Random vibration and statistical linearization

Random vibration and statistical linearization

J. B. Roberts
School of Engineering and Applied Sciences
University of Sussex
UK

P. D. Spanos
George R. Brown School of Engineering
L. B. Ryon Chair in Engineering
Rice University
Texas
USA

JOHN WILEY & SONS
Chichester · New York · Brisbane · Toronto · Singapore

Other Wiley Editorial Offices

John Wiley & Sons, Inc., 605 Third Avenue,
New York, NY 10158-0012, USA

Jacaranda Wiley Ltd, G.P.O. Box 859, Brisbane,
Queensland 4001, Australia

John Wiley & Sons (Canada) Ltd, 22 Worcester Road,
Rexdale, Ontario M9W 1L1, Canada

John Wiley & Sons (SEA) Pte Ltd, 37 Jalan Pemimpin 05-04.
Block B, Union Industrial Building, Singapore 2057

Library of Congress Cataloging in Publication Data:

Roberts, J. B. (John Brian)
Random vibration and statistical linearization / J. B. Roberts, P. D. Spanos.
p. cm.
Includes bibliographical references.
ISBN 0 471 91699 4
1. Random vibration—Statistical methods. I. Spanos, P. D. (Pol D.) II. Title.
TA355.R55 1990
620.3—dc20 89-24807
CIP

British Library Cataloguing in Publication Data:

Roberts, J. B. (John Brian)
Random vibration and statistical linearization.
1. Random vibration. Mathematics
I. Title II. Spanos, Pol. D.
531 .32
ISBN 0 471 91699 4

Typeset by Thomson Press (India) Ltd, New Delhi
Printed and bound in Great Britain by
Biddles Ltd, Guildford and King's Lynn

Contents

Preface

The study of random vibration problems, using the concepts of stochastic process theory, is a relatively new engineering discipline. Interest in this field has grown rapidly in the last few decades, due to the need to design structures and machinery which can operate reliably when subjected to random environmental loads. Examples include the response of buildings to wind loading and earthquakes, the dynamic behaviour of marine structures in waves, the vibration of vehicles travelling over rough ground and the excitation of aircraft and missile vibration by atmospheric turbulence and jet noise.

In many practical applications the system of concern has non-linearities, which must be taken into account if one is to predict its performance in a realistic way. For such non-linear problems the classical linear theory, which can be found in a number of textbooks, is not directly applicable and new techniques are required. Much of the recent research effort in the field of random vibration has been directed towards developing suitable methods for analysing non-linear systems.

Of the various possible approaches which are available, the method of statistical, or equivalent, linearization has, over the last three decades, proved to be the most useful approximate technique. Its value lies in the fact that, unlike many other methods, it can readily be used to deal with complex systems having many degrees of freedom, and with complex types of excitation. Moreover, it can cope with hysteretic elements and is capable of yielding results even when the input is non-stationary. Unlike several other methods which relate to linear theory, statistical linearization usually gives reasonably good results when there are strong non-linear effects.

Many important advances in the general method have been made recently. However, with the exception of some recent, very brief, review papers, knowledge of the current 'state of the art' can only be gleaned from a large number of scientific journals and conference proceedings, which are widely scattered in the literature. It is the aim of this book to present, for the first time, a coherent and reasonably self-contained account of the general method and its various recent developments, together with numerous examples drawn from a wide variety of engineering problems. Thus, this book provides a comprehensive account of the statistical linearization method in a form which, it is hoped,will be directly useful to engineers faced with non-linear random vibration problems.

Further, the book can be used, with supplementary material reflecting the individual instructor's preference, for an introductory course in random vibration.

We acknowledge with pleasure and gratitude the help, both direct and indirect, from colleagues and research students over a number of years. The dedicated typing on many sections of this book by Ms L. Anderson and Mrs L. Richardson of Rice University, and Ms D. Staples of the University of Sussex, is highly appreciated. Thanks are also due to Mrs P. Cherry and Mrs J. Robertson for tracing most of the figures. Dr. J. F. Dunne, of the University of Sussex, has read this book at the proof stage; the authors are grateful for his help in locating errors and for his useful comments. Finally, we wish to express our thanks to our publisher, John Wiley, and particularly to Ellen Taylor for her patience and encouragement during the preparation of this book.

J. B. Roberts
P. D. Spanos
November 1989

Dedications

JBR *To Margaret, Mark and Matthew*

PDS *To my family for loving and challenging me*

Dedication

Chapter 1

Introduction

1.1 RANDOM VIBRATION

Methods of predicting the vibration response of mechanical and structural systems to fluctuating external forces have grown rapidly in importance, in engineering design, over the last century. The continuing tendency to reduce the weight and cost of structures and to increase the power-to-weight ratio of machines, engines and vehicles has brought to the fore the need to predict vibration response levels in many fields of engineering. High vibration amplitudes are almost invariably undesirable, due to factors such as accompanying high noise levels, increase in wear and fretting of joints and, often most importantly, the possibility of component failure leading to catastrophic system failure. Such a failure may occur at the instant the stress levels exceed safe working limits or as a result of accumulated damage through metal fatigue.

For many years engineers were mainly concerned with periodic vibration—i.e. with situations where the excitation and response of the system are simple periodic functions of time. Vibration of this kind typically arises through lack of perfect balance in rotating or reciprocating machinery and can be transmitted through support structures and foundations to neighbouring systems. Following the extensive pioneering work of Rayleigh (1877), a very comprehensive body of classical theory now exists for analysing this type of vibration. The theory is particularly well developed for linear systems and may be found in a number of standard textbooks (e.g. Den Hartog (1965), Bishop *et al.* (1965) and McCallion (1973)).

In the mid 1950s, however, a new type of vibration problem arose in the aerospace industry which could not be solved by the classical methods. It was discovered that aircraft fuselage panels in the neighbourhood of jet engines were reaching such high levels of vibration response, due to acoustic excitation from the jet exhausts, that fatigue cracks could develop and spread quite rapidly. Studies revealed that the vibration response of these panels was extremely complex, mirroring the very complex nature of the spatial and temporal variations of pressure on the panel surfaces (e.g. see Clarkson and Mead, 1973). Not only was this kind of excitation, and response, non-periodic, and highly irregular, but it also lacked repeatibility. Thus, two successive experiments, carried out under identical conditions, would yield vibration amplitude response against time histories which differed significantly in detail, although they had the same overall behaviour, on 'average'. It became evident that it was not feasible to tackle such a problem on the conventional 'deterministic' basis. A new 'statistical', or more accurately a 'probabilistic', approach was required in which the excitation and response was described in terms of statistical parameters, such as the mean square of the vibration amplitudes. Similar problems arose, at about the same time, in designing aircraft to withstand buffeting due to atmospheric turbulence (e.g. see Press and Houbolt, 1955) and in assessing the reliability of payloads in rocket-propelled vehicles (e.g., see Bendat *et al.*, 1962). In both these applications the vibration response was found to be so irregular and complex that a probabilistic, rather than deterministic, approach proved to be far more fruitful.

In the probabilistic approach to vibration, subsequently developed by structural engineers, both the excitation and the response are modelled as 'stochastic', or 'random' processes, which can be specified in terms of a fairly small number of statistical parameters, and functions, such as the power spectrum. Such processes can be viewed as an infinite 'ensemble' of possible sample functions, or 'realizations'. It was soon discovered that much previous work in statistical communication theory, undertaken about a decade earlier, notably by Rice (1944), could be adapted fairly easily to formulate a solution methodology for the case of linear systems responding to random excitation. In addition, there was already a considerable body of mathematical literature on stochastic processes, which originated from early work by physicists, notably Einstein (1905), on the theory of Brownian motion. Thus, for linear systems at least, it was found that a relatively simple, and complete, theory of probabilistic structural dynamics could be developed through an adaptation of existing knowledge. In the case of Gaussian excitation this theory enables all the statistical parameters of the response to be directly related to the corresponding parameters of the excitation.

It was realized fairly quickly that this linear theory of 'random vibration' was applicable not only in the aerospace field but in a wide variety of other engineering disciplines. Thus, ships and other offshore structures responding to wave excitation (e.g., see St Denis and Pierson, 1953), civil engineering structures such as tall buildings and bridges responding to earthquakes (e.g. see Vanmarke, 1976) and wind excitation (e.g., see Davenport and Novak, 1976) and land-based vehicles such as cars and trains responding to irregularities in ground and track surfaces (e.g. see Schiehlen, 1985) can all be analysed using the concepts and results of linear random vibration theory. Application of the theory was greatly assisted in the 1970s by the development of digital processing techniques and the use of fast Fourier transform (FFT) algorithms, which enabled experimental data to be processed rapidly, and efficiently, to yield estimates of the required statistical parameters. Moreover, the cost of implementing these techniques has fallen dramatically, in the last few years, due to the introduction of inexpensive, but very powerful, microcomputers and associated data acquisition hardware.

This rapid and accelerating growth, during the last three decades, has resulted in a very extensive literature on linear random vibration, covering both theoretical and practical aspects. There are now available a number of text-books which give a good overview of the subject (Crandall and Mark, 1963; Robson, 1963; Lin, 1967; Elishakoff, 1983; Augusti *et al.*, 1983; Newland, 1984; Bolotin, 1969, 1984; Nigam, 1983; Piszczek and Niziol, 1986; Yang, 1986; Krée and Soize, 1986). In addition, mention should be made of several very useful, fairly recent reviews (Crandall and Zhu, 1983; Wedig, 1984; Spanos and Lutes, 1986).

1.2 IMPORTANCE OF NON-LINEARITIES

In engineering applications the use of a linear model for the system under consideration leads to fairly simple, and often useful, results. If the excitation processes have a Gaussian distribution then, according to the linear theory, the response processes are also Gaussian. This enables one to compute various statistics of the response relevant to reliability, such as level crossing rates, in terms of a few statistical parameters.

It is most important to bear in mind, however, that no real system is exactly linear. In mechanical and structural systems non-linearities can arise in various forms, and usually become progressively more significant as the amplitude of vibration increases. Since systems must usually be designed to withstand, with a specified probability, the severest possible levels of excitation which they can encounter, during their operational life-time, it is often vital to properly account for the effect of non-linearities. Failure to do so can result, at best, in an excessively conservative design or, at worst, in a system liable to catastrophic failure.

An important example of the occurrence of non-linearity occurs in the problem of designing earthquake resistant buildings and other structures. Here structural components usually exhibit a very significant hysteretic behaviour, when subjected to high levels of dynamic loading. Such hysteresis is normally highly non-linear in character and thus the overall system behaviour can only be represented with satisfactory accuracy by using a non-linear, differential model. In combination with a suitable stochastic representation of seismic excitation, such a non-linear system model provides the basis for formulating realistic analysis and design procedures.

Other examples of non-linearity abound in engineering and will be discussed in some detail in the following chapter. Here it is worth mentioning briefly fluid loading forces, which occur when structures oscillate in contact with a fluid. These forces usually depend on the relative velocity between the structure and the fluid, in a highly non-linear manner. Thus, as a specific example, for a ship rolling in irregular waves, the damping moment is found to depend on the roll angular velocity, $\dot{\phi}$, according to the linear-plus-quadratic relationship $a\dot{\phi} + b\dot{\phi}|\dot{\phi}|$, where a and b are constants; the second, quadratic term is usually significantly larger, in magnitude, than the first, linear term. A similar, non-linear damping model has been found to be appropriate in many other situations, e.g. fluid loading on vibrating heat exchanger tubes and on vertical support columns in offshore gravity platforms, and the fluid damping forces provided by car shock absorbers. It is also worth mentioning the effect of membrane forces in transversely loaded plates and shells, leading to significantly non-linear force–deformation characteristics, the non-linear nature of the dissipative and restoring forces arising in many dynamically loaded materials, and the extremely non-linear behaviour of the Coulomb frictional forces which arise when dry contacting surfaces are in relative movement.

1.3 NON-LINEAR RANDOM VIBRATION PROBLEMS

If a non-linear mathematical model of the system under consideration is adopted, together with a random process model of the excitation, then one is faced with the problem of predicting the system response. Since the excitation is described in a probabilistic fashion, in terms of various statistical functions and parameters (e.g. mean, mean-square and power spectrum) one would ideally like to be able to predict those statistical functions and parameters for the response which lead to a full probabilistic specification.

In its simplest form, where the system has only one 'input' and only one 'output' is of interest, this prediction problem can be represented as in Figure 1.1. The excitation process, $X(t)$, is 'transformed' into a response process, $Y(t)$, by the non-linear system and one wishes to compute the statistical characteristics of $Y(t)$ from a knowledge of the statistical character of $X(t)$ and the non-linear dynamic model of the system. The dynamic model will invariably take the form of an ordinary differential relationship between $X(t)$ and $Y(t)$ (see Chapter 2). Since both $X(t)$ and $Y(t)$ are stochastic in nature, one is actually faced with the problem of solving a non-linear 'stochastic differential equation' (Arnold, 1973; Soong, 1973).

Often the problem of concern is more complicated than indicated by Figure 1.1. One may have a number of inputs and outputs to consider; in this

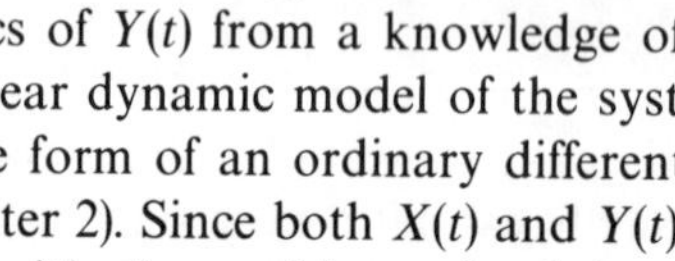

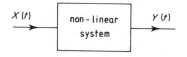

Figure 1.1
Transformation of $X(t)$ to $Y(t)$ by a non-linear system

case the dynamic model is in the form of a set of coupled ordinary differential equations. Here it is necessary to specify not only the statistical characteristics of each individual input or output process but also the statistical relationships between the various processes. At a higher level of complexity, the excitation may be spatially as well as temporally dependent, as in the case, for example, of acoustic excitation of aircraft panels. Here one can proceed by adopting a partial differential model of the system, with the attendant considerable analytical difficulties, or, alternatively, one can simplify the problem on the basis of a suitable lumped mass approximation. The latter has the effect of reducing the system model to a set of ordinary differential equations. One then has a spatially discrete rather than a spatially continuous model with the excitation consisting of a set of statistically dependent random processes, each of which depends only on time.

1.4 METHODS OF SOLUTION

Non-linear stochastic differential equations are much harder to solve than their linear counterparts. This situation is not surprising, in view of the similar situation which exists in deterministic vibration theory. There, a comprehensive linear theory exists but no correspondingly general theoretical framework exists for treating non-linear problems.

Here the various approaches which have been developed over the years will be briefly discussed, starting with the method which will be the main theme of this book. For further discussion of the background and development of these various methods the reader may consult a number of relevant review articles (Bendat *et al.*, 1962; Caughey, 1971; Lemaitre 1971; Osinki, 1971; Iwan 1974; Crandall, 1977a; Roberts, 1981a, 1981b; Spanos, 1976, 1978, 1981; Roberts, 1984. To, 1984, 1987; Spanos and Lutes, 1986; Roberts and Dunne, 1988).

1.4.1 Statistical linearization

A natural method of attacking non-linear problems is to replace the governing set of non-linear differential equations by an equivalent set of linear equations; the difference between the sets being minimized in some appropriate sense. This technique has been used extensively for studying deterministic non-linear problems for many years (e.g. see Bogoliubov and Mitropolsky, 1963) and is often referred to as the 'describing function method' in the electrical engineering literature. An adaptation of the approach to deal with stochastic problems was first developed by Booton (see Booton *et al.*, 1953; Booton, 1954 and Kazakov, 1954, 1955) and used as a tool in control engineering. Subsequent developments in this field have been described by Sawaragi *et al.* (1962), Kazakov (1965a,

1965b), Gelb and Van Der Velde (1968), Atherton (1975) and Sinitsyn (1976). Independently, the method (now variously known as 'statistical linearization', 'equivalent linearization' or 'stochastic linearization') was proposed by Caughey (1963) as a means of solving non-linear stochastic problems in structural dynamics.

As a simple illustration of the basic idea it is convenient to consider the following very general form for the stochastic differential equation of a non-linear oscillator (or single degree of freedom system).

$$g(\mathbf{Y}) = X(t) \tag{1.1}$$

Here $\mathbf{Y}(t)$ is the vector

$$\mathbf{Y} = [Y, \dot{Y}, \ddot{Y}]^{\mathrm{T}} \tag{1.2}$$

where $Y(t)$ is the displacement response, $\dot{Y}(t)$ the velocity and $\ddot{Y}(t)$ the acceleration. $g(\mathbf{Y})$ denotes an arbitrary, non-linear function of $\mathbf{Y}$ and $X(t)$ is the input. The notation here conforms with Figure 1.1 and $X(t)$ and $Y(t)$ are taken to be modelled as stochastic processes. Thus, the non-linear system 'transforms' $X(t)$ into the process $Y(t)$.

The method of statistical linearization consists of replacing equation (1.1) with the equivalent linear form

$$m\ddot{Y} + c\dot{Y} + kY = X(t) \tag{1.3}$$

where m, c and k are parameters, which will be time-dependent if the statistical nature of $X(t)$ varies with time. The 'equation error' between equations (1.1) and (1.3) is clearly given by

$$\varepsilon = g(\mathbf{Y}) - m\ddot{Y} - c\dot{Y} - kY \tag{1.4}$$

This quantity can be minimized in some convenient way. The usual method is to minimize the mean-square of ε, with respect to the parameters m, c and k. This process will yield a set of equations involving the average values of various functions of $\mathbf{Y}$.

To evaluate these averages exactly it is necessary to know the probability laws governing $\mathbf{Y}(t)$. These are almost invariably unknown; indeed if they are known then the solution of the original non-linear problem is known exactly and there is no point in resorting to the use of statistical linearization. This difficulty is usually resolved by approximating $\mathbf{Y}(t)$ as a Gaussian vector process (see Chapter 3). The justification behind this step lies in the fact that the response of the equivalent linear system will be exactly Guassian if the excitation process, $X(t)$, is Gaussian, and approximately so if $X(t)$ is non-Gaussian. Once the Gaussian assumption for $\mathbf{Y}(t)$ is invoked then, using standard linear theory, a closed system of non-linear equations can be derived for the parameters m, c and k. These equations will be algebraic if the parameters are constant, and differential if they are time-dependent; they are, in either case, easily solved by numerical means and, in certain special cases, may be solved analytically.

Numerous applications of this technique for studying the response of non-linear oscillators to random excitation have been described in the literature (Roberts, 1981b, and Spanos, 1981a).

In several areas of application, such as earthquake engineering, the non-linearity function is hysteretic in character—i.e. it depends on the history of the motion, rather than simply on the instantaneous motion. In this case, as originally shown by Caughey (1963), the averaging principle of Krylov and Bogoliubov (1937) can be applied. This involves the assumption that the response motion, $Y(t)$, resembles a sinusoid with slowly varying amplitude and phase. Thus the validity of the method is limited to situations where the response is narrow-band in nature (i.e. to cases where the energy dissipation per cycle is relatively small). The evaluation of the average quantities, which occur in the expressions for the various linear parameters, can then be accomplished by averaging over one cycle of oscillation, on the basis that the amplitude and phasc of $Y(t)$ is reasonably constant over such a time period. In conjunction with rsults obtained from the equivalent linear system, non-linear equations for the required parameters may be easily derived. This technique was first applied to oscillators with bilinear hysteresis (Caughey, 1960a) and its accuracy has been systematically studied by Lutes and coworkers (Iwan and Lutes, 1968; Lutes, 1970a; Takemiya, 1973; Lutes and Takemiya, 1974; Takemiya and Lutes, 1977).

The energy dissipation in hysteretic structures can, however, often be relatively large. In these circumstances the response process has a wide-band character and an 'averaging over a cycle' method is inappropriate. A method of overcoming this difficulty is to represent the hysteretic force in the form of a first-order non-linear differential equation. It can be shown (e.g., see Suzuki and Minai 1987) that a wide variety of shapes of hysteretic loop can be satisfactorily modelled on this basis. A combination of a differential hysteresis model with the normal equation of motion for the system yields a non-linear third-order system model, from which an equivalent linear model can be constructed by means of a slight extension of the method employed for non-hysteretic, non-linear oscillators. Its relevance to the design of earthquake resistant structures has prompted much research activity on this topic in recent years (Roberts and Dunne, 1988; Wen, 1986; 1989).

A great advantage of the statistical linearization method is that it can easily be generalized to cope with multi-degree of freedom systems, including those where hysteretic elements are incorporated. The earliest extensions of the theory in this direction were given by Caughey (1963) and Kazakov (1965a, 1965b)). Subsequently there have been a number of theoretical advances in this area (Roberts, 1981b). The technique has no difficulty in dealing with non-white excitations and can be further generalized to cope with non-stationary excitations and responses (Spanos, 1981a and Roberts, 1981b).

From the perspective of treatable physical problems, the statistical linearization method has proved a useful analytical tool across a very wide spread of engineering applications (Spanos, 1981a). Recent examples include the analyses of the sliding motion of a structure on a randomly moving surface (Constantinou and Tadjbakhsh, 1984; Noguchi, 1985), the response of offshore structures to waves (Spanos and Agarwal, 1984; Gumestad and Connor, 1983; Leira, 1987; Grigoriu and Allbe, 1986), the large amplitude response of clamped plates and skin structures to acoustic loading (Mei and Paul, 1986; Mei and Wolfe, 1986; Maymon, 1984), the sloshing of liquids in tanks subjected to earthquake excitation (Sakata *et al.*, 1984), the motion of vehicles traversing rough ground (Harrison and Hammond, 1986) and the effect of non-linear soil-structure interaction on the dynamic response of buildings (Chu, 1985).

1.4.2 Moment closure

Equations for the moments of the response such as the mean, mean square and mean cube, can be derived fairly readily from the equations of motion. From these moments, or related quantities known as cumulants (or semi-invariants) and quasi-moments (Stratonovitch, 1964), it is possible to derive estimates of the probability distribution of the response, using a variety of analytical expansions of the Gaussian distribution.

A notorious difficulty with this approach, when dealing with non-linear systems, is that the moments are generally governed by an infinite hierachy of coupled equations. Thus, to obtain a solution it is necessary to introduce a 'closure approximation', to obtain a soluble set of equations. The simplest level of closure is to assume that the response is Gaussian. It can be shown that this approach leads to results which are identical to those obtained from the statistical linearization approach. However, it should be noted that with the Gaussian closure method an equivalent linear set of equations is not explicitly invoked.

Improvements in accuracy can be obtained, in principle at least, by resorting to a higher, non-Gaussian level of closure. For example, all cumulants above a certain order, n, where $n > 2$, can be assumed to be negligible (the choice $n = 2$ corresponding to Gaussian closure). The basic principle of non-Gaussian closure is now well established (see Dashevskii, 1967; Dashevskii and Lipster, 1967; Nakamizo, 1970; Assaf and Zirkle, 1976; Crandall 1977b, 1978, 1980, 1985; Beaman and Hedrick, 1981).

A characteristic feature of this approach is that the complexity of the moment equations dramatically increases as the order of closure increases. However, results for non-linear oscillators have been obtained, for n values up to 6, using both cumulant closure (Wu and Lin, 1984; Lin and Wu, 1984) and quasi-moment closure (Bover, 1978). These studies demonstrate that a significant improvement in accuracy can be obtained by progressing beyond simple Gaussian closure. The method has recently been applied by Ibrahim and coworkers (Ibrahim,

1985; Ibrahim and Soundararayan, 1985; Ibrahim *et al.*, 1985) to a study of the response of systems with non-linear inertial terms to random excitation.

1.4.3 Equivalent non-linear equations

An interesting, alternative generalization of statistical linearization has been proposed by Caughey (1986a). The idea is to replace the original set of non-linear differential equations by an equivalent non-linear set, where the latter belong to a class of problems which can be solved exactly. Unfortunately, this class is, at present, very limited, and thus the range of applicability of the technique is correspondingly restricted. To date, results have only been obtained for oscillators with non-linearity in damping and stiffness (Caughey, 1986a; Cai and Lin, 1988a; Zhu and Yu, 1989). Here, however, it has been demonstrated that the method is very effective as a means of predicting the probability distribution of the response, with reasonable accuracy. Thus, unlike the situation with the normal statistical linearization method, one can obtain information on the departure of the response distribution from the Gaussian form, due to non-linear effects. This information is of prime importance in reliability assessments, as will be demonstrated in Chapter 9.

1.4.4 Perturbation and functional series

If the non-linearities in the system are sufficiently small, the perturbation method of solution can be used. This approach has been well known in deterministic vibration theory for many years (Stoker, 1950) and was generalized to the case of stochastic excitation by Crandall (1963). The basic idea is to expand the solution to the non-linear set of equations in terms of a small scaling parameter, λ say, which characterizes the magnitude of the non-linear terms. The first term in the expansion is simply the linear response which is the response when all the non-linearities in the system are removed. The subsequent terms express the influence of non-linearity. As with perturbation schemes in general, the calculations are usually lengthy and rapidly become more tedious as the order of λ increases. In practice, results are usually obtained only to the first order in λ (e.g., see Manning (1975) and Soni and Surrendran (1975)).

Functional series methods offer an alternative approach to developing an expansion based on the linear solution. A recent example of the application of such methods is the work of Orabi and Ahmadi (1987). They used a Wiener–Hermite expansion and presented a formal procedure for deriving the deterministic equations governing the kernel functions arising in the expansion (see also Roy and Spanos (1990)).

A common difficulty with all expansion methods lies in establishing the regimes of convergence, in the appropriate parameter space. Frequently it is found that, due to a combination of poor convergence properties and excessive computational requirements, it is only possible to obtain reliable results if there is a very small degree of non-linearity.

1.4.5 Markov methods

A totally different approach to the solution of non-linear random vibration problems is based on the theory of continuous Markov processes (otherwise known as 'diffusion processes'). Since the early pioneering work of physicists on Brownian motion it has been well known that the response of dynamic systems to wide-band random excitation can, in many cases, be accurately modelled in terms of multi-dimensional Markov processes. The state transition probability function for such a process is governed by a linear partial differential equation, known as the Fokker–Planck–Kolmogorov (FPK) equation, or forward diffusion equation (Arnold, 1973; Soong, 1973). This equation, governing the diffusion of probability 'mass' in state space, is closely analogous to the diffusion equations which govern the diffusion of heat, or mass, in thermo-fluid mechanics problems. It is possible to relate the 'drift' and 'diffusion' coefficients in the FPK equation directly to the parameters in the dynamic equations of motion of the system under consideration. Thus, within the limitations of the modelling, which essentially involves approximating the actual excitation processes as white noises, the theoretical framework of Markov process theory offers, in principle, a direct approach to the exact treatment of non-linear random vibration problems. For cases where the excitations can not be adequately approximated as white noises, it is possible to introduce 'pre-filters' which operate on white noises to generate excitation processes for the system, with the required power spectra. Then Markov process theory is applicable to a higher dimensional combined system consisting of the pre-filters in series with the original system.

Unfortunately, however, the class of non-linear random vibration problems for which the appropriate FPK equation can be solved exactly is still quite limited, despite a number of very recent theoretical advances (see Caughey, 1986b; Yong and Lin, 1987; Dimentberg, 1988; Cai and Lin, 1988b; Langley, 1988a; Lin *et al.*, 1988; Lin and Cai, 1988; Scheurkogel and Elishakoff, 1988; Zhu, 1989). The broadest reported class of single degree of freedom structural systems, for which the stationary solution of the associated FPK equation can be determined, requires that the mass, damping and stiffness of the oscillator be a function of displacement and velocity, of a very particular form. It is unfortunate that this class does not include important models of vibrating structural systems, such as, for example, oscillators with linear-plus-quadratic damping, of the type commonly encountered when fluid loading is present. As far as non-stationary, or transient analytical solutions are concerned, these are extremely scarce, or non-existent. For multi-degree of freedom systems the available exact solutions, in the non-linear case, are confined to the stationary response of a highly restricted class of systems (Dimentberg 1988).

For lightly damped oscillators, with wide-band random excitation, a useful approximate analytical technique, known as 'stochastic averaging' has been widely used in recent years; for a review see Roberts and Spanos, 1986. This method enables the basic two-dimensional Markov process governing the response to be replaced, approximately, by a one-dimensional Markov process governing an envelope amplitude process, $a(t)$. The appropriate FPK equation for $a(t)$ can be easily solved analytically to yield simple expressions for the stationary probability distribution of the amplitude process. By considering an associated phase process, approximate analytical expressions for the joint distribution of the response displacement and velocity can be derived, from which a wide variety of statistics, such as level crossing rates, can be calculated. The reduction in dimension of the governing FPK equation, from two to one, also considerably simplifies the computation of non-stationary, or transient solutions and is of great value in estimating reliability statistics of the 'first passage' type (see Roberts, 1986). There are, however, considerable difficulties in applying this method to systems with more than one degree of freedom.

For more complicated systems, or when pre-filters must be used (resulting in an increase in the dimension of the total system), it is necessary to resort to numerical, or combined analytical and numerical, techniques for solving the governing FPK equations. These approaches tend to involve considerable memory requirements, and excessive amounts of computer time, when the system is of high dimension.

1.4.6 Monte Carlo simulation

An alternative method for estimating, within any desired confidence level, the exact response statistics of randomly excited non-linear systems is based on random computation experiments, popularly known as Monte Carlo simulation (e.g. see Shinozuka, 1972; Rubinstein, 1981). Currently digital simulation is used almost exclusively; for a recent review article see Spanos and Mignolet, 1989. The theoretical foundation of Monte Carlo studies is associated with the fact that the stochastic differential equation governing the motion of the system can be interpreted as an infinite set of deterministic differential equations (Soong, 1973). For each member of this set, the input is a sample function of the excitation process, and the output is the corresponding sample function of the response process.

The backbone of any digital simulation study is an algorithm which provides a set of pseudo-random numbers belonging to a population with a specified probability density function. Proper processing of this set of numbers can yield the values of sample functions of random process excitations, with pre-selected frequency content and temporal variation of intensity, at successive discrete equi-spaced times. Upon generating a single sample of the random excitation the response may be computed by any of the commonly available subroutines for the numerical integration of differential equations. Then, another sample of the excitation can be generated and the computed values of the structural response used to update its statistics. Obviously this approach is applicable for the estimation of both stationary and non-stationary response statistics.

Evidently, the larger the number of the simulated records is, the smaller the expected deviation of the obtained numerical values from the theoretical values of the response statistics should be. Unfortunately, the number of sample records which are necessary for the estimation of the response statistics, within commonly acceptable engineering confidence levels, is of the order of five hundred. This fact makes the computational cost of simulation quite significant, especially for multi-degree of freedom vibratory systems. It must be noted, however, that the necessity of a large number of records can often be eliminated if interest is confined to stationary response statistics. In this case, under certain conditions (Lin, 1967), ergodicity with respect to a particular statistical moment can be assumed. This assumption allows the determination of this specific ensemble statistical moment by using its temporal counterpart, which is calculated by using a single sample function of the response. For example, this approach can be appropriate for dynamic analysis involving probabilistic models of sea waves or atmospheric turbulence. However, a similar approach is inappropriate for stochastic models of inherently non-stationary and, therefore, non-ergodic, physical processes such as earthquakes.

1.5 ROLE OF STATISTICAL LINEARIZATION

Most of the approximate methods of analysis, discussed in the preceding section, are best suited to single degree of freedom systems, with stationary random excitation. For multi-degree of freedom systems, which are prevalent in most engineering applications, these methods are very difficult to apply; they tend to involve severe analytical complexity, often combined with excessive computational requirements, in terms of core storage or execution time. The single exception is the method of statistical linearization, which enables results to be obtained with relative ease, even in situations where multi-degree of freedom systems subjected to non-stationary random excitations are of concern.

The high degree of flexibility of the statistical linearization method is intimately connected with its basic, inherent limitation. Specifically, it only yields estimates of the first and second moments of the response (e.g., mean, mean square and power spectrum). Due to the in-built assumption that the response is Gaussian it does not lead to predictions concerning the influence of non-linearities on the distribution of the response. It is thus ineffective in providing sound reliability estimates based on the probability of the system failing in some prescribed manner, during a fixed interval of time. These estimates are particularly sensitive to the precise shape of the probability distribution of the response, in its extreme 'tails'. To obtain reliability estimates for multi-degree of freedom systems there are really only two avenues of attack. One can either attempt to solve an appropriate FPK equation, by some numerical means, or one can use a digital simulation technique. Both methods tend to be extremely costly, in terms of computational effort, and thus have limited value, from a design viewpoint.

In designing structures and machines to withstand complex environmental loads there are usually two basic stages. In the first, a systematic study is undertaken of the influence of various disposable system parameters on the overall level of response. For example, one is often interested in how a variety of possible structural modifications can affect the system's response level. On the basis of such a study an optimization of the system, with regard to certain parameters, is often the ultimate objective. For this purpose a detailed knowledge of the probability distribution of the response is not required. It is sufficient to use simple response level indices, such as the mean and mean-square characteristics of the response. In the second stage, once an optimal design configuration for the system has been established, a reliability study is carried out in which statistics relating to the system's safety are estimated, for a class of random excitations which are likely to be encountered during the system's lifetime.

The method of statistical linearization, whilst inappropriate for the second stage, for the reasons already given, is ideally suited for the first stage. This explains its high degree of popularity amongst designers, across a very wide range of engineering disciplines. Through a judicious use of this technique the computational cost involved in the second stage, where currently simulation is most commonly used, can be very substantially reduced. It has been estimated (Spanos, 1981b) that the computational efficiency of the statistical linearization method can be of the order of one hundred to one thousand times greater than that of Monte Carlo simulation. This range is representative for structural systems with the amount of damping commonly encountered in most engineering applications. Clearly, the significance of the computational superiority of statistical linearization will increase with the number of response samples generated in a simulation study.

1.6 SCOPE OF BOOK

This book is principally concerned with the statistical linearization method, and its application to a variety of non-linear engineering problems. It is implicitly assumed throughout that the systems of concern have been discretized, in some suitable fashion, so that their motion can be represented in terms of a finite number, n, of coordinates. Thus the most general type of systems which will be considered will be multi-degree of freedom systems with n degrees of freedom. As in the deterministic situation, this approach of modelling real continuous systems in terms of a finite number of degrees of freedom appears to be the most direct and practical way of analysing the response of complex structures.

The scope of this book is also deliberately restricted to systems in which there is no parametric, or multiplicative, excitation present. For systems with parametric excitation, stochastic stability or bifurcation is often of principal concern and it generally accepted that statistical linearization is unsuitable for studying this aspect of dynamic response. It is noted however, that for the case of white noise excitation, some improvement in accuracy, over that obtained from the standard statistical linearization procedure, can be obtained by using a different approach, in which the drift term in the diffusion equation is linearized, and the square of the diffusion term is replaced by a second-degree polynomial (Bruckner and Lin, 1987; see also Spanos and Agarwal (1984)). For a recent study of parametric random vibration problems in general the reader is referred to the book by Ibrahim (1985).

Within the framework of the above constraints, this book aims to present a comprehensive coverage of statistical linearization, and related methods, with emphasis on their application in engineering. Particular attention is paid to the analysis of systems with hysteretic elements, and with non-stationary excitation, in view of the criticality of these aspects in important fields such as earthquake engineering, but the numerous examples given, throughout the text, are drawn from a variety of engineering disciplines.

1.7 PLAN OF BOOK

The first part of the book (up to, and including, Chapter 4) is intended to provide the reader with a background to the statistical linearization technique. In Chapter 2 the question of formulating suitable equations of motion is addressed. Using a Lagrangian approach the general form of the equations of motion for discrete, multi-degree of freedom systems is derived. This is followed by a description of the various kinds of non-linearity which can be encountered in applications, and methods of representing these non-linearities in terms of analytical expressions, suitable for incorporating into the equations of motion.

This is followed, in Chapter 3, by a brief account of those aspects of probability theory, and stochastic process theory, which are relevant to statistical linearization methods. Also included here is a description of expansion series for the probability distribution of multi-dimensional random variables; this material is required later, in Chapter 9, when an extension of the statistical linearization method, based on non-Gaussian closure, is discussed.

Chapter 4 contains a fairly thorough exposition of the various classical methods for analysing linear systems, including time–domain and frequency–domain relationships, combined with modal analysis. Methods based on a state variable formulation, such as complex modal analysis, are also introduced, and illustrated by means of simple examples.

The next four chapters (5 to 8, inclusive) deal with the analysis of non-linear random vibration problems, by the statistical linearization method. Chapter 5 is concerned solely with single degree of freedom systems, which provide a means for introducing the basic concepts in a simple manner, with the minimum of mathematical complexity. The discussion here starts with the linearization of simple, single elements, with zero memory, and progressively builds up to a consideration of oscillators with various forms of non-linearity, subjected to stationary random excitation, of both the wide-band and narrow-band kinds.

Chapter 6, which may be regarded as the core of the book, presents the method for linearizing multi-degree of freedom systems. Questions of uniqueness and existence are examined and the mechanization of the method is discussed in some detail.

In Chapter 7 it is shown that the statistical linearization method can be extended, fairly readily, to cope with problems where the excitation and response processes are non-stationary in character. The main complication here is that the parameters in the equivalent linear model are no longer time invariant. It is shown that state variable based methods of analysis can be applied very easily when the excitation is a modulated white noise and, with additional complexity, in cases where the excitation in non-white.

Systems with hysteretic elements are analysed in some depth in Chapter 8. Here two basic approaches are examined, namely the averaging method referred to earlier in this Chapter, and the use of differential equations, for modelling hysteresis loops. Specific results from both methods are obtained for the special case of an oscillator with bilinear hysteresis. It is shown that complex problems, involving multi-degree of freedom systems with a number of degrading hysteretic elements, can also be dealt with through suitable extensions of the analysis procedures.

A basic limitation of the statistical linearization method, already referred to, is that it is necessary to approximate the probability distribution of the response by the Gaussian form. At the beginning of Chapter 9, it is shown that this approach, which is equivalent to Gaussian closure, can be generalized to higher order, non-Gaussian, closure and results are obtained for a specific non-linear oscillator. This is followed by an exposition of the method of equivalent non-linear equations, referred to earlier. This method can be regarded as another approach to relaxing the Gaussian response assumption, inherent in the normal statistical linearization technique. The chapter concludes with a quite different approach to linearization, which also avoids the need to assume that the response is Gaussian. This involves generating sample functions of the response process and fitting an equivalent linear model to this data, using parametric identification techniques. One advantage of this approach is that it can be applied to experimental data, in cases where the exact nature of the non-linearities which are present are not known *a priori*.

The book concludes with an assessment of the accuracy of statistical linearization, and the related methods considered in Chapter 9. This is achieved partly through a comparison of results in cases where an exact solution of the governing FPK equation is available and partly through a comparison with digital simulation results, in a representative selection of particular cases.

Chapter 2

General equations of motion and the representation of non-linearities

2.1 INTRODUCTION

The majority of engineering systems can be modelled, to a first approximation, in terms of linear differential equations of motion, if the amplitude of motion is relatively small. Clearly, since the treatment of non-linear equations is a good deal more complicated than the treatment of linear ones, the adoption of a linear model is a quite desirable step in the analysis of any particular system under investigation. There are cases, however, in which a linearized mathematical model of the physical problem does not account adequately for the quantitative, or even the qualitative, behaviour of the problem. A typical example of the former case is the large amplitude vibration of elastic systems. An example of the latter case involves the existence of sub-harmonics and ultra-harmonics in the response of periodically excited non-linear systems.

From an alternative perspective, in comparing linear and non-linear systems, the latter exhibits two distinct features. First, the superposition principle is invalid for non-linear systems. For example, if the amplitude of excitation of a non-linear system is doubled, its response amplitude is not necessarily doubled. Second, a linear system has one and only one position of equilibrium. A non-linear system can have more than one equilibrium state, depending on the condition of operation.

In this chapter the general form of the equations of motion of discrete, lumped-mass systems is derived, using a Lagrangian approach. This is followed by a description of some representative non-linearities encountered in vibration problems. The goal is to expose the reader in a natural way to the considerations which may lead to the formulation of a non-linear mathematical model in analysing a particular engineering problem.

2.2 THE GENERAL EQUATIONS OF MOTION

If a Lagrangian approach is adopted, the configuration of an n degree of freedom system can be expressed in terms of an n-vector, $\mathbf{q}$, of 'generalized coordinates', $q_1, q_2, \ldots, q_n$ (e.g. see Whittaker, 1937). Thus, assuming for simplicity that any constraints on the system do not depend explicitly on time, the rectangular coordinates x_i, y_i, z_i of the ith particle of mass in the system can be expressed as

$$\begin{aligned} x_i &= x_i(q_1, q_2, \ldots, q_n) \\ y_i &= y_i(q_1, q_2, \ldots, q_n) \\ z_i &= z_i(q_1, q_2, \ldots, q_n) \end{aligned} \tag{2.1}$$

For a virtual, infinitesimally small displacement of the system, expressible as $\delta\mathbf{q}$, and assuming that the system is 'holonomic' (see Whittaker, 1937), the virtual work done may be written as

$$\delta W = \mathbf{Q}^{\mathrm{T}}\delta\mathbf{q} \tag{2.2}$$

where $\mathbf{Q}$ is the n-vector of 'generalized forces', $Q_1, Q_2, \ldots, Q_n$. The elements of $\mathbf{Q}$ relate to all the various kinds of forces acting on the system, both internal and external. Using D'Alembert's principle, the 'inertial forces' required to accelerate the masses in the system may also be incorporated into $\mathbf{Q}$. It is convenient to distinguish the various contributions to $\mathbf{Q}$ as follows

$$\mathbf{Q} = \mathbf{Q}^{\mathrm{I}} + \mathbf{Q}^{\mathrm{in}} + \mathbf{Q}^{\mathrm{ex}} \tag{2.3}$$

where $\mathbf{Q}^{\mathrm{I}}$, $\mathbf{Q}^{\mathrm{in}}$ and $\mathbf{Q}^{\mathrm{ex}}$ are, respectively, the generalized inertial, internal and external force vectors. Thus $\mathbf{Q}^{\mathrm{I}}$ derives from the inertial forces, $\mathbf{Q}^{\mathrm{in}}$ from the non-inertial internal forces, such as restoring and damping forces, and $\mathbf{Q}^{\mathrm{ex}}$ from the externally applied forces.

With inertial forces included in $\mathbf{Q}$, one can use the methods of statics to formulate the equations of motion. In particular, the principle of virtual work may be used; this states that, during a virtual displacement, $\delta\mathbf{q}$,

$$\delta W = 0 \tag{2.4}$$

On combining equations (2.2) to (2.4) it is evident that, if equation (2.4) is to be satisfied for any arbitrary $\delta\mathbf{q}$, then one must have $\mathbf{Q} = \mathbf{0}$, or, alternatively,

$$\mathbf{Q}^{\mathrm{I}} + \mathbf{Q}^{\mathrm{in}} + \mathbf{Q}^{\mathrm{ex}} = \mathbf{0} \tag{2.5}$$

This equation represents a set of n equations of motion; i.e.

$$Q_i^{\mathrm{I}} + Q_i^{\mathrm{in}} + Q_i^{\mathrm{ex}} = 0 \quad i = 1, 2, \ldots, n \tag{2.6}$$

In general these equations will be differential in nature, and coupled.

The inertial force component, $\mathbf{Q}^{\mathrm{I}}$, can be derived directly from the scalar 'kinetic energy function', T, of the systems. The symbol T represents the total, instantaneous kinetic energy of the system. It is given by the expression

$$T = \sum_i m_i(\dot{x}_i^2 + \dot{y}_i^2 + \dot{z}_i^2) \tag{2.7}$$

where m_i is the mass of the ith particle and the summation is over all particles in the system. It can be shown that (e.g. see Whittaker, 1937)

$$Q_i^{\mathrm{I}} = -\frac{\mathrm{d}}{\mathrm{d}t}\left(\frac{\partial T}{\partial \dot{q}_i}\right) + \frac{\partial T}{\partial q_i} \quad i = 1, 2, \ldots, n \tag{2.8}$$

Hence, on combining equations (2.6) and (2.8) one has

$$\frac{\mathrm{d}}{\mathrm{d}t}\left(\frac{\partial T}{\partial \dot{q}_i}\right) - \frac{\partial T}{\partial q_i} = Q_i^* \quad i = 1, 2, \ldots n \tag{2.9}$$

where Q_i^* are the elements of the vector $\mathbf{Q}^*$ defined by

$$\mathbf{Q}^* = \mathbf{Q}^{\mathrm{in}} + \mathbf{Q}^{\mathrm{ex}} \tag{2.10}$$

Equation (2.9) is generally known as Lagrange's equation.

The internal forces acting on a system may be classified as either conservative, or non-conservative. Conservative forces are such that the work done by them, in a displacement of the system, from one configuration to the next, depends only on the initial and final configurations of the system, and not on the *path* taken between the two configurations. Gravitational forces and elastic restoring forces are examples of forces which are conservative in nature. Non-conservative forces are associated with path-dependent work and normally involve energy dissipation. Thus it is convenient to decompose $\mathbf{Q}^{\mathrm{in}}$ as follows

$$\mathbf{Q}^{\mathrm{in}} = \mathbf{Q}^{\mathrm{c}} + \mathbf{Q}^{\mathrm{nc}} \tag{2.11}$$

where $\mathbf{Q}^{\mathrm{c}}$ is derived from conservative forces and $\mathbf{Q}^{\mathrm{nc}}$ from non-conservative, dissipative forces.

Conservative forces can be expressed in terms of a scalar 'potential energy function', V, which depends only on $q_1, q_2, \ldots, q_n$. It may be defined as the work done by conservative forces during a displacement of the system from an original configuration, $\mathbf{q}$, to some standard configuration, $\mathbf{q}^*$ say. Considering a small virtual displacement, $\delta\mathbf{q}$, again, it is evident that the virtual work done by the conservative forces, during such an infinitesimal displacement, can be written as

$$-\sum_{i=1}^{n} \frac{\partial V}{\partial q_i} \delta q_i \tag{2.12}$$

and hence the ith component of $\mathbf{Q}^{\mathrm{c}}$ is given by

$$Q_i^{\mathrm{c}} = -\frac{\partial V}{\partial q_i} \tag{2.13}$$

Combining this result with equations (2.9) to (2.11) one obtains another version of Lagrange's equation, as follows

$$\frac{\mathrm{d}}{\mathrm{d}t}\left(\frac{\partial T}{\partial \dot{q}_i}\right) - \frac{\partial T}{\partial q_i} + \frac{\partial V}{\partial q_i} - Q_i^{\mathrm{nc}} = Q_i^{\mathrm{ex}} \quad i = 1, 2, \ldots, n \tag{2.14}$$

In the special case where the non-conservative forces arise solely from linear, viscous damping effects, the components of $\mathbf{Q}^{\mathrm{nc}}$ can be derived from Rayleigh's scalar dissipation function, D, where D is equal to one half of the total rate of energy dissipation. One can show that (Whittaker, 1937)

$$Q_i^{\mathrm{nc}} = -\frac{\partial D}{\partial \dot{q}_i} \tag{2.15}$$

and hence the complete set of system equations can be expressed as

$$\frac{\mathrm{d}}{\mathrm{d}t}\left(\frac{\partial T}{\partial \dot{q}_i}\right) - \frac{\partial T}{\partial q_i} + \frac{\partial V}{\partial q_i} + \frac{\partial D}{\partial \dot{q}_i} = Q_i^{\mathrm{ex}} \quad i = 1, 2, \ldots, n \tag{2.16}$$

The inertial contributions to the equations of motion, as derived from the first two terms on the left-hand side of equation (2.16), can be expressed in explicit form. First it is noted, from equation (2.1), that the velocity components for the ith particle can be written as

$$\dot{x}_i = \frac{\partial x_i}{\partial q_1}\dot{q}_i + \frac{\partial x_i}{\partial q_2}\dot{q}_2 + \cdots + \frac{\partial x_i}{\partial q_n}\dot{q}_n \tag{2.17}$$

and similarly for $\dot{y}_i$ and $\dot{z}_i$. Thus the velocities are homogeneous, linear functions of the generalized velocities, $\dot{q}_1, \dot{q}_2, \ldots, \dot{q}_n$. The differential coefficients, such as $\partial x_i/\partial q_1$, are, however, not constant but will depend, in general, on $q_1, q_2, \ldots, q_n$. Squaring equation (2.17), and the companion expressions for $\dot{y}_i$ and $\dot{z}_i$, and substituting into equation (2.7), it can be seen that T must be a homogeneous function of second degree in $\dot{q}_1, \dot{q}_2, \ldots, \dot{q}_n$. Thus, T can be expressed as

$$T = \tfrac{1}{2}(m_{11}\dot{q}_1^2 + m_{22}\dot{q}_2^2 + \cdots + m_{nn}\dot{q}_n^2 + 2m_{12}\dot{q}_1\dot{q}_2 + \cdots) \tag{2.18}$$

where m_{ij}, will, in general, be functions of the elements of $\mathbf{q}$. Equation (2.18) can be expressed compactly as

$$T = \tfrac{1}{2}\dot{\mathbf{q}}^{\mathrm{T}}\mathbf{M}\dot{\mathbf{q}} \tag{2.19}$$

where

$$\mathbf{M} = \begin{bmatrix} m_{11} & m_{12} & \cdots & m_{1n} \\ \vdots & & & \\ m_{n1} & m_{n2} & \cdots & m_{nn} \end{bmatrix} \tag{2.20}$$

The $n \times n$ matrix $\mathbf{M}$ is usually called the 'inertia matrix' (or 'mass matrix'). Substituting equation (2.18) into the first two terms of equation (2.14) one obtains

$$\frac{\mathrm{d}}{\mathrm{d}t}\left(\frac{\partial T}{\partial \dot{q}_i}\right) - \frac{\partial T}{\partial q_i} = \sum_{s=1}^{n} m_{is}\ddot{q}_s + \sum_{l=1}^{n}\sum_{m=1}^{n}\begin{bmatrix} l\ m \\ i \end{bmatrix}\dot{q}_l\dot{q}_m \tag{2.21}$$

where the 'Christoffel symbol' $\begin{bmatrix} l & m \\ & i \end{bmatrix}$, denotes the expression

$$\frac{1}{2}\left(\frac{\partial m_{li}}{\partial q_m}+\frac{\partial m_{mi}}{\partial q_l}-\frac{\partial m_{lm}}{\partial q_i}\right)$$

A combination of equation (2.21) with equation (2.14) will generate, for a specific n degree of freedom system, a complete set of n coupled, non-linear differential equations of motion. For the specific case of a single degree of freedom system, where $n = 1$, one finds that the single general equation of motion is of the form

$$m_{11}(q_1)\ddot{q}_1+\frac{1}{2}\frac{dm_{11}(q_1)}{dq_1}\dot{q}_1^2+\frac{\partial V}{\partial q_1}-Q_1^{nc}=Q_1^{ex} \tag{2.22}$$

Clearly here, as in the more general n degree of freedom case, non-linearities may occur in the inertially dependent terms, derived from $m_{11}(q_1)$, the conservative force term, $\partial V/\partial q_1$ and the non-conservative, dissipative term, Q_1^{nc}.

It is noted that, throughout the above analysis, the word 'displacement' can be interpreted to mean either a linear, translatory displacement, or a rotational displacement. Similarly, the word 'force' can be interpreted to mean either a force or a moment. Where both translational and rotational displacements occur in a system there is a correspondence between forces and translational displacements, and moments and rotational displacements, in the vectors $\mathbf{Q}$ and $\mathbf{q}$.

2.2.1 Small vibrations

For most of the systems of concern in this book, vibration occurs about a single static equilibrium position, as the result of externally applied, fluctuating forces. Since the dissipative forces will be zero when the system is at rest, it follows from equation (2.14) that, in this condition (i.e., with $Q_i^{ex} = 0$, for $i = 1, 2, \ldots, n$), one must have

$$\frac{\partial V}{\partial q_i}=0 \quad i=1,2,\ldots,n \tag{2.23}$$

where the derivatives are evaluated at the equilibrium position. One can, without loss of generality, measure displacements $q_1, q_2, \ldots, q_n$ from the static equilibrium configuration; thus $\mathbf{q} = \mathbf{0}$ represents this configuration. Then, from equation (2.23) one has

$$\left(\frac{\partial V}{\partial q_i}\right)_{q_i=0}=0 \quad i=1,2,\ldots,n \tag{2.24}$$

If the vibration of the system, as a result of non-zero Q_i^{ex} $(i = 1, 2, \ldots, n)$, involves only *small* displacements of the system, about the static equilibrium configuration, $\mathbf{q} = \mathbf{0}$ then it is often possible to obtain a very useful, linear approximation of the equations of motion.

Considering, for simplicity, the case of a single degree of freedom system, initially, one has, in general, from equation (2.18),

$$T = \tfrac{1}{2} m_{11}(q_1)\dot{q}_1^2 \tag{2.25}$$

The inertial coefficient, $m_{11}(q_1)$ can be expanded as a McLaurin series about the static equilibrium position, $q_1 = 0$. Thus

$$m_{11}(q) = m_{11}(0) + q_1\left(\frac{\mathrm{d}m_{11}}{\mathrm{d}q_1}\right)_{q_1=0} + \text{higher order terms} \tag{2.26}$$

If q_1 is sufficiently small then the first term in this expansion will be dominant, and one can set

$$m_{11}(q) = m_{11}(0) \equiv m_{11} \quad \text{(a constant)} \tag{2.27}$$

to a good approximation.

A similar expansion can be applied to the potential energy function $V(q_1)$. Then

$$V(q_1) = V(0) + q_1\left(\frac{\mathrm{d}V}{\mathrm{d}q_1}\right)_{q_1=0} + \frac{1}{2}q_1^2\left(\frac{\mathrm{d}^2V}{\mathrm{d}q_1^2}\right)_{q_1=0} + \text{higher order terms} \tag{2.28}$$

Now $V(0) = 0$ if, without loss of generality, one measures potential energy with reference to the static equilibrium position. Furthermore, one has, from equation (2.24), $\mathrm{d}V/\mathrm{d}q_1 = 0$ at $q_1 = 0$. Hence, the first non-zero term in the expansion of equation (2.28) is of order q_1^2, i.e. is quadratic in q_1. For sufficiently small q_1 this quadratic term becomes dominant and one can approximate as follows

$$V(q_1) = \tfrac{1}{2} k_{11} q_1^2 \tag{2.29}$$

where

$$k_{11} = \left(\frac{\mathrm{d}^2V}{\mathrm{d}q_1^2}\right)_{q_1=0} \tag{2.30}$$

This approximation is illustrated graphically in Figure 2.1. It is noted that, for stability, it is essential that k_{11} be positive. This corresponds to a conservative force which tends to restore the system to its static equilibrium position.

If the approximations indicated by equations (2.27) and (2.29) are substituted into the general equation of motion for a single degree of freedom, given by equation (2.22), then one obtains

$$m_{11}\ddot{q}_1 + k_{11}q_1 - Q_1^{\text{nc}} = Q_1^{\text{ex}} \tag{2.31}$$

The first two terms on the left-hand side of the equation are seen to be linear.

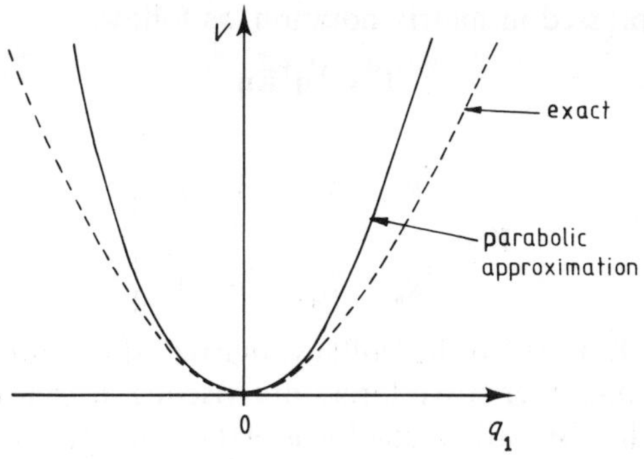

Figure 2.1
Approximate, parabolic representation of the potential energy function in the neighbourhood of the equilibrium position

Thus the approximate quadratic forms for T and V, given by equation (2.25), (with m_{11} a constant) and equation (2.29) correspond to a linear approximation of the inertial and conservative restoring force contributions to the equation of motion. If one further assumes that the dissipative term, Q_1^{nc}, arises from viscous damping effects then the dissipation function D is also quadratic in form, i.e.

$$D = \tfrac{1}{2} c_{11} \dot{q}_1^2 \tag{2.32}$$

where c_{11} is a constant. Applying equation (2.15) one has

$$Q_1^{nc} = -c_{11} \dot{q}_1 \tag{2.33}$$

and hence, from equation (2.22)

$$m_{11} \ddot{q}_1 + c_{11} \dot{q}_1 + k_{11} q_1 = Q_1^{ex} \tag{2.34}$$

This equation of motion is now completely linear and is, in fact, a general form of the linearized equation of motion for systems of the single degree of freedom type.

These considerations may be easily extended to the multi-degree of freedom situation. Thus, if all the elements in $\mathbf{q}$ are small, the coefficients m_{ij} $(i, j = 1, 2, \ldots, n)$ in equation (2.18) may be approximated by their constant values at $\mathbf{q} = \mathbf{0}$. Thus, in equation (2.19), the elements in $\mathbf{M}$ may be treated as constants. Moreover, $V(\mathbf{q})$ may be expanded about $\mathbf{q} = \mathbf{0}$, using a multi-dimensional equivalent of equation (2.28). Again the first non-zero terms in this expansion are quadratic in $q_1, q_2, \ldots, q_n$. It follows that, for small $\mathbf{q}$, the following homogeneous quadratic form for V may be employed

$$V = \tfrac{1}{2}(k_{11} q_1^2 + k_{22} q_2^2 + \cdots + k_{nn} q_n^2 + 2k_{12} q_1 q_2 + \cdots) \tag{2.35}$$

where k_{ij} $(i, j = 1, 2, \ldots, n)$ are constants. This form, by analogy with equation

(2.19), may be expressed in matrix notation, as follows

$$V = \tfrac{1}{2}\mathbf{q}^{\mathrm{T}}\mathbf{K}\mathbf{q} \tag{2.36}$$

where

$$\mathbf{K} = \begin{bmatrix} k_{11} & k_{12} & \cdots & k_{1n} \\ \vdots & & & \\ k_{n1} & k_{n2} & \cdots & k_{nn} \end{bmatrix} \tag{2.37}$$

The $n \times n$ matrix, **K**, is called the 'stiffness matrix'. If one further supposes that the dissipative, non-conservative forces are viscous in character then (e.g. see Whittaker, 1937) the dissipation function is of the homogeneous quadratic form

$$D = \tfrac{1}{2}(c_{11}\dot{q}_1^2 + c_{22}\dot{q}_2^2 + \cdots + c_{nn}\dot{q}_n^2 + 2c_{12}\dot{q}_1\dot{q}_2 + \cdots) \tag{2.38}$$

where c_{ij} $(i, j = 1, 2, \ldots, n)$ are constants. The quantity D may be alternatively represented as

$$D = \tfrac{1}{2}\dot{\mathbf{q}}^{\mathrm{T}}\mathbf{C}\dot{\mathbf{q}} \tag{2.39}$$

where

$$\mathbf{C} = \begin{bmatrix} c_{11} & c_{12} & \cdots & c_{1n} \\ \vdots & & & \\ c_{n1} & c_{n2} & \cdots & c_{nn} \end{bmatrix} \tag{2.40}$$

The matrix **C** is called the 'damping matrix'. Substitution of the quadratic forms for T, V and D, given by equation (2.18) (with m_{ij} constants), equations (2.35) and (2.38), respectively, into equation (2.16) results in the following compact matrix representation of the general, linearized equations of motion for an n degree of freedom system

$$\mathbf{M}\ddot{\mathbf{q}} + \mathbf{C}\dot{\mathbf{q}} + \mathbf{K}\mathbf{q} = \mathbf{Q} \tag{2.41}$$

where

$$\mathbf{Q} = \begin{bmatrix} Q_1^{\mathrm{ex}} \\ \vdots \\ Q_n^{\mathrm{ex}} \end{bmatrix} \tag{2.42}$$

now denotes the n-vector of generalized external forces. It can be shown that **M**, **C** and **K** are, in general, all symmetric matrices (e.g. see Bishop *et al.*, 1965).

2.2.2 Large vibrations

In many practical applications it is convenient to separate out the linear and non-linear contributions to the equation of motion. Clearly the non-linear contributions can be expected to become increasingly significant as the overall amplitude of vibration increases.

To distinguish between linear and non-linear contributions one can write

$$T = \tfrac{1}{2}\dot{\mathbf{q}}^{\mathrm{T}}\mathbf{M}\dot{\mathbf{q}} + T_{\mathrm{n}}(\mathbf{q}, \dot{\mathbf{q}}, \ddot{\mathbf{q}}) \tag{2.43}$$

$$V = \tfrac{1}{2}\mathbf{q}^{\mathrm{T}}\mathbf{K}\mathbf{q} + V_{\mathrm{n}}(\mathbf{q}) \tag{2.44}$$

where T_{n} and V_{n} are the non-linear components of T and V, respectively. The dissipative terms, Q_i^{nc}, in equation (2.14) can be similarly decomposed into linear and non-linear contributions. The linear contribution may, from equations (2.15) and (2.38), be written as

$$Q_i^{\mathrm{nc}} = -\sum_{s=1}^{n} c_{is}\dot{q}_s \tag{2.45}$$

and hence, in general, one can write

$$Q_i^{\mathrm{nc}} = -\sum_{s=1}^{n} c_{is}\dot{q}_s + Q_{in}^{\mathrm{nc}}(\mathbf{q}, \dot{\mathbf{q}}, \ddot{\mathbf{q}}) \tag{2.46}$$

where Q_{in}^{nc} is the non-linear component of Q_i^{nc}. If these expressions are substituted into equation (2.14) one finds that a general form of the non-linear equations of motion, suitable for large amplitude motion, is as follows

$$\mathbf{M}\ddot{\mathbf{q}} + \mathbf{C}\dot{\mathbf{q}} + \mathbf{K}\mathbf{q} + \boldsymbol{\Psi}(\mathbf{q}, \dot{\mathbf{q}}, \ddot{\mathbf{q}}) = \mathbf{Q}(t) \tag{2.47}$$

where $\boldsymbol{\Psi}(\mathbf{q}, \dot{\mathbf{q}}, \ddot{\mathbf{q}})$ is some non-linear function of $\mathbf{q}$ and its derivatives.

It should be noted that it has been implicitly assumed here that the dissipative non-linearity depends only on the instantaneous motion, as described by $\mathbf{q}, \dot{\mathbf{q}}$ and $\ddot{\mathbf{q}}$. It will be shown later in this chapter that dissipative forces frequently are hysteretic in nature—in these circumstances, the zero-memory non-linearity function $\boldsymbol{\Psi}(\mathbf{q}, \dot{\mathbf{q}}, \ddot{\mathbf{q}})$ is inappropriate.

In the remainder of this chapter various sources of non-linearity, which can occur in engineering vibration problems, will be briefly discussed.

2.3 NON-LINEAR CONSERVATIVE FORCES

In most mechanical and structural systems, forces which tend to return the system to its static equilibrium position arise from one or more of sources such as

gravity field;
hydrostatic pressures on floating bodies;
internal stresses generated in deformed solids.

Here it will be shown, through the use of simple examples, how non-linearities can arise in restoring forces derived from each of the above sources. In all cases the forces involved are conservative in nature, or at least approximately so in the case of deformed solids.

2.3.1 Motion in a gravitational field

Consider the motion of the simple pendulum shown in Figure 2.2. The pendulum consists of a mass m (approximated here as a point-mass) which is attached to the lower end of a light, rigid rod, of length l. The upper end of the rod is free to pivot about a fixed point, at 0.

If attention is restricted to oscillation of the pendulum in a vertical plane,then one has a single degree of freedom system, approximately. An appropriate coordinate to define the motion is the angle Φ which the rod makes with the vertical through 0; the latter is, of course, the static equilibrium position. Clearly Φ can be identified as an appropriate generalized coordinate, i.e. $\Phi = q_1$.

The kinetic energy of the system is given by

$$T = \tfrac{1}{2}m(l\dot{q}_1)^2 \tag{2.48}$$

and the (gravitational) potential energy by

$$V = mgl(1 - \cos q_1) \tag{2.49}$$

It is seen that T is of the quadratic form which leads to a linear inertial term in the equation of motion. However V is only approximately quadratic, when q_1 is small. Thus, expanding $\cos q_1$ in powers of q_1 one has

$$V = mgl\left(\frac{q_1^2}{2} - \frac{q_1^4}{24}\right) + \text{higher order terms} \tag{2.50}$$

For sufficiently small q_1 one can use the quadratic form

$$V = \frac{mgl}{2} q_1^2 \tag{2.51}$$

as a 'linear' approximation for V.

Substituting the 'exact' expressions for T and V into equation (2.14), and

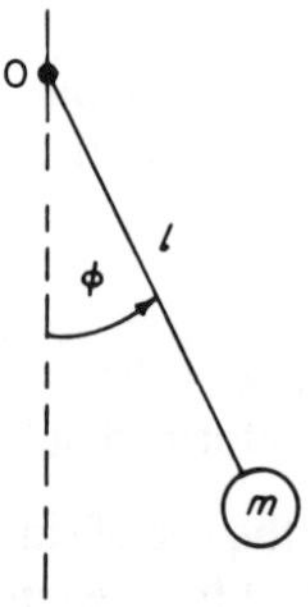

Figure 2.2
The simple pendulum

assuming that dissipative forces are negligible, the following equation of motion is obtained for free oscillations

$$\ddot{q}_1 + \frac{g}{l}\sin q_1 = 0 \tag{2.52}$$

Here the non-linear restoring term, proportional to sin q_1, renders the complete system non-linear.

For small amplitude vibration, using equation (2.51) in place of equation (2.49), the linear equation of motion is found to be

$$\ddot{q}_1 + \frac{g}{l} q_1 = 0 \tag{2.53}$$

This approximation can be improved by including the next term in the McLaurin expansion for V (see equation (2.28)). Hence, for moderate amplitudes of oscillation one has the non-linear equation

$$\ddot{q}_1 + \frac{g}{l}\left(q_1 - \frac{q_1^3}{6}\right) = 0 \tag{2.54}$$

Figure 2.3 shows a comparison between the exact restoring characteristic,

$$\lambda(q_1) = \frac{g}{l}\sin q_1 \tag{2.55}$$

the linear approximation, contained in equation (2.53), and the linear-plus-cubic approximation contained in equation (2.54). It is evident that, in this example, the non-linearity is of the 'softening' kind, i.e. the slope of the characteristic, which is a measure of the 'stiffness' associated with the restoring force, tends to decrease as the amplitude increases.

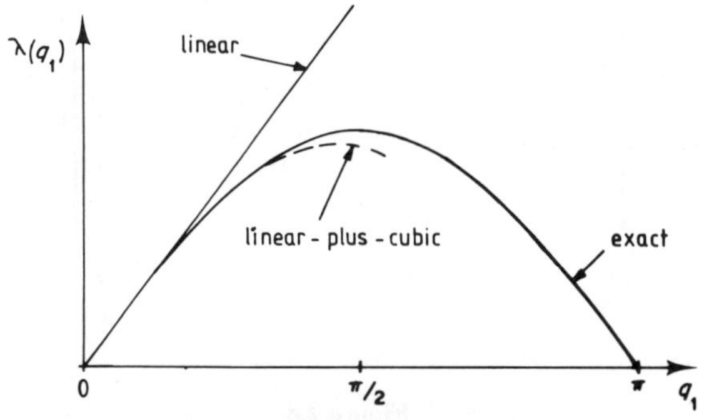

Figure 2.3
Restoring force characteristic of the simple pendulum

2.3.2 Restoring moments for floating bodies

Consider an effectively rigid floating body, symmetrical about a vertical plane. This may be, for example, a ship, or an offshore drilling platform. In the static equilibrium position the body will be as sketched in the sectional view of Figure 2.4a. Here the buoyancy force, B, is equal to the weight of the body, W, and the lines of action of these two forces are coincident, in the plane of symmetry, both passing through the centre of gravity of the body, at G.

Suppose that the body is given a rotational displacement, Φ, as shown in Figure 2.4b, such that the mass of liquid displaced by the body remains unaltered. This can be achieved by applying a pure couple to the body. The buoyancy force remains equal to B, but in the displaced position the lines of action of B and W are no longer coincident. The buoyancy force will intersect the plane of symmetry at a point, M, known as the 'metacentre'; for small displacements M is effectively a fixed point. Clearly, for stability it is essential that the metacentre lies above the centre of gravity. The distance between G and M, GM, usually called the 'metacentric height', is obviously a parameter of prime importance in the design of floating bodies.

The distance between the lines of action of W and B, give by GZ in Figure 2.4b, is known as the 'righting lever', and the restoring moment is directly proportional to this distance. For small rotational displacements, one has $GZ \sim GM\,\Phi$ so that the righting lever is linearly related to the rotational displacement. However, for larger displacements GZ varies in a distinctly non-linear manner with Φ, the behaviour depending on the shape of the body. Figure 2.5 shows a typical GZ—Φ characteristic for a fishing trawler (see Brook, 1986). It can be seen that, for small angles of rotation ($\Phi < 10°$), a linear approximation is reasonably

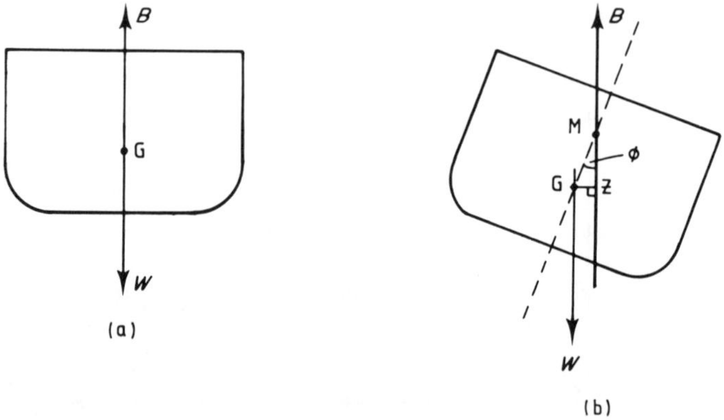

Figure 2.4
Forces acting on a floating body. (a) The static equilibrium position. (b) Effect of rotational displacement

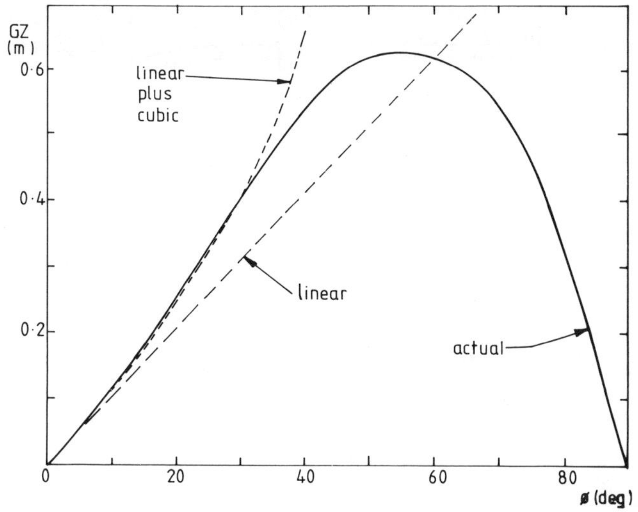

Figure 2.5
A typical GZ characteristic for a fishing trawler (reproduced from Brook (1986) by permission of the Royal Institution of Naval Architects, London)

accurate. For roll angles up to 30°, a 'hardening' linear-plus-cubic representation is more accurate, as Figure 2.5 shows. At very large angular displacements the GZ characteristic is of the 'softening kind', GZ reducing to zero at 90°. The condition $GZ = 0$, corresponds to incipient capsize; since there is no effective restoring moment, a small perturbation will cause the body to roll over.

2.3.3 Elastic restoring forces

Many components in mechanical and structural systems produce a restoring force, by virtue of some degree of elastic deformation. Some typical components with this 'spring-like' property are sketched in Figure 2.6. The variation of applied static force, which is equal and opposite to the restoring force, with resulting deformation depends, of course, primarily on the constitutive properties of the material of which the component is made. In many cases, particularly if the deformation is small, the force–displacement relationship obeys Hooke's law, i.e. is linear, as sketched in Figure 2.7. However, at large displacements, deviations from the linear characteristics often occur. Such deviations may be of either the softening or hardening type (see Figure 2.7).

The word 'elastic' in this context means that the material does not permanently deform when subjected to external loads. Thus if a material behaves elastically, it will return to its original state when the external loads are removed. Evidently

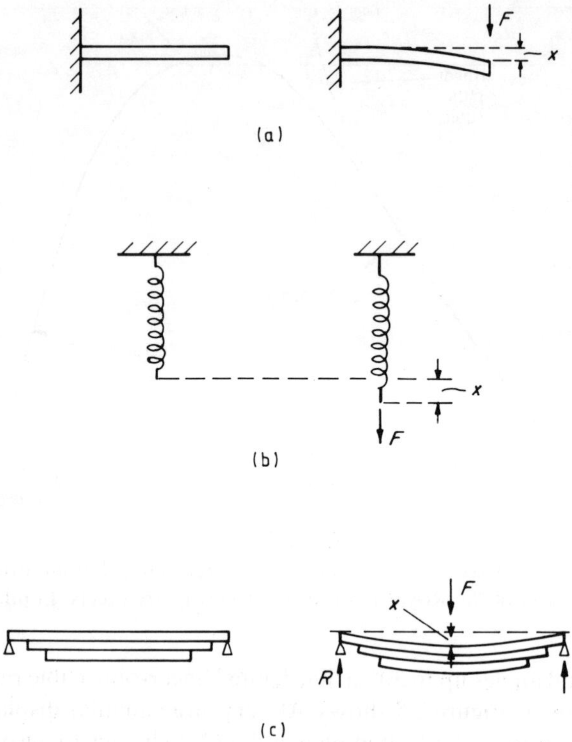

Figure 2.6
Typical deformable elements (a) Cantilevered beam. (b) Helical spring. (c) Leaf spring

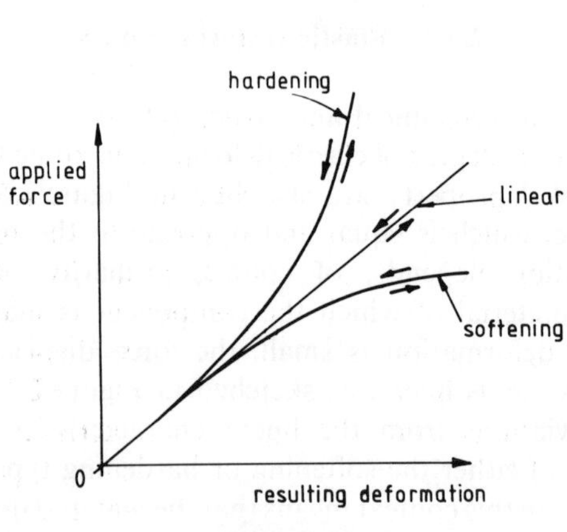

Figure 2.7
Force–displacement characteristics for elastic elements

elastic restoring forces are of the conservative type. Some materials are such that they remain approximately elastic, even though their constitutive property as expressed by their stress–strain curves is distinctly non-linear. For example, Figure 2.8 shows some typical non-linear stress–strain curves obtained experimentally for various types of rubber. Such materials occur frequently in isolation mountings for various types of machinery.

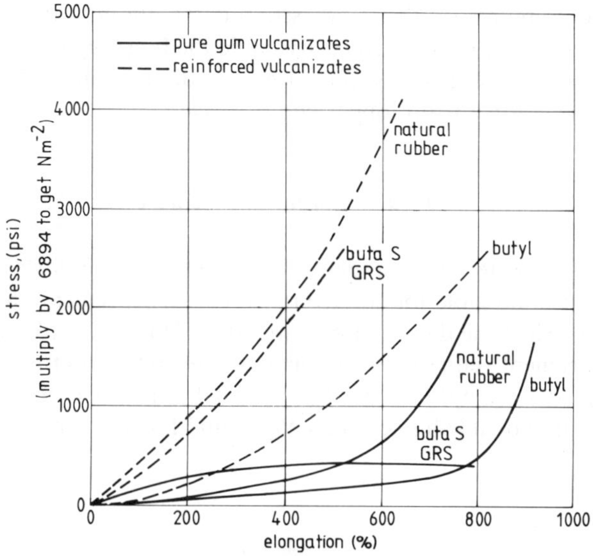

Figure 2.8

Stress–strain curves for some rubber vulcanizates (reproduced from Schmidt and Marlies (1948) by permission of McGraw-Hill)

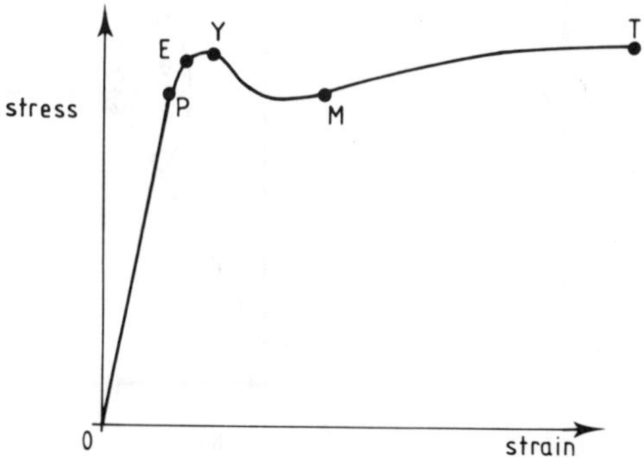

Figure 2.9

Stress–strain curve for mild steel. Yield point $Y \rightarrow M$ region of high ductility, $M \rightarrow T$ strain hardening, T ultimate tensile strength

For many metallic materials the loading range for which non-linear elastic behaviour occurs is quite limited, and, beyond this, permanent deformation results, which is associated with plastic behaviour. Figure 2.9 sketches the stress–strain characteristic of mild steel. In the range O to P the material behaves linearly, and elastically, to a very close approximation. From P to E elastic behaviour still prevails, but the material now behaves non-linearly. Beyond E, the 'elastic limit', some degree of permanent, plastic deformation occurs. Several other significant points on the stress–strain curve are indicated in Figure 2.9.

As discussed later in this chapter, dynamic loading, beyond the elastic limit of the material will result in significant energy dissipation, through hysteresis, and the restoring force is no longer conservative.

2.3.4 Non-linear elasticity

Even when the material of a deformed body obeys a linear stress–strain relationship, it is possible for the overall force–displacement characteristic to be non-linear, when the deformations are large. This is due to the fact that the strain–displacement equations become significantly non-linear, for large deformations. An example of such non-linear behaviour is the case of a plate supported by immovable hinges at its edges, and loaded transversely. For plate

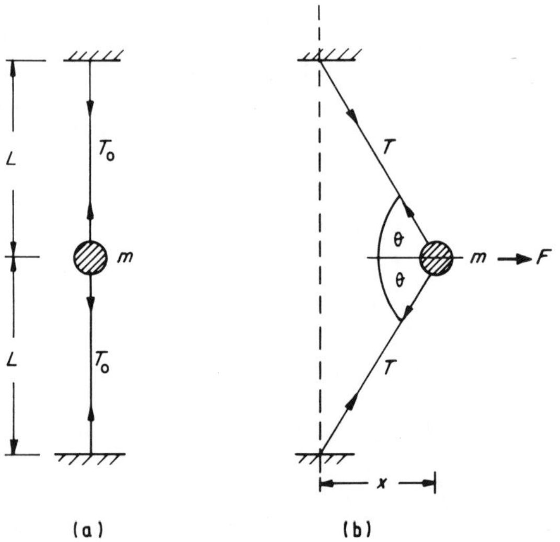

Figure 2.10
Simple mass–spring system. (a) Static-equilibrium position. (b) Deformed position as a result of applied force

deflections that are of the same order as the plate thickness, membrane strains, which depend non-linearly on plate deformation, become significant, causing a non-linear load–deflection relationship, of the hardening kind (e.g. see Lin, 1967). The general theory for large deformations of elastic bodies is well developed, and may be found in a number of text books (e.g. see Timoshenko and Woinowsky–Krieger, 1959; Stoker, 1968).

Here this type of non-linearity will be illustrated through an analysis of the very simple system shown in Figure 2.10. It consists of a mass m, centrally mounted on a uniform string, of total length $2L$. It will be assumed that the string behaves linearly, so that the tension, T, is proportional to the elongation, δ. Thus, $T = k\delta$ where k, the 'spring stiffness', is a constant. In the static equilibrium position the system is as shown in Figure 2.10a, and there is an initial tension in the string, of magnitude T_0. It will be assumed that $T_0 \gg mg$, so that gravitational forces can be neglected.

If a transverse force, F, is applied to the mass, a transverse deflection, x will occur, as shown in Figure 2.10b. The relationship between F and x is easily found by considering the force equilibrium of the stationary mass. Thus

$$2T\cos\theta = F \tag{2.56}$$

where the tension T is given by

$$T = T_0 + k[(L^2 + x^2)^{1/2} - L] \tag{2.57}$$

Hence

$$F = 2\{T_0 + k[(L^2 + x^2)^{1/2} - L]\}\frac{x}{(L^2 + x^2)^{1/2}} \tag{2.58}$$

This is clearly a non-linear relationship between F and x.

When x is very small ($x \ll L$) equation (2.58) yields the linear asymptote

$$F \to \frac{2T_0}{L}x \tag{2.59}$$

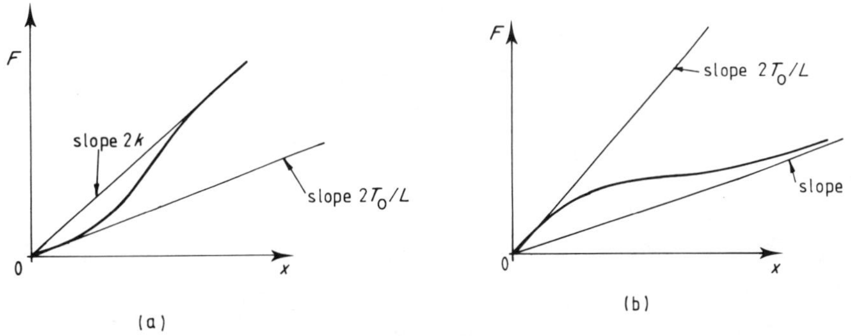

Figure 2.11
Force–displacement relationship. (a) $k > T_0/L$, (b) $k < T_0/L$

For larger values of x an expansion of $(L^2 + x^2)^{1/2}$ leads to

$$F = \frac{2T_0}{L}x + \frac{1}{L^2}\left(k - \frac{T_0}{L}\right)x^3 \tag{2.60}$$

correct to order x^3. This shows that for moderate displacements the non-linearity in the F–x relationship may be of either the hardening or softening kind, depending on whether $k > T_0/L$ or $k < T_0/L$, respectively. For very large x then, from equation (2.58), one has the linear asymptotic relationship

$$F = 2kx \tag{2.61}$$

The overall F–x relationship is sketched in Figure 2.11, for the two cases (a) $k > T_0/L$, (b) $k < T_0/L$.

Another related example of non-linear behaviour occurs in the mooring of ships and offshore platforms. Figure 2.12 shows, schematically, a simple mooring arrangement for a floating platform, consisting of two cables, AB and CD, attached to the ocean floor at A and D. If the vessel undergoes a horizontal displacement, x, then the net effect of the tensions in the cables is to provide a restoring force, F, returning the vessel to its equilibrium position. However, even if the tension–extension relationship for the cables is linear, the net F–x relationship is usually decidedly non-linear. Figure 2.13 shows a typical mooring system characteristic, together with the linear approximation for small amplitudes.

Another non-linear phenomenon can also be illustrated very easily by using a model similar to that in Figure 2.10. Figure 2.14 shows a very simple model of a shallow arch, comprising two linear springs, of stiffness k, pinned to rigid supports at A and B, and pinned together at C. In the static equilibrium position, the springs make an angle α to the horizontal.

With a vertical load, F, applied at C, the arch will deform to a new position, as shown in Figure 2.14b. The relationship between the deformation, as

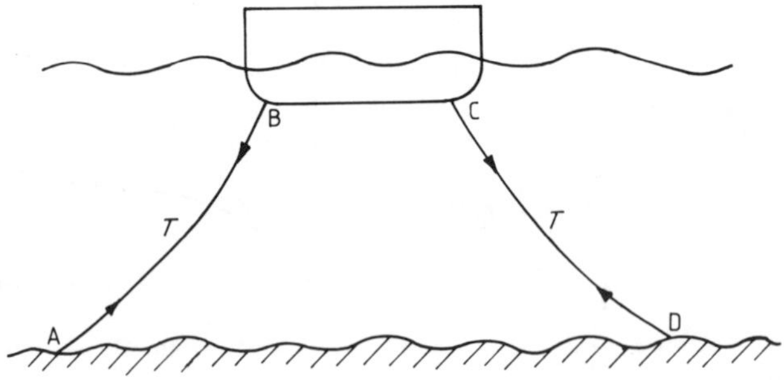

Figure 2.12
Simple mooring arrangement for a floating vessel

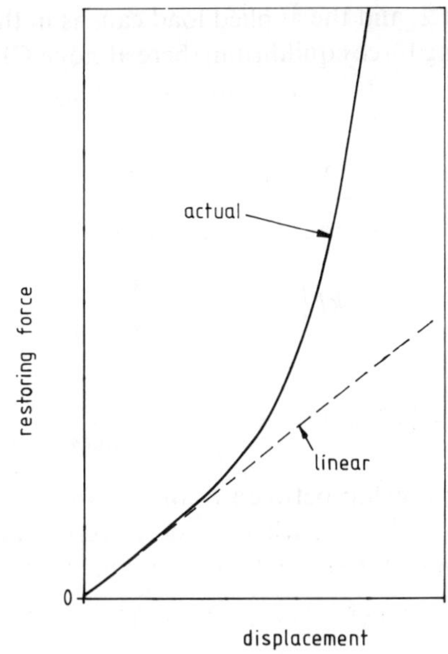

Figure 2.13
Typical mooring characteristic

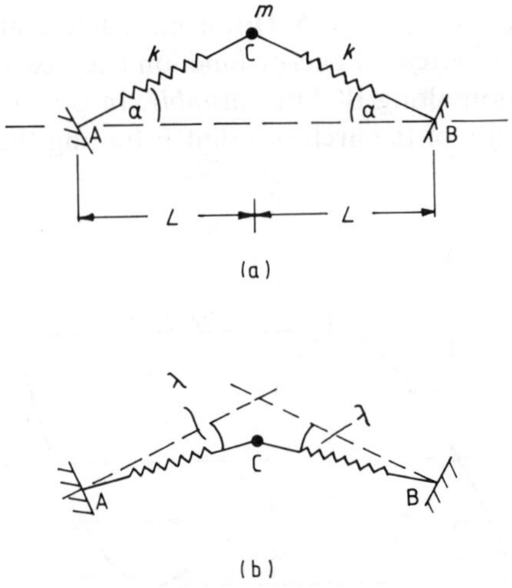

Figure 2.14
Model of a shallow arch. (a) One static equilibrium position. (b) Deformed position, as a result of applied force

measured by the angle λ, and the applied load can, as in the preceding example, be found by considering force equilibrium (here at node C). Assuming, as before, that gravitational forces are comparatively negligible, one has, for vertical equilibrium at C,

$$2F_s \sin(\alpha - \lambda) = F \tag{2.62}$$

and here the force in each spring, F_s, is given by

$$F_s = kL\left(\frac{1}{\cos\alpha} - \frac{1}{\cos(\alpha - \lambda)}\right) \tag{2.63}$$

Thus

$$F = 2kL\sin(\alpha - \lambda)\left(\frac{1}{\cos\alpha} - \frac{1}{\cos(\alpha - \lambda)}\right) \tag{2.64}$$

This non-linear relationship between F and λ can be simplified somewhat if one restricts attention to the case where α and λ are *small* angles. Then, making the usual approximations for sin and cos one finds that

$$F = kL(\alpha - \lambda)[\alpha^2 - (\alpha - \lambda)^2] = kL(2\alpha^2\lambda - 3\alpha\lambda^2 - \lambda^3) \tag{2.65}$$

For small λ, this reduces to the linear relationship

$$F = (2kL\alpha^2)\lambda \tag{2.66}$$

Figure 2.15 sketches the variation of F with λ. As the load increases, the deflection increases from A to B, along the softening characteristic shown. Equation (2.64) indicates that further deformation occurs, with F reducing, along the dotted path BCD in Figure 2.15. However, it can easily be shown, through a consideration of the potential energy function (e.g. see Thompson and Hunt, 1973), that the position along BCD is *unstable*. In practice, loading beyond B causes a 'snap-through' of the arch to point F (noting that there will be some

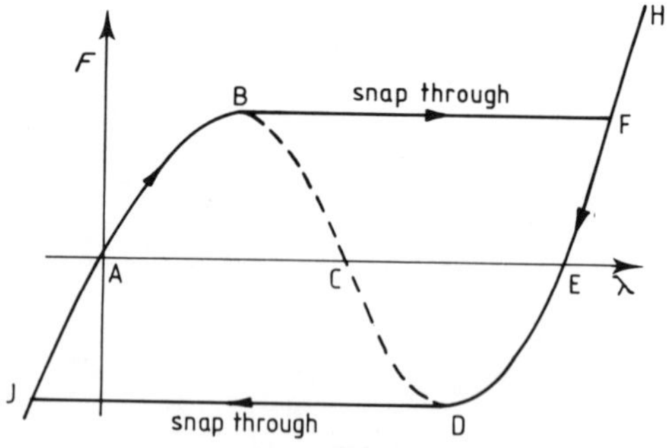

Figure 2.15
Force–displacement characteristic for the arch shown in Figure 2.14

transient vibration before the system comes to rest at F). Further loading will be along the 'hardening' characteristic, FH. If the loading is reversed there will be a snap-through, in the other direction, from D to J.

2.3.5 Geometric non-linearities

Mass–spring systems can be arranged, geometrically, so that the overall force–displacement is strongly non-linear.

As an example, consider the system shown in Figure 2.16a. Clearly the linear springs with stiffness k_2 become activated only when the horizontal displacement, x, is such that $|x| > a$, where a is the gap between the oscillating body of mass m and the springs. The equation of motion of the system can be written in the following manner

$$m\ddot{x} + f(x) = 0 \tag{2.67}$$

where

$$f(x) = 2k_1x \quad |x| < a \tag{2.68}$$

and

$$f(x) = 2k_1a + 2(x - a)(k_1 + k_2) \quad |x| > a \tag{2.69}$$

The piecewise linear force–displacement characteristic for this system is shown in Figure 2.16b.

The type of non-linearity shown in Figure 2.16 occurs in many applications where one wishes to limit amplitudes of vibration by introducing high stiffness 'buffers', or 'barriers'. Thus, for example, the flexural vibration of heat exchanger tubes is often limited through the introduction of a buffer, as illustrated in Figure 2.17a. Similarly fenders on dock-side walls often act in parallel with

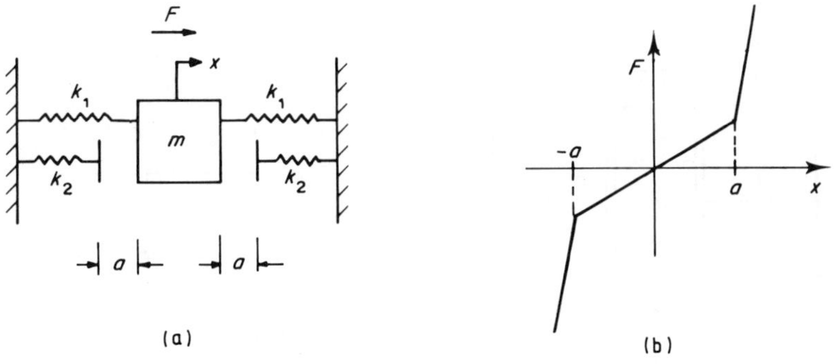

Figure 2.16
Effect of clearances in mass–spring systems. (a) Mass–spring system. (b) Force–displacement characteristic

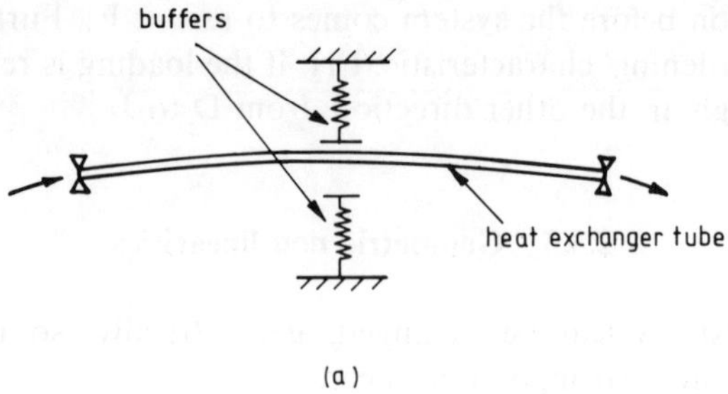

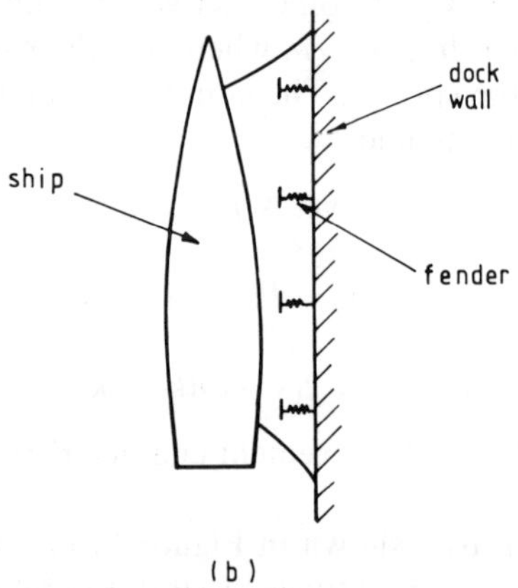

Figure 2.17
Examples of non-linear restrainers. (a) Heat-exchanger tube with stops. (b) Ship with mooring lines and fenders

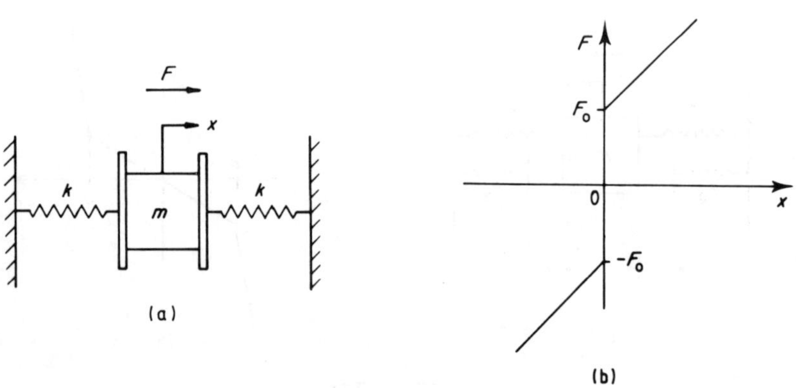

Figure 2.18
Set-up spring with stops. (a) Mass–spring system. (b) Force–displacement characteristic

mooring lines to produce an overall non-linear restoring characteristic (see Figure 2.17b).

Piecewise non-linear restoring characteristics can be obtained with various geometric configurations of linear springs. Another example is shown schematically in Figure 2.18. Here the 'set-up springs with stops' is such that the springs have some initial compressive force, F_0. Thus the mass cannot move until the applied force overcomes the initial compression.

2.4 NON-LINEAR DISSIPATIVE FORCES

The description of dissipative forces is usually inherently bound-up with the methods of estimating their magnitude, from experimental data. Frequently they are modelled on the basis of the free-decay behaviour of single degree of freedom systems and it is, therefore, appropriate to initiate the discussion here with an analysis of this particular case.

A general form of the equation of motion of a single degree of freedom system, with non-conservative, dissipative forces, was given earlier (see equation (2.22)). Assuming for simplicity that the inertial term is linear, this equation, for the case of free decay, may be written as

$$m\ddot{q} + \frac{\partial V}{\partial q} = Q^{\mathrm{nc}} \tag{2.70}$$

where the subscripts on m, q and Q^{nc} have been omitted. If the total energy of the system is defined as the sum of kinetic and potential energies, i.e.

$$E = \frac{m\dot{q}^2}{2} + V(q) \tag{2.71}$$

then it is possible to rewrite equation (2.70) very simply, in terms of E. Thus

$$\dot{E} = -L(E) \tag{2.72}$$

where

$$L(E) = -Q^{\mathrm{nc}}\dot{q} \tag{2.73}$$

The function $L(E)$ in equation (2.72) may be called a 'loss function' and represents the rate of energy loss, due to dissipative forces. It is measurable, if one has a knowledge of $V(q)$. Thus, supposing a free decay record is obtained, as sketched in Figure 2.19a. Successive peak amplitudes give a set of numerical values, $A_1, A_2, \ldots$ etc., as shown. These may be converted into energy values, using the relationship

$$E_i = V(A_i) \quad i = 1, 2, \ldots \tag{2.74}$$

and hence the variation of E_i with time may be plotted, as sketched in

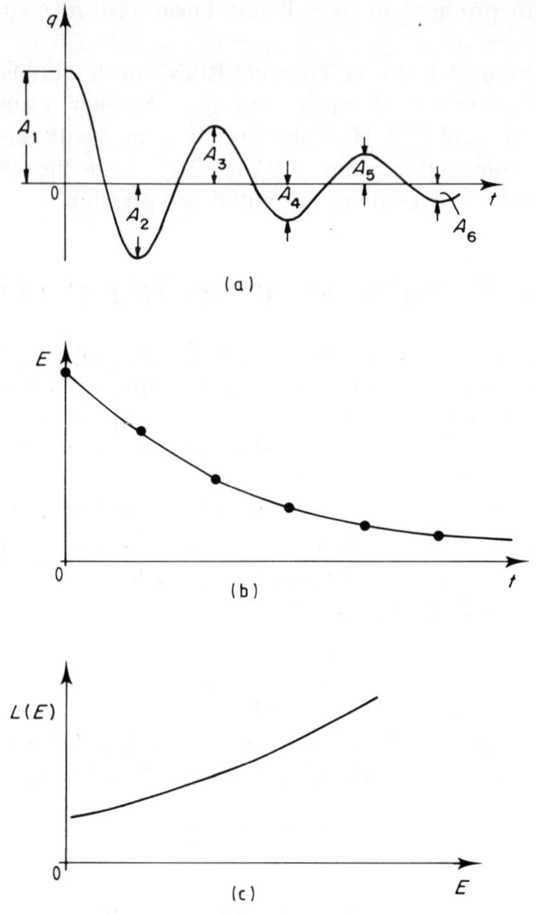

Figure 2.19
Determination of the loss function from free-decay data. (a) Free-decay curve. (b) Variation of energy level with time. (c) The loss function

Figure 2.19b. If the energy loss per cycle is fairly low then a good estimate of the gradient, $\dot{E}$, may be obtained, by fitting a curve through the points in Figure 2.19b. Hence one may obtain the loss function, as sketched in Figure 2.19c (see Roberts, 1985).

A useful reference case is that of linear damping and restoring forces. The appropriate equation of motion here is, from equation (2.34) (with $Q_1^{\text{ex}} = 0$, and omitting subscripts)

$$m\ddot{q} + c\dot{q} + kq = 0 \tag{2.75}$$

or

$$\ddot{q} + 2\zeta\omega_{\text{n}}\dot{q} + \omega_{\text{n}}^2 q = 0 \tag{2.76}$$

where

$$\omega_{\mathrm{n}} = \left(\frac{k}{m}\right)^{1/2} \tag{2.77}$$

is the undamped natural frequency and

$$\zeta = \frac{c}{2(km)^{1/2}} \tag{2.78}$$

is a non-dimensional damping factor. From equation (2.73) one has

$$L(E) = 2\zeta\omega_{\mathrm{n}} m\dot{q}^2 \tag{2.79}$$

If the damping is reasonably light ($\zeta \ll 1$) then one can write

$$\begin{aligned} q &= A\cos\omega_{\mathrm{n}}t \\ \dot{q} &= -\omega_{\mathrm{n}}A\sin\omega_{\mathrm{n}}t \end{aligned} \tag{2.80}$$

to a good approximation where A varies slowly, and is virtually constant over one cycle of oscillation, of period $T = 2\pi/\omega_{\mathrm{n}}$. Substituting equation (2.80) into equation (2.79) and, averaging over one cycle, one has, to a good approximation

$$L(E) = (2\zeta\omega_{\mathrm{n}})\frac{kA^2}{2} = 2\zeta\omega_{\mathrm{n}}E \tag{2.81}$$

A combination of equation (2.72) with equation (2.81) shows that, in the linear lightly damped case, E decays exponentially with time, i.e.

$$E = E_0\exp(-2\zeta\omega_{\mathrm{n}}t) \tag{2.82}$$

where E_0 is some initial value of E.

Equation (2.81) shows that the loss function, $L(E)$, increases linearly with E, in the case of a linear oscillator. In the non-linear case it is sometimes useful to define a non-linear damping factor, analogous to ζ, as follows

$$Q(E) = \frac{L(E)}{2\omega(E)E} \tag{2.83}$$

where $\omega(E)$ is the frequency of undamped oscillations obtained by solving

$$m\ddot{q} + \frac{\partial V}{\partial q} = 0 \tag{2.84}$$

Thus, in the linear case one evidently has

$$Q(E) = \zeta \quad \text{a constant} \tag{2.85}$$

and non-linearity in damping is evidenced by a variation of $Q(E)$ with amplitude level, as measured by E, or A, where $E = V(A)$.

If the dissipative force is non-linear, but of a known parametric form, then an explicit form for the loss function can be derived, by generalizing the preceding analysis for the linear case. For example, suppose that it is known that

$$Q^{\mathrm{nc}} = -a\dot{q} - b\dot{q}^3 \tag{2.86}$$

Here the dissipative force is a linear-plus-cubic function of the instantaneous velocity. From equations (2.86) and (2.73) one has

$$L(E) = a\dot{q}^2 + b\dot{q}^4 \tag{2.87}$$

Now, if the damping is light, a good approximation is to replace q in (2.87) by q', where q' is the solution for the undamped case as described by equation (2.84). One can then average over one cycle, $T(E) = 2\pi/\omega(E)$ to produce a loss function of the general form

$$L(E) = aH(E) + bG(E) \tag{2.88}$$

where $H(E)$ and $G(E)$ are known functions. One can estimate the parameters a and b by fitting the parametric form given by equation (2.88) to an experimentally determined $L(E)$ function, deduced by the procedure sketched in Figure 2.19. Details of this method of determining the parameters in explicit models of damping are given by Roberts (1985).

If the damping is not light then the parameters in an explicit model of the dissipative forces can be determined by using the powerful methodology of parametric identification (see Chapter 9 for an application of this general approach). A specific technique for estimating the parameters in non-linear damping models, from free-decay data, has been described by Gawthrop (1984). This approach has the advantage that it may be readily extended to deal with the case of forced excitation (e.g. see Gawthrop *et al.*, 1988).

2.4.1 Internal damping in materials

None of the materials which occur in mechanical and structural systems are perfectly elastic, i.e. the restoring forces generated as a result of deformations are not perfectly conservative. Inelastic behaviour is manifested in the form of a hysteresis loop, if the material is subjected to a sinusoidally varying load, Q'. Typical hysteresis loops are sketched in Figure 2.20. Here the instantaneous applied load is plotted against the resulting displacement, over a complete cycle. The area enclosed by the loops

$$E = \oint Q'\,\mathrm{d}q = \oint Q'\dot{q}\,\mathrm{d}t \tag{2.89}$$

where integration is over a complete cycle, is a measure of the energy dissipated as a result of internal friction within the material.

In some cases the hysteresis loops are extremely slim and the energy dissipated is consequently very small. This occurs, for example, when a nominally elastic

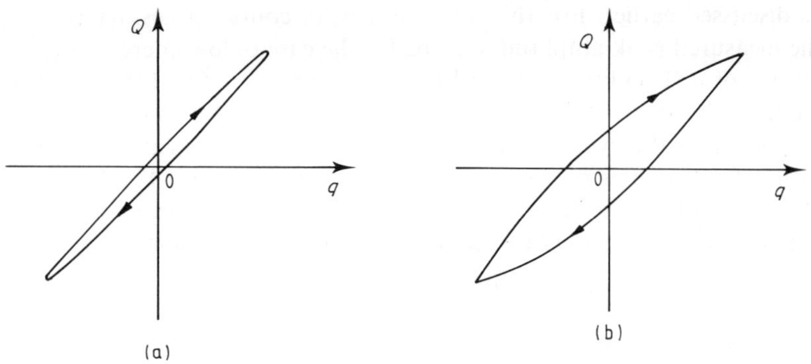

Figure 2.20
Typical hysteresis loops. (a) Linear. (b) Non-linear

material, such as mild steel, is loaded within its elastic range. In this situation the hysteresis loops are usually elliptical in shape, as shown in Figure 2.20a. For materials loaded at high levels inelastic behaviour often becomes more apparent, and the hysteresis loops become fatter, and usually have the characteristic pointed shape shown in Figure 2.20b. For example, if mild steel is loaded beyond its elastic limit (see Figure 2.9)) then plastic deformation will occur, accompanied by a sharp increase in energy dissipation, under cyclic loading conditions. Composite materials can also exhibit significant energy dissipation, as exhibited by fat hysteresis loops.

A wide variety of micromechanisms which contribute to energy dissipation in cyclically loaded materials have been identified (e.g. see Lazen, 1968). However, except in very special cases it is not possible to quantitatively predict the gross, macroscopic behaviour, as manifested by the hysteresis loops, from physical models of the microscopic behaviour. From a practical viewpoint, therefore, one must usually rely on experimental measurements of energy loss, for materials of engineering interest.

The loss function $L(E)$, defined earlier (see equation (2.73)) can be estimated by subjecting a material to a sinusoidal load, measuring the area enclosed by the hysteresis loop, as given by equation (2.89), and then dividing by the period of the cycle, T. Thus

$$L(E) = \frac{1}{T}\int_0^T Q'\dot{q}\,\mathrm{d}t \tag{2.90}$$

Strictly this gives the average energy loss per cycle, but for most cases the energy loss per cycle is a small fraction of the total energy, E, in that cycle, and the latter may be treated as approximately constant over the period of the cycle.

Alternatively, $L(E)$ may be estimated by performing a free-decay test, and determining $\dot{E}$ from the rate of decay of the peaks in the oscillatory response, as discussed earlier. For this purpose it is, of course, necessary to relate E to the measured peak amplitudes. Here, for the case of low energy dissipation, the notion of a 'backbone' to the hysteresis loop is useful. Figure 2.21 shows a family of hysteresis loops, which could be obtained by using various levels of amplitude in the cyclic loading process. The backbone can be defined as the locus of extremities of the loops, and denoted $G(q)$, say. The hysteretic force, $Q'(q, t)$, can then be decomposed into a sum of a non-hysteretic, conservative component and a purely dissipative term, $h(\dot{q}, t)$, say. Thus

$$Q'(q,t) = G(q) + h(q,t) \tag{2.91}$$

The potential energy

$$V(q) = \int_0^q G(\xi)\,d\xi \tag{2.92}$$

may be used to determine the conversion between E and A. Thus, as before, since amplitude peaks are associated with zero velocity, one has, from equation (2.71), $E = V(A)$. The function $G(q)$ can usually be determined experimentally, either directly, by means of a static loading test, or indirectly from a knowledge of the variation of the natural frequency of oscillation, $\omega(E)$, with amplitude. If this frequency is reasonably independent of amplitude, such that $\omega(E) \sim \omega_n$, a constant, the implication is that $G(q)$ may be treated as linear. Then $G(q) = kq$ and $E = kA^2/2$.

If the hysteresis loop is elliptical, as in Figure 2.20a, then it can be represented, exactly, as the result of a combination of a linear spring and a viscous dashpot. Thus

$$Q'(q,t) = kq + c\dot{q} \tag{2.93}$$

where k is the spring stiffness and c a viscous damping coefficient. Thus, similar to equation (2.81) one finds that the loss function is proportional to E (or,

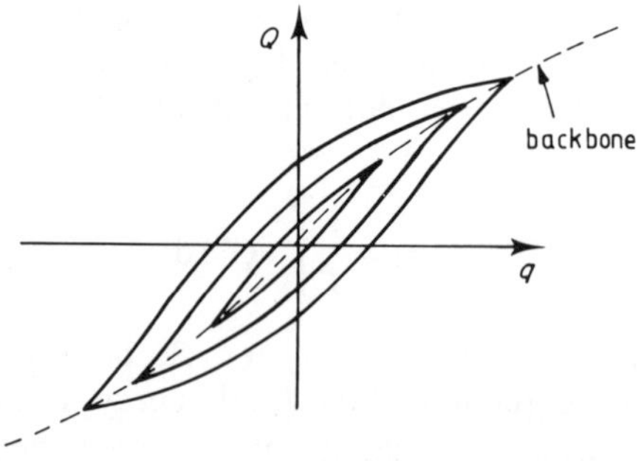

Figure 2.21
The backbone for hysteresis loops

alternatively, proportional to the amplitude squared) and that the function $Q(E)$ is given by (see equation (2.85))

$$Q(E) = \zeta = \frac{c\omega}{2k} \tag{2.94}$$

where ω is the frequency of oscillation. Equation (2.94) shows that, according to the spring–dashpot model, $Q(E)$ varies linearly with ω, for fixed values of c and k. In fact, experimental results for material damping indicate that $Q(E)$ is usually fairly insensitive to the frequency of oscillation, as sketched in Figure 2.22. Fortunately, in most applications, it is only necessary to model the damping accurately when ω is in the vicinity of a resonant frequency, ω_n. Thus, by choosing c and k so that $Q(E)$, according to equation (2.94) agrees with the experimentally determined values of $Q(E)$, for $\omega \sim \omega_n$, one can obtain a damping model which is adequate for most cases. It is noted that an alternative, linear hysteretic damping model can be employed (e.g. see Bishop *et al.*, 1965) which results in a frequency independent $Q(E)$ function. However, as usually formulated, this model is only applicable in the case of sinusoidal excitation. A generalization of this hysteretic damping model, to allow non-sinusoidal motion to be analysed, has recently been given by Bishop and Price (1986).

Non-elliptical loops, such as that shown in Figure 2.20b, are such that $L(E)$ is no longer proportional to E and they cannot be produced by combinations of linear elements. However, if the value of $Q(E)$ is small then the decomposition of equation (2.91) is still appropriate, where now $G(q)$ will usually be non-linear. By choosing a suitable, parametric form for $h(q, t)$ one can often obtain a good fit between the model and an experimentally determined $Q(E)$ against A characteristic, by following the general procedure outlined in the previous section. For example, one can assume, for simplicity, that $G(q)$ is linear, i.e.

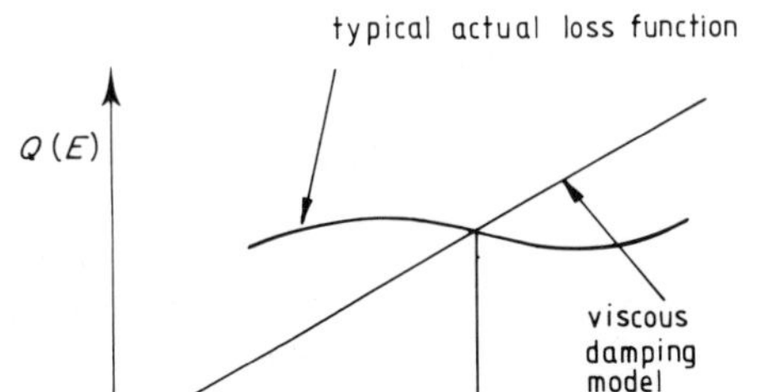

Figure 2.22
Comparison between the variation of $Q(E)$ with frequency, for a typical material, and the viscous damping model result

$G(q) = kq$ and choose

$$h(q, t) = a\dot{q} + b\dot{q}^{n-1} \tag{2.95}$$

where $n = 4, 6, \ldots$ Then, using the approximation for $\dot{q}$ given by equation (2.80) (which is valid when the damping is light) one obtains $L(E)$ in the form

$$L(E) = \alpha A^2 + \beta A^n \tag{2.96}$$

where α and β are constants. Since $E = kA^2/2, Q(E)$ is obtained from equations (2.83) and (2.96) in the form

$$Q(E) = \lambda\left[1 + \left(\frac{A}{A^*}\right)^{n-2}\right] \tag{2.97}$$

where λ and A^* are also constants. Appropriate values of λ and A^* can be found by fitting equation (2.97) to an experimental $Q(E)$ versus A characteristic, in some suitable, optimal way.

A particular example of the above modelling approach has been given by Roberts and Yousri (1978). Two small cantilevered beams, of constant rectangular cross-section, were tested, using the free-decay method. The larger beam was steel, whilst the smaller one was of a copper alloy. The dimensions of the beams, and their natural frequencies, are given in Figure 2.23. The natural frequencies were found to be insensitive to the tip amplitude, A, over the range measured, indicating that a linear $G(q)$ was appropriate, for both beams. Figure 2.24 shows the estimated $Q(E)$ function, plotted against A, for both

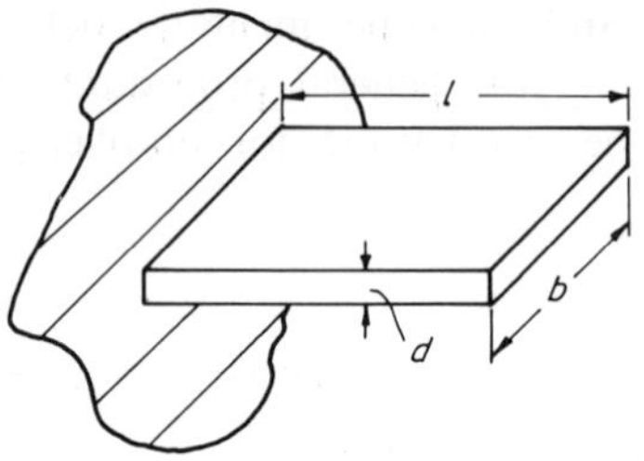

Figure 2.23

	l (mm)	b (mm)	d (mm)	Natural frequency (Hz)
Steel beam	57.0	12.8	0.64	158
Copper alloy beam	32.0	10.3	0.53	281

Dimensions of the two cantilevers (reproduced from Roberts and Yousri (1978) by permission of the American Society of Mechanical Engineers)

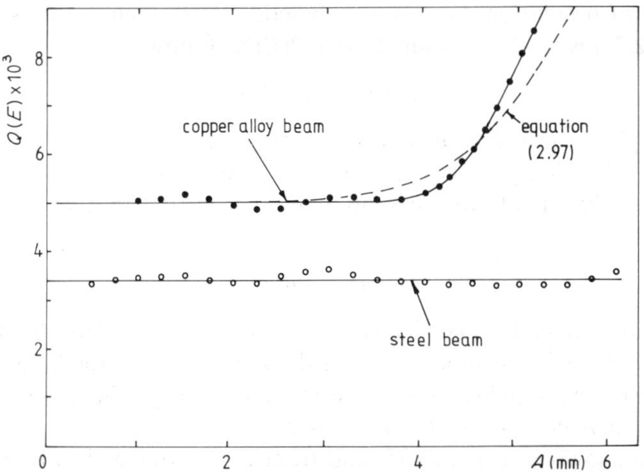

Figure 2.24
Variation of $Q(E)$ with tip amplitude, as deduced from free-decay data (reproduced from Roberts and Yousri (1978) by permission of the American Society of Mechanical Engineers)

beams, as deduced from the free-decay data. It is observed that, for tip amplitudes up to 6 mm, the steel beam has a linear characteristic and the appropriate damping factor, ζ, is about 0.0034. In contrast, the copper alloy beam only behaves linearly for tip amplitudes less than 4 mm and for $A > 4$ mm the damping function rises sharply. This non-linear behaviour is represented reasonably well by equation (2.97), if the parameters are given the values $n = 8$, $A^* = 5.8$ mm and $\lambda = 0.0050$. The dotted line shows the variation of $Q(E)$ with A according to this relationship.

Clearly, measured loss functions, $L(E)$, will depend on the geometry of the vibrating structure, and on the mode of vibration, as well as on the constitutive properties of the material from which the structure is built. The material property can be isolated by defining a specific damping function, D_s, as the energy loss per unit volume of material, per cycle. This parameter will be a function of the stress amplitude of a cycle, σ. Thus

$$D_s = f(\sigma) \tag{2.98}$$

If D_s against σ information is available, it can be converted into the total energy loss per cycle for the whole structure, $P(E)$ say. This is achieved by first calculating the stress distribution in the structure, from a knowledge of the relevant mode shape. This stress distribution can then be converted into a specific energy loss distribution, using equation (2.98), and finally, integrating over the whole volume of the structure, one obtains $P(E)$. If $T(E) = 2\pi/\omega(E)$ is the period of oscillation, known as a function of E, then $L(E)$, as defined by equation (2.73) is readily obtained from $P(E)$ as follows

$$L(E) = P(E)/T(E) \tag{2.99}$$

Some typical variations of D_s with σ are shown in Figure 2.25, for various metals, on a log–log plot. Up to a critical stress limit, σ_L, indicated by points P, a straight line relationship is observed, suggesting that D_s varies with σ according to

$$D_s = \rho\sigma^s \tag{2.100}$$

where ρ and s are constants. The value of s is observed to vary between 2 and 3, and the value of σ_L is found to be of the same order as the fatigue strength, σ_f, at 2×10^7 cycles, indicated by E. For $\sigma > \sigma_L$ the specific damping function is found to increase sharply, in most cases.

The data shown in Figure 2.25, and from other sources has been replotted in Figure 2.26. Here the range of normalized D_s values is indicated, as a function of the non-dimensional stress σ/σ_f (from Lazan, 1968). The broken line shows a geometric mean curve, which is well represented by the single equation

$$D_s = \left(\frac{\sigma}{\sigma_f}\right)^{2.3} + 6\left(\frac{\sigma}{\sigma_f}\right)^8 \tag{2.101}$$

With little loss of accuracy, the exponent of 2.3 in the first term of the above can be replaced by 2.0. This term can then be identified as linear, since the resulting $Q(E)$ function, for a particular structure, will be independent of the

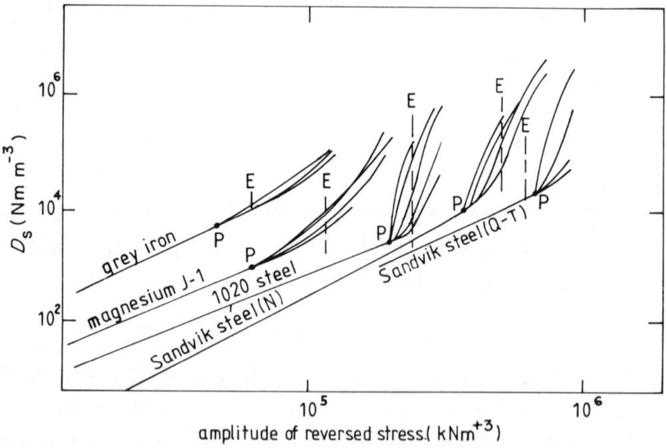

Figure 2.25
Typical variation of specific damping energy, D_s, with stress level (reproduced from Lazan (1968) by permission of Pergamon Press)

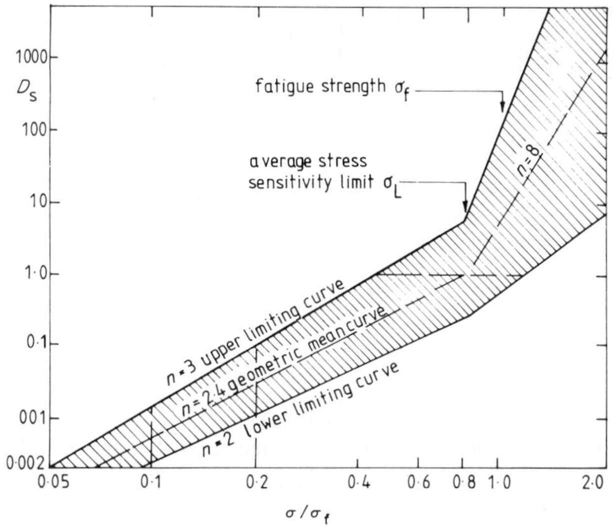

Figure 2.26
Variation of D_s, ratio of reversed stress fatigue strength, with σ/σ_f for a range of materials (reproduced from Lazan (1968) by permission of Pergamon Press)

amplitude of oscillation. The complete $Q(E)$ function for a particular structure, derived from equation (2.101), with 2.3 replaced by 2.0, will be of the form given by equation (2.97), where $n = 8$. It follows that, in cases where the damping is light ($Q(E)$ small compared with unity) a model of the form given by equations (2.91) and (2.95) may be used (as suggested by Crandall *et al.*, 1964).

2.4.2 Mathematical representation of hysteresis loops

For slim hysteresis loops the modelling decomposition given by equation (2.91), which is based on the concept of a backbone, is useful. However, when the deformations are large the area enclosed by the hysteresis loops can become correspondingly large, and a more accurate model of such loops is required. 'Fat' hysteresis loops commonly occur, for example, when composite materials are cyclically loaded, at large amplitudes. Figure 2.27 shows the typical hysteretic behaviour of a structural element made of reinforced concrete.

A wide variety of different shapes of hysteretic loops can be obtained by using the simple differential model originally proposed by Bouc (1967) and subsequently generalized by Wen (1980). In this model the applied force, here denoted z, required to produce a displacement, q, is governed by the

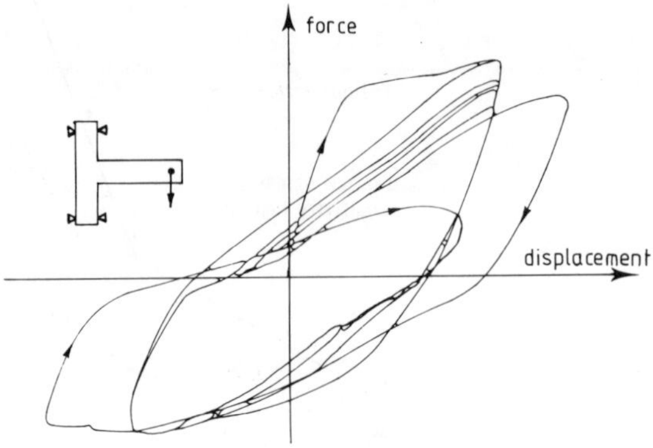

Figure 2.27
Typical hysteresis looping for a reinforced concrete structure subjected to dynamic loading (reproduced from Casciati (1982) by permission of Elsevier Science Publishers)

following single differential relationship

$$\dot{z} = -\gamma|\dot{q}|z|z|^{n-1} - v\dot{q}|z|^n + A\dot{q} \tag{2.102}$$

Here the quantities γ, v, A and n are 'loop parameters', which control the shape and magnitude of the hysteresis loop. Some typical loops obtained, for various sets of parameters, are shown in Figure 2.28.

If z, as described by equation (2.102) is combined with a linear, conservative force, then a more general representation of a hysteretic force is as follows

$$Q' = rq + (1-r)z \tag{2.103}$$

r is usually referred to as the 'rigidity ratio'.

Through an appropriate choice of parameters in this differential model it is often possible to obtain a very good representation of actual, measured hysteresis loops. An example of this may be found in the work of Pivovarov and Vinogradov (1987) who studied the hysteretic behaviour of flexing stranded cables, of the type to be found in Stockbridge dampers. Figure 2.29a shows a family of three experimentally determined hysteresis loops, relating to three different levels of sinusoidal excitation. The corresponding loops, obtained from the differential model, by using suitable chosen values of the loop parameters, are shown in Figure 2.29b.

Differential models of hysteresis, of the above 'curvilinear' type, have many advantages in vibration analysis. The chief of these is that they can be readily combined with system differential equations to yield an overall differential model. For example, consider the case of a single degree of freedom system, with the general differential equation of motion given by equation (2.22). If the non-conservative force, Q_1^{nc} in this equation is modelled according to

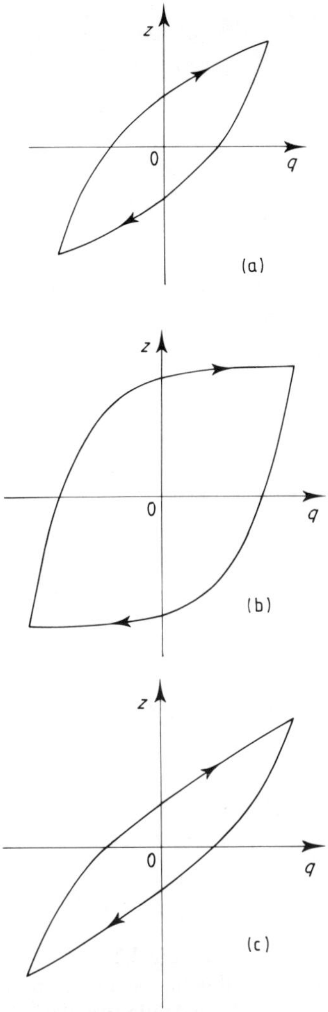

Figure 2.28
Typical hysteresis loops generated by the curvilinear model. (a) $\gamma = 0.5$, $\nu = -0.5$, $A = 1$, $n = 1$. (b) $\gamma = 0.9$, $\nu = 0.1$, $A = 1$, $n = 1$. (c) $\gamma = 0.25$, $\nu = -0.75$, $A = 1$, $n = 1$

equations (2.102) and (2.103) then one has, noting that Q' is opposite to the internal force, Q_1^{nc}

$$m(q)\ddot{q} + \frac{1}{2}\frac{\mathrm{d}m(q)}{\mathrm{d}q}\dot{q}^2 + \frac{\partial V}{\partial q} + rq + (1-r)z = Q^{\mathrm{ex}} \qquad (2.104)$$

where z is given by equation (2.102). Equations (2.102) and (2.104), together,

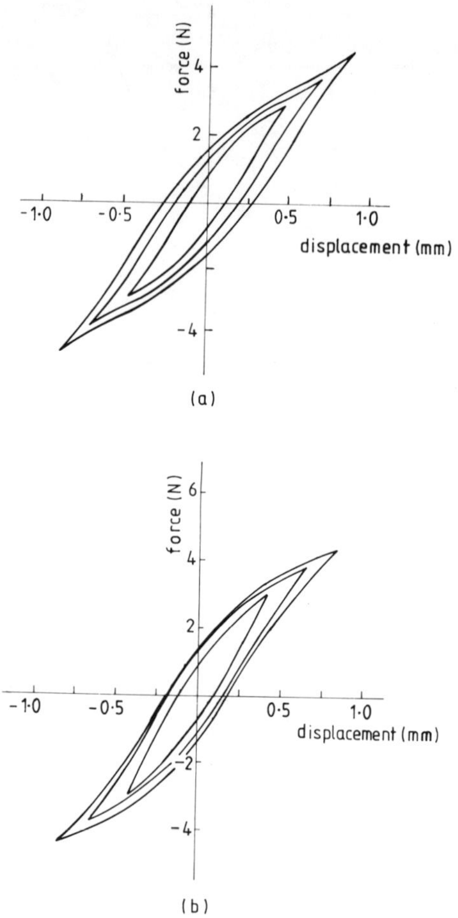

Figure 2.29
Hysteresis loops for a stranded cable in flexure (reproduced from Pivovarov and Vinogradov (1987), by permission of Academic Press). (a) Experimental. (b) Theoretical, curvilinear model

constitute an extended third-order differential model; as shown later, in Chapter 8, such extended differential models can be analysed by using techniques identical to those normally applied to non-hysteretic systems.

Another main advantage of this type of model is that it can be easily generalized. For example, as shown later in Chapter 8, by allowing the loop parameters to become time dependent one can allow for the effect of system degradation, which usually occurs in structures undergoing large deformations.

As shown by Suzuki and Minai (1987), the curvilinear model is only one example of a wide variety of hysteretic constitutive laws which can be expressed in differential form. Another example, namely bilinear hysteresis, will be described shortly, in the context of the following discussion of interface damping.

2.4.3 Interface damping

Frictional forces frequently arise, in mechanical and structural systems from the relative motion of two dry contacting surfaces. A simple example is shown in Figure 2.30a. Here a block, of mass M, rests on a rough horizontal surface. Experimental work shows that the frictional force, F_k, which resists relative motion, is proportional to the normal reaction force, N, between the two surfaces; thus

$$F_k = \mu_k N \tag{2.105}$$

where μ_k is the 'dynamic coefficient of friction'.

The subscript k, in equation (2.105), signifies that kinetic considerations apply in the modelling of the friction force. Actually, to a first approximation,

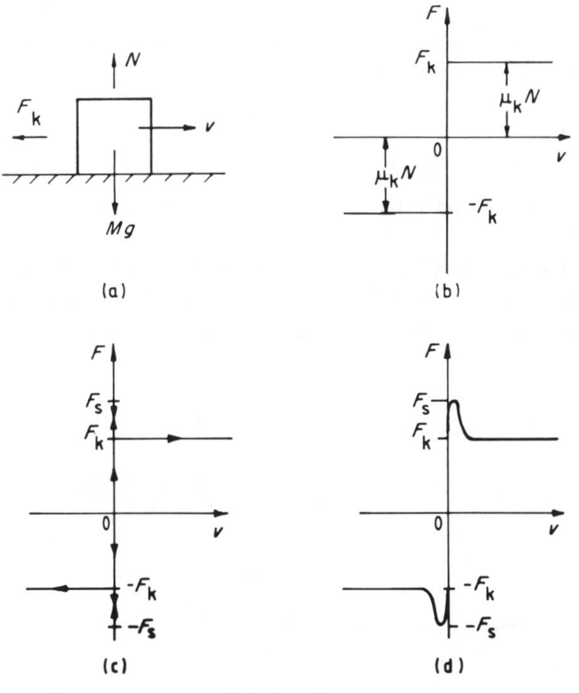

Figure 2.30
Frictional damping characteristics. (a) Block sliding on a plane surface.(b) Idealized dynamic friction law. (c) Effect of static friction. (d) Actual friction law

experimental studies reveal that F_k and μ_k remain constant, during sliding. Thus, an idealized model of 'dynamic friction' appears as shown in Figure 2.30b, where F_k is plotted against the velocity, v, of the relative motion. This model is usually called 'Coulomb friction'; evidently the discontinuity in F_k at $v = 0$ means that Coulomb damping is highly non-linear.

It is found that the kinetic frictional force, F_k, is smaller than the maximum static frictional force, F_s, which the system can sustain before sliding occurs. That is

$$F_k < F_s = \mu_s N \tag{2.106}$$

where μ_s is the 'static coefficient of friction'. Thus a more accurate model of friction would imply that there are two discontinuities at the origin of Figure 2.30b; a jump from zero to F_s, before motion can occur, followed by a jump down to F_k, once motion starts. This is illustrated in Figure 2.30c.

Actual frictional force versus velocity characteristics do not, of course, tend to exhibit such discontinuities. A more realistic model should involve a continuous function of the velocity which exhibits a sharp peak at the vicinity of zero velocity and varies slowly beyond a critical velocity, as sketched in Figure 2.30d. In fact, mathematical models exhibiting this behaviour have been reported in the literature (Belokobylskii and Prokopov, 1982). A typical example of such a model, used in connection with modelling the frictional forces experienced by oil drill strings, is as follows

$$F(v) = A\exp(-Bv^2) + C\tan^{-1}(Rv) \quad v > 0 \tag{2.107}$$

Here the values of the parameters A, B, C and R depend on the nature of the contacting surfaces. The force, $F(v)$, as given by equation (2.107), involves two components. The first involves a negative exponential, the initial value of which, at $v = 0$, is the static frictional force, F_s, i.e. $A = F_s$. This component decays away to zero as v becomes large. In the second component of $F(v)$ the arctangent

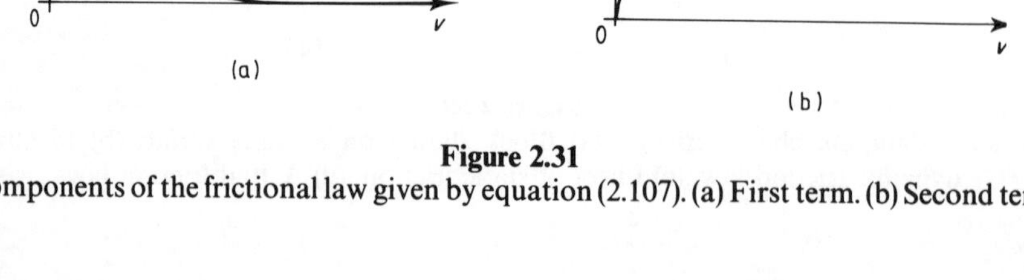

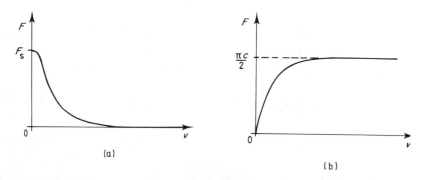

Figure 2.31
Components of the frictional law given by equation (2.107). (a) First term. (b) Second term

is a continuous approximation to the sign function. The parameter R denotes a constant with a numerically high value which controls the slope at the origin. The slope increases with increasing R. These two components are sketched graphically in Figure 2.31.

Dynamic systems which incorporate a frictional damping element will clearly dissipate energy, when subject to cyclic loading. This energy dissipation will be associated with hysteresis loops of a non-linear (i.e., non-elliptical) kind. A very simple example of this is furnished by the single degree of freedom system sketched in Figure 2.32a. It consists of a mass m, attached to a rigid foundation through a linear spring, of stiffness αk and a spring, of stiffness $(1-\alpha)k$ in series with a frictional damping element. It will be supposed that the frictional element can be modelled as an ideal Coulomb damper, and that the mass is originally in a position of static equilibrium.

Suppose that the mass is moved to right, and consider the force $F_1(t)$ exerted by the lower, combined elements, on the mass (see Figure 2.32a). If $F(t) < F_k$, where F_k is the Coulomb friction force, then there will be no slipping in the damper element and F_1 will be equal to $(1-\alpha)kq$, where q is the displacement of the mass. Thus the $F_1 - q$ characteristic will be the straight line, AB, shown in Figure 2.32b, of slope $(1-\alpha)k$. At B, $F_1 = F_k = (1-\alpha)kq^*$. If q exceeds the critical value q^* then the slipping will begin in the damping element and F_1 will remain at its critical value of F_k. Therefore, the force–displacement characteristic will follow the straight horizontal line BC.

If the motion of the mass is now reversed in direction, the force in the lower spring falls below the value F_k and slipping in the damping element ceases. Thus the force F_1 decreases linearly with displacement, along the line CD, of slope $(1-\alpha)k$. At point D, where the mass displacement is $q_c - 2q^*$, the compressive force in the lower spring becomes equal to F_k, and the damper slips again. Thus, the path DE is followed. If, at E, the motion is reversed again, then the linear characteristic EF is followed, and so on. If the mass is subjected to a cyclic motion, the force (F_1) displacement (q) characteristic will eventually become a repeatable, closed loop, as indicated in Figure 2.32c.

If one considers the combined force $h(t) = F_1(t) + F_2(t)$ exerted by the springs and damper, on the mass, then it is clear that, throughout any motion, $F_2(t) = \alpha kq$. Consequently one can write

$$h(t) = \alpha kq + F_1 \qquad (2.108)$$

and the $h(t)$ characteristic, corresponding to that for F_1 shown in Figure 2.32b, will appear as shown in Figure 2.32d. The closed loop eventually attained under cyclic loading conditions is shown in Figure 2.32e; this may be compared with Figure 2.32c. On the 'slipping paths' the slope is now αk, whereas on the 'no slipping' paths the slope is simply k, representing the combined spring stiffness.

Systems with hysteretic characteristics of the type shown in Figure 2.32d are said to exhibit 'bilinear hysteresis'. A special type of such hysteresis, corres-

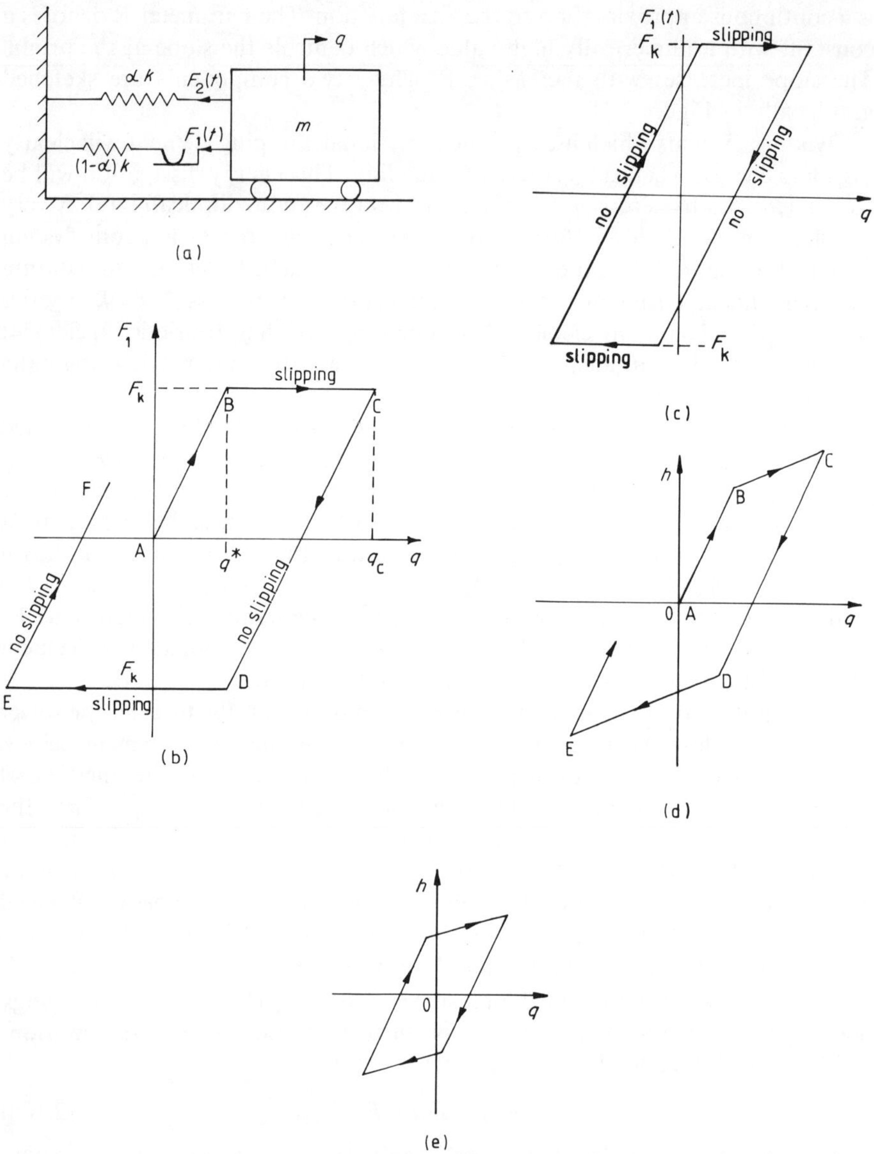

Figure 2.32

Mass–Spring system with Coulomb damping. (a) Schematic diagram of system. (b) Variation of F_1 with displacement. (c) Variation of F_1 with displacement during cyclic motion. (d) Variation of total restoring force with displacement during cyclic motion. (e) Closed loop attained under cyclic loading ponding to $\alpha = 0$, is elasto–plastic hysteresis, as shown in Figure 2.32b. The physical model shown in Figure 2.32a shows that general bilinear hysteresis can be decomposed into a sum of an ordinary elastic restoring term and a purely elasto–plastic hysteretic term. The latter can be expressed in differential equation form, as shown later (in Chapter 8).

It is noted that there are a wide variety of alternative models of hysteresis, involving slipping elements (e.g. see Suzuki and Minai, 1987) which can also be modelled in terms of differential equations.

2.4.4 Flow induced forces

In many engineering systems forces exerted by surrounding fluids are not negligible. In these situations the mathematical modelling of the fluid induced load becomes crucial as it is the basis for an analytical or numerical determination of the systems dynamic behaviour. For example, Morison's equation (Morison *et al.*, 1950) is widely used to model the wave forces experienced by members of offshore structures when the effects of reflection or diffraction are secondary. If the structural member velocity is considered negligible, the wave force per unit length of the member may be estimated from Morison's equation. This has the form

$$F(t) = \tfrac{1}{2}\rho\pi d^2 C_M \dot{v} + \tfrac{1}{2}\rho d C_D v|v| \tag{2.109}$$

where $v(t)$ is the wave-induced water particle velocity, relative to the structural member, $\dot{v}$ the corresponding acceleration, ρ the fluid density and d the equivalent diameter of the member. Further, the coefficients C_M, which has an approximate value equal to 2 and C_D, which has an approximate value equal to 1 are the inertia and drag coefficients respectively. The first term of the right-hand side of equation (2.109) is linear with respect to acceleration, and represents the non-dissipative inertia force. The second term is non-linear with respect to velocity, and represents the dissipative drag force.

The mathematical expression given by equation (2.109) is strictly applicable only to stationary members. There are cases, however, where the flexibility of the structural members cannot be ignored without compromising the accuracy of the results obtained. Thus, it has been found necessary to revise equation (2.109) for flexible structures. The revised equation expresses the force per unit length, in the following form

$$F(t) = \tfrac{1}{2}\rho\pi d^2 \dot{v} + \tfrac{1}{2}C_I\rho\pi d^2(\dot{v} - \ddot{x}) + \tfrac{1}{2}\rho d C_D|v - \dot{x}|(v - \dot{x}) \tag{2.110}$$

where

$$C_I = 1 - C_M \tag{2.111}$$

and the symbols $\dot{x}$ and $\ddot{x}$ denote the structural velocity and acceleration, respectively (see Blevins, 1977 and Chen, 1977).

There are a number of applications where the dissipative fluid loading is well represented by a linear-plus-quadratic model, of the following form

$$F(t) = \alpha v + \beta |v| v| \tag{2.112}$$

where α and β are constants and v is the relative flow velocity. This model is particularly relevant when flow separation occurs and/or the flow is turbulent. The second, non-linear component on the right-hand side of equation (2.112) can become dominant, in these circumstances.

One example occurs in connection with the shock absorbers which are used in automobiles to reduce dynamic loading. Basically, a shock absorber consists of a piston moving inside a closed cylinder, as illustrated in Figure 2.33. The cylinder is filled with oil and the two chambers are interconnected by a channel with cross-section denoted here by α. Following Hagedorn and Wallaschek (1987), let $p_1(t)$ and $p_2(t)$ be the pressures in the left-hand, and right-hand containers, respectively, and $F(t)$ be the force applied to the piston. If one allows for friction between the piston and the chamber, and models this as ideal Coulomb friction, then an appropriate equation of motion for the piston is as follows

$$m\ddot{q} = F(t) + A(p_1 - p_2) - F^* \operatorname{sgn}(\dot{q}) \tag{2.113}$$

Here F^* is Coulomb friction force and sgn is the signum function (sgn $x = 1$ if $x > 0$, and -1 if $x < 0$). The symbol A stands for the effective area of the piston.

Assuming incompressible flow, the volume flow rate, V, through the connecting chamber is proportional to $\dot{q}$, i.e.

$$\dot{V} = A\dot{q} \tag{2.114}$$

Now the force, $A\Delta p = A(p_2 - p_1)$, relating to the pressure drop Δp, can be modelled by equation (2.112), where v can be identified with $\dot{V}$ (see Hagedorn and Wallaschek, 1987). The linear term originates from viscous flow through the interconnecting channel; the second, non-linear term derives from throttle losses in this channel. Combining results one obtains the equation of motion

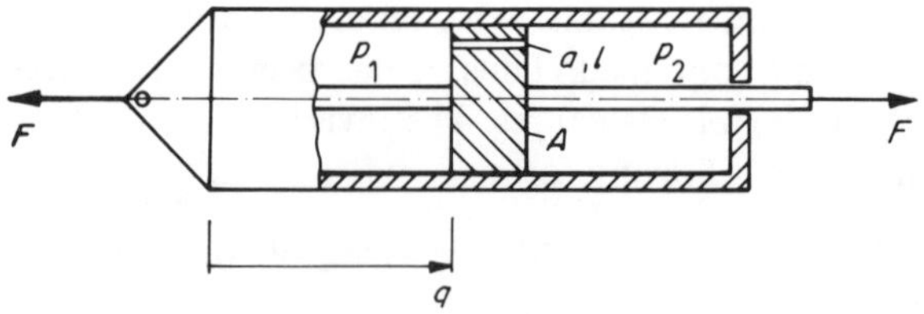

Figure 2.33
Schematic diagram of a shock absorber (reproduced from Hagedorn and Wallaschek (1987) by permission of Springer-Verlag)

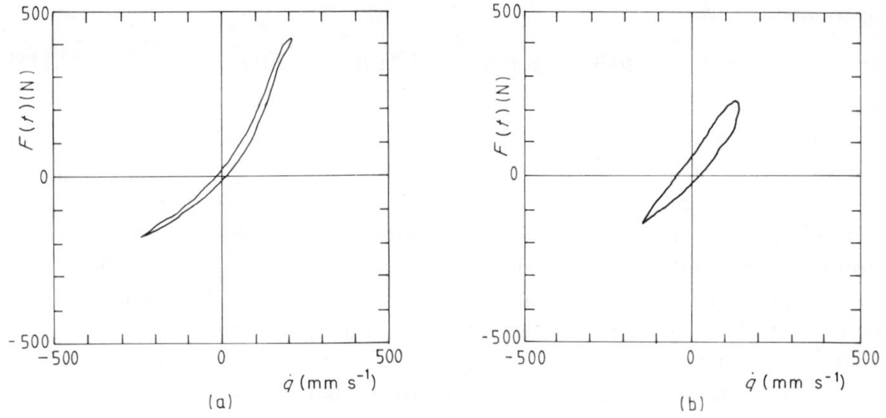

Figure 2.34
Experimental force-velocity relationships for a typical shock absorber (reproduced from Hagedorn and Wallaschek (1987) by permission of Springer-Verlag). (a) $f = 1.67$ Hz, (b) $f = 8.35$ Hz

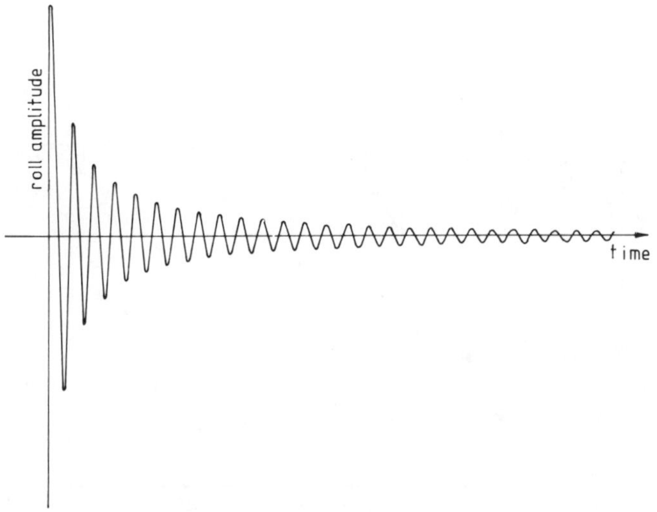

Figure 2.35
Typical experimental free decay curve for a ship rolling in calm water (reproduced from Roberts (1985) by permission of the Society of Naval Architects and Marine Engineers). Variation of roll angle with time

as follows

$$m\ddot{q} + \alpha'\dot{q} + \beta'|\dot{q}|\dot{q} + F^* \operatorname{sgn}(\dot{q}) = F(t) \tag{2.115}$$

where $\alpha' = A\alpha$, $\beta' = A^2\beta$.

Figure 2.34 (from Hagedorn and Wallaschek, 1987), shows some experimental measurements of the force $F(t)$, plotted against velocity $\dot{q}$, for a typical shock absorber, undergoing simple harmonic motion. At a relatively low frequency of motion of 1.67 Hz (see Figure 2.34a) the force–velocity characteristic is almost non-hysteretic, but distinctly non-linear. This suggests that inertial effects, are in this case, negligible and that non-linear terms, of the kind appearing in equation (2.115) are of primary importance. At a higher frequency of excitation Figure 2.34b pronounced hysteresis in the force–velocity characteristic is observed, which is probably attributable to inertial effects.

Another example where the modelling represented by equation (2.112) is

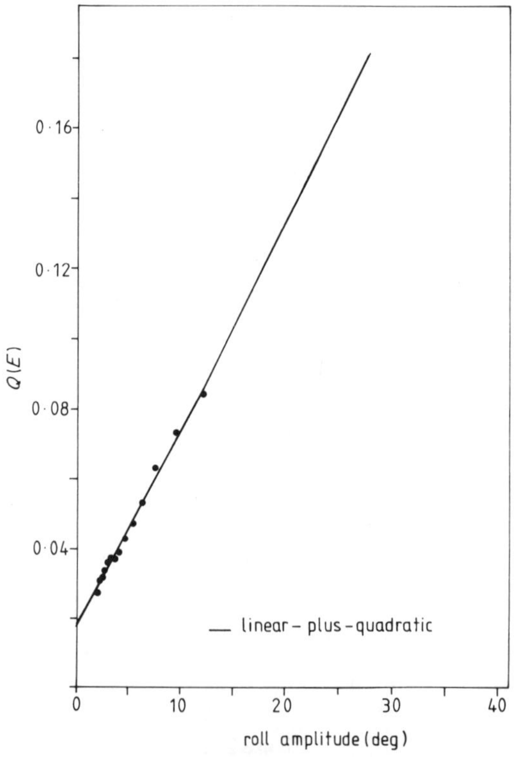

Figure 2.36
Variation of the damping function $Q(E)$ with roll amplitude, as deduced from a free decay curve, similar to that shown in Figure 2.35. Comparison with linear-plus-quadratic model

appropriate occurs in the field of ship dynamics. It was demonstrated over one hundred years ago, by Froude (1955) that the dissipative, damping moment experienced by a ship when undergoing a rolling motion, would be represented as a combination of linear and quadratic damping. Here the quadratic term relates to the non-linear effects of vortex shedding from the hull, which is a significant effect except for very small angles of roll. Figure 2.35 shows a typical free-decay record obtained from a scale model of a ship, undergoing transient rolling motion in a tank of water (see Roberts, 1985). The very rapid decay of amplitude observed at the start of the decay, where the roll amplitude is high may be compared here with the very slow rate of decay observed in the 'tail' of the decay, when the amplitude is low. This is clear evidence of the influence of non-linearity in the dissipative fluid forces. Figure 2.36 shows the variation of the function $Q(E)$ with roll amplitude, as deduced from the peak amplitudes in data similar to that in Figure 2.35 (see equation (2.83) for a definition of $Q(E)$). Strong non-linear behaviour is evidenced by the fact that $Q(E)$ is not independent of amplitude, but increases sharply as the amplitude increases. The full line in Figure 2.36 is a 'best fit' of the linear-plus-quadratic model of roll damping. This fits the data extremely well. The linear component of the damping, given by the intercept at zero roll amplitude, is seen to be here relatively small.

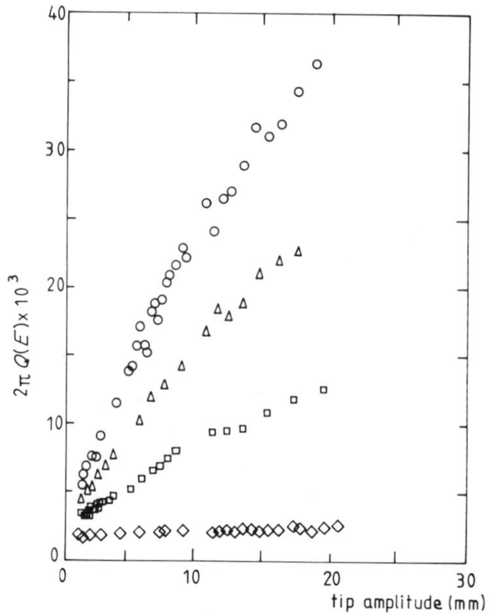

Figure 2.37
Variation of $Q(E)$ with amplitude, at various ambient pressure levels, for a cantileyered beam (reproduced from Baker *et al.* (1967) by permission of Pergamon Press) (◇ < 1 mm Hg, □ 220 mm Hg, △ 447 mm Hg, ○ 742 mm Hg)

At high roll amplitudes, the result in Figure 2.36 indicates that the damping is principally quadratic in nature.

As a final example of fluid dissipative effects some results of Baker *et al.* (1967) are presented. They measured the damping of thin cantilevered beams by processing free decay data. By performing the experiment inside a chamber, which could be evacuated, they were able to isolate the effects of structural, or material damping, from those arising from the motion of the cantilever through the surrounding air. Figure 2.37 shows how the damping function, $Q(E)$ varies with tip amplitude for various air pressures, ranging from a near-vacuum pressure of less than 1 mm of mercury to normal atmospheric pressure. Under near-vacuum conditions the damping is due entirely to material effect, and it can be seen that $Q(E)$ is virtually independent of tip amplitude, suggesting that a linear damping model is appropriate. As the ambient air pressure increases, however, the effect of 'air damping' becomes increasingly significant, and, for large tip amplitudes, is dominant at normal atmospheric pressure. The linear variation of $Q(E)$ with amplitude suggests that, as in Figure 2.36, a linear-plus-quadratic damping force, of the form of equation (2.112) is appropriate. This is indeed confirmed by the detailed calculations of Baker *et al.* (1967).

Chapter 3

Probability theory and stochastic processes

3.1 INTRODUCTION

The application of probabilistic methods to the solution of vibration problems in engineering does, of course, require a basic understanding of the fundamentals of probability theory. In this chapter those principles of modern probability theory which are relevant to the development of the technique of statistical linearization, and related methods, as means of solving non-linear problems in random vibration, are summarized.

To begin with, the notion of a random event and the specification of a probability associated with such an event is introduced. This leads naturally to a discussion of random variables, and their probabilistic specification, and hence to the concept of a stochastic process, as a generalization of a random variable. The description of stochastic processes in the time domain and the frequency domain, by means of covariance and spectral density functions, respectively, is outlined. Certain special classes of stochastic process are described, including the very important class known as Gaussian, or normal, processes.

It is fortunate that the nature of this book allows certain areas of probability theory, such as the theory of Markov chains and Markov processes, which feature prominently in many modern text books on probability theory and stochastic processes, to be omitted entirely. However, it must be admitted that the discussion in this chapter, being necessarily fairly brief, is no more than an introduction to the subject. The reader who requires a fuller treatment should consult some of the many available treatises on probability theory and stochastic processes. (e.g. Kolmogorov, 1956; Doob, 1953; Loeve, 1977; Gnedenko, 1962; Yaglom, 1962; Parzen, 1962; Feller, 1971; Cooper and McGillem, 1986; Chung, 1979; Grimmett and Stirzaker, 1982; Papoulis, 1984).

3.2 RANDOM EVENTS AND PROBABILITY

The phenomena with which this book is concerned obey 'probabilistic laws', which may be expressed in the following manner: 'given a set of specified conditions then an event "A" will occur, with probability $P(\mathrm{A})$'. A may be termed a 'random event' since it is neither certain nor impossible, but has some chance of happening. The quantity $P(\mathrm{A})$ is a numerical measure of that chance, and is a number such that $0 \leqslant P(\mathrm{A}) \leqslant 1$. Higher values of $P(\mathrm{A})$ are associated with greater chances of A occurring. The event A is 'certain' if $P(\mathrm{A}) = 1$ and 'impossible' if $P(\mathrm{A}) = 0$.

In many situations it is necessary to consider the relationships between a number of events, A, B,..., N. The following definitions are then useful

(a) $P(\mathrm{AB} \cdots \mathrm{N})$ is the probability that all the events A, B,..., N occur.
(b) $P(\mathrm{A} + \mathrm{B} + \cdots + \mathrm{N})$ is the probability that at least one of the events occurs.
(c) $P(\mathrm{A} \mid \mathrm{B})$ is the 'conditional probability' that the event A occurs, given that B has occurred.

With this notation, the two most important axioms which govern probability (due to Kolmogorov, 1956) may be written as

$$P(\mathrm{A} + \mathrm{B} + \cdots + \mathrm{N}) \leqslant P(\mathrm{A}) + P(\mathrm{B}) + \cdots + P(\mathrm{N}) \tag{3.1}$$

$$P(\mathrm{AB}) = P(\mathrm{A} \mid \mathrm{B}) P(\mathrm{B}) \tag{3.2}$$

The events A, B,..., N are called 'mutually exclusive' if only one of them can occur; in this case the inequality in equation (3.1) becomes an equality. The two events A and B are said to be 'independent' if $P(\mathrm{A} \mid \mathrm{B}) = P(\mathrm{A})$; i.e. if $P(\mathrm{AB}) = P(\mathrm{A}) P(\mathrm{B})$. This definition is easily extended to the general case; thus the events A, B,..., N are independent if

$$P(\mathrm{AB} \cdots \mathrm{N}) = P(\mathrm{A}) P(\mathrm{B}) \cdots P(\mathrm{N}) \tag{3.3}$$

If at least one of the events A, B,..., N is bound to occur then they form a 'complete' set, and

$$P(\mathrm{A} + \mathrm{B} + \cdots + \mathrm{N}) = 1 \tag{3.4}$$

3.3 RANDOM VARIABLES

For a complete set of exclusive events, each event can be characterized by a *number*, η, called a random variable. For example, if a die is rolled there are six exclusive events, each of which is characterized by an integer number (1 to 6).

Die rolling is an example of the occurrence of a discrete random variable, where η takes only a finite (or countably finite) number of distinct values. Random variables may also be of the continuous kind, where η can assume any real number in a range which may stretch from $-\infty$ to $+\infty$. In this case the underlying set of events associated with the random variable is infinitely large in number. Continuous random variables result from the measurement of continuous physical quantities and are of primary interest in this book. Henceforth, for conciseness, only random variables of the continuous type will be treated here. The slightly modified treatment required when dealing with discrete random variables may be found in numerous textbooks on probability theory.

3.3.1 Probability distributions

In many cases there is more than one random variable of interest. If there are n random variables, $\eta_1, \eta_2, \ldots, \eta_n$, then a vector random variable, $\boldsymbol{\eta}$, may be defined, which has a 'joint distribution function'

$$F(\mathbf{x}) = P(\boldsymbol{\eta} \leqslant \mathbf{x}) \tag{3.5}$$

Here $\boldsymbol{\eta} \leqslant \mathbf{x}$ means that each element of $\boldsymbol{\eta}$ is not greater than the corresponding element of $\mathbf{x}$.

The joint distribution function has the following properties: $F(\mathbf{x}) \to 0$ as any element of $\mathbf{x}$ approaches zero and $F(\mathbf{x}) \to 1$ as all the elements of $\mathbf{x}$ approach $+\infty$. Moreover $F(\mathbf{x})$ increases monotonically in all variables.

Associated with the distribution function $F(\mathbf{x}) = F(x_1, x_2, \ldots, x_n)$ is the joint density function, defined by

$$f(\mathbf{x}) = f(x_1, x_2, \ldots, x_n) = \frac{\partial^n F(x_1, x_2, \ldots, x_n)}{\partial x_1\, \partial x_2 \cdots \partial x_n} \tag{3.6}$$

It can be easily shown, using the axioms in the preceding section, that the joint density function $f(\mathbf{x})$ possesses the following properties

(a) It is non-negative, i.e. $f(\mathbf{x}) \geqslant 0$.
(b) The probability that the vector random variable $\boldsymbol{\eta}$ falls in some region R is equal to

$$P(\boldsymbol{\eta} \in R) = \iint_R \cdots \int f(\mathbf{x})\, \mathrm{d}\mathbf{x} \tag{3.7}$$

In particular, if the region R is infinite then the above integral equals unity.

A knowledge of $f(\mathbf{x})$ allows the density functions for the individual elements of $\boldsymbol{\eta}$ to be computed. For example, $f_1(x_1)$, the density function for η_1, can be found by integrating $f(\mathbf{x})$ with respect to $x_2, x_3, \ldots, x_n$, from $-\infty$ to $+\infty$ (this follows from equation (3.7)). Hence,

$$f_1(x_1) = \int_{-\infty}^{\infty} \int_{-\infty}^{\infty} \cdots \int_{-\infty}^{\infty} f(\mathbf{x})\, \mathrm{d}x_2\, \mathrm{d}x_3 \cdots \mathrm{d}x_n \tag{3.8}$$

and similarly for $f_2(x_2), f_3(x_3)$, etc., the density functions for η_2, η_3, etc.

A particularly important case occurs when the random variable elements in $\boldsymbol{\eta}$ are statistically independent. In accordance with the previous definition of independence of events (see equation (3.3)), it can be said that the random variables $\eta_1, \eta_2, \ldots, \eta_n$ are independent if the relation

$$P(\boldsymbol{\eta} \leqslant \mathbf{x}) = P(\eta_1 \leqslant x_1) P(\eta_2 \leqslant x_2) \cdots P(\eta_n \leqslant x_n) \tag{3.9}$$

holds for arbitrary values of $x_1, x_2, \ldots, x_n$. In terms of distribution functions, equation (3.9) may be written as

$$F(\mathbf{x}) = F_1(x_1)F_2(x_2)\cdots F_n(x_n) \tag{3.10}$$

where $F_k(x_k)$ denotes the distribution function of the random variable η_k. Alternatively, in terms of density functions

$$f(\mathbf{x}) = f_1(x_1)f_2(x_2)\cdots f_n(x_n) \tag{3.11}$$

for independent random variables. It follows that, for independent variables, a knowledge of the individual density functions $f_k(x_k)$ $(k = 1, 2, \ldots, n)$ is sufficient to construct the joint density $f(\mathbf{x})$. This is *not* true in the more general case of dependent variables.

It is noted that although care has been taken here, in the notation, to distinguish between a random variable vector, $\boldsymbol{\eta}$, and the argument, $\mathbf{x}$, of its distribution or density function, this is frequently unnecessary. In later chapters of this book, where there is no risk of confusion, the same symbol will be used; thus, for example, $f(\mathbf{x})$ will be used to denote the density function of a random variable vector $\mathbf{x}$.

3.3.2 Transformation of random variables

If η and ρ are two scalar random variables, related through the algebraic equation

$$\rho = g(\eta) \tag{3.12}$$

then the density function for ρ, $f(y)$ say, can be readily related to the density function of η, $f(x)$ say. In the general case where $g(x)$ is multi-valued, and x_i $(i = 1, 2, \ldots, m)$ are the m values of x which correspond to $y = g(x)$, then it can be shown that (Papoulis, 1984)

$$f(y) = \sum_{i=1}^{m} f(x_i)|g'(x_i)| \tag{3.13}$$

where $g'(x) = \mathrm{d}g/\mathrm{d}x$.

This result can be readily generalized to the case of vector random variables. For example, if

$$\boldsymbol{\rho} = \mathbf{g}(\boldsymbol{\eta}) \tag{3.14}$$

where $\boldsymbol{\rho}$ and $\boldsymbol{\eta}$ are vectors, and the elements of $\mathbf{g}$ are single-valued functions, then the density function $f(\mathbf{x})$ for $\boldsymbol{\eta}$ is related to the density function $f(\mathbf{y})$ for $\boldsymbol{\rho}$ as follows (e.g., see Papoulis, 1984):

$$f(\mathbf{y}) = f(\mathbf{x})|\mathbf{J}|_{\mathbf{x}=\mathbf{g}(\mathbf{y})} \tag{3.15}$$

where $|\mathbf{J}|$ is the determinant of the Jacobian matrix

$$\mathbf{J} = \begin{bmatrix} \dfrac{\partial g_1}{\partial y_1} & \dfrac{\partial g_1}{\partial y_2} & \cdots & \dfrac{\partial g_1}{\partial y_n} \\ \vdots & & & \\ \dfrac{\partial g_n}{\partial y_1} & \dfrac{\partial g_n}{\partial y_2} & \cdots & \dfrac{\partial g_n}{\partial y_n} \end{bmatrix}. \tag{3.16}$$

3.4 EXPECTATION OF RANDOM VARIABLES

If $\boldsymbol{\eta}$ is a vector random variable, with n components, and $\mathbf{g}(\boldsymbol{\eta})$ is a matrix function of $\boldsymbol{\eta}$, then the expectation of $\mathbf{g}(\boldsymbol{\eta})$ is defined by

$$E\{\mathbf{g}(\boldsymbol{\eta})\} = \int_{-\infty}^{\infty}\int_{-\infty}^{\infty}\cdots\int_{-\infty}^{\infty} \mathbf{g}(\mathbf{x})f(\mathbf{x})\,\mathrm{d}\mathbf{x} \tag{3.17}$$

A very important special case is where $\mathbf{g}(\boldsymbol{\eta}) = (\boldsymbol{\eta}-\mathbf{m})(\boldsymbol{\eta}-\mathbf{m})^{\mathrm{T}}$. Here $\mathbf{m}$ is the vector of the mean values of $\boldsymbol{\eta}$, i.e.

$$\mathbf{m} = E\{\boldsymbol{\eta}\} \tag{3.18}$$

and $(\boldsymbol{\eta}-\mathbf{m})^{\mathrm{T}}$ denotes the transpose of $\boldsymbol{\eta}-\mathbf{m}$. The expectation

$$\mathbf{V} = E\{(\boldsymbol{\eta}-\mathbf{m})(\boldsymbol{\eta}-\mathbf{m})^{\mathrm{T}}\} \tag{3.19}$$

is called the covariance (or variance) matrix of $\boldsymbol{\eta}$.

The diagonal elements of $\mathbf{V}$ are simply the variances

$$\operatorname{var}\{\eta_i\} = \sigma_i^2 = E\{(\eta_i - m_i)\}^2 \quad i = 1, 2, \ldots, n \tag{3.20}$$

of η_i where σ_i are the standard deviations of η_i. The off-diagonal elements

$$\operatorname{cov}\{\eta_i, \eta_j\} = E\{(\eta_i - m_i)(\eta_j - m_j)\} \quad i, j = 1, 2, \ldots, n \quad i \neq j \tag{3.21}$$

are called the *covariances* of the variables η_i and η_j. If the elements of $\boldsymbol{\eta}$ are statistically independent random variables then $f(\mathbf{x})$ is given by equation (3.11) and it follows, fairly readily, that the covariances are all zero. Thus, for a set of independent random variables the variance matrix $\mathbf{V}$ is diagonal. However, the converse is not generally true, i.e. a diagonal $\mathbf{V}$ matrix does not imply that elements of $\boldsymbol{\eta}$ are statistically independent.

Another very important special case of equation (3.17) is where

$$\mathbf{g}(\boldsymbol{\eta}) = g(\boldsymbol{\eta}) = \exp(\mathrm{i}\boldsymbol{\omega}^{\mathrm{T}}\boldsymbol{\eta}) \tag{3.22}$$

where $g(\boldsymbol{\eta})$ is a scalar function and $\boldsymbol{\omega}$ is an n-vector of variables $\omega_1, \omega_2, \ldots, \omega_n$.

The scalar function

$$\Theta(\boldsymbol{\omega}) = E\{\exp(\mathrm{i}\boldsymbol{\omega}^{\mathrm{T}}\boldsymbol{\eta})\} \tag{3.23}$$

is known as the 'characteristic function'. From equations (3.17) and (3.23) one has

$$\Theta(\boldsymbol{\omega}) = \int_{-\infty}^{\infty}\int_{-\infty}^{\infty}\cdots\int_{-\infty}^{\infty} \exp(\mathrm{i}\boldsymbol{\omega}^{\mathrm{T}}\mathbf{x}) f(\mathbf{x})\,\mathrm{d}\mathbf{x} \tag{3.24}$$

It is evident from this relationship that $\Theta(\boldsymbol{\omega})$ is the multi-dimensional Fourier transform of $f(\mathbf{x})$; it can thus be used in place of $f(\mathbf{x})$ as a means of defining the probability distribution of $\boldsymbol{\eta}$.

The general moments

$$m_s(\eta_j, \eta_k, \ldots, \eta_m) = E\{\eta_j \eta_k \cdots \eta_m\} \tag{3.25}$$

where $j, k, \ldots, m$ can assume any integer values between 1 and n, and s, the 'order' of the moment, is equal to the number of arguments in the function, can be derived directly from the characteristic function, as follows

$$m_s(\eta_j, \eta_k, \ldots, \eta_m) = (-\mathrm{i})^s \left(\frac{\partial^s \Theta(\boldsymbol{\omega})}{\partial\omega_j \partial\omega_k \cdots \partial\omega_m} \right)_{\boldsymbol{\omega}=\mathbf{0}} \tag{3.26}$$

It is evident, from this definition, that m_s are symmetrical functions of their arguments, i.e. their value does not depend on the order in which the arguments appear. For example, for $n = 2$ and $s = 2$,

$$m_2(\eta_1, \eta_2) = m_2(\eta_2, \eta_1) \tag{3.27}$$

Using equation (3.26), $\Theta(\boldsymbol{\omega})$ can be expanded about $\boldsymbol{\omega} = \mathbf{0}$ as a multi-dimensional Taylor series, as follows

$$\Theta(\boldsymbol{\omega}) = 1 + \sum_{s=1}^{\infty} \frac{\mathrm{i}^s}{s!} \sum_{j,k,\ldots,m=1}^{n} m_s(\eta_j, \eta_k, \ldots, \eta_m)\omega_j\omega_k\cdots\omega_m \tag{3.28}$$

In view of the symmetry of the moment functions a different notation is possible, and useful, in the case where $s \geqslant n$. One can define

$$m_{k_1 \cdots k_n} = m_s(\underbrace{\eta_1, \ldots, \eta_1}_{k_1 \text{ times}}, \underbrace{\eta_2, \ldots, \eta_2}_{k_2 \text{ times}}, \ldots, \underbrace{\eta_n, \ldots, \eta_n}_{k_n \text{ times}}) \tag{3.29}$$

where

$$s = k_1 + k_2 + \cdots + k_n \tag{3.30}$$

Thus, from equation (3.26)

$$m_{k_1 \cdots k_n} = (-\mathrm{i})^s \left(\frac{\partial^s \Theta(\boldsymbol{\omega})}{\partial\omega_1^{k_1} \partial\omega_2^{k_2} \cdots \partial\omega_n^{k_n}} \right)_{\boldsymbol{\omega}=\mathbf{0}} \tag{3.31}$$

A set of useful functions, closely related to the moments m_s, are the cumulants.

Here κ_s denotes a cumulant of order s, which can be derived from the characteristic function by taking derivatives of its natural logarithm. Thus

$$\kappa_s(\eta_j, \eta_k, \ldots, \eta_m) = (-\mathrm{i})^s \left(\frac{\partial^s \ln \Theta(\boldsymbol{\omega})}{\partial\omega_j \partial\omega_k \cdots \partial\omega_m} \right)_{\boldsymbol{\omega}=0} \tag{3.32}$$

From this definition it is clear that κ_s, like m_s, are symmetrical functions of their arguments. Similar to the expansion (3.28), $\ln \Theta(\boldsymbol{\omega})$ can be expanded, about $\boldsymbol{\omega} = 0$, as a Taylor series involving the cumulants. By noting that $\ln \Theta(\mathbf{0}) = 0$ and taking exponentials of both sides of the expansion of $\ln \Theta(\boldsymbol{\omega})$, one obtains

$$\Theta(\boldsymbol{\omega}) = \exp\left(\sum_{s=1}^{\infty} \frac{\mathrm{i}^s}{s!} \sum_{j,k,\ldots,m=1}^{n} \kappa_s(\eta_j, \eta_k, \ldots, \eta_m)\omega_j\omega_k \cdots \omega_m \right) \tag{3.33}$$

By analogy with equation (3.29), when $s \geqslant n$ a cumulant $\kappa_{k_1 \cdots k_n}$ can be defined in terms of κ_s, where s is given by equation (3.30) and the argument η_1 appears k_1 times, η_2 appears k_2 times, and so on.

By equating the two expansions for $\Theta(\boldsymbol{\omega})$, given by equation (3.28) and (3.33), it is possible to obtain a set of relationships between m_s and κ_s. For the first four moments one has (Stratonovitch, 1964)

$$m_1(\eta_j) = \kappa_1(\eta_j) \tag{3.34}$$

$$m_2(\eta_j, \eta_k) = \kappa_2(\eta_j, \eta_k) - \kappa_1(\eta_j)\kappa_1(\eta_k) \tag{3.35}$$

$$\begin{aligned} m_3(\eta_j, \eta_k, \eta_l) = {} & \kappa_3(\eta_j, \eta_k, \eta_l) + 3\{\kappa_1(\eta_j)\kappa_2(\eta_k, \eta_l)\}_s \\ & + \kappa_1(\eta_j)\kappa_1(\eta_k)\kappa_1(\eta_l) \end{aligned} \tag{3.36}$$

$$\begin{aligned} m_4(\eta_j, \eta_k, \eta_l, \eta_m) = {} & \kappa_4(\eta_j, \eta_k, \eta_l, \eta_m) + 3\{\kappa_2(\eta_j, \eta_k)\kappa_2(\eta_l, \eta_m)\}_s \\ & + 4\{\kappa_1(\eta_j)\kappa_3(\eta_k, \eta_l, \eta_m)\}_s + 6\{\kappa_1(\eta_j)\kappa_1(\eta_k)\kappa_2(\eta_l, \eta_m)\}_s \\ & + \kappa_1(\eta_j)\kappa_1(\eta_k)\kappa_1(\eta_l)\kappa_1(\eta_m) \end{aligned} \tag{3.37}$$

In these expressions the symbol $\{\ \ \}_s$ indicates a 'symmetrizing operation', with respect to all the arguments. This involves taking the arithmetic mean of all *different permuted* terms similar to the one shown within the braces. For example

$$\begin{aligned} \{\kappa_2(\eta_j, \eta_k)\kappa_2(\eta_l, \eta_m)\}_s = {} & \tfrac{1}{3}[\kappa_2(\eta_j, \eta_k)\kappa_2(\eta_l, \eta_m) + \kappa_2(\eta_j, \eta_l)\kappa_2(\eta_k, \eta_m) \\ & + \kappa_2(\eta_j, \eta_m)\kappa_2(\eta_k, \eta_l)] \end{aligned} \tag{3.38}$$

The coefficient preceding each symmetrizing operation turns out to be exactly equal to the total number of terms being averaged. This is a useful cross-check when applying equations (3.36) and (3.37). It is possible to derive relationships between m_s and κ_s, for $s > 4$, but the complexity of the expressions increases rapidly as s increases. Ibrahim (1985) has given relationships for κ_s, in terms of m_s, for $s = 1, 2, \ldots, 8$.

An inspection of equation (3.34) reveals that κ_1 is equal to the mean of its argument, i.e.

$$\kappa_1(\eta_j) = E\{\eta_j\} = m_j \tag{3.39}$$

Using this result, and equation (3.35), the second-order cumulant can be written as

$$\kappa_2(\eta_j, \eta_k) = m_2(\eta_j, \eta_k) - m_j m_k = E\{(\eta_j - m_j)(\eta_k - m_k)\} \tag{3.40}$$

On comparing this with equation (3.21) it is evident that $\kappa_2(\eta_j, \eta_k)$ is simply the covariance of the variables η_j, η_k. Thus

$$\kappa_2(\eta_j, \eta_k) = \text{cov}\{\eta_j, \eta_k\} \tag{3.41}$$

As pointed out earlier, κ_2 will vanish if there is no statistical dependency between η_j and η_k.

A similar interpretation can be applied to the third-order cumulants. From equations (3.34) to (3.36) it follows that one can express κ_3 in the following alternative way

$$\kappa_3(\eta_j, \eta_k, \eta_l) = E\{(\eta_j - m_j)(\eta_k - m_k)(\eta_l - m_l)\} \tag{3.42}$$

Therefore, κ_3 can be regarded as a 'triple covariance function', and will vanish when at least one of the three random variables, η_j, η_k and η_l, is statistically independent of the other two. Similar interpretations can be applied to the higher order cumulants, but their physical significance is less well defined.

In many applications the means of all the elements in $\boldsymbol{\eta}$ are zero. Specifically

$$m_1(\eta_j) = \kappa(\eta_j) = m_j = 0 \quad j = 1, 2, \ldots, n \tag{3.43}$$

In this special case relationships such as equations (3.35) to (3.37) simplify considerably. For moments up to sixth order one has

$$m_2(\eta_j, \eta_k) = \kappa_2(\eta_j, \eta_k) \tag{3.44}$$

$$m_3(\eta_j, \eta_k, \eta_m) = \kappa_3(\eta_j, \eta_k, \eta_m) \tag{3.45}$$

$$m_4(\eta_j, \eta_k, \eta_m) = \kappa_4(\eta_j, \eta_k, \eta_m) + 3\{\kappa_2(\eta_j, \eta_k)\kappa_2(\eta_l, \eta_m)\}_s \tag{3.46}$$

$$m_5(\eta_j, \eta_k, \eta_l, \eta_m, \eta_n) = \kappa_5(\eta_j, \eta_k, \eta_l, \eta_m, \eta_n) + 10\{\kappa_2(\eta_j, \eta_k)\kappa_3(\eta_l, \eta_m, \eta_n)\}_s \tag{3.47}$$

$$\begin{aligned} m_6(\eta_j, \eta_k, \eta_l, \eta_m, \eta_n, \eta_o) = \kappa_6(\eta_j, \eta_k, \eta_l, \eta_m, \eta_n, \eta_o) &+ 15\{\kappa_2(\eta_j, \eta_k)\kappa_4(\eta_l, \eta_m, \eta_n, \eta_o)\}_s \\ &+ 15\{\kappa_2(\eta_j, \eta_k)\kappa_2(\eta_l, \eta_m)\kappa_2(\eta_n, \eta_o)\}_s \\ &+ 10\{\kappa_3(\eta_j, \eta_k, \eta_l)\kappa_3(\eta_m, \eta_n, \eta_o)\}_s \end{aligned} \tag{3.48}$$

3.5 THE GAUSSIAN DISTRIBUTION

In many practical problems the random variables of concern have a probability distribution which can be closely approximated by the ideal form known as the 'normal', or 'Gaussian' distribution. The theoretical basis for this observation is the so-called Central Limit Theorem, which states that if $\xi_1, \xi_2, \ldots, \xi_n$ are n independent random variables, with *arbitrary* distributions, then the sum $\eta = \xi_1 + \xi_2 + \cdots + \xi_n$ of these variables is a random variable which tends to that of a Gaussian distribution as n becomes large. In practice the random variables of interest frequently arise from a large number of unrelated effects and thus, by virtue of the Central Limit Theorem, it can be expected that they will have distributions which are close to the Gaussian form.

For a vector random variable, $\boldsymbol{\eta}$, the Gaussian form of the density function is

$$f(\mathbf{x}) = \frac{1}{(2\pi)^{n/2}|\mathbf{V}|^{1/2}} \exp[-\tfrac{1}{2}(\mathbf{x}-\mathbf{m})^{\mathrm{T}}\mathbf{V}^{-1}(\mathbf{x}-\mathbf{m})] \tag{3.49}$$

where $\mathbf{V}$ is the covariance matrix of $\boldsymbol{\eta}$ and $\mathbf{m}$ is the mean vector, equal to $E\{\boldsymbol{\eta}\}$. $|\mathbf{V}|$ denotes the determinant of $\mathbf{V}$. Here $f(\mathbf{x})$ is completely determined by $\mathbf{m}$ and $\mathbf{V}$. It is noted that if $\mathbf{V}$ is diagonal, the elements of $\boldsymbol{\eta}$ are uncorrelated and $f(\mathbf{x})$ reduces to the form

$$f(\mathbf{x}) = f_1(x_1) f_2(x_2) \cdots f_n(x_n) \tag{3.50}$$

where $f_i(x_i)$ is the Gaussian form for the ith component, i.e.

$$f_i(x_i) = \frac{1}{(2\pi)^{1/2}\sigma_i} \exp\left(\frac{(x_i - m_i)^2}{\sigma_i}\right) \tag{3.51}$$

where m_i is the mean of η_i and σ_i is the standard deviation of x_i. Thus, in the particular case of a joint normal distribution, absence of correlation *does* imply statistical independence, and vice versa.

The characteristic function for the Gaussian distribution can be found by combining equations (3.24) and (3.49). On performing the integration one obtains

$$\Theta(\boldsymbol{\omega}) = \exp(i\mathbf{m}^{\mathrm{T}}\boldsymbol{\omega} - \tfrac{1}{2}\boldsymbol{\omega}^{\mathrm{T}}\mathbf{V}\boldsymbol{\omega}) \tag{3.52}$$

or

$$\Theta(\boldsymbol{\omega}) = \exp\left(i \sum_{j=1}^{n} m_j \omega_j - \frac{1}{2} \sum_{j,k=1}^{n} v_{jk}\omega_j\omega_k\right) \tag{3.53}$$

where m_j are the elements of $\mathbf{m}$, and v_{jk} are the elements of $\mathbf{V}$. As previously shown, m_j are equal to the first-order cumulants $\kappa_1(\eta_j)$, (see equation (3.39)). Moreover

$$v_{jk} = E\{\eta_j - m_j)(\eta_k - m_k)\} \tag{3.54}$$

are equal to the second-order cumulants, $\kappa_2(\eta_j, \eta_k)$ (see equation (3.40)). Hence, in terms of cumulants, equation (3.53) can be written as

$$\Theta(\omega) = \exp\left(i \sum_{j=1}^{n} \kappa_1(\eta_j)\omega_j - \frac{1}{2} \sum_{j,k=1}^{n} \kappa_2(\eta_j, \eta_k)\omega_j\omega_k\right) \tag{3.55}$$

Comparing this with the general expansion for the characteristic function, in terms of cumulants, given by equation (3.33), it is seen that the Gaussian distribution is such that all the cumulants of order greater than two are zero. Thus, one can, in fact, *define* a Gaussian multi-dimensional distribution as one for which only the first and second order cumulants are non-zero.

3.5.1 Properties of Gaussian random variables

An extremely important property of Gaussian random variables is that any linear combination of such variables is also Gaussian. Thus, if $\boldsymbol{\xi} = \mathbf{A}\cdot\boldsymbol{\eta}$ where $\mathbf{A}$ is a linear transformation matrix and $\boldsymbol{\eta}$ is a Gaussian vector variable, then $\boldsymbol{\xi}$ is another Gaussian vector variable. The far reaching implications of this result will be discussed later.

Another important result for a Gaussian vector $\boldsymbol{\eta}$, which will be used later, is summarized in the following formula (Kazakov, 1965a)

$$E\{f(\boldsymbol{\eta})\boldsymbol{\eta}\} = E\{\boldsymbol{\eta}\boldsymbol{\eta}^{\mathrm{T}}\}E\{\boldsymbol{\nabla} f(\boldsymbol{\eta})\} \tag{3.56}$$

where $\boldsymbol{\nabla}$ is the gradient operator defined by

$$\boldsymbol{\nabla} = \left[\frac{\partial}{\partial\eta_1}, \frac{\partial}{\partial\eta_2}, \ldots, \frac{\partial}{\partial\eta_n}\right]^{\mathrm{T}} \tag{3.57}$$

3.5.2 Expansions of the Gaussian distribution

For non-Gaussian distributions, where the cumulants of order three, or greater, are non-zero, it is useful to have an expansion for $f(\mathbf{x})$, the density function for an n-vector $\boldsymbol{\eta}$, in terms of the Gaussian density function, here denoted $f_{\mathrm{G}}(\mathbf{x})$, and defined by equation (3.49).

In principle such an expansion can be derived by applying an inverse Fourier transform to the series expansion for $\Theta(\boldsymbol{\omega})$, given by either equation (3.28) or equation (3.33). This inverse transform is expressed by

$$f(x) = \frac{1}{(2\pi)^n}\int_{-\infty}^{\infty}\int_{-\infty}^{\infty}\cdots\int_{-\infty}^{\infty} \exp(-i\boldsymbol{\omega}^{\mathrm{T}}\mathbf{x})\Theta(\boldsymbol{\omega})\,\mathrm{d}\boldsymbol{\omega} \tag{3.58}$$

However, this approach leads to severe difficulties, since the required integrals cannot be calculated directly.

These difficulties can be reduced considerably by introducing the concept of quasi-moments (see Kuznetsov et al, 1965; Stratonovitch 1964). These quantities will be denoted

$$b_s(\eta_j, \eta_k, \ldots, \eta_m)$$

where, as for m_s and κ_s, s denotes the order of the quasi-moment and the number of arguments is equal to s.

Quasi-moments may be defined by the following relationship with the cumulants

$$\exp\left(\sum_{s=3}^{\infty}\sum_{j,k,\ldots,m=1}^{n}\frac{\mathrm{i}^s}{s!}\kappa_s(\eta_j,\eta_k,\ldots,\eta_m)\omega_j\omega_k\cdots\omega_m\right)$$

$$=1+\sum_{s=3}^{\infty}\frac{\mathrm{i}^s}{s!}b_s(\eta_j,\eta_k,\ldots,\eta_m)\omega_j\omega_k\cdots\omega_m \tag{3.59}$$

On comparing the expansion for the characteristic function, $\Theta(\boldsymbol{\omega})$, given by equation (3.33), with equation (3.59), and noting that, for a Gaussian distribution, $\Theta(\boldsymbol{\omega})$ is given by equation (3.53), one can write

$$\Theta(\boldsymbol{\omega})=\Theta_G(\boldsymbol{\omega})\left(1+\sum_{s=3}^{\infty}\frac{\mathrm{i}^s}{s!}b_s(\eta_j,\eta_k,\ldots,\eta_m)\omega_j\omega_k\cdots\omega_m\right) \tag{3.60}$$

In this equation $\Theta_G(\omega)$ is the characteristic function of a Gaussian vector, $\boldsymbol{\eta}_G$, with the same first- and second-order cumulants as the non-Gaussian vector $\boldsymbol{\eta}$.

Using equation (3.59) it is possible to derive a set of relationships between b_s and κ_s. In fact the relationship between m_s and κ_s (see equation (3.28) and (3.33)) is very similar to that between b_s and κ_s, and is identical if one sets $m_1=m_2=0$, or $\kappa_1=\kappa_2=0$. Hence, in equations (3.36) and (3.37), if m_3 and m_4 are replaced by κ_3 and κ_4, respectively, and κ_1 and κ_2 are set to zero, it follows that

$$b_3=\kappa_3 \quad b_4=\kappa_4 \tag{3.61}$$

Further calculations show that (Stratonovitch, 1964)

$$b_5=\kappa_5 \tag{3.62}$$

$$b_6=\kappa_6+10\{\kappa_3\kappa_3\}_s \tag{3.63}$$

$$b_7=\kappa_7+35\{\kappa_3\kappa_4\}_s \tag{3.64}$$

Thus, up to sixth order, the quasi-moments are identical to the cumulants. All quasi-moments are, like m_s and κ_s, symmetrical functions of their arguments. Bover (1978) has derived expressions for b_s in terms of m_s, in the one-dimensional case, for $s=3,4,\ldots,10$.

Returning to the problem of finding an expression for a non-Gaussian $f(\mathbf{x})$, it is now possible to combine equation (3.58) with equation (3.60). Using the fact that $f_G(\mathbf{x})$ is related to $\Theta_G(\boldsymbol{\omega})$ through equation (3.58), it follows that

$$\left(-\frac{\partial}{\partial x_j}\right)\left(-\frac{\partial}{\partial x_k}\right)\cdots\left(-\frac{\partial}{\partial x_m}\right)f_G(\mathbf{x})$$

$$=\int_{-\infty}^{\infty}\int_{-\infty}^{\infty}\cdots\int_{-\infty}^{\infty}\mathrm{i}^s\omega_j\omega_k\cdots\omega_m\exp(-\mathrm{i}\boldsymbol{\omega}^{\mathrm{T}}\mathbf{x})\Theta_G(\boldsymbol{\omega})\,\mathrm{d}\boldsymbol{\omega} \tag{3.65}$$

Hence the following expression for $f(\mathbf{x})$, in terms of $f_G(\mathbf{x})$, is obtained

$$f(\mathbf{x}) = \left[1 + \sum_{s=3} \frac{(-1)^s}{s!} \sum_{j,k,\ldots,m=1}^{n} b_s(\eta_j, \eta_k, \ldots, \eta_m) \left(\frac{\partial}{\partial x_j}\right)\left(\frac{\partial}{\partial x_k}\right)\cdots\left(\frac{\partial}{\partial x_m}\right)\right] f_G(\mathbf{x}) \tag{3.66}$$

Equation (3.66) can be written, alternatively, in terms of generalized Hermite polynomials, defined by

$$H_{jk\cdots m}(\mathbf{x}) = \exp(\theta(\mathbf{x}))\left(-\frac{\partial}{\partial x_j}\right)\left(-\frac{\partial}{\partial x_k}\right)\cdots\left(-\frac{\partial}{\partial x_m}\right)\exp(-\theta(\mathbf{x})) \tag{3.67}$$

where

$$\theta(\mathbf{x}) = \tfrac{1}{2}\mathbf{x}^{\mathrm{T}}\mathbf{V}^{-1}\mathbf{x} \tag{3.68}$$

Thus, in place of equation (3.66) one can write (Stratonovitch, 1964)

$$f(\mathbf{x}) = \left(1 + \sum_{s=3}^{\infty} \frac{1}{s!} \sum_{j,k,\ldots,m=1}^{n} b_s(\eta_j, \eta_k, \ldots, \eta_m) H_{jk\cdots m}(\mathbf{x}-\mathbf{m})\right) f_G(\mathbf{x}) \tag{3.69}$$

In the one-dimensional case where $\boldsymbol{\eta} = \eta$ is scalar, having a standard deviation, σ, and mean, m, equation (3.69) reduces to

$$f(x) = \left[1 + \sum_{s=3}^{\infty} \frac{1}{s!}\frac{b_s}{\sigma^s} H_s\left(\frac{x-m}{\sigma}\right)\right] f_G(x) \tag{3.70}$$

where H_s are ordinary Hermite polynomials, defined by

$$H_s\left(\frac{x}{\sigma}\right) = \sigma^s H_{\underbrace{11\cdots 1}_{s \text{ times}}}(x) \tag{3.71}$$

or

$$H_s(x) = \exp\left(\frac{x^2}{2}\right)\left(-\frac{\mathrm{d}}{\mathrm{d}x}\right)^s \exp\left(-\frac{x^2}{2}\right) \tag{3.72}$$

and

$$b_s = b_s(\underbrace{x, x, \ldots, x}_{s \text{ times}}) \tag{3.73}$$

The symbol $f_G(x)$ in equation (3.70) represents the one-dimensional Gaussian density function, defined by

$$f_G(x) = \frac{1}{(2\pi)^{1/2}\sigma}\exp\left[-\frac{1}{2}\left(\frac{x-m}{\sigma}\right)^2\right] \tag{3.74}$$

Expansion (3.70) is often referred to as Edgeworth series (see Cramer, 1946).

The first six Hermite polynomials are as follows:

$$\begin{aligned} &H_1(x) = x && H_4(x) = x^4 - 6x^2 + 3 \\ &H_2(x) = x^2 - 1 && H_5(x) = x^5 - 10x^3 + 15x \\ &H_3(x) = x^3 - 3x && H_6(x) = x^6 - 15x^4 + 45x^2 - 15 \end{aligned} \tag{3.75}$$

For the two-dimensional case, where $n = 2$, equation (3.69) can be written as

$$f(x_1, x_2) = \left[1 + \sum_{s=3}^{\infty} \frac{1}{s!} \sum_{l+m=s} b_{lm} H_{lm}(x_{01}, x_{02}) \right] f_G(x_1, x_2) \tag{3.76}$$

where, similar to equation (3.29)

$$b_{lm} = b_{l+m}(\underbrace{x_1, \ldots, x_1}_{l\text{ times}}, \underbrace{x_2, \ldots, x_2}_{m\text{ times}}) \tag{3.77}$$

and, similar to equation (3.71)

$$H_{lm} = H_{1\cdots 1\ 2\cdots 2}(x_{01}, x_{02}) \tag{3.78}$$

Here

$$x_{01} = x_1 - m_1 \quad x_{02} = x_2 - m_2 \tag{3.79}$$

Expressions for H_{lm} up to third order ($l + m = 3$) have been derived by Stratonovitch (1964).

3.6 THE CONCEPT OF A STOCHASTIC PROCESS

Earlier the idea of a random variable as a number which characterizes an event, belonging to a complete set of events was introduced. Every time an experiment is performed (i.e., the specified set of conditions is realized), a sample value of the random variable is obtained. As a natural extension, it may be imagined that a *function of time*, rather than just a number, is obtained every time an experiment is carried out. The sample function obtained will, in general, be different every time the experiment is performed, and there is thus an infinite set, or 'ensemble' of possible sample functions, or 'realizations'. This infinite set will be referred to as a stochastic (or random) process, denoted $\eta(t)$.

As a simple example of a stochastic process, suppose that the flow velocity in an air jet, as a function of time, is measured at a particular location in the flow, and in a particular direction. If the air flow is turbulent then a record, $\eta_1(t)$, as shown in Figure 3.1, might be obtained. If the experiment is now repeated, keeping all the conditions which are under control the same, another record, $\eta_2(t)$, as shown in Figure 3.1, would be obtained. After n repetitions of the experiment, n records, as indicated in Figure 3.1, would be generated. Each

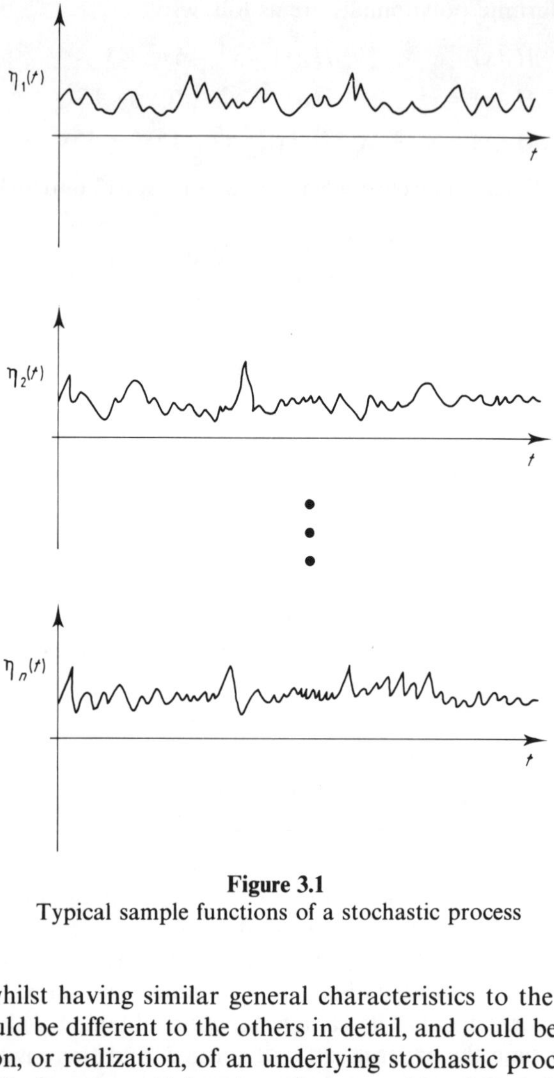

Figure 3.1
Typical sample functions of a stochastic process

record, $\eta_i(t)$, whilst having similar general characteristics to the other records measured, would be different to the others in detail, and could be regarded as a sample function, or realization, of an underlying stochastic process, $\eta(t)$.

Although only functions of time have been considered here, it is noted that in some problems it may be convenient to consider functions of time and space, i.e. to define a stochastic process, $\eta(t, \mathbf{x})$ where $\mathbf{x}$ is a vector spatial variable. In this book, attention will be restricted to functions of time only.

3.6.1 The complete probabilistic specification

Consider a stochastic process, $\eta(t)$, at the set of time instants $t_1, t_2, \ldots, t_n$. Then $\eta(t_1)$, $\eta(t_2), \ldots, \eta(t_n)$ are a set of random variables and constitute a vector random variable, denoted ξ.

The stochastic process $\eta(t)$, is specified completely, in a probabilistic sense, if the distribution function,

$$\mathbf{F}(\mathbf{x}) = P(\boldsymbol{\xi} < \mathbf{x}) \tag{3.80}$$

is known for any set of time instants, $t_1, t_2, \ldots, t_n$, where n is any integer. Alternatively, the joint density function, $f(\mathbf{x})$ is required, for any set of time instants.

In terms of density functions, there is a hierarchy of descriptive functions, which will be denoted as

$f_1(x_1;t_1)$	first-order density function for $\eta(t_1)$
$f_2(x_1;t_1,x_2;t_2)$	second-order density function for $\eta(t_1)$ and $\eta(t_2)$
$f_3(x_1;t_1,x_2;t_2,x_3;t_3)$	third-order density function for $\eta(t_1)$, $\eta(t_2)$ and $\eta(t_3)$

and so on. It is noted that a knowledge of the nth-order density function, for all possible values of its arguments, implies a complete knowledge of all lower order density functions (of order less than n). For example

$$f_1(x_1;t_1) = \int_{-\infty}^{\infty} f_2(x_1;t_1,x_2;t_2)\,\mathrm{d}x_2 \tag{3.81}$$

for any value of t_2.

3.6.2 The Gaussian process

A special, and very important case, is that of the Gaussian, or normal process. Here $\boldsymbol{\xi}^{\mathrm{T}} = [\eta(t_1), \eta(t_2), \ldots, \eta(t_n)]$ has, for any set of time instants $t_1, t_2, \ldots, t_n$, a multi-variate Gaussian distribution, as given by equation (3.49). Thus, a knowledge of $\mathbf{m}$, the mean vector, and $\mathbf{V}$, the covariance matrix, for any set of time instants, is sufficient to define a Gaussian process completely. It is immediately evident that the complete specification of a Gaussian process requires far less information than is required for a general process.

Suppose that a 'mean function', $m(t)$, is defined, such that

$$m(t) = E\{\eta(t)\} \tag{3.82}$$

and a 'covariance function', $w(t_1, t_2)$, is defined by the relation

$$w(t_1,t_2) = E\{[\eta(t_1) - m(t_1)][\eta(t_2) - m(t_2)]\} = R(t_1,t_2) - m(t_1)m(t_2) \tag{3.83}$$

where

$$R(t_1,t_2) = E\{\eta(t_1)\eta(t_2)\} \tag{3.84}$$

is the 'correlation function' for $\eta(t)$. It is evident that, if $m(t)$ and $w(t_1, t_2)$ (or $R(t_1, t_2)$) are known, for all values of their arguments, then the quantities $\mathbf{m}$ and $\mathbf{V}$ in equation (3.49) can be found, for any set of time instants. A Gaussian process is, therefore, completely specified by the mean function, $m(t)$, and the covariance function $w(t_1, t_2)$.

By virtue of the Central Limit Theorem, referred to earlier, many real processes which occur in engineering applications can be expected to be such that they can be idealized as Gaussian processes, to a close approximation. This is particularly true for excitation processes occurring from the natural environment, such as wind and wave forces.

3.6.3 Stationary processes

Frequently the generating mechanism behind a stochastic process does not change its character significantly with time, over periods of time which are much longer than the time-scale of fluctuations in the realizations of the process. It is then reasonable to model the stochastic process as a stationary process.

We define a stochastic process as stationary (in the strict sense) if its statistics are not affected by a shift in the time origin. Thus, if $\eta(t)$ is stationary then the two processes $\eta(t)$ and $\eta(t+s)$ have the same probabilistic description, for any value of s. It follows that the nth-order density function is such that

$$f_n(x_1;t_1,x_2;t_2,\dots,x_n;t_n)=f_n(x_1;t_1+s,x_2;t_2+s,\dots,x_n;t_n+s) \tag{3.85}$$

for any value of s.

As a special case, for $n=1$, f_1 is independent of t. From this it follows that the mean value, m, is a constant, and that all other statistics derived from $f_1(x)$ are also constants. In particular, the standard deviation, σ, is a constant.

For $n=2$, equation (3.85) shows that f_2 depends only on the time difference, $\tau=t_2-t_1$. An important consequence of this is that the covariance function $w(t_1, t_2)$ and the correlation function $R(t_1,t_2)$ depend only on τ i.e.

$$w(\tau)=E\{[\eta(t)-m][\eta(t+\tau)-m]\}=R(\tau)-m^2 \tag{3.86}$$

where

$$R(\tau)=E\{\eta(t)\eta(t+\tau)\} \tag{3.87}$$

Usually information on all the joint density functions of a stochastic process is unavailable and it is not possible to specify whether the process is stationary in the strict sense. A rather less stringent definition, which is of more practical value, is to define a process as stationary 'in the wide sense' (or 'weakly' stationary) if the mean value is a constant and the covariance function is independent only on the time difference, t_2-t_1. It is important to note here that, in the special case of a Gaussian process, stationarity in the wide sense also implies stationarity in the strict sense.

The covariance function, $w(\tau)$, of a stationary stochastic process has certain simple properties. it is an even function with a maximum at the origin and such that $w(\tau)\to 0$ as $|\tau|\to\infty$.

The last property follows from the fact that $\eta(t_1)$ and $\eta(t_2)$ will tend to become uncorrelated, as $|\tau|=|t_2-t_1|$ becomes large, for all real processes.

The rate at which $w(\tau)\rightarrow 0$ as $|\tau|$ increases is a measure of the 'time scale' of the process. An integral time scale τ_{cor}, may be defined by

$$\tau_{\text{cor}} = \frac{1}{w(0)} \int_0^{\infty} \tau w(\tau)\,\mathrm{d}\tau \tag{3.88}$$

If τ_{corr} is small we have a 'wide-band' process, whereas if τ_{cor} is large the process is narrow-band. A special, idealized case is where $\tau_{\text{cor}} = 0$; the process is then uncorrelated, and is usually referred to as white noise.

3.7 DIFFERENTIATION OF STOCHASTIC PROCESSES

The derivative, $\dot{\eta}(t)$, of a stochastic process, $\eta(t)$, can be defined formally as

$$\dot{\eta}(t) = \frac{\mathrm{d}\eta}{\mathrm{d}t} = \lim_{\varepsilon\rightarrow 0} \left| \frac{\eta(t+\varepsilon) - \eta(t)}{\varepsilon} \right| \tag{3.89}$$

However, since $\eta(t)$ can only be described probabilistically, it follows that the limit in equation (3.89) can not be interpreted in the normal, deterministic sense. A probabilistic interpretation is required.

For the present purposes, it is convenient to adopt a 'mean square' interpretation. The process $\eta(t)$ is said to be differentiable, in a 'mean square sense', (m.s. sense, for short) if a derivative process, $\dot{\eta}(t)$, can be found, such that

$$\lim_{\varepsilon\rightarrow 0} E\left\{\left(\frac{\eta(t+\varepsilon) - \eta(t)}{\varepsilon} - \dot{\eta}(t)\right)^2\right\} = 0 \tag{3.90}$$

All real processes which arise in practice are such that their realizations are differentiable in the normal sense. In these circumstances, one can be certain that derivatives of these processes will exist, in a m.s. sense. However, more caution is required when dealing with idealized models of real processes.

Sufficient conditions for a process to be differentiable n times, in a m.s. sense, are that (e.g. see Parzen, 1962) the mean function is differentiable n times, and that

$$\frac{\partial^{2n} w(t_1, t_2)}{\partial t_1^n \partial t_2^n}$$

exists, and is continuous.

In the special case of a process which is stationary in the wide sense, and differentiable in the m.s. sense, it may be shown that the derivative process is also stationary, in a m.s. sense. In these circumstances

$$E\{\dot{\eta}(t_1), \eta(t_2)\} = -\frac{\mathrm{d}}{\mathrm{d}\tau} R(\tau) \tag{3.91}$$

$$E\{\dot{\eta}(t_1), \dot{\eta}(t_2)\} = -\frac{\mathrm{d}^2}{\mathrm{d}\tau^2} R(\tau) \tag{3.92}$$

where $R(\tau)$ is the correlation function for $\eta(t)$. Similar relations exist for the covariance function.

3.8 INTEGRATION OF STOCHASTIC PROCESSES

A formal definition of the integral of a stochastic process, $\eta(t)$, over an interval $a < t < b$, is as follows

$$\int_a^b \eta(t)\,\mathrm{d}t = \lim_{n\to\infty} \sum_{i=1}^{n} \eta(t_i)\Delta t_1 \tag{3.93}$$

where $\Delta t_i = t_i - t_{i-1}$, $t_1 = a$ and $t_n = b$. Here it is understood that the maximum value of Δt_i tends to zero as n becomes large. This is the usual 'approximating sum' definition of an integral, which is familiar in the context of deterministic functions. When $\eta(t)$ is a stochastic process the above limiting operation requires a probabilistic interpretation.

Here, as in the case of differentiation, a mean square (m.s.) interpretation of integration will be adopted. It can be shown that a stochastic process is integrable, in the m.s. sense, (i.e. the approximating sum of equation (3.93) has a limit, in the m.s. sense) provided that the function $E\{\eta(t_1)\eta(t_2)\}$ is Riemann integrable over the range $a \leqslant s,\ t \leqslant b$.

For m.s. integrable processes it can be proved that the linear operations of integration and expectation may be commuted; this is a result which may be intuitively anticipated. Hence, for example, one finds that

$$E\left\{\int_a^b \eta(t)\,\mathrm{d}t\right\} = \int_a^b E\{\eta(t)\}\,\mathrm{d}t = \int_a^b m(t)\,\mathrm{d}t \tag{3.94}$$

and

$$\operatorname{cov}\left\{\int_a^b \eta(t_1)\,\mathrm{d}t_1 \int_c^d \eta(t_2)\,\mathrm{d}t_2\right\} = \int_a^b \mathrm{d}t_1 \int_c^d \mathrm{d}t_2 w(t_1, t_2) \tag{3.95}$$

Relationships of this kind are extremely useful when dealing with linear transformation of stochastic processes, as will be demonstrated in the following chapter.

3.9 ERGODICITY

In many experimental situations only one realization, $\eta_i(t)$ say, of a stationary stochastic process is available. It is then natural to estimate quantities such as the mean, m, and the correlation function, $R(\tau)$, from *time averaging* operations on $\eta_i(t)$.

If the available realization is of duration T, a natural estimate of the mean, m, is as follows

$$m_{iT} = \frac{1}{T}\int_0^T \eta_i(t)\,\mathrm{d}t \tag{3.96}$$

Clearly m_{iT} can be regarded as a sample value of the random variable m_T, where

$$m_T = \frac{1}{T}\int_0^T \eta(t)\,\mathrm{d}t \tag{3.97}$$

and the stochastic integral is here interpreted in a m.s. sense.

If, in the limit, as $T \to \infty$

$$\lim_{T\to\infty} m_T = E\{\eta(t)\} = m \tag{3.98}$$

where, again, the limit is understood in a m.s. sence, then $\eta(t)$ is said to be *ergodic in first moment*, or *ergodic in mean.* Applying the mean-square convergence criterion one has, similar to equation (3.90), the requirement that

$$\lim_{T\to\infty} E\left\{\left(\frac{1}{T}\int_0^T \eta(t)\,\mathrm{d}t - m\right)^2\right\} = 0 \tag{3.99}$$

for first moment ergodicity. It follows (e.g. see Lin, 1967) that in addition to the requirement that $\eta(t)$ be stationary (in the wide sense), one must also satisfy the condition

$$\lim_{T\to\infty} \frac{1}{T}\int_0^T w(\tau)\,\mathrm{d}\tau = 0 \tag{3.100}$$

Similarly, the correlation function, $R(\tau)$, can be estimated from a single realization as follows

$$R_{iT}(\tau) = \frac{1}{(T-\tau)}\int_0^{T-\tau} \eta_i(t)\eta_i(t+\tau)\,\mathrm{d}t \tag{3.101}$$

and $R_{iT}(\tau)$ may be regarded as a sample value of the random variable

$$R_T(\tau) = \frac{1}{(T-\tau)}\int_0^{T-\tau} \eta(t)\eta(t+\tau)\,\mathrm{d}t \tag{3.102}$$

If

$$\lim_{T\to\infty} R_T(\tau) = E\{\eta(t)\eta(t+\tau)\} = R(\tau) \tag{3.103}$$

where, as before, the integration and limiting operation are understood in a m.s. sense, then $\eta(t)$ is said to be *ergodic in second moments* or *ergodic in correlation.* Necessary and sufficient conditions for ergodicity in correlation can be readily derived (e.g. see Lin, 1967).

In practical applications, where the processes of concern are stationary, or nearly so, it is very common to implicitly assume ergodicity, by the act of estimating the mean and the correlation function from a single sample function, according to equations (3.96) and (3.101). This assumption almost invariably leads to meaningful results, when dealing with real processes.

3.10 SPECTRAL DECOMPOSITION

For a stationary stochastic process, $\eta(t)$, the concept of the power spectral density function (or power spectrum), denoted $S(\omega)$, is particularly useful, $S(\omega)$ may be defined as the Fourier transform of the correlation function, $R(\tau)$; thus

$$S(\omega) = \frac{1}{2\pi}\int_{-\infty}^{\infty} R(\tau)\exp(\mathrm{i}\omega\tau)\,\mathrm{d}\tau \tag{3.104}$$

Since $R(\tau)$ is an even function, this can be written as

$$S(\omega) = \frac{1}{2\pi}\int_{-\infty}^{\infty} R(\tau)\cos\omega\tau\,\mathrm{d}\tau \tag{3.105}$$

The inverse of equation (3.104) is

$$R(\tau) = \int_{-\infty}^{\infty} S(\omega)\exp(-\mathrm{i}\omega\tau)\,\mathrm{d}\omega \tag{3.106}$$

and since $S(\omega)$ is also an even function (from equation (3.105)), it follows that equation (3.106) can be expressed alternatively as

$$R(\tau) = \int_{-\infty}^{\infty} S(\omega)\cos\omega\tau\,\mathrm{d}\omega \tag{3.107}$$

Setting $\tau = 0$ in equation (3.107) one has

$$R(0) = E\{\eta^2(t)\} = \int_{-\infty}^{\infty} S(\omega)\,\mathrm{d}\omega \tag{3.108}$$

Therefore, the total area under the power spectral density curve gives the total mean square of the process.

The function $S(\omega)$ can be interpreted as a spectral decomposition of the mean square. Thus, if a frequency band, $\omega, \omega + \delta\omega$ is considered, then the contribution to $E\{\eta^2\}$ from these harmonic components in the process which lie in this frequency band is $S(\omega)\,\delta\omega$, if $\delta\omega$ is small.

The power spectral density function can also be related directly to the Fourier transform of a stationary stochastic process, $\eta(t)$. Here, since realizations of $\eta(t)$

are not periodic, and do not decay to zero, is is not possible to use normal Fourier series, or even Fourier transform methods. Instead, the theory of 'generalized harmonic analysis', due to Weiner, is required (Papoulis, 1984). According to this theory one can write

$$\eta(t) = \int_{-\infty}^{\infty} \exp(i\omega t)\, dZ(\omega) \tag{3.109}$$

where $Z(\omega)$ is also a random process. The above 'stochastic integral' can be interpreted in a mean square (m.s.) sense. However, it should be noted that $Z(\omega)$ is not m.s. differentiable; in fact $dZ(\omega)$ is of order $(d\omega)^{1/2}$.

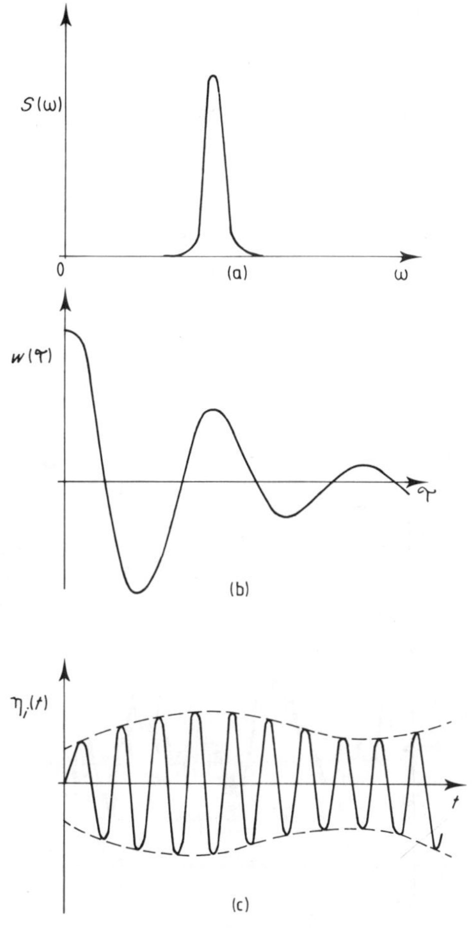

Figure 3.2
Characteristics of a narrow-band stochastic process. (a) Power spectrum. (b) Correlation function. (c) Typical sample function

It can be shown (e.g., see Doob, 1953) that $Z(\omega)$ is an orthogonal process, when $\eta(t)$ is stationary, i.e., for two non-overlapping intervals $d\omega$ and $d\omega'$, the increments $dZ(\omega)$ and $dZ(\omega')$ are uncorrelated. If two such intervals *do* overlap then the quantity $E\{|dZ(\omega)|^2\}$ is non-zero, and represents the average, or expected, power of the process within the frequency interval $d\omega$. This quantity is of order $d\omega$, and it is thus natural to define a power spectral density function, $S(\omega)$, as

$$S(\omega) = \frac{E\{|dZ(\omega)|^2\}}{d\omega} \tag{3.110}$$

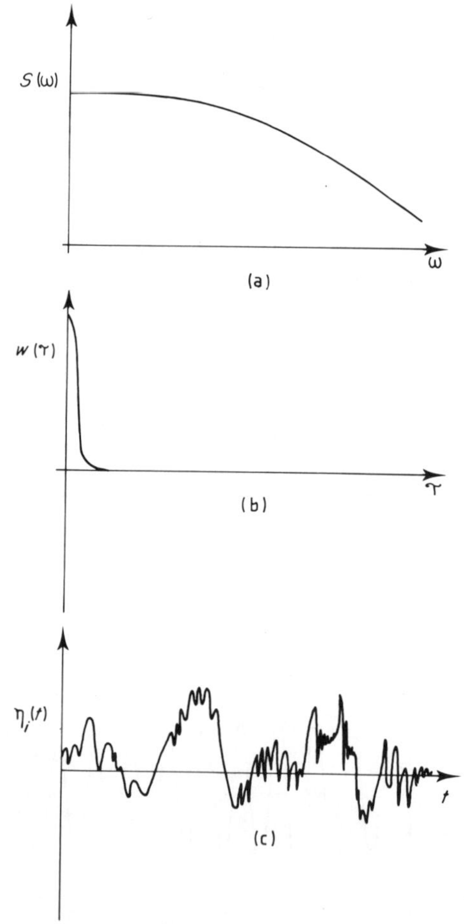

Figure 3.3
Characteristics of a wide-band stochastic process. (a) Power spectrum. (b) Correlation function. (c) Typical sample function

It is fairly easy to demonstrate that this definition is fully in accord with the definition given earlier by equation (3.104). Further, it lends itself to an extension, leading to the concept of time dependent, or evolutionary, spectra for non-stationary processes (Priestley, 1981).

If the correlation time scale, τ_{cor}, is small, then the power of the process, as measured by the mean square, will be spread over a wide range of frequencies—hence the description 'wide-band', referred to earlier. On the other hand, if τ_{cor} is large then the power may be limited to a narrow-band of frequencies, centred at $\omega = 0$. More generally, a narrow-band process will have a single, sharp peak in the neighbourhood of some non-zero frequency, ω_0 say. In this case the corresponding covariance function will establish a pronounced oscillating characteristic, the oscillation frequency being close to ω_0. The effect of a non-zero mean, m, is to introduce a delta function 'spike' at $\omega = 0$, of area magnitude m^2, i.e. a term representable by $m^2\delta(0)$. Typical spectra, correlation functions and sample functions are sketched in Figure 3.2 and 3.3.

A special limiting case is that of ideal white noise. Here, if the mean is non-zero

$$S(\omega) = S_0 + m^2\delta(0) \tag{3.111}$$

where S_0 is a constant; the spectrum thus appears as shown in Figure 3.4a. From equation (3.111), the corresponding covariance function is found to be

$$w(\tau) = 2\pi S_0 \delta(\tau) \tag{3.112}$$

where δ is Dirac's delta function (see Figure 3.4b). For this process $\tau_{cor} = 0$ and it is evident, from equation (3.112), that the mean square is infinite. Clearly no process can have an infinite power, but nonetheless, the idealization of real, wide-band processes, as white noises can be very useful in analysis, provided that care is taken in interpreting results derived from this idealization.

An expression for the spectral density, $S_{\dot{\eta}}(\omega)$ of the derivative process, $\dot{\eta}(t)$, can be found in terms of the spectral density $S_\eta(\omega)$, of $\eta(t)$. From equations

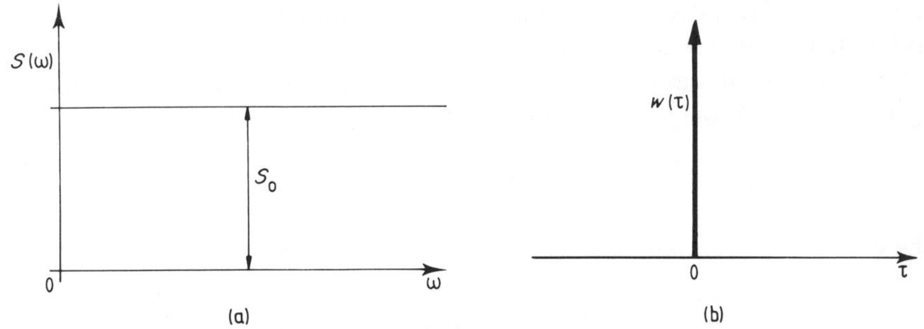

Figure 3.4
Characteristics of ideal white noise. (a) Power spectrum. (b) Correlation function

(3.92) and (3.104) it is found that

$$S_{\dot{\eta}}(\omega) = \omega^2 S_\eta(\omega) \tag{3.113}$$

Similarly, for the second derivative, $\ddot{\eta}(t)$

$$S_{\ddot{\eta}}(\omega) = \omega^4 S_\eta(\omega) \tag{3.114}$$

and so on.

3.11 SPECIFICATION OF JOINT PROCESSES

Frequently the relationship between two or more stochastic processes needs to be specified. For example, when considering a dynamic system with an input process, $\eta(t)$, and an output process, $\xi(t)$, the statistical relationships which exist between $\eta(t)$ and $\xi(t)$ are often of interest.

A two-dimensional process consists of two stochastic processes, $\eta(t)$ and $\xi(t)$. It is determined fully, in a probabilistic sense, if the joint distribution of the random variables $\eta(t_1), \eta(t_2), \ldots, \eta(t_n), \xi(t'_1), \xi(t'_2), \ldots, \xi(t'_m)$ is known for any set of the instants $t_1, t_2, \ldots, t_n$ and $t'_1, t'_2, \ldots, t'_m$. If this joint distribution is of the Gaussian form, for any set of time instants, then the joint process is said to be Gaussian. This implies that the individual processes are themselves Gaussian.

For joint Gaussian processes a complete probabilistic specification can be obtained by defining

(1) The mean functions of $\eta(t)$ and $\xi(t)$
(2) The covariance functions of $\eta(t)$ and $\xi(t)$
(3) The cross-covariance function, $w_{\eta\xi}(t_1, t_2)$, where

$$w_{\eta\xi}(t_1, t_2) = E\{[\eta(t_1) - m_\eta(t_1)][\xi(t_2) - m_\xi(t_2)]\} \tag{3.115}$$

Two processes $\eta(t)$ and $\xi(t)$ are jointly stationary if their joint statistics are the same as the joint statistics of $\eta(t+s)$ and $\xi(t+s)$. Joint stationarity implies that both individual processes are stationary.

For jointly stationary processes, $\eta(t)$ and $\xi(t)$, the cross-covariance function $w_{\eta\xi}(t_1, t_2)$ depends only on the time difference $\tau = t_2 - t_1$, i.e.

$$w_{\pi\xi}(\tau) = E\{[\eta(t_1) - m_\eta][\xi(t_2) - m_\xi]\} = R_{\eta\xi}(\tau) - m_\eta m_\xi \tag{3.116}$$

where

$$R_{\eta\xi}(\tau) = E\{\eta(t_1)\xi(t_2)\} \tag{3.117}$$

is the cross-correlation function for $\eta(t)$ and $\xi(t)$. Associated with $w_{\eta\xi}(\tau)$ one has the cross-spectral density function $S_{\eta\xi}(\omega)$, defined by

$$S_{\eta\xi}(\omega) = \frac{1}{2\pi}\int_{-\infty}^{\infty} R_{\eta\xi}(\tau)\exp(\mathrm{i}\omega\tau)\,\mathrm{d}\tau \tag{3.118}$$

It is noted that the cross-correlation is not an even function of τ, and hence the cross-spectral density, $S_{\eta\xi}(\omega)$, is in general a complex quantity possessing both real and imaginary components. Fourier inversion of equation (3.118) gives

$$R_{\eta\xi}(\tau) = \int_{-\infty}^{\infty} S_{\eta\xi}(\omega)\exp(-\mathrm{i}\omega\tau)\,\mathrm{d}\omega \tag{3.119}$$

These concepts are readily extendable to the more general case of an n-dimensional stochastic process, consisting of n stochastic processes, $\eta_1(t), \eta_2(t), \ldots, \eta_n(t)$. For such a process the covariance matrix $\mathbf{w}_{\boldsymbol{\eta}}(t_1, t_2)$ is

$$\begin{bmatrix} w_{\eta_1\eta_2}(t_1,t_2) & w_{\eta_1\eta_2}(t_1,t_2) & \cdots & w_{\eta_1\eta_n}(t_1,t_2) \\ \vdots & & & \\ w_{\eta_n\eta_1}(t_1,t_2) & w_{\eta_n\eta_2}(t_1,t_2) & \cdots & w_{\eta_n\eta_n}(t_1,t_2) \end{bmatrix} \tag{3.120}$$

For a jointly stationary process, all the elements of the above matrix depend only on $\tau = t_2 - t_1$, and a corresponding spectral density matrix (or power spectrum matrix), $\mathbf{S}_\eta(\omega)$ exists, as follows

$$\begin{bmatrix} S_{\eta_1\eta_1}(\omega) & S_{\eta_1\eta_2}(\omega) & \cdots & S_{\eta_1\eta_n}(\omega) \\ \vdots & & & \\ S_{\eta_n\eta_1}(\omega) & S_{\eta_n\eta_2}(\omega) & \cdots & S_{\eta_n\eta_n}(\omega) \end{bmatrix} \tag{3.121}$$

where

$$S_{\eta_i\eta_j}(\omega) = \frac{1}{2\pi}\int_{-\infty}^{\infty} R_{\eta_i\eta_j}(\tau)\exp(\mathrm{i}\omega\tau)\,\mathrm{d}\tau \quad 1 \leqslant i,j \leqslant n \tag{3.122}$$

and

$$R_{\eta_i\eta_n}(\tau) = E\{\eta_i(t)\eta_j(t+\tau)\} = w_{\eta_i\eta_j}(\tau) + m_{\eta_i}m_{\eta_j} \tag{3.123}$$

are the elements of the correlation matrix $\mathbf{R}_{\boldsymbol{\eta}}(t_1, t_2)$.

Chapter 4

Elements of linear random vibration theory

4.1 INTRODUCTION

In the preceding chapter it was shown that a stochastic process can be described in a probabilistic fashion; fully in terms of distribution functions and partially in terms of statistical moments, such as the mean and covariance functions. When dealing with random vibration problems, both the excitation and response processes are modelled as stochastic processes, and the central problem may be stated as follows: to predict various statistical parameters of the response processes from a knowledge of

(a) a probabilistic specification of the excitation processes, and
(b) the equations of motion of the system.

In this chapter the basic elements of deterministic linear vibration theory will be reviewed and it will be shown how these can be combined with the theory of stochastic processes (discussed in the previous chapter) to yield stochastic input-output relationships. The application of the theory to lumped-parameter systems, with a finite number of degrees of freedom, will be discussed and the particular case of a single degree of freedom system will be considered in some detail. It will be implicitly assumed throughout that the systems of concern are time-invariant, and stable.

4.2 GENERAL INPUT–OUTPUT RELATIONSHIPS

Suppose one has a system with r inputs, $x_1(t)$, $x_2(t), \ldots, x_r(t)$, persisting from $t = -\infty$, and s outputs, $y_1(t), y_2(t), \ldots, y_s(t)$, as shown in Figure 4.1. If $\mathbf{x}(t)$ is the r-vector of inputs and $\mathbf{y}(t)$ is the s-vector of outputs, then the time-domain, convolution integral relationship between $\mathbf{x}(t)$ and $\mathbf{y}(t)$ can be expressed as

$$\mathbf{y}(t) = \int_0^\infty \mathbf{h}(v)\mathbf{x}(t-v)\,\mathrm{d}v \tag{4.1}$$

where $\mathbf{h}(t)$ is an $r \times s$ matrix of impulse response functions, i.e.

$$\mathbf{h}(t) = [h_{ij}(t)] \tag{4.2}$$

The element h_{ij} of $\mathbf{h}$ is here defined as the response at the ith output to a unit

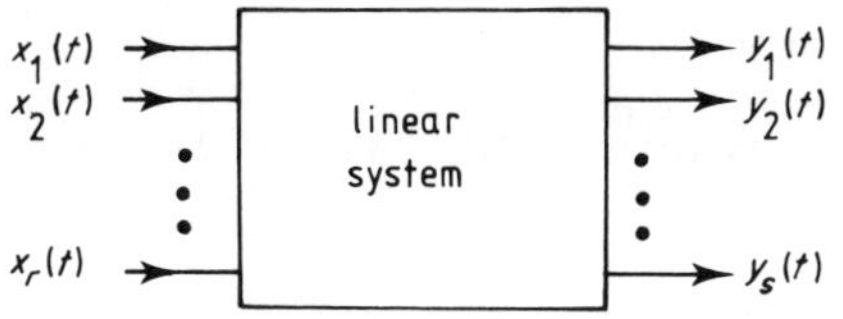

Figure 4.1
Block diagram of a linear system with r inputs and s outputs

ideal impulse, represented by Dirac's delta function $\delta(t)$, applied at the jth input, at $t = 0$ (see Figure 4.2a and 4.2b). The quantities h_{ij} must satisfy the condition of 'physical realizability', i.e. $h_{ij}(t) = 0$ for $t < 0$. A typical variation of h_{ij} with time, for a lightly damped system, is sketched in Figure 4.2c.

If the elements of $\mathbf{x}(t)$ and $\mathbf{y}(t)$ are all absolutely integrable equation (4.1) can be transformed into the frequency domain. This leads to

$$\mathbf{Y}(\omega) = \boldsymbol{\alpha}(\omega)\mathbf{X}(\omega) \tag{4.3}$$

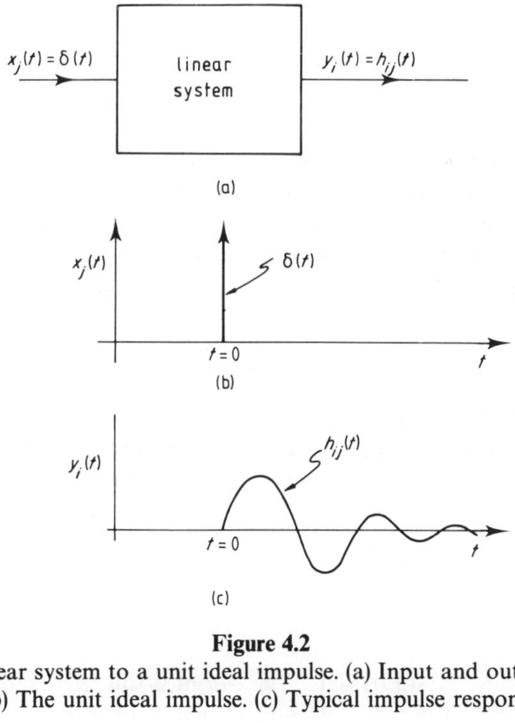

Figure 4.2
Response of a linear system to a unit ideal impulse. (a) Input and output of the system. (b) The unit ideal impulse. (c) Typical impulse response

where

$$\mathbf{X}(\omega) = \int_{-\infty}^{\infty} \mathbf{x}(t) \exp(\mathrm{i}\omega t)\,\mathrm{d}t \quad \mathbf{Y}(\omega) = \int_{-\infty}^{\infty} \mathbf{y}(t) \exp(\mathrm{i}\omega t)\,\mathrm{d}t \tag{4.4}$$

and

$$\boldsymbol{\alpha}(\omega) = \int_{0}^{\infty} \mathbf{h}(t) \exp(\mathrm{i}\omega t)\,\mathrm{d}t \tag{4.5}$$

Here $\boldsymbol{\alpha}(\omega)$ is an $r \times s$ matrix of frequency response functions, i.e.

$$\boldsymbol{\alpha}(\omega) = [\alpha_{ij}(\omega)] \tag{4.6}$$

The components of $\boldsymbol{\alpha}(\omega)$ can be determined from the steady state response to sinusoidal inputs. Thus, if

$$x_j(t) = \exp(\mathrm{i}\omega t) \tag{4.7}$$

then

$$y_i(t) = \alpha_{ij}(\omega) \exp(\mathrm{i}\omega t) \tag{4.8}$$

This result can be deduced directly from equation (4.3) if the Fourier transform of a sine wave is represented as a delta function.

4.3 STOCHASTIC INPUT–OUTPUT RELATIONSHIPS

It will now be shown that the deterministic input–output relationships, described in Section 4.2, can be used to develop stochastic input–output relationships.

If $\mathbf{x}(t)$ and $\mathbf{y}(t)$ are both stochastic processes then equation (4.1) is still valid. However this relationship is now stochastic in nature, and the integration must be interpreted in a probabilistic (here mean square) sense (see Chapter 3).

Denote

$$\mathbf{m}_x(t) = E\{\mathbf{x}(t)\} \tag{4.9}$$

the mean vector of $\mathbf{x}(t)$, and $\mathbf{m}_y(t)$ the mean vector of $\mathbf{y}(t)$. Then on taking expectations of equation (4.1), and using the fact that the linear operations of integration and expectation may be commuted (see Section 3.8), it is found that

$$\mathbf{m}_y(t) = \int_{0}^{\infty} \mathbf{h}(v)\mathbf{m}_x(t - v)\,\mathrm{d}v \tag{4.10}$$

or

$$\mathbf{m}_y(t) = \int_{-\infty}^{\infty} \mathbf{h}(t - u)\mathbf{m}_x(u)\,\mathrm{d}u \tag{4.11}$$

If $\mathbf{x}(t)$ is jointly stationary then $\mathbf{m}_x(t) = \mathbf{m}_x$ (a constant) and it follows, from equation (4.10), that $\mathbf{m}_y$ is also independent of t.

It is noted, however, that it has been assumed that the excitation persists from $t = -\infty$. In practice the excitation will be 'switched on', at $t = 0$, say; there will then be transient response, during which $\mathbf{m}_y(t)$ will vary with time, according to (from equation (4.10))

$$\mathbf{m}_y(t) = \mathbf{m}_x \int_0^t \mathbf{h}(v)\,\mathrm{d}v \tag{4.12}$$

As t becomes large the integral approaches a limiting value, asymptotically, and $\mathbf{m}_y(t)$ approaches the steady-state value

$$\mathbf{m}_y = \mathbf{m}_x \int_0^\infty \mathbf{h}(v)\,\mathrm{d}v \tag{4.13}$$

The correlation matrix of the response process can be related to the corresponding correlation matrix of the input process by using equations (3.123) and (4.1) and once again interchanging the expectation and integration operations. Hence it is found that

$$\mathbf{R}_y(t_1, t_2) = \int_{-\infty}^{\infty}\int_{-\infty}^{\infty} \mathbf{h}(t_1 - u_1)\mathbf{R}_x(u_1, u_2)\mathbf{h}^{\mathrm{T}}(t_2 - u_2)\,\mathrm{d}u_1\,\mathrm{d}u_2 \tag{4.14}$$

Similarly, the covariance matrix, $\mathbf{w}_y(t_1, t_2)$, of the response is related to the corresponding covariance matrix of the excitation process by the relation

$$\mathbf{w}_y(t_1, t_2) = \int_{-\infty}^{\infty}\int_{-\infty}^{\infty} \mathbf{h}(t_1 - u_1)\mathbf{w}_x(u_1, u_2)\mathbf{h}^{\mathrm{T}}(t_2 - u_2)\,\mathrm{d}u_1\,\mathrm{d}u_2 \tag{4.15}$$

If $\mathbf{x}(t)$ is jointly stationary (in the wide sense) then $\mathbf{R}_x(t_1, t_2)$ depends only on the time difference $\tau = t_2 - t_1$, and it follows from equation (4.14) that $\mathbf{R}_y(t_1, t_2)$ is also dependent only on τ. Thus, from equation (4.14), with a change of variables

$$\mathbf{R}_y(\tau) = \int_0^\infty\int_0^\infty \mathbf{h}(v_1)\mathbf{R}_x(\tau + v_1 - v_2)\mathbf{h}^{\mathrm{T}}(v_2)\,\mathrm{d}v_1\,\mathrm{d}v_2 \tag{4.16}$$

and, similarly,

$$\mathbf{w}_y(\tau) = \int_0^\infty\int_0^\infty \mathbf{h}(v_1)\mathbf{w}_x(\tau + v_1 - v_2)\mathbf{h}^{\mathrm{T}}(v_2)\,\mathrm{d}v_1\,\mathrm{d}v_2 \tag{4.17}$$

It may be deduced from this result, and the corresponding result for the mean vector $\mathbf{m}_y(t)$, that if $\mathbf{x}(t)$ is jointly stationary then $\mathbf{y}(t)$ is also jointly stationary (in the wide sense).

Once again, it should be noted that, if the excitation is switched on, at $t = 0$, the above results are strictly valid only asymptotically, as $t \to \infty$. During the non-stationary, transient response $\mathbf{R}_y$ and $\mathbf{w}_y$ will not depend simply on $t_2 - t_1$, and equation (4.14) and (4.15) must be used.

When both $\mathbf{x}(t)$ and $\mathbf{y}(t)$ are stationary equation (4.16) can be transformed into the frequency domain, to yield a comparatively simple result. Thus, if

$$\mathbf{S}_x(\omega) = \frac{1}{2\pi}\int_{-\infty}^{\infty} \mathbf{R}_x(\tau)\exp(\mathrm{i}\omega\tau)\,\mathrm{d}\tau \tag{4.18}$$

is the spectral density matrix for $\mathbf{x}(t)$, and $\mathbf{S}_y(\omega)$ is, similarly, the spectral density matrix of $\mathbf{y}(t)$, then the application of a Fourier transformation to equation (4.16) yields the relationship

$$\mathbf{S}_y(\omega) = \boldsymbol{\alpha}(\omega)\mathbf{S}_x(\omega)\boldsymbol{\alpha}^{\mathrm{T}*}(\omega) \tag{4.19}$$

where $\boldsymbol{\alpha}(\omega)$ is the matrix of frequency response functions and the symbols, $*$ and T denote transposition and conjugation, respectively. In the scalar case this result reduces to the well known result

$$\mathbf{S}_y(\omega) = |\alpha(\omega)|^2\mathbf{S}_x(\omega) \tag{4.20}$$

Equation (4.19) is undoubtedly the most important relationship in linear vibration theory. It gives a simple, and direct relationship between the power spectra of the input and output processes.

It was pointed out in Chapter 3 that any linear combination of Gaussian random variables is also Gaussian. Since the convolution integral can be regarded as the limit of a summation process, in the time domain, it follows that the response of a linear system to Gaussian excitation is also Gaussian. Thus, in the general case, if the r-vector input $\mathbf{x}(t)$ is jointly Gaussian then the s-vector output process, $\mathbf{y}(t)$ is also jointly Gaussian.

Since joint Gaussian processes are completely specified in terms of the mean vector, $\mathbf{m}(t)$, and the covariance matrix $\mathbf{w}(t_1, t_2)$, it follows that the foregoing relationships between $\mathbf{m}_x(t)$ and $\mathbf{m}_y(t)$, and between $\mathbf{w}_x(t_1, t_2)$ and $\mathbf{w}_y(t_1, t_2)$, are completely sufficient to obtain a full probabilistic specification of the response of a linear system to Gaussian excitation. In addition, it follows that, if the excitation to a linear system is Gaussian and stationary, in the strict sense, then the steady-state response is also Gaussian and stationary, in the strict sense.

For non-Gaussian excitation (and hence non-Gaussian response), $\mathbf{m}(t)$ and $\mathbf{w}(t_1, t_2)$ are only partial descriptions of the input and output processes. The analysis given here can be extended to obtain higher-order moment relationships, which give some information on the way linear systems transform non-Gaussian processes, but the calculations tend to be very cumbersome (Stratonovitch, 1964). However, a general observation is worth making here. If a linear system is lightly damped, so that it has a long 'relaxation time', then the output processes will be more Gaussian than the input processes. This is because a linear transformation is essentially a summation operation. By virtue of the Central Limit Theorem, such an operation will tend to produce processes with a Gaussian distribution. As the damping in the system decreases the effective 'length' of the summation increases and the tendency of the output processes toward Gaussianity will be increased. A specific example of this effect has been discussed by Roberts (1966).

4.4 ANALYSIS OF LUMPED PARAMETER SYSTEMS

In the analysis of real mechanical and structural systems there are considerable advantages in adopting a lumped parameter model, in which the mass is 'lumped' into a finite number of rigid masses. The number of degrees of freedom, n, is equal to the number of lumped masses.

A general form of the equations of motion of a lumped-parameter n degree of freedom system is as follows (see Chapter 2)

$$\mathbf{M}\ddot{\mathbf{q}} + \mathbf{C}\dot{\mathbf{q}} + \mathbf{K}\mathbf{q} = \mathbf{Q} \tag{4.21}$$

where $\mathbf{M}$, $\mathbf{C}$ and $\mathbf{K}$ are symmetric, $n \times n$ matrices. The symbol $\mathbf{M}$ represents the inertia matrix, $\mathbf{C}$ is the damping matrix and $\mathbf{K}$ is the stiffness matrix. $\mathbf{q}$ is an n vector containing the n (generalized) displacements of the system and $\mathbf{Q}$ is an n vector containing the n (generalized) forces, corresponding to $\mathbf{q}$.

4.4.1 Response prediction

To predict the response, $\mathbf{q}(t)$, of a linearized lumped mass system to an input, $\mathbf{Q}(t)$, either the frequency domain relationship, involving the matrix of frequency-response functions, $\boldsymbol{\alpha}(\omega)$ (see equation (4.3)), or, alternatively, the convolution integral involving the matrix of impulse response functions, $\mathbf{h}(t)$ (see equation (4.1)) may be used.

To find $\boldsymbol{\alpha}(\omega)$ one can consider the system excited by harmonic forcing, of frequency ω. A general form for $\mathbf{Q}$ is then $\mathbf{Q} = \mathbf{Q}_0 \exp(\mathrm{i}\omega t)$ where $\mathbf{Q}_0$ is an amplitude vector. The steady-state response of the system will be of the form $\mathbf{q} = \boldsymbol{\alpha}(\omega)\mathbf{Q}$ (see equation (4.7) and (4.8)) and hence, on substituting these expressions into equation (4.21) the following expression for $\boldsymbol{\alpha}(\omega)$ is obtained

$$\boldsymbol{\alpha}(\omega) = [-\omega^2\mathbf{M} + \mathrm{i}\omega\mathbf{C} + \mathbf{K}]^{-1} \tag{4.22}$$

The matrix of impulse response functions, $\mathbf{h}(t)$, can be found by performing the inverse Fourier transformation to equation (4.5), i.e.

$$\mathbf{h}(t) = \frac{1}{2\pi}\int_{-\infty}^{\infty} \boldsymbol{\alpha}(\omega)\exp(-\mathrm{i}\omega t)\,\mathrm{d}\omega \tag{4.23}$$

As an alternative to the use of Fourier transforms, Laplace transforms may be employed. Thus, if $\mathbf{T}(s)$ is the transfer function between the Laplace transform of $\mathbf{q}(t)$ and the Laplace transform of $\mathbf{Q}(t)$ then

$$\mathbf{T}(s) = [s^2\mathbf{M} + s\mathbf{C} + \mathbf{K}]^{-1} \tag{4.24}$$

and $\mathbf{h}(t)$ can be found by computing the inverse Laplace transform of $\mathbf{T}(s)$. Thus

$$\mathbf{h}(t) = L^{-1}[\mathbf{T}(s)] \tag{4.25}$$

The Laplace transform approach is useful when either the input or the output functions, or both, are not absolutely integrable.

Yet another approach to the calculation of $\mathbf{h}(t)$ is possible, using a state-space formulation of the equations of motion, as will be shown later in this chapter.

4.4.2 Free undamped motion

The evaluation of $\boldsymbol{\alpha}(\omega)$ (and hence $\mathbf{h}(t)$) may be simplified in many cases by carrying out a modal analysis. An essential first step in this approach is to consider the solution for free, undamped vibrations. In the absence of damping and external forces the general, linearized equations of motion reduce to

$$\mathbf{M}\ddot{\mathbf{q}} + \mathbf{K}\mathbf{q} = \mathbf{0} \tag{4.26}$$

A possible solution of equation (4.26) is of the form $\mathbf{q} = \mathbf{q}_0 \exp(\mathrm{i}\omega t)$ where $\mathbf{q}_0$ is an amplitude vector. Substituting this expression in to equation (4.26) gives

$$(-\omega^2\mathbf{M} + \mathbf{K})\mathbf{q}_0 = \mathbf{0} \tag{4.27}$$

For a non-trivial solution of this equation it is necessary that

$$|-\omega^2\mathbf{M} + \mathbf{K}| = 0 \tag{4.28}$$

where $|\ |$ denotes a determinant. This is an algebraic equation of nth order in ω^2, the roots of which are called characteristic values, or eigenvalues. In general there are n real, positive roots, denoted by $\omega_1^2, \omega_2^2, \ldots, \omega_n^2$. It is convenient to order them such that $\omega_1 < \omega_2 < \cdots < \omega_n$. ω_1 is the first 'natural frequency' of oscillation, ω_2 is the second natural frequency, and so on.

For each natural frequency there is a corresponding 'mode shape', or 'eigenvector'—this is the value of $\mathbf{q}_0$ in equations (4.27). Thus, for vibration at the jth natural frequency, where $1 \leqslant j \leqslant n$

$$\mathbf{q} = \mathbf{q}_0^{(j)} \exp(\mathrm{i}\omega_j t) \tag{4.29}$$

where $\mathbf{q}_0^{(j)}$ is the jth eigenvector. Only the ratios of the elements of an eigenvector can be determined from the equations of motion; the absolute values are dependent on the initial conditions. For this reason it is convenient to introduce normalized eigenvectors, $\boldsymbol{\lambda}^{(j)}$, such that

$$\mathbf{q}_0^{(j)} = \gamma_j \boldsymbol{\lambda}^{(j)} \tag{4.30}$$

where γ_j is the first element of $\mathbf{q}_0^{(j)}$ and the first element of $\boldsymbol{\lambda}^{(j)}$ is unity. The vectors, $\boldsymbol{\lambda}^{(j)}$, may be assembled together to form the $n \times n$ 'modal matrix', where

$$\boldsymbol{\lambda} = [\lambda_{ij}] = [\boldsymbol{\lambda}^{(j)}] \tag{4.31}$$

The general solution of free vibration is, by virtue of the principle of superposition, an arbitrary sum of modal solutions, such as that given by equation (4.29). Using the modal matrix, $\boldsymbol{\lambda}$, the general solution may be written as

$$\mathbf{q} = \boldsymbol{\lambda}\mathbf{p} \tag{4.32}$$

where

$$\mathbf{p} = \begin{bmatrix} \gamma_1 \exp(\mathrm{i}\omega_1 t) \\ \gamma_2 \exp(\mathrm{i}\omega_2 t) \\ \vdots \\ \gamma_n \exp(\mathrm{i}\omega_n t) \end{bmatrix} \tag{4.33}$$

Here γ_j represents the 'amplitude' of the jth mode; to allow for arbitrary phase variations, γ_j will generally be complex. They are determined by the initial conditions.

4.4.3 Classical modal analysis

Equation (4.32) expresses a linear transformation between the original (arbitrarily chosen) coordinate vector $\mathbf{q}$ and another, alternative coordinate vector, $\mathbf{p}$. The elements of $\mathbf{p}$ are called 'principal', or 'normal' coordinates.

When expressed in terms of $\mathbf{p}$, the equations of motion for free, undamped vibration become uncoupled. Thus, on combining equation (4.26) with equation (4.32), and premultiplying throughout by $\boldsymbol{\lambda}^{\mathrm{T}}$, the following is obtained

$$\mathbf{L}\ddot{\mathbf{p}} + \mathbf{N}\mathbf{p} = \mathbf{0} \tag{4.34}$$

where

$$\mathbf{L} = \boldsymbol{\lambda}^{\mathrm{T}}\mathbf{M}\boldsymbol{\lambda} \quad \text{and} \quad \mathbf{N} = \boldsymbol{\lambda}^{\mathrm{T}}\mathbf{K}\boldsymbol{\lambda} \tag{4.35}$$

By virtue of the Principle of Orthogonality (Bishop *et al.*, 1965), the $n \times n$ matrices $\mathbf{L}$ and $\mathbf{N}$ are both *diagonal*, i.e. the equations of motion (4.34) may be written as

$$l_i\ddot{p}_i + n_i p_i = 0 \quad i = 1, 2, \ldots, n \tag{4.36}$$

where l_i and n_i are, respectively, the diagonal elements of $\mathbf{L}$ and $\mathbf{N}$. The quantities l_i $(i = 1, 2, \ldots, n)$ are the modal masses and n_i are the modal stiffness coefficients.

Alternatively, equation (4.36) may be written as

$$\ddot{p}_i + \omega_i^2 p_i = 0 \quad i = 1, 2, \ldots, n \tag{4.37}$$

where

$$\omega_i^2 = \frac{n_i}{l_i} = \frac{\boldsymbol{\lambda}^{(i)\mathrm{T}}\mathbf{K}\boldsymbol{\lambda}^{(i)}}{\boldsymbol{\lambda}^{(i)\mathrm{T}}\mathbf{M}\boldsymbol{\lambda}^{(i)}} \tag{4.38}$$

Note that ω_i are the natural frequencies of the system, as defined previously.

If the more general case where damping and external forces are also present is now considered, and equation (4.21) is transformed from $\mathbf{q}$ to $\mathbf{p}$ coordinates, the equations of motion

$$\mathbf{L}\ddot{\mathbf{p}} + \mathbf{D}\dot{\mathbf{p}} + \mathbf{N}\mathbf{p} = \mathbf{P} \tag{4.39}$$

are obtained where $\mathbf{L}$ and $\mathbf{N}$ are given by equation (4.35) and

$$\mathbf{D} = \boldsymbol{\lambda}^{\mathrm{T}}\mathbf{C}\boldsymbol{\lambda} \quad \mathbf{P} = \boldsymbol{\lambda}^{\mathrm{T}}\mathbf{Q} \tag{4.40}$$

In general $\mathbf{D}$ will *not* be a diagonal matrix. However, if the damping is everywhere light, a very good approximation, in many cases, can be obtained by neglecting the off-diagonal elements of $\mathbf{D}$. Equation (4.39) then represents a set of n uncoupled equations, which may be written as

$$l_i\ddot{p}_i + d_i\dot{p}_i + n_i p_i = P_i \quad i = 1, 2, \ldots, n \tag{4.41}$$

where d_i are the diagonal elements of $\mathbf{D}$. Each equation in the above set is identical in form to that of a single degree of freedom system.

For harmonic forcing, at frequency ω, $\mathbf{p}$ is related to $\mathbf{P}$ by the relation

$$\mathbf{p} = \boldsymbol{\Lambda}(\omega)\mathbf{P} \tag{4.42}$$

where $\boldsymbol{\Lambda}(\omega)$ is a diagonal frequency response function matrix, given by

$$\boldsymbol{\Lambda}(\omega) = \begin{bmatrix} \alpha_1'(\omega) & 0 & \cdots & 0 \\ 0 & \alpha_2'(\omega) & \cdots & 0 \\ \vdots & \vdots & & \vdots \\ 0 & 0 & \cdots & \alpha_n'(\omega) \end{bmatrix} \tag{4.43}$$

and

$$\alpha_i'(\omega) = \frac{1}{(-\omega^2 l_i + i\omega d_i + n_i)} \tag{4.44}$$

is the modal frequency response function, for the ith mode.

$\boldsymbol{\Lambda}(\omega)$ may be related simply to the frequency response function connecting the original coordinate vector $\mathbf{q}$ to the original force vector, $\mathbf{Q}$. From equations (4.32), (4.40) and (4.42) it is found that

$$\boldsymbol{\alpha}(\omega) = \boldsymbol{\lambda}\boldsymbol{\Lambda}(\omega)\boldsymbol{\lambda}^{\mathrm{T}} \tag{4.45}$$

By using the fact that the model matrix, $\boldsymbol{\lambda}$, is comprised of the eigenvectors $\boldsymbol{\lambda}^{(j)}$

$(j=1,2,\ldots,n)$ equation (4.45) may be expressed as

$$\boldsymbol{\alpha}(\omega)=\sum_{i=1}^{n}\boldsymbol{\lambda}^{(i)}\boldsymbol{\lambda}^{(i)\mathrm{T}}\alpha_i(\omega)=\sum_{r=1}^{n}\frac{\boldsymbol{\lambda}^{(r)}\boldsymbol{\lambda}^{(r)\mathrm{T}}}{l_r(\omega_r^2-\omega^2+2\mathrm{i}\zeta_r\omega_r\omega)} \tag{4.46}$$

where ζ_i is the ith modal damping factor, defined by

$$\zeta_i=\frac{d_i}{2(l_in_i)^{1/2}} \tag{4.47}$$

Equation (4.46) is a useful series expression for $\boldsymbol{\alpha}(\omega)$, which avoids the necessity for carrying out the inverse matrix multiplication indicated in equation (4.22). For the case of no damping (all $\zeta_i=0$) it is an exact result. In many applications the series may be truncated to only the first few terms, with little loss of accuracy.

The impulse response matrix, $\mathbf{h}(t)$, corresponding to $\boldsymbol{\alpha}(\omega)$, may be evaluated by applying a Fourier transform to equation (4.46), to give

$$\mathbf{h}(t)=\sum_{i=1}^{n}\boldsymbol{\lambda}^{(i)}\boldsymbol{\lambda}^{(i)\mathrm{T}}h_i(t) \tag{4.48}$$

where

$$h_i(t)=\frac{1}{2\pi}\int_{-\infty}^{\infty}\alpha_i(\omega)\exp(-\mathrm{i}\omega t)\,\mathrm{d}\omega \tag{4.49}$$

is the modal impulse response function. Alternatively, one can find $h_i(t)$ from the inverse Laplace transform of the modal transfer function

$$T_i(s)=\frac{1}{l_i(s^2+2\zeta_i\omega_is+\omega_i^2)} \tag{4.50}$$

(obtained by a Laplace transformation of equation (4.41)). For the normal case of underdamped vibration in the ith mode, where $\zeta_i<1$, one finds that

$$h_i(t)=\frac{1}{l_i\omega_i(1-\zeta_i^2)^{1/2}}\exp(-\zeta_i\omega_it)\sin[(1-\zeta_i^2)^{1/2}\omega_it] \tag{4.51}$$

4.4.4 State variable formulation

The general linearized equations of motion for an n degree of freedom system, as expressed by equation (4.21), can be re-cast into the 'state variable' form by defining a $2n$ 'state vector', $\mathbf{z}(t)$, as follows

$$\mathbf{z}(t)=\begin{bmatrix}\mathbf{q}\\ \dot{\mathbf{q}}\end{bmatrix} \tag{4.52}$$

A first-order matrix equation of motion may then be written as

$$\dot{\mathbf{z}}=\mathbf{G}\mathbf{z}+\mathbf{f} \tag{4.53}$$

where

$$\mathbf{G} = \begin{bmatrix} \mathbf{0} & \mathbf{I} \\ -\mathbf{M}^{-1}\mathbf{K} & -\mathbf{M}^{-1}\mathbf{C} \end{bmatrix} \tag{4.54}$$

and

$$\mathbf{f} = \begin{bmatrix} \mathbf{0} \\ \mathbf{M}^{-1}\mathbf{Q} \end{bmatrix}. \tag{4.55}$$

In the particular case of free vibration where $\mathbf{f} = \mathbf{0}$, equation (4.53) is homogeneous and has the general solution

$$\mathbf{z}(t) = \exp(\mathbf{G}t)\mathbf{z}(0) \tag{4.56}$$

where $\mathbf{z}(0)$ is the vector of initial conditions and the exponential of a square matrix, $\boldsymbol{\alpha}$ say, is defined by

$$\exp(\boldsymbol{\alpha}) = \mathbf{I} + \boldsymbol{\alpha} + \frac{1}{2!}\boldsymbol{\alpha}^2 + \frac{1}{3!}\boldsymbol{\alpha}^3 \cdots \tag{4.57}$$

The expression $\exp(\mathbf{G}t)$ is known as the 'transition matrix' for the system. From a knowledge of this matrix the response to a non-zero forcing, $\mathbf{f}$, can be computed using the equation (Reid, 1983)

$$\mathbf{z}(t) = \exp(\mathbf{G}t)\mathbf{z}(0) + \int_0^t \exp[\mathbf{G}(t-\tau)]\mathbf{f}(\tau)\,\mathrm{d}\tau \tag{4.58}$$

If the system is initially at rest at $t = 0$, $\mathbf{z}(0) = \mathbf{0}$, and equation (4.58) reduces to

$$\mathbf{z}(t) = \int_0^t \exp[\mathbf{G}(t-\tau)]\mathbf{f}(\tau)\,\mathrm{d}\tau \tag{4.59}$$

or

$$\mathbf{z}(t) = \int_0^t \exp(\mathbf{G}v)\mathbf{f}(t-v)\,\mathrm{d}v \tag{4.60}$$

Equation (4.60) (or equation (4.59)) is simply the convolution integral relating $\mathbf{z}(t)$ to $\mathbf{f}(t)$; hence $\exp(\mathbf{G}v)$ is the appropriate impulse response matrix for those input and output vectors. If the transition matrix is partitioned as

$$\exp(\mathbf{G}v) = \begin{bmatrix} \mathbf{a}(v) & \mathbf{b}(v) \\ \mathbf{c}(v) & \mathbf{d}(v) \end{bmatrix} \tag{4.61}$$

where the submatrices are all $n \times n$, it follows from equations (4.55), (4.60) and (4.61) that

$$\mathbf{q}(t) = \int_0^t \mathbf{b}(v)\mathbf{M}^{-1}\mathbf{Q}(t-\tau)\,\mathrm{d}\tau \tag{4.62}$$

A comparison between equations (4.1) and (4.62) now reveals that

$$\mathbf{h}(t) = \mathbf{b}(t)\mathbf{M}^{-1} \tag{4.63}$$

where $\mathbf{h}(t)$ is the impulse response matrix relating $\mathbf{q}$ to $\mathbf{Q}$.

To facilitate the computation of $\exp(\mathbf{G}t)$, and hence $\mathbf{h}(t)$, it is convenient to first compute the eigenvalues, $\lambda_1, \lambda_2, \ldots, \lambda_{2n}$ of the $2n \times 2n$ matrix $\mathbf{G}$. If

$$\mathbf{T} = [\mathbf{d}_1, \mathbf{d}_2, \ldots, \mathbf{d}_n] \tag{4.64}$$

is the $2n \times 2n$ 'complex modal matrix' (see, for example, Fang and Wang, 1986a, b; Faravelli and Casciati, 1987) whose columns are the eigenvectors, or 'complex modes', of $\mathbf{G}$, and

$$\boldsymbol{\eta} = \operatorname{diag}(\lambda_1, \lambda_2, \ldots, \lambda_n) \tag{4.65}$$

is the $2n \times 2n$ diagonal matrix containing the eigenvalues. Then it can be shown that

$$\exp(\mathbf{G}t) = \mathbf{T}\exp(\boldsymbol{\eta}t)\mathbf{T}^{-1} \tag{4.66}$$

where

$$\exp(\boldsymbol{\eta}t) = \operatorname{diag}(\exp(\lambda_1 t), \exp(\lambda_2 t), \ldots, \exp(\lambda_n t)) \tag{4.67}$$

This result enables the elements of the impulse response matrix to be easily evaluated, once the eigenvalues and eigenvectors of $\mathbf{G}$ are determined.

The eigenvalues of $\mathbf{G}$ are either real, or occur in complex conjugate pairs. They are often referred to as the 'poles' of the system. The dynamic character of a system can be determined, to a large extent, from the position of the poles in the complex plane. This is evident from the fact that the general solution for free vibration may be expressed as

$$\mathbf{z}(t) = \sum_{i=1}^{2n} k_i \exp(\lambda_i t)\mathbf{d}_i \tag{4.68}$$

where k_i are constants, dependent upon the initial conditions. If λ_i are all real, and negative, then the free vibration decays exponentially. If there is a pair of complex conjugate roots then the oscillatory free vibration is possible. In the special case of an undamped system all the eigenvalues occur in complex pairs, and are entirely imaginary, i.e.

$$\begin{array}{llll} \lambda_1 & = \mathrm{i}\omega_1 & \lambda_2 & = -\mathrm{i}\omega_1 \\ \lambda_3 & = \mathrm{i}\omega_2 & \lambda_4 & = -\mathrm{i}\omega_2 \\ \cdots & \cdots & \cdots & \cdots \\ \lambda_{2n-1} & = \mathrm{i}\omega_n & \lambda_{2n} & = -\mathrm{i}\omega_n \end{array} \tag{4.69}$$

where ω_i are the natural frequencies of the system, as previously defined.

4.4.5 Complex modal analysis

As pointed out earlier, the application of classical modal analysis, described in Section 4.4.3, to damped linear systems does not lead, in general, to a completely decoupled set of equations. Such decoupling occurs, in fact, if the damping is zero, or for special cases, such as the damping matrix, **C**, being proportional to the stiffness matrix, **K** (Clough and Penzien, 1975). In other words, it is not possible, in general, to diagonalise all three system matrices (**M**, **C** and **K**) by using a transformation based on the undamped normal modes. However, it *is* possible to obtain a completely decoupled set of equations if one allows the modes to be complex (i.e., to have both real and imaginary components). As shown in the previous section, there are, associated with the $2n \times 2n$ state-space system matrix **G**, a set of complex eigenvalues, λ_i $(i = 1, 2, \ldots, 2n)$ and a corresponding set of complex modes, $\mathbf{d}_i$ $(i = 1, 2, \ldots, 2n)$.

The complex modal matrix, **T**, formed from the complex modes (see equation (4.64)) can be used as an appropriate transformation matrix. Introducing the transformed state vector **v**, where

$$\mathbf{z} = \mathbf{T}\mathbf{v} \tag{4.70}$$

and substituting into equation (4.53) one finds that

$$\dot{\mathbf{v}} = \mathbf{T}^{-1}\mathbf{G}\mathbf{T}\mathbf{v} + \mathbf{T}^{-1}\mathbf{f} \tag{4.71}$$

Now this equation can be simplified by using the fact that the eigenvectors are orthogonal with respect to **G** (Reid, 1983). Therefore

$$\mathbf{T}^{-1}\mathbf{G}\mathbf{T} = \boldsymbol{\eta} \tag{4.72}$$

where $\boldsymbol{\eta}$ is the diagonal matrix containing the $2n$ eigenvalues (see equation (4.65)). Combining equations (4.71) and (4.72) one obtains

$$\dot{\mathbf{v}} = \boldsymbol{\eta}\mathbf{v} + \mathbf{g} \tag{4.73}$$

where

$$\mathbf{g} = \mathbf{T}^{-1}\mathbf{f} \tag{4.74}$$

This equation is of exactly the same form as the original matrix equation for **z** (see equation (4.53)). However, since $\boldsymbol{\eta}$ is diagonal, equation (4.73) represents, in fact, a set of $2n$ *uncoupled* equations; i.e. one can write, alternatively,

$$\dot{v}_i = \lambda_i v_i + g_i \quad i = 1, 2, \ldots, 2n \tag{4.75}$$

where v_i, g_i are, respectively, the ith elements of **v** and **g**.

Assuming, for simplicity, that the system is initially at rest, the solution to equation (4.73) can be written as (c.f. equation (4.60))

$$\mathbf{v}(t) = \int_0^t \exp(\boldsymbol{\eta}u)\mathbf{g}(t-u)\,\mathrm{d}u \tag{4.76}$$

where the transition matrix, $\exp(\boldsymbol{\eta}t)$, is now diagonal (see equation (4.67)). Alternatively, one can write

$$v_i(t) = \int_0^t \exp(\lambda_i u) g_i(t-u)\,\mathrm{d}u \quad i = 1, 2, \ldots, 2n \tag{4.77}$$

Thus, the impulse response function, $h_i(t)$, for the ith mode, is simply

$$h_i(t) = \exp(\lambda_i t) \quad t > 0 \tag{4.78}$$

It is noted that, once $\mathbf{v}(t)$ is computed, according to equation (4.76), then it is a simple matter to transform back to the original state vector, $\mathbf{z}$, using equation (4.70).

When the damping is sufficiently light, all the eigenvalues of $\mathbf{G}$ will be in complex conjugate pairs. In this case it is possible to simplify the above analysis somewhat, as discussed by Fang and Wang (1986b).

4.5 STOCHASTIC RESPONSE OF LINEAR SYSTEMS

It will now be shown how the general theory developed earlier in this chapter may be used to find specific results for randomly excited, lumped-parameter systems.

For simplicity, at this stage attention will be restricted to situations where the excitation processes are stationary and Gaussian. It follows, from the argument given earlier, that the response processes will also be stationary, once transients have decayed away, and Gaussian. For this class of problems, both input and output processes are completely specified by their means and by covariance functions, or power spectral density functions. The task is thus to compute the means and covariance functions, or spectral densities, of the response processes from a knowledge of the corresponding means and covariance functions, or spectral densities, of the excitation processes, and the dynamic characteristics of the system.

More complicated situations, where the excitation is non-stationary including the case of non-stationary response to stationary excitation, 'switched on' at $t = 0$, will be considered later in this book.

4.5.1 Single degree of freedom systems

Many real engineering systems can be modelled adequately as single degree of freedom (SDOF) linear systems.

As a simple example, consider the one-storey building structure shown in Figure 4.3a. Suppose the vibration of the structure, in the vertical plane shown, as a result of an external, fluctuating horizontal force $F(t)$ (due to wind

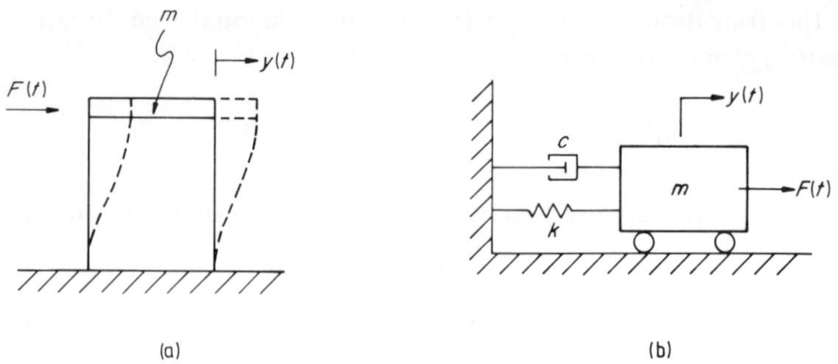

Figure 4.3
Single degree of freedom system. (a) One-storey building structure. (b) Equivalent mass-spring-damper model

turbulence), is considered. For generality the mean of $F(t)$ will be assumed to be non-zero. Let y be the absolute horizontal displacement of the mass, m, of the roof. If the structure is idealized as a rigid mass supported by light (massless), flexible walls then an equivalent SDOF model of the structure can be determined, as shown in Figure 4.3b. Here the flexibility of the walls is represented by a linear spring, of stiffness k, and the damping characteristics of the structure are modelled by the inclusion of a linear damper, in parallel with the spring, with a damping coefficient c.

Application of Newton's laws to the model shown in Figure 4.3b results in the following equation of motion

$$m\ddot{y} + c\dot{y} + ky = F(t) \tag{4.79}$$

This is the same form as the general linear, matrix equation of motion given by equation (4.21). Here $n = 1$ and **M**, **C**, **K**, **q** and **Q** are scalar.

If equation (4.79) is divided throughout by the mass, m, one can write

$$\ddot{y} + 2\zeta\omega_{\mathrm{n}}\dot{y} + \omega_{\mathrm{n}}^2 y = f(t) \tag{4.80}$$

where ζ, the critial damping factor, is given by

$$\zeta = \frac{c}{2(km)^{1/2}} \tag{4.81}$$

and

$$f(t) = F(t)/m \tag{4.82}$$

Equation (4.80) is the standard form of SDOF systems, and thus represents the equation of motion of any system of this type.

The frequency response function, $\alpha(\omega)$, relating $y(t)$ to $f(t)$ may be found by considering harmonic excitation (see Section 4.4.1). It is found that

$$\alpha(\omega) = \frac{1}{\omega_n^2 - \omega^2 + 2i\zeta\omega\omega_n} \tag{4.83}$$

The corresponding transfer function, $T(s)$, is found by replacing $i\omega$ by s and an inverse Laplace transformation of $T(s)$ yields $h(t)$, the impulse response function. Thus, assuming that $\zeta < 1$

$$h(t) = \frac{\exp(-\zeta\omega_n t)}{\bar{\omega}} \sin \bar{\omega} t \quad t > 0 \tag{4.84}$$

where

$$\bar{\omega} = (1 - \zeta^2)^{1/2}\omega_n \tag{4.85}$$

is the damped natural frequency.

The mean, m_y, of the response process, $y(t)$ is related to the mean, m_f, of the excitation by the scalar form of equation (4.13). Hence one finds that

$$m_y = m_f/\omega_n^2 \tag{4.86}$$

This result can, alternatively, be obtained directly by simply taking expectations of the terms in the equation of motion (4.80).

If $w_f(\tau)$ and $w_z(\tau)$ are, respectively, the covariance functions of the excitation process, $f(t)$, and the response process, $y(t)$, then, from the scalar form of equations (4.17) and (4.84), one can relate $w_y(\tau)$ to $w_f(\tau)$ as follows

$$w_y(\tau) = \frac{1}{\bar{\omega}^2} \int_0^\infty \int_0^\infty \exp[-\zeta\omega_n(v_1 + v_2)] \sin[\bar{\omega} v_1] \sin[\bar{\omega} v_2] w_f(\tau + v_1 - v_2)\, dv_1\, dv_2 \tag{4.87}$$

Here the double integration must, in general, be evaluated numerically. However, in certain special cases, where $w_f(\tau)$ can be represented in a simple form, it is possible to evaluate $w_y(\tau)$ exactly, by analytical integration.

The most important special case is where the correlation time scale of $f(t)$ is so small that this process may be approximated as an ideal white noise. Then $w_f(\tau)$ may then be written as

$$w_f(\tau) = S_0 \delta(\tau) \tag{4.88}$$

where S_0 is the constant level of the spectrum of $f(t)$. On substituting equation (4.88) into the covariance relationship given by equation (4.87) and using the properties of the delta function, it is found that

$$w_y(\tau) = \frac{2\pi S_0}{\bar{\omega}^2} \int_0^\infty \exp[-\zeta\omega_n(\tau + 2v)] \sin[\bar{\omega} v] \sin[\bar{\omega}(\tau + v)]\, dv \tag{4.89}$$

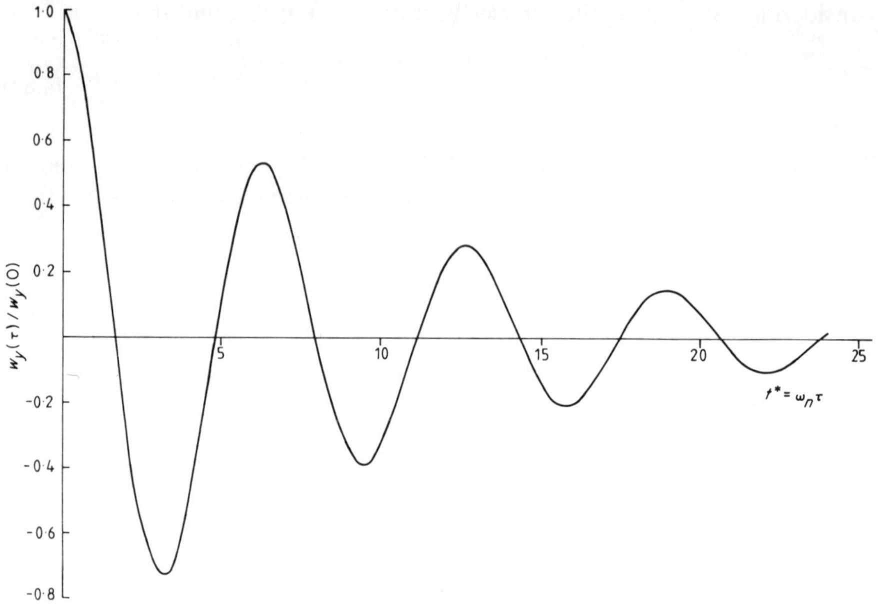

Figure 4.4
Covariance function of the response, in the case of light damping

The single integral can now be evaluated by standard methods to obtain the result

$$w_y(\tau) = \frac{\pi S_0}{2\zeta\omega_n^3}\exp(-\zeta\omega_n|\tau|)\left(\cos\bar{\omega}\tau + \frac{\zeta\omega_n}{\bar{\omega}}\sin\bar{\omega}|\tau|\right) \tag{4.90}$$

For light damping ($\zeta < 1$) the covariance function $w_y(\tau)$, as given by equation (4.90), has the appearance of an oscillating function, with peaks which decay exponentially as τ increases. Figure 4.4 shows a typical variation of $w_y(\tau)$ with τ, for $\tau \geqslant 0$ and for $\zeta = 0.1$. In line with the general properties of covariance functions of stationary processes (see Section 3.8.3), $w_y(\tau)$ is an even function with a maximum value at the origin. Setting $\tau = 0$, this maximum value is the square of the standard deviation, σ_y, of $y(t)$. Thus, from equation (4.90)

$$\sigma_y^2 = \frac{\pi S_0}{2\zeta\omega_n^3} \tag{4.91}$$

Once the covariance function, $w_y(\tau)$, of $y(t)$ is determined then the covariance of derivatives of $y(t)$, and the cross-covariance between $y(t)$ and its derivatives may be readily founded by using equations similar to equation (3.91) and (3.92). For example, for the velocity process, $\dot{y}(t)$, it is found that

$$w_{\dot{y}}(\tau) = -\frac{d^2}{d\tau^2}w_y(\tau) \tag{4.92}$$

Application of this relation to equation (4.90) yields the following results for white noise excitation

$$w_{\dot{y}}(\tau)=\frac{\pi S_0}{2\zeta\omega_{\mathrm{n}}}\exp(-\zeta\omega_{\mathrm{n}}|\tau|)\left(\cos\bar{\omega}\tau-\frac{\zeta\omega_{\mathrm{n}}}{\bar{\omega}}\sin\bar{\omega}|\tau|\right) \tag{4.93}$$

It is important to note here that a consequence of adopting a simple, idealized model for the excitation process may be that derivatives of $y(t)$, beyond a certain order, do not exist (in a mean square sense). For example, with a white noise model for $f(t)$, the derivatives of $y(t)$ of second and higher order do not exist. In particular, the acceleration process, $\ddot{y}(t)$, does not exist. This follows from the fact that the third derivative of $w_y(\tau)$ is discontinuous at $\tau=0$. Hence application of the general formula

$$w_{\ddot{y}}(\tau)=-\frac{d^4}{d\tau^4}w_y(\tau) \tag{4.94}$$

does not lead to a sensible answer for the covariance of $\ddot{y}(t)$. In fact $\ddot{y}(t)$ has the same character as the input process, $f(t)$, i.e. it is a totally uncorrelated process, with an infinite mean square value. This conclusion is one example of the care which must be exercised in interpreting results derived from idealized models of real processes. In the present example an evaluation of the statistics of the acceleration process will require a more accurate model of the excitation process, in which the correlation time scale is non-zero.

From $w_y(\tau)$ the correlation function $R(\tau)$ can be found, and hence, by a Fourier transformation, the spectral density $S_y(\omega)$ of $y(t)$ can be obtained. However, $S_y(\omega)$ can be found more easily by working directly in the frequency domain. Using the general input–output relationship given by equation (4.20), together with the specific form of the frequency response function given by equation (4.83), the equation

$$S_y(\omega)=\frac{S_f(\omega)}{|\omega_{\mathrm{n}}^2-\omega^2+2\mathrm{i}\zeta\omega\omega_{\mathrm{n}}|^2} \tag{4.95}$$

is obtained for the power spectrum of $y(t)$.

Figure 4.5 shows an interpretation of equation (4.95), in graphical form. For light damping, $|\alpha(\omega)|^2$ will peak sharply at $\omega\sim\omega_{\mathrm{n}}$. If $f(t)$ is a wide-band process, with a slowly varying spectrum (as shown) then it is evident that a white noise approximation

$$S_f(\omega)=S_0+m_f^2\delta(0) \tag{4.96}$$

where S_0 is a constant, will yield a good approximation to $S_f(\omega)$. For very light damping an appropriate value for S_0 is $S_f(\omega_{\mathrm{n}})$ (see Figure 4.5a).

The standard deviation of $y(t)$ can be determined directly from $S_y(\omega)$, rather

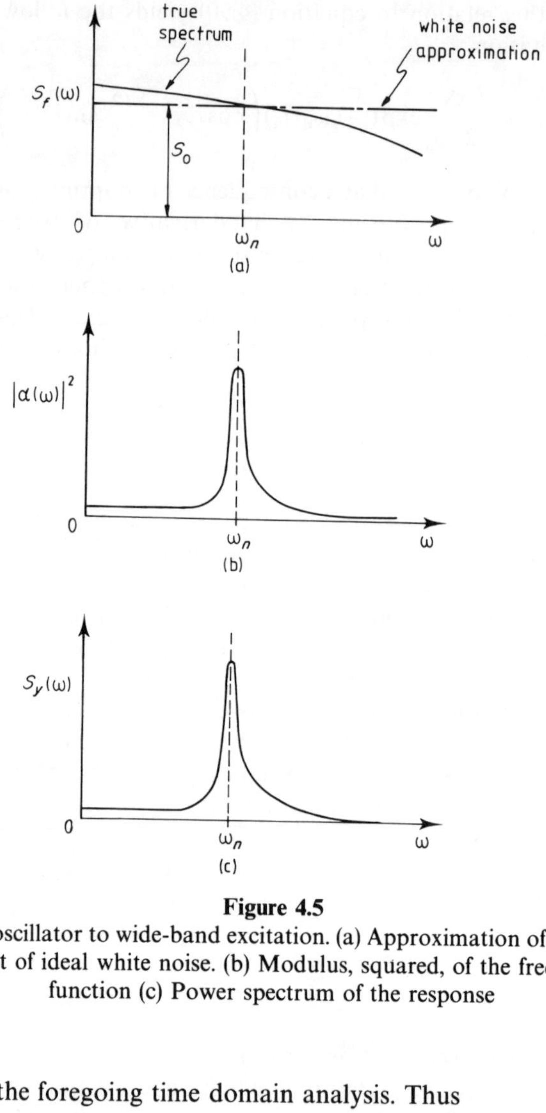

Figure 4.5
Response of an oscillator to wide-band excitation. (a) Approximation of the input power spectrum by that of ideal white noise. (b) Modulus, squared, of the frequency response function (c) Power spectrum of the response

than by using the foregoing time domain analysis. Thus

$$\sigma_y^2 = E\{y^2\} - m_y^2 = \int_{-\infty}^{\infty} S_y(\omega)\,\mathrm{d}\omega - m_y^2 \tag{4.97}$$

With the white noise approximation for $S_y(\omega)$, equation (4.95) and (4.97) combine to give (using equation (4.86))

$$\sigma_y^2 = S_0 \int_{-\infty}^{\infty} \frac{\mathrm{d}\omega}{|\omega_n^2 - \omega^2 + 2\mathrm{i}\zeta\omega\omega_n|^2} \tag{4.98}$$

Here the integral can be evaluated by the method given in Appendix B (see equation (B.28)); the result agrees, of course, with that given by equation (4.91).

The spectra of derivatives of $y(t)$ (where these exist in an m. s. sense) can be found by using equations such as equations (3.113) and (3.114). Thus, for example,

$$S_{\dot{y}}(\omega) = \frac{\omega^2 S_f(\omega)}{|\omega_{\mathrm{n}}^2 - \omega^2 + 2\mathrm{i}\zeta\omega\omega_{\mathrm{n}}|^2} \tag{4.99}$$

With the white noise approximation, the area under the spectrum for $S_{\dot{y}}(\omega)$ is finite, indicating that the process $\dot{y}(t)$ exists. However, the area under the spectrum of $\ddot{y}(t)$, given by

$$S_{\ddot{y}}(\omega) = \omega^4 S_y(\omega) \tag{4.100}$$

is infinite, indicating that $\ddot{y}(t)$ does not exist. This is in agreement with the conclusion reached through a study of the covariance function for $y(t)$. To obtain the statistics of $\ddot{y}(t)$ it is necessary to model the excitation process more accurately. In particular it is necessary for the spectrum of $S(\omega)$ to 'roll off' at high frequencies, at a sufficiently fast rate to ensure that the area under the spectrum of $S_{\ddot{y}}(\omega)$ is finite. The spectra of all real processes are, of course, such that the spectral level drops to zero for sufficiently high frequencies.

4.5.2 Two degree of freedom systems

The analysis given in the preceding section can be extended fairly easily to deal with systems with more than one degree of freedom.

As a simple example of a two degree of freedom system, consider the two storey building structure shown in Figure 4.6a. Here the structure may be idealized as two, rigid masses, m_1 and m_2, which move horizontally as the result of applied, fluctuating forces, $F_1(t)$ and $F_2(t)$, due to turbulent wind. The masses may be assumed to be interconnected by light, massless, flexible walls. An equivalent two degrees of freedom model of the structure is then as shown in Figure 4.6b. Here y_1 and y_2 are the horizontal mass displacements. The symbols k_1 and k_2 represent the wall stiffnesses and the damping properties of the structure are modelled by the inclusion of two linear dampers, with coefficients c_1 and c_2.

The equations of motion for the model shown in Figure 4.5b can be written in the matrix form of equation (4.21). Here

$$\mathbf{M} = \begin{bmatrix} m_1 & 0 \\ 0 & m_2 \end{bmatrix} \quad \mathbf{C} = \begin{bmatrix} (c_1 + c_2) & -c_2 \\ -c_2 & c_2 \end{bmatrix} \quad \mathbf{K} = \begin{bmatrix} (k_1 + k_2) & -k_2 \\ -k_2 & k_2 \end{bmatrix} \tag{4.101}$$

$$\mathbf{q} = \begin{bmatrix} y_1 \\ y_2 \end{bmatrix} \quad \mathbf{Q} = \begin{bmatrix} F_1(t) \\ F_2(t) \end{bmatrix} \tag{4.102}$$

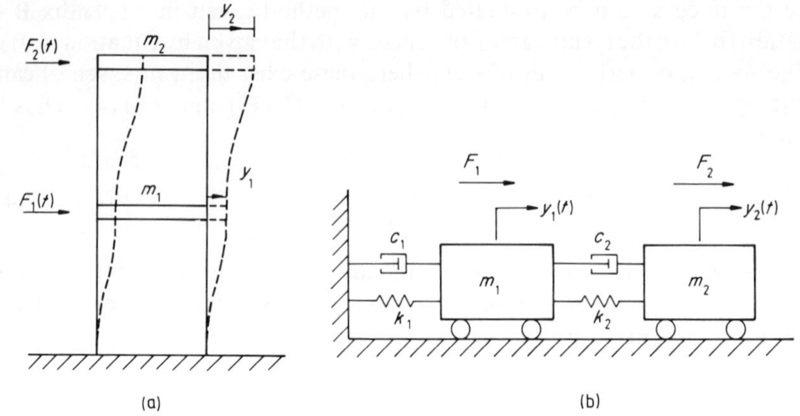

Figure 4.6
Two degree of freedom system. (a) Two-storey building structure. (b) Equivalent mass–spring–damper model

In this problem there are two inputs ($F_1(t)$ and $F_2(t)$) and two outputs, ($y_1(t)$ and $y_2(t)$). Thus, the frequency response function matrix, $\boldsymbol{\alpha}(\omega)$, is 2×2 and may be found from equation (4.22). The Fourier transform of equation (4.23) then yields an expression for $\mathbf{h}(t)$, the impulse response matrix. If the 2×2 power spectrum matrix, $\mathbf{S}_F(\omega)$, of the joint input process $[F_1(t), F_2(t)]$ is specified then the power spectrum matrix, $\mathbf{S}_y(\omega)$ of the joint output process $[y_1(t), y_2(t)]$, can be computed by using equation (4.19). Alternatively the time domain input–output relationship of equation (4.17), may be used.

In general, the frequency domain approach is simpler, and more direct. For the particular case of white noise excitation it is possible to obtain exact results by this method (see Appendix B). As a simple, specific example, consider the problem of computing the standard deviation of $y_1(t)$, in the special case where $F_1(t) = 0$ and $F_2(t)$ can be modelled as a white noise, with zero mean. The spectral matrix $\mathbf{S}_F(\omega)$ is then of the form

$$\mathbf{S}_F(\omega) = \begin{bmatrix} 0 & 0 \\ 0 & S_0 \end{bmatrix} \tag{4.103}$$

where S_0 is the constant, spectral level of $F_2(t)$. Application of equation (4.19) then gives

$$S_{y1}(\omega) = S_0 |\alpha_{12}(\omega)|^2 \tag{4.104}$$

where $\alpha_{12}(\omega)$ is given by (from equations (4.101) to (4.103), and (4.22))

$$\alpha_{12}(\omega) = \frac{(k_2 + \mathrm{i}\omega c_2)}{\Delta} \tag{4.105}$$

and Δ is the determinant of the matrix $\boldsymbol{\alpha}(\omega)$. This quantity is of the form

$$\Delta = \sum_{r=0}^{4} A_r(i\omega)^r \tag{4.106}$$

where A_r are constants, which can be expressed in terms of the system parameters. Thus

$$A_0 = k_1 k_2 \tag{4.107}$$

$$A_1 = c_1 k_2 + c_2 k_1 \tag{4.108}$$

$$A_2 = m_2(k_1 + k_2) + c_2(c_1 + c_2) - m_1 k_2 + c_2^2 \tag{4.109}$$

$$A_3 = m_1 c_2 + m_2(c_1 + c_2) \tag{4.110}$$

$$A_4 = m_1 m_2 \tag{4.111}$$

The evaluation of the standard deviation, σ_{y1} of $y_1(t)$, from

$$\sigma_{y1}^2 = \int_{-\infty}^{\infty} S_{y1}(\omega)\,d\omega = S_0 \int_{-\infty}^{\infty} |\alpha_{12}(\omega)|^2\,d\omega \tag{4.112}$$

can now be accomplished by using the integral result given in Appendix B.

On comparing the integral in equation (4.112) with the standard form given by (B.1) one can write

$$\sigma_{y1}^2 = S_0 I_4 \tag{4.113}$$

where, from equation (B.26),

$$I_4 = \frac{\pi}{\lambda_4} \frac{\begin{vmatrix} \xi_3 & \xi_2 & \xi_1 & \xi_0 \\ -\lambda_4 & \lambda_2 & -\lambda_0 & 0 \\ 0 & -\lambda_3 & \lambda_1 & 0 \\ 0 & \lambda_4 & -\lambda_2 & \lambda_0 \end{vmatrix}}{\begin{vmatrix} \lambda_3 & -\lambda_1 & 0 & 0 \\ -\lambda_4 & \lambda_2 & -\lambda_0 & 0 \\ 0 & -\lambda_3 & \lambda_1 & 0 \\ 0 & \lambda_4 & -\lambda_2 & \lambda_0 \end{vmatrix}} \tag{4.114}$$

In the present problem, a comparison of equations (4.105) and (4.106) with equations (B2) and (B3) shows that, here,

$$\xi_1 = k_2^2 \quad \xi_2 = c_2^2 \quad \xi_3 = 0 \tag{4.115}$$

and

$$\lambda_r = A_r \quad r = 0, 1, \ldots, 4 \tag{4.116}$$

An alternative procedure to the above, exact approach is to use an approximate modal analysis. Here, if the equations of motion are expressed in principal coordinates, and the off-diagonal elements in the transformed damping matrix are neglected, then the frequency response function matrix, $\boldsymbol{\alpha}(\omega)$, relating $F_1(t)$ and $F_2(t)$ to $y_1(t)$ and $y_2(t)$, can be expressed in a modal expression form as given by equation (4.46).

Using the above specific example, where $F_1(t) = 0$ and $F_2(t)$ is a white noise, as an illustration, equation (4.112) for σ_{y1} is again applicable and the appropriate modal form for $\sigma_{12}(\omega)$ is (from equation (4.46))

$$\alpha_{12}(\omega) = \sum_{r=1}^{2} k_r \alpha_r(\omega) \tag{4.117}$$

where

$$\alpha_r(\omega) = \frac{1}{\omega_r^2 - \omega^2 + 2i\zeta_r \omega \omega_r} \qquad r = 1, 2 \tag{4.118}$$

The symbols k_1 and k_2 represent constants. ω_1 and ω_2 are the undamped natural frequencies and ζ_1 and ζ_2 are the modal damping factors. The constants k_1 and k_2 are, in fact the 'modal participation factors', found from an eigenvalue analysis of the equations of motion for free, undamped vibration, as explained earlier.

On combining equations (4.112) to (4.118) the following expression for σ_{y1} is obtained

$$\sigma_{y1}^2 = k_1^2 S_0 \int_{-\infty}^{\infty} |\alpha_1(\omega)|^2 \, d\omega + k_2^2 \int_{-\infty}^{\infty} |\alpha_2(\omega)|^2 \, d\omega + 2k_1 k_2 S_0 \int_{-\infty}^{\infty} [\alpha_1(\omega)\alpha_2^*(\omega)] \, d\omega \tag{4.119}$$

The first two integrals on the right-hand side of equation (4.119) are of the form obtained for a single degree of freedom system (see equation (4.98)). Using the result in Appendix B one finds that (see equation (B28))

$$\int_{-\infty}^{\infty} |\alpha_1(\omega)|^2 \, d\omega = \frac{\pi}{2\zeta_1 \omega_1^3} \tag{4.120}$$

$$\int_{-\infty}^{\infty} |\alpha_2(\omega)|^2 \, d\omega = \frac{\pi}{2\zeta_2 \omega_2^3} \tag{4.121}$$

The third integral on the right-hand side of equation (4.119) can also be found by using the result given in Appendix B. For this purpose t is necessary to multiply both the numerator and the denominator of the integral by the factor $(\omega_1^2 - \omega^2 - 2i\zeta_1 \omega \omega_1)(\omega_2^2 - \omega^2 + 2i\zeta_2 \omega \omega_2)$. This converts the required integral into the standard form given by equation (B1), i.e.

$$\int_{-\infty}^{\infty} \alpha_1(\omega)\alpha_2^*(\omega) \, d\omega = I_4 \tag{4.122}$$

where I_4 is once again given by equation (4.114) and here

$$\xi_0 = \omega_1^2\omega_2^2 \tag{4.123}$$

$$\xi_1 = -\omega_1^2 - \omega_2^2 - 4\zeta_1\zeta_2\omega_1\omega_2 \tag{4.124}$$

$$\xi_2 = 1 \tag{4.125}$$

$$\xi_3 = 0 \tag{4.126}$$

Also

$$\lambda_0 = \omega_1^2\omega_2^2 \tag{4.127}$$

$$\lambda_1 = 2\zeta_1\omega_1^2\omega_2 + 2\zeta_2\omega_2\omega_1^2 \tag{4.128}$$

$$\lambda_2 = \omega_1^2 + \omega_2^2 \tag{4.129}$$

$$\lambda_3 = 2\zeta_1\omega_1 + 2\zeta_2\omega_2 \tag{4.130}$$

$$\lambda_4 = 1 \tag{4.131}$$

After some algebra, the following result emerges

$$\int_{-\infty}^{\infty} \alpha_1(\omega)\alpha_2^*(\omega)\,\mathrm{d}\omega = \frac{4\pi(\omega_1\zeta_1 + \omega_2\zeta_2)}{[(\omega_1^2 + \omega_2^2 + 2\zeta_1\zeta_2\omega_1\omega_2)^2 - 4\bar{\omega}_1^2\bar{\omega}_2^2]} \tag{4.132}$$

where

$$\bar{\omega}_1 = \omega_1(1 - \zeta_1^2)^{1/2} \quad \bar{\omega}_2 = \omega_2(1 - \zeta_2^2)^{1/2} \tag{4.133}$$

are the damped natural frequencies of the two modes. On combining equations (4.119) to (4.133), an explicit expression for σ_{y1}^2 is obtained.

4.5.3 Multi-degree of freedom systems

For multi-degree of freedom (MDOF) systems with several inputs and outputs the frequency-domain input–output formula given by equation (4.19) may be used to find the power spectrum matrix of the vector response process. Then, the Fourier inversion

$$\mathbf{R}_y(\tau) = \int_{-\infty}^{\infty} \mathbf{S}_y(\omega)\exp(-\mathrm{i}\omega\tau)\,\mathrm{d}\omega \tag{4.134}$$

gives the correlation matrix, and hence the covariance matrix, for the response.

Although $\boldsymbol{\alpha}(\omega)$, and hence $\mathbf{S}_y(\omega)$, can usually be found analytically, for a specific lumped parameter system, the Fourier inversion indicated by equation (4.134) will be difficult to carry out analytically, in a closed form, when the system is complex. For systems with many degrees of freedom it is often possible to obtain results of sufficient accuracy by using the modal expansion for $\boldsymbol{\alpha}(\omega)$ given by equation (4.46). When this expansion is employed an analytical evaluation of $\mathbf{w}_y(\tau)$ is possible in those cases where $\mathbf{S}_x(\omega)$, the input spectral matrix, is of a simple analytical form. In particular, if the input processes can be modelled as white noises, fairly simple analytical expressions for $\mathbf{w}_y(\tau)$ can be found.

As a simple illustration, consider the special case of a MDOF system with a single (scalar) input $x(t)$ and a single (scalar) output $y(t)$. If $x(t)$ is modelled as a zero-mean white noise, with spectral level S_0, then equation (4.19) reduces to

$$S_y(\omega) = S_0 |\alpha(\omega)|^2 \tag{4.135}$$

where $\alpha(\omega)$ is the appropriate frequency response function. From equation (4.46) a modal expansion for $\alpha(\omega)$ will be of the form

$$\alpha(\omega) = \sum_{r=1}^{n} \frac{k_r}{(\omega_r^2 - \omega^2 + 2\mathrm{i}\zeta_r \omega\omega_r)} \tag{4.136}$$

where k_r are the appropriate modal participation factors, ω_r are the undamped natural frequencies and ζ_r the modal damping factors ($r = 1, 2, \ldots, n$).

From equations (4.134) to (4.136), using the fact that the mean of $y(t)$ is zero, the covariance function for $y(t)$ is

$$w_y(\tau) = S_0 \sum_{r=1}^{n} \sum_{s=1}^{n} k_r k_s I_{rs}(\tau) \tag{4.137}$$

where

$$I_{rs}(\tau) = \int_{-\infty}^{\infty} \frac{\exp(\mathrm{i}\omega\tau)\,\mathrm{d}\omega}{(\omega_r^2 - \omega^2 + 2\mathrm{i}\zeta_r\omega\omega_r)(\omega_s^2 - \omega^2 - 2\mathrm{i}\zeta_s\omega\omega_s)} \tag{4.138}$$

The integral I_{rs} may be evaluated by the method of contour integration to obtain the result, for $\tau \geqslant 0$

$$I_{rs} = 2\pi \exp(-\omega_r\zeta_r\tau)\left(2\omega_r(\omega_r\zeta_r + \omega_s\zeta_s)\cos\bar{\omega}_r\tau - \frac{[(\omega_r^2 - \omega_s^2) - (\omega_r\zeta_r + \omega_s\zeta_s)^2]\sin\bar{\omega}_r\tau}{\bar{\omega}_r[(\omega_r^2 + \omega_s^2 + 2\zeta_r\zeta_s\omega_r\omega_s)^2 - 4\bar{\omega}_r^2\bar{\omega}_s^2]}\right) \tag{4.139}$$

This result is a generalization of the specific results given earlier for one and two degree of freedom systems. For example, setting $n = 1$, equations (4.137) to (4.139) reduce to the result given by equation (4.90). As a special case of the above general result τ can be set to zero to obtain an expression for the mean square of $y(t)$. Hence

$$w(0) = 4\pi S_0 \sum_{r=1}^{n} \sum_{s=1}^{n} \frac{k_r k_s(\omega_r\zeta_r + \omega_s\zeta_s)}{[(\omega_r^2 + \omega_s^2 + 2\zeta_r\zeta_s\omega_r\omega_s)^2 - 4\bar{\omega}_r^2\bar{\omega}_s^2]} \tag{4.140}$$

which may be compared with equations (4.119) to (4.133), for the case where $n = 2$, and the mean of the input (and output) is zero.

4.5.4 State variable analysis

An alternative approach to the computation of statistical response parameters is to form expectations by direct manipulations on the equations of motion. This is most conveniently carried out by using the state-variable formulation of these equations, as described in Section 4.4.5.

As an example, for a MDOF system, with a $2n$ state vector $\mathbf{z}(t)$, as defined by equation (4.52), the expectation of $\mathbf{z}(t)$ can be found by simply taking expectations of all the terms in the state-variable form of the equation of motion, as given by equation (4.53). This gives

$$\dot{\mathbf{m}}_z = \mathbf{G}\mathbf{m}_z + \mathbf{m}_f \tag{4.141}$$

where

$$\mathbf{m}_z = E\{\mathbf{z}(t)\} \quad \mathbf{m}_f = E\{\mathbf{f}(t)\} \tag{4.142}$$

Equation (4.141) may be solved to find $\mathbf{m}_z$, as a function of time. If $\mathbf{m}_f = 0$ then the solution for $\mathbf{m}_z$ is (compare with equation (4.56))

$$\mathbf{m}_z = \exp(\mathbf{G}t)\mathbf{m}_z(0) \tag{4.143}$$

and $\mathbf{m}_z \to \mathbf{0}$ as $t \to \infty$.

A differential equation for the covariance matrix

$$\underset{\sim}{\mathbf{V}} = E\{[\mathbf{z}(t) - \mathbf{m}_z(t)][\mathbf{z}(t) - \mathbf{m}_z(t)]^{\mathrm{T}}\} \tag{4.144}$$

may also be found from the state variable form of the equation of motion. Thus, from equations (4.53) and (4.141),

$$\dot{\boldsymbol{\lambda}} = \mathbf{G}\boldsymbol{\lambda} + \boldsymbol{\eta}(t) \tag{4.145}$$

where

$$\boldsymbol{\lambda}(t) = \mathbf{z}(t) - \mathbf{m}_z(t) \quad \boldsymbol{\eta}(t) = \mathbf{f}(t) - \mathbf{m}_f(t) \tag{4.146}$$

From a combination of equation (4.144) to (4.146) the following differential equations for $\mathbf{V}$ can be deduced

$$\dot{\mathbf{V}} = \mathbf{G}\mathbf{V}^{\mathrm{T}} + \mathbf{V}\mathbf{G}^{\mathrm{T}} + E\{\boldsymbol{\eta}\boldsymbol{\lambda}^{\mathrm{T}}\} + E\{\boldsymbol{\lambda}\boldsymbol{\eta}^{\mathrm{T}}\} \tag{4.147}$$

Here the last two terms may be evaluated by using the general solution for $\mathbf{z}(t)$, given by equation (4.58). Hence

$$\boldsymbol{\lambda}(t) = \exp(\mathbf{G}(t))\boldsymbol{\lambda}(0) + \int_0^t \exp(\mathbf{G}(t-\tau))\boldsymbol{\eta}(\tau)\,\mathrm{d}\tau \tag{4.148}$$

and it follows that

$$E\{\boldsymbol{\lambda}\boldsymbol{\eta}^{\mathrm{T}}\} = \exp(\mathbf{G}(t))E\{\boldsymbol{\lambda}(0)\boldsymbol{\eta}^{\mathrm{T}}\} + \int_0^t \exp(\mathbf{G}(t-\tau))E\{\boldsymbol{\eta}(\tau)\boldsymbol{\eta}^{\mathrm{T}}(t)\}\,\mathrm{d}\tau \tag{4.149}$$

If the initial conditions are deterministic, or at least uncorrelated with the excitation process $\boldsymbol{\eta}(t)$, then $E\{\boldsymbol{\lambda}(0)\boldsymbol{\eta}^{\mathrm{T}}\}$. Moreover

$$E\{\boldsymbol{\eta}(t)\boldsymbol{\eta}^{\mathrm{T}}(\tau)\} = \mathbf{w}_{\eta}(t, \tau) \tag{4.150}$$

where $\mathbf{w}_{\eta}(t, \tau)$ is the covariance matrix for $\boldsymbol{\eta}(t)$. Hence

$$E\{\boldsymbol{\lambda}\boldsymbol{\eta}^{\mathrm{T}}\} = \int_0^t \exp(\mathbf{G}(t-\tau))\mathbf{w}_{\eta}(t, \tau)\,\mathrm{d}\tau \tag{4.151}$$

Similarly, it is found that

$$E\{\boldsymbol{\eta}\boldsymbol{\lambda}^{\mathrm{T}}\} = \int_0^t \exp(\mathbf{G}(t-\tau))\mathbf{w}_{\eta}^{\mathrm{T}}(t, \tau)\,\mathrm{d}\tau \tag{4.152}$$

Hence, on collecting results, the equation

$$\dot{\mathbf{V}} = \mathbf{G}\mathbf{V}^{\mathrm{T}} + \mathbf{V}\mathbf{G}^{\mathrm{T}} + \int_0^t \exp(\mathbf{G}(t-\tau))[\mathbf{w}_{\eta}(t, \tau) + \mathbf{w}_{\eta}^{\mathrm{T}}(t, \tau)]\,\mathrm{d}\tau \tag{4.153}$$

is obtained. In the special case where the elements of $\boldsymbol{\eta}(t)$ are stationary white noises, equation (4.153) may be significantly simplified. One has

$$\mathbf{w}_{\eta}(t, \tau) = \mathbf{D}\delta(t-\tau) \tag{4.154}$$

where $\mathbf{D}$ is a real, symmetric, non-negative matrix of constants. Substituting this form into equation (4.153) it is found that the integral then is equal to $\mathbf{D}$. Hence

$$\dot{\mathbf{V}} = \mathbf{G}\mathbf{V}^{\mathrm{T}} + \mathbf{V}\mathbf{G}^{\mathrm{T}} + \mathbf{D} \tag{4.155}$$

is the appropriate differential equation for $\mathbf{V}(t)$, in this case.

To illustrate the use of equation (4.155), the problem of computing the response of the standard, SDOF system to white noise with zero mean can be considered. Here the equation of motion, as given by equation (4.80), may be written in the state variable form of equation (4.53), where

$$\mathbf{z} = \begin{bmatrix} y \\ \dot{y} \end{bmatrix} \quad \mathbf{f} = \begin{bmatrix} 0 \\ f(t) \end{bmatrix} \tag{4.156}$$

and

$$\mathbf{G} = \begin{bmatrix} 0 & 1 \\ -\omega_{\mathrm{n}}^2 & -2\zeta\omega_{\mathrm{n}} \end{bmatrix} \tag{4.157}$$

If $f(t)$ is a white noise, with a correlation function

$$w_f(\tau) = I\delta(\tau) = 2\pi S_0\delta(\tau) \tag{4.158}$$

where S_0 is the (constant) spectral level for $f(t)$, then

$$\mathbf{D} = \begin{bmatrix} 0 & 0 \\ 0 & 2\pi S_0 \end{bmatrix} \tag{4.159}$$

If attention is restricted to the case of the steady-state, stationary response, where the elements of

$$\mathbf{V} = \begin{bmatrix} v_{11} & v_{12} \\ v_{21} & v_{22} \end{bmatrix} \tag{4.160}$$

are constant, then $\dot{\mathbf{V}} = 0$ in equation (4.155). In this case the following matrix equation for $\mathbf{V}$ is obtained

$$\begin{bmatrix} 0 & 1 \\ -\omega_n^2 & -2\zeta\omega_n \end{bmatrix} \begin{bmatrix} v_{11} & v_{21} \\ v_{12} & v_{22} \end{bmatrix} + \begin{bmatrix} v_{11} & v_{12} \\ v_{21} & v_{22} \end{bmatrix} \begin{bmatrix} 0 & -\omega_n^2 \\ 1 & -2\zeta\omega_n \end{bmatrix} + \begin{bmatrix} 0 & 0 \\ 0 & 2\pi S_0 \end{bmatrix} = \begin{bmatrix} 0 & 0 \\ 0 & 0 \end{bmatrix} \tag{4.161}$$

Solving of this equation gives

$$\begin{aligned} v_{12} &= v_{21} = 0 \\ v_{11} &= \sigma_y^2 = \frac{\pi S_0}{2\zeta\omega_n^3} \end{aligned} \tag{4.162}$$

and

$$v_{22} = \sigma_{\dot{y}}^2 = \frac{\pi S_0}{2\zeta\omega_n} \tag{4.163}$$

This is in agreement with the result found earlier for this case, using different methods.

4.5.5 Analysis using complex modes

The state variable formulation forms a convenient basis for the application of complex modal analysis, as shown earlier, in Section 4.4.5. In this approach the statistical moments of the transformed state vector **v**, defined by equation (4.70), are considered initially, and then converted into the corresponding moments for the original state vector **z**. As before, the input is here assumed to be stationary; it follows that both **v** and **z** are stationary vector processes.

The mean of the response may be evaluated by first taking expectations of equation (4.76). Hence

$$\mathbf{m}_v = \mathbf{m}_g \int_0^t \exp(\boldsymbol{\eta} u)\, du \tag{4.164}$$

where

$$\mathbf{m}_v = E\{\mathbf{v}(t)\} \quad \mathbf{m}_g = E\{\mathbf{g}(t)\} \tag{4.165}$$

Since $\exp(\boldsymbol{\eta} t)$ is a diagonal matrix (see equation (4.67)), the ith element of $\mathbf{m}_v$, denoted m_v^i, can be written as

$$m_v^i = m_g^i \int_0^t \exp(\lambda_i u)\,du = \frac{m_g^i}{\lambda_i}[\exp(\lambda_i t) - 1] \tag{4.166}$$

Now, for a stable system, *all* the eigenvalues, λ_i, will lie on the left-hand side of the Argand plane, i.e. all λ_i will have negative real parts. It follows that

$$\exp(\lambda_i t) \to 0 \quad \text{as } t \to \infty \tag{4.167}$$

for every i. Hence

$$m_v^i \to -m_g^i/\lambda_i \quad \text{as } t \to \infty \tag{4.168}$$

or, in vector form,

$$\mathbf{m}_v \to -\boldsymbol{\eta}^{-1}\mathbf{m}_g \quad \text{as } t \to \infty \tag{4.169}$$

This result can be obtained rather more directly by simply taking expectations of equation (4.73), and setting $\dot{\mathbf{m}}_v = \mathbf{0}$.

To convert $\mathbf{m}_v$ into $\mathbf{m}_z$ one simply takes expectations of the transform relationship given by equation (4.70). Thus

$$\mathbf{m}_z = \mathbf{T}\mathbf{m}_v \tag{4.170}$$

Also, from equation (4.74)

$$\mathbf{m}_g = \mathbf{T}^{-1}\mathbf{m}_f \tag{4.171}$$

Combining equations (4.169) to (4.171) yields

$$\mathbf{m}_z = -\mathbf{T}\boldsymbol{\eta}^{-1}\mathbf{T}^{-1}\mathbf{m}_f \tag{4.172}$$

or

$$\mathbf{m}_z = -\mathbf{G}^{-1}\mathbf{m}_f \tag{4.173}$$

if the orthogonality relationship given by equation (4.72) is used. Equation (4.173) can, of course, be derived directly from the expectation of equation (4.53). Thus, setting $\dot{\mathbf{m}}_z = \mathbf{0}$ in equation (4.141), equation (4.173) is obtained immediately.

The covariance matrix for the state vector, $\mathbf{v}$, in the case of stationary response, can also be related fairly easily to the corresponding covariance matrix of $\mathbf{g}(t)$. Since $\mathbf{v}(t)$ and $\mathbf{g}(t)$ are, in general, both complex, it is convenient to define the following complex covariance matrix for $\mathbf{v}(t)$

$$\mathbf{w}_v(\tau) = E\{[\mathbf{v}(t) - \mathbf{m}_v][\mathbf{v}(t+\tau) - \mathbf{m}_v]^{*\mathrm{T}}\} \tag{4.174}$$

and similarly for $\mathbf{g}(t)$.

Using the convolution integral relationship of equation (4.76), the following covariance input–output relationship can be obtained which is similar to that of equation (4.17), for real inputs and outputs

$$\mathbf{w}_v(\tau) = \int_0^\infty \int_0^\infty \mathbf{h}(v_1)\mathbf{w}_g(\tau + v_1 - v_2)\mathbf{h}^{*\mathrm{T}}(v_2)\,dv_1\,dv_2 \tag{4.175}$$

where

$$\mathbf{h}(t) = \exp(\boldsymbol{\eta} t) = \mathrm{diag}\,[h_i(t)] \tag{4.176}$$

and $h_i(t)$ are given by equation (4.78). Due to the simple, diagonal nature of $\mathbf{h}(t)$, in this case, the elements of $\mathbf{w}_v(\tau)$ can be written as

$$w_v^{ij}(\tau) = \int_0^\infty \int_0^\infty h_i(v_1) w_g^{ij}(\tau + v_1 - v_2) h_j^*(v_2)\,\mathrm{d}v_1\,\mathrm{d}v_2 \tag{4.177}$$

where

$$\mathbf{w}_v(\tau) = [w_v^{ij}(\tau)] \quad \mathbf{w}_g(\tau) = [w_g^{ij}(\tau)] \tag{4.178}$$

Once $\mathbf{w}_v(\tau)$ is determined, the corresponding covariance matrix,

$$\mathbf{w}_z(\tau) = E\{[\mathbf{z}(t) - \mathbf{m}_z][\mathbf{z}(t+\tau) - \mathbf{m}_z]^{\mathrm{T}}\} \tag{4.179}$$

for $\mathbf{z}(t)$ can be found by using the transformation relationship given by equation (4.70), again. It is easy to show that

$$\mathbf{w}_z(\tau) = \mathbf{T}\mathbf{w}_v(\tau)\mathbf{T}^{*\mathrm{T}} \tag{4.180}$$

Similarly, the covariances of $\mathbf{f}(t)$ and $\mathbf{g}(t)$ are related through

$$\mathbf{w}_g(\tau) = \mathbf{T}^{-1}\mathbf{w}_f(\tau)(\mathbf{T}^{-1})^{*\mathrm{T}} \tag{4.181}$$

In the special case of white noise excitation, the analysis is very much simplified. Thus, if $\mathbf{f}(t)$ is a white noise vector process, with a correlation function given by equation (4.154), then, from equation (4.181), the covariance matrix of $\mathbf{g}(t)$ is given by

$$\mathbf{w}_g(\tau) = \mathbf{R}\delta(\tau) \tag{4.182}$$

where

$$\mathbf{R} = [R_{ij}] = \mathbf{T}^{-1}\mathbf{D}(\mathbf{T}^{-1})^{*\mathrm{T}} \tag{4.183}$$

Substituting this expression for $\mathbf{w}_g(\tau)$ into equation (4.177) gives

$$w_v^{ij}(\tau) = R_{ij} \int_0^\infty h_i(v) h_j^*(\tau + v)\,\mathrm{d}v \tag{4.184}$$

Now, using expression (4.78) for $h_i(t)$ one has

$$w_r^{ij}(\tau) = R_{ij}\exp(\lambda_i\tau) \int_0^\infty \exp[(\lambda_i + \lambda_j^*)v]\,\mathrm{d}v \tag{4.185}$$

$$= -\frac{R_{ij}\exp(\lambda_i\tau)}{(\lambda_i + \lambda_j^*)} \quad \tau > 0 \tag{4.186}$$

For the evaluation at negative τ, the relationship

$$\mathbf{w}_v(-\tau) = \mathbf{w}_v^*(\tau) \tag{4.187}$$

is useful. At zero lag one has the very simple result

$$w_v^{ij}(0) = -\frac{R_{ij}}{(\lambda + \lambda_j^*)} \tag{4.188}$$

It is also possible to obtain fairly simple results, by this method, for certain types of non-white excitation (Fang and Wang, 1986a, 1986b). For example, if

$$\mathbf{w}_f(\tau) = \mathbf{D}\rho(\tau) \tag{4.189}$$

where, as before, $\mathbf{D}$ is a real, symmetric matrix of constants, and $\rho(\tau)$ is a symmetric function of τ then, from equation (4.181),

$$\mathbf{w}_g(\tau) = \mathbf{R}\rho(\tau) \tag{4.190}$$

and

$$w_v^{ij}(\tau) = R_{ij} \int_0^\infty \int_0^\infty h_i(v_1)\rho(\tau + v_1 - v_2)h_j^*(v_2)\,\mathrm{d}v_1\,\mathrm{d}v_2 \tag{4.191}$$

This double integral can be evaluated fairly easily if $\rho(\tau)$ is of exponential form. Thus, for example, if

$$\rho(\tau) = \exp(-\alpha|\tau|) \tag{4.192}$$

where α is some real, positive constant, corresponding to first-order filtered noise, then

$$w_v^{ij}(0) = -\frac{R_{ij}}{(\lambda_i + \lambda_j^*)}\left(\frac{1}{\alpha - \lambda_i} + \frac{1}{\alpha - \lambda_j^*}\right) \tag{4.193}$$

This equation can be used to find a result for second-order filtered noise. Thus, if

$$\rho(\tau) = c\exp(q|\tau|) + c^*\exp(q^*|\tau|) \tag{4.194}$$

where c and q are complex constants, then, from equation (4.193)

$$w_v^{ij}(0) = \frac{R_{ij}}{(\lambda_i + \lambda_j^*)}\left[c\left(\frac{1}{\lambda_i + q} + \frac{1}{\lambda_j^* + q}\right) + c^*\left(\frac{1}{\lambda_i + q^*} + \frac{1}{\lambda_j^* + q^*}\right)\right] \tag{4.195}$$

Example

To illustrate the application of complex modal analysis, the case of a linear SDOF system, excited by zero-mean white noise, will once again be considered. Here, $\mathbf{G}$ is given by equation (4.157), $\mathbf{z}$ and $\mathbf{f}$ are given by equation (4.156) and the correlation function of $f(t)$ is of the delta function form given by equation (4.158). Moreover, $\mathbf{w}_f(\tau)$ is of the form of equation (4.189), where $\mathbf{D}$ is given by equation (4.159) and $\rho(\tau) = \delta(\tau)$.

To find the eigenvalues and the eigenvectors of $\mathbf{G}$ one simply substitutes the trial solution

$$\mathbf{z} = \mathbf{d} \exp(\lambda t) \tag{4.196}$$

into the homogeneous version of the state-vector equation of motion (4.53). Hence, with $\mathbf{f}(t) = \mathbf{0}$, one obtains

$$(\mathbf{I}\lambda - \mathbf{G})\mathbf{d} = \mathbf{0} \tag{4.197}$$

For a non-trival solution, an eigenvalue, λ, must satisfy the characteristic equation

$$|\mathbf{I}\lambda - \mathbf{G}| = 0 \tag{4.198}$$

which is, in this particular case,

$$\left|\begin{bmatrix} \lambda & 0 \\ 0 & \lambda \end{bmatrix} - \begin{bmatrix} 0 & 1 \\ -\omega_n^2 & -2\zeta\omega_n \end{bmatrix}\right| = 0 \tag{4.199}$$

or

$$\begin{vmatrix} \lambda & -1 \\ \omega_n^2 & (\lambda + 2\zeta\omega_n) \end{vmatrix} = 0 \tag{4.200}$$

Expanding the determinant leads immediately to the characteristic equation

$$\lambda^2 + 2\zeta\omega_n\lambda + \omega_n^2 = 0. \tag{4.201}$$

On solving this quadratic equation for λ, the following two eigenvalues of $\mathbf{G}$ are obtained

$$\begin{aligned} \lambda_1 &= \omega_n[-\zeta + \mathrm{i}(1-\zeta^2)^{1/2}] \\ \lambda_2 &= \lambda_1^* = \omega_n[-\zeta - \mathrm{i}(1-\zeta^2)^{1/2}] \end{aligned} \tag{4.202}$$

Clearly, if $\zeta < 1$, these two eigenvalues comprise a complex conjugate pair. If $\zeta > 1$, both eigenvalues are real. In the subsequent analysis it will be assumed that the eigenvalues are complex, i.e, $\zeta < 1$.

The eigenvectors $\mathbf{d}_1$ and $\mathbf{d}_2$, corresponding to λ_1 and λ_2, can be found from equation (4.197); specifically

$$(\mathbf{I}\lambda_i - \mathbf{G})\mathbf{d}_i = 0; \quad i = 1, 2 \tag{4.203}$$

Hence it is found that

$$\mathbf{d}_1 = \begin{bmatrix} 1 \\ \lambda_1 \end{bmatrix}, \quad \mathbf{d}_2 = \begin{bmatrix} 1 \\ \lambda_1^* \end{bmatrix}, \tag{4.204}$$

where, for consistency with the approach adopted in the classical modal method, described earlier, the eigenvectors are normalized so that the first element is unity. On combining $\mathbf{d}_1$ and $\mathbf{d}_2$, the complex modal matrix

$$\mathbf{T} = \begin{bmatrix} 1 & 1 \\ \lambda_1 & \lambda_1^* \end{bmatrix} \tag{4.205}$$

is obtained. The inverse of $\mathbf{T}$ is given by

$$\mathbf{T}^{-1} = \frac{1}{(\lambda_1^* - \lambda_1)} \begin{bmatrix} \lambda_1^* & -1 \\ -\lambda_1 & 1 \end{bmatrix} \tag{4.206}$$

Since the excitation is white, the basic result by equation (4.188) can be applied if attention is restricted, for simplicity, to the case $\tau = 0$. Hence it is necessary to evaluate the matrix, $\mathbf{R}$, defined by equation (4.183). On carrying out the appropriate matrix multiplications one obtains the result

$$\mathbf{R} = \frac{2\pi S_0}{|\lambda_1^* - \lambda_1|^2} \begin{bmatrix} 1 & -1 \\ -1 & 1 \end{bmatrix} \tag{4.207}$$

On using this expression in equation (4.186), the following equation for $\mathbf{w}_v(0)$ is obtained

$$\mathbf{w}_v(0) = -\frac{2\pi S_0}{|\lambda_1^* - \lambda_1|^2} \begin{bmatrix} \dfrac{1}{\lambda_1 + \lambda_1^*} & -\dfrac{1}{2\lambda_1} \\ -\dfrac{1}{2\lambda_1^*} & \dfrac{1}{\lambda_1 + \lambda_1^*} \end{bmatrix} \tag{4.208}$$

The final step in the analysis is to convert $\mathbf{w}_v(0)$ to $\mathbf{w}_z(0)$, according to equation (4.180) (with $\tau = 0$). Specifically

$$\mathbf{w}_z(0) = -\frac{2\pi S_0}{|\lambda_1^* - \lambda_1|^2} \begin{bmatrix} 1 & 1 \\ \lambda_1 & \lambda_1^* \end{bmatrix} \begin{bmatrix} \dfrac{1}{\lambda_1 + \lambda_1^*} & -\dfrac{1}{2\lambda_1} \\ -\dfrac{1}{2\lambda_1^*} & \dfrac{1}{\lambda_1 + \lambda_1^*} \end{bmatrix} \begin{bmatrix} 1 & \lambda_1^* \\ 1 & \lambda_1 \end{bmatrix} \tag{4.209}$$

Hence

$$w_z^{11}(0) = E\{y^2\} = \sigma_y^2 = -\frac{2\pi S_0}{|\lambda_1^* - \lambda_1|^2} \left(\frac{2}{\lambda_1 + \lambda_1^*} - \frac{1}{2\lambda_1} - \frac{1}{2\lambda_1^*} \right) \tag{4.210}$$

$$w_z^{22}(0) = E\{\dot{y}^2\} = \sigma_{\dot{y}}^2 = -\frac{2\pi S_0}{|\lambda_1^* - \lambda_1|^2} \left(\frac{2\lambda_1 \lambda_1^*}{\lambda_1 + \lambda_1^*} - \frac{\lambda_1}{2} - \frac{\lambda_1^*}{2} \right) \tag{4.211}$$

Substituting the expression for λ_1 given by equation (4.202) into equations (4.210) and (4.211) yields

$$w_z^{11}(0) = \frac{2\pi S_0}{4\zeta\omega_n^3} = \frac{w_z^{22}(0)}{\omega_n^2} \tag{4.212}$$

in agreement with the results obtained earlier, by other methods. One also finds, as expected, that $w_z^{12}(0) = w_z^{21}(0) = 0$.

It is evident from this simple example that complex modal analysis can lead to cumbersome calculations, when performed by hand. However, the method is well suited for programming on a digital computer, where complex arithmetic can be handled without difficulty. In this respect the method offers an attractive basis for developing general purpose computer routines.

In many situations, only a relatively small number of complex modes will contribute to the response. In these cases it is possible to reduce the computational effort involved by neglecting all but the significant modes.

Chapter 5

Statistical linearization for simple systems with stationary response

5.1 INTRODUCTION

In this chapter the method of statistical linearization is introduced by first applying it to the problem of determining the response of simple, non-linear elements of the 'zero-memory' type. For such elements exact input–output relationships can be established and thus the accuracy of the statistical linearization approximation can be assessed, in specific cases.

It is then shown how the statistical linearization technique can be extended to a consideration of single degree of freedom systems, containing one or more non-linear elements. A study of this particular class of system, which is the simplest, and one of the most important, to be found in engineering applications (see Chapter 4), enables the basic procedure for dealing with systems with memory to be established, without undue algebraic complexity. As will be demonstrated in the following chapter, the analytical techniques developed here can be readily generalized to deal with more complex systems, with more than one degree of freedom.

To keep matters as simple as possible it will be assumed, at the outset, that the non-linearities which occur are not of the 'hysteretic' type, i.e. do not depend on the entire history of the system response, but only on the instantaneous motion. Generalizations of the statistical linearization method, to cope with hysteretic non-linearities, will be considered later (in Chapter 8). It will also be assumed, throughout this chapter, that the excitation and response processes are stationary. Generalizations of the method, which are appropriate when the excitation is non-stationary, will be considered in Chapter 7.

5.2 NON-LINEAR ELEMENTS WITHOUT MEMORY

Initially, the response of a simple non-linear element to random excitation will be considered. This element is assumed to be of the 'zero-memory' type, such that it transforms an input, $x(t)$, into an output, $y(t)$, according to the algebraic, non-differential, relationship

$$y(t) = g[x(t)] \tag{5.1}$$

Thus $y(t)$ depends only on the instantaneous value of the input at time t. The function $g(x)$ is the element non-linearity function.

For such a simple type of non-linearity the probability distribution of the output, $y(t)$, may be related to that of the input, $x(t)$, fairly directly, without recourse to approximate techniques (see equation (3.13)).

5.2.1 Statistical linearization procedure

An approximate solution to the problem of determining the response of a zero-memory, non-linear element to random excitation can be found by the statistical linearization method. Clearly this approach is not necessary for such a simple problem, since (as pointed out in the preceding section) the exact solution can be found for any particular non-linearity function. However, its application here is very useful for demonstrating the basic principle of the statistical linearization method. Moreover, it will be seen that the analysis for a single element can be extended to cope with non-linear systems, governed by differential systems, which contain such an element, interconnected with other elements.

In the general case the mean values of both the input and the output processes will be non-zero. It is convenient to decompose $x(t)$ as follows

$$x(t) = x_0(t) + m_x \tag{5.2}$$

where m_x is the mean value of $x(t)$. Clearly $x_0(t)$ is a zero-mean process.

A suitable linear element for the purposes of approximating the actual non-linear element, is one which produces an output

$$y_e = ax_0 + b \tag{5.3}$$

where a and b are constants. Here, and henceforth, the time argument will be dropped from x and y. The parameters a and b can be chosen to minimize the difference

$$\varepsilon = y - y_e = g(x) - ax_0 - b \tag{5.4}$$

between the non-linear element output and the linear element output. If ε is suitably minimized the system governed by equation (5.4) will be referred to as the 'equivalent linear system'. It approximates the actual non-linear system governed by equation (5.1). The statistics of y_e, computed from equation (5.4), may be used as approximations to the statistics of the output of the non-linear element. This is the essence of the statistical linearization method.

Various minimization criteria have been suggested but the one based on minimizing the expected value of ε^2, $E\{\varepsilon^2\}$, has been found to give, in general, as good, if not better, results than those obtained using other possible criteria (Iwan and Patula, 1972; Smith, 1966). Thus, adopting this particular criterion here, the values of a and b are sought which minimize

$$E\{\varepsilon^2\} = E\{[g(x) - ax_0 - b]^2\} \tag{5.5}$$

The quantity $E\{\varepsilon^2\}$ can be minimized with respect to a and b by solving the equations

$$\frac{\partial}{\partial a}E\{\varepsilon^2\} = \frac{\partial}{\partial b}E\{\varepsilon^2\} = 0 \tag{5.6}$$

This procedure leads to

$$a = \frac{E\{x_0 g(x)\}}{E\{x_0^2\}} = \frac{E\{x_0 y\}}{\sigma_x^2} \tag{5.7}$$

and

$$b = E\{g(x)\} = E\{y\} \tag{5.8}$$

It can readily be verified that these values of a and b do indeed minimize, rather than maximize, the value of $E\{\varepsilon^2\}$, as given by equation (5.5).

Equation (5.8) shows that the optimum value of b is equal to the expected value of the output of the non-linear element, i.e. to the true mean value of y. From equation (5.3) one also finds, on taking expectations, that

$$b = E\{y_e\} \tag{5.9}$$

Thus, the optimum linear element is such that

$$E\{y\} = E\{y_e\} \tag{5.10}$$

The quantity $E\{x_0 g(x)\} = E\{x_0 y\}$, appearing in the above expression for a, is the cross-correlation for the input process, $x_0(t)$, and the output process, $y(t)$, of the non-linear element, evaluated at zero time lag. It may be computed from the first order density function of $x(t)$, $f_x(x)$, using the expression

$$E\{x_0 g(x)\} = \int_{-\infty}^{\infty} (x - m_x) g(x) f_x(x)\, \mathrm{d}x \tag{5.11}$$

For the special case where $x(t)$ is a Gaussian process, a simpler expression for the parameter a can be found, by using a result for Gaussian random variables given earlier in Chapter 3 (see equation (3.56)). From this general relationship one obtains

$$E\{x_0 g(x_0 + m)\} = \sigma_x^2 E\left\{\frac{\mathrm{d}g}{\mathrm{d}x}\right\} \tag{5.12}$$

and a combination of equations (5.7) and (5.12) leads to

$$a = E\left\{\frac{\mathrm{d}g}{\mathrm{d}x}\right\} \tag{5.13}$$

Therefore a is simply equal to the expectation of the gradient of the function $g(x)$. It is often easier to calculate a from equation (5.13), rather than from equation (5.7).

Once a and b have been determined, the distribution of the output, y_e, of the equivalent linear system can be readily determined from equation (5.3). For example, in the simple case where the input and output means are zero, the first-order density function is given by

$$f_y(y) = \frac{1}{a} f_x(x) \tag{5.14}$$

Similar relationships exist between the higher order density functions of the input and output processes. It is noted that the response of the equivalent linear system to a Gaussian input will also be Gaussian (see Chapter 3). Therefore the statistical linearization technique, when applied to systems with Gaussian inputs, involves approximating the true, non-Gaussian response distribution by a Gaussian distribution. Often the distribution of the input is not known with any precision and it is then common practice to assume that this process is Gaussian, in the absence of further information.

5.2.2 Optimum linearization

In the preceding section the equivalent linear system was assumed to be a linear, zero-memory element. However, it is possible to adopt a more general form for the equivalent linear system, which allows this system to have a finite memory. As shown in Chapter 4, the most general linear input–output relationship is the convolution integral, given by

$$y_e(t) = \int_{-\infty}^{t} h(t-u)x(u)\,du + b \tag{5.15}$$

This may be used as an alternative to equation (5.3). The problem is then to determine the optimum impulse response function, $h(u)$, which minimizes the difference between $y = g(x)$ and y_e. It can be shown (e.g., see Atherton, 1975) that, for the following two classes of input process, equation (5.3) *does* represent the optimum linear filter:

(a) Sinusodial processes, consisting of sums of sinusoids of the same amplitude and frequency, but randomly varying phases.
(b) Gaussian processes

In applications involving isolated elements, the input is often Gaussian, to a good approximation, and it is then reasonable to assume that the simple zero-memory linear element, governed by equation (5.5), is close to an optimal substitute.

5.2.3 Examples

To illustrate the statistical linearization procedure, it will now be applied to two specific non-linear elements, with the following non-linearity functions:

(a) Cubic non-linearity

$$g(x) = x^3 \tag{5.16}$$

(b) Quadratic non-linearity

$$g(x) = x^2 \tag{5.17}$$

In each case it will be assumed that the input is Gaussian, i.e. that the first-order density function for $x(t)$ is given by (see Section 3.5)

$$f_x(x) = \frac{1}{(2\pi)^{1/2}\sigma_x} \exp\left(-\frac{(x - m_x)^2}{2\sigma_x^2} \right) \tag{5.18}$$

where m_x and σ_x are, respectively, the mean and standard deviation of $x(t)$.

Example a

An exact expression for $f_y(y)$ is readily found from equation (3.13). This gives

$$f_y(y) = \frac{1}{3(2\pi)^{1/2}\sigma_x y^{2/3}} \exp\left(-\frac{(y^{1/3} - m_x)^2}{2\sigma_x^2} \right) \tag{5.19}$$

Hence, by integration, the following exact expressions for m_y and σ_y, may be found

$$m_y = \sigma_x^3(3\lambda + \lambda^3) \tag{5.20}$$

and

$$\sigma_y = \sigma_x^3(15 + 36\lambda^2 + 9\lambda^4)^{1/2} \tag{5.21}$$

where

$$\lambda = \frac{m_x}{\sigma_x} \tag{5.22}$$

From equations (5.13) and (5.8) the following expressions are obtained for the coefficients a and b in the equivalent linear system (see Appendix A, Table A2)

$$a = 3\sigma_x^2(1 + \lambda^2) \quad b = \sigma_x^3(3\lambda + \lambda^3) \tag{5.23}$$

Here, as in general, b is equal to the exact mean value of $y(t)$, m_y. This distribution of the response of the equivalent linear system is, of course, Gaussian with a mean value equal to m_y, as given by equation (5.20). The standard deviation appropriate to this Gaussian distribution is

$$\sigma_y = 3\sigma_x^3(1 + \lambda^2) \tag{5.24}$$

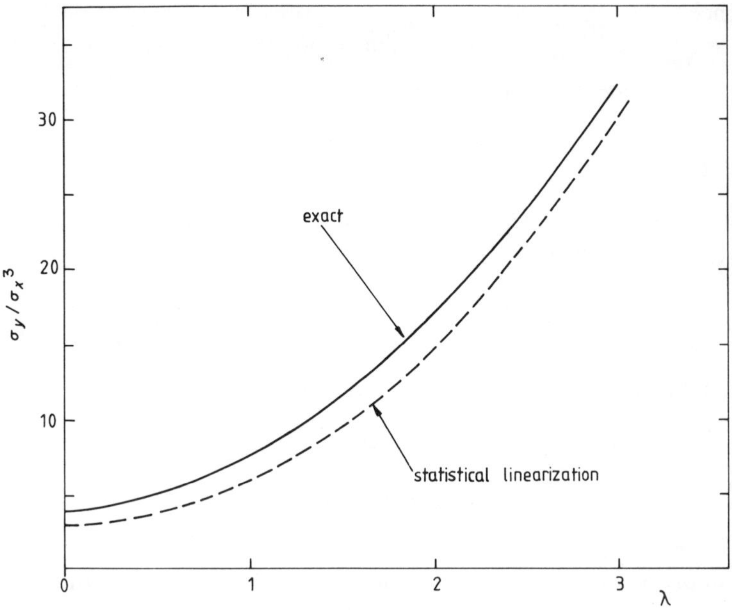

Figure 5.1
Variation of σ_y/σ_x^3 with $\lambda = m_x/\sigma_x$ according to the exact theory and a comparison with the result obtained by statistical linearization. Cubic non-linearity function

Figure 5.1 compares the variation of the exact value of σ_y/σ_x^3 with λ against the corresponding statistical linearization approximation. The latter is seen to underestimate the true value of σ_y, to some extent, but the percentage error in the approximation tends to decrease as λ increases. In fact, from equation (5.24)

$$\sigma_y \to 3\sigma_x^3\lambda^2 \quad \text{as } \lambda \to \infty \tag{5.25}$$

which is the same asymptotic limit given by the exact expression (5.21). This trend is due to the fact that when λ is large the ratio of m_y to σ_y is also large; thus, from equations (5.20) and (5.21),

$$v \equiv \frac{m_y}{\sigma_y} \to \frac{\lambda}{3} \quad \text{as } \lambda \to \infty \tag{5.26}$$

Large values of v imply that the disperson of y values around the value of m_y is relatively small, so that the element non-linearity function can be linearized to a good approximation. It is therefore not surprising that the equivalent linear system is a good approximation for large λ values. In fact, if the non-linearity function is linearized about the value $x = m_x$, for small changes δx and δy, it is found that

$$\delta y = (3m_x^2)\delta x \tag{5.27}$$

and hence

$$\sigma_y = 3m_x^2\sigma_x = 3\sigma_x^3\lambda^2 \tag{5.28}$$

This is in agreement with equation (5.25).

Example b

Here $y = x^2$ is not a single-valued function. From equation (3.13) it is found that $f_y(y)$ is given, exactly, by

$$f_y(y) = \frac{1}{2(2\pi)^{1/2} y\sigma_x}\left[\exp\left(-\frac{[y^{1/2} - m_x]^2}{2\sigma_x^2}\right) + \exp\left(-\frac{[y^{1/2} + m_x]^2}{2\sigma_x^2}\right)\right] \tag{5.29}$$

and the following expressions for m_y and σ_y are obtained

$$m_y = \sigma_x^2(1 + \lambda^2) \tag{5.30}$$

and

$$\sigma_y = \sigma_x^2(2 + 4\lambda^2)^{1/2} \tag{5.31}$$

where $\lambda = m_x/\sigma_x$, as before.

The appropriate expressions for the coefficients a and b, in the equivalent

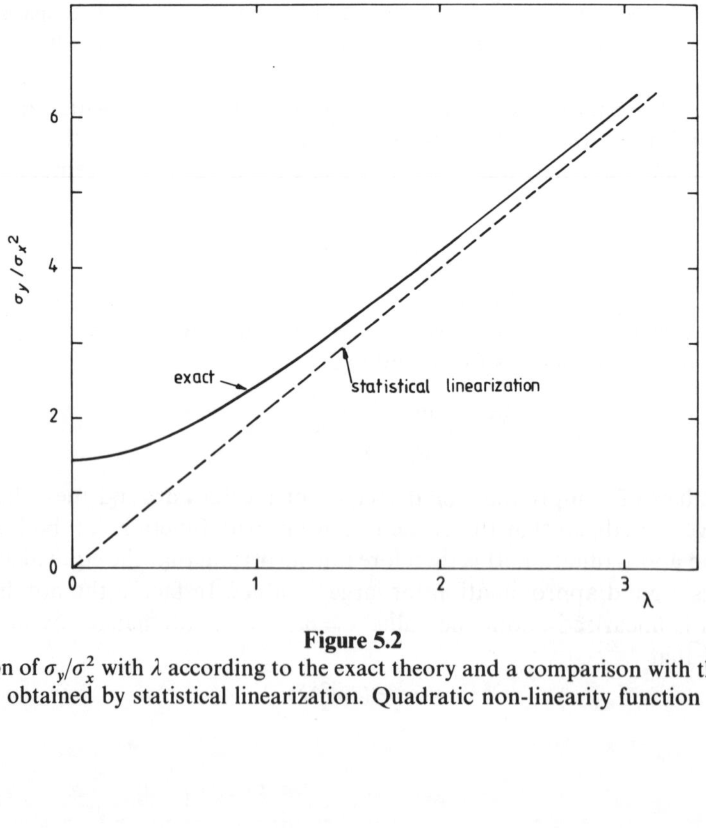

Figure 5.2

Variation of σ_y/σ_x^2 with λ according to the exact theory and a comparison with the result obtained by statistical linearization. Quadratic non-linearity function linear system are found, from equations (5.13) and (5.8), to be (see also Appendix A, Table A2)

$$a = 2\sigma_x \lambda \quad b = \sigma_x^2(1 + \lambda^2) \tag{5.32}$$

Again, b is equal to the exact value of m_y, which follows from the general requirement that $E\{y\} = E\{y_e\}$. The statistical linearization approximation for σ_y is found to be

$$\sigma_y = 2\lambda\sigma_x^2 \tag{5.33}$$

It is interesting to note that if $m_x = 0$ the above expression indicates that $\sigma_y = 0$. This is in direct contradiction with the exact result (5.31) and, indeed, with simple physical reasoning. Thus the statistical linearization procedure, as described here, must be regarded as unsatisfactory when applied to completely symmetrical non-linearity functions, such as $y = x^2$, driven by inputs with zero mean. This difficulty can be overcome by resorting to a different minimization criterion, but such a step is likely to bring difficulties in other applications. However, for non-zero mean inputs, the statistical linearization technique is useful, and will give a non-zero estimate of σ_y.

Figure 5.2 shows the variation of σ_y/σ_x^2 with λ, according to the exact result (5.31) and to the statistical linearization result (5.33). As in the preceding example, the relative error of the statistical linearization result reduces as λ increases.

5.3 OSCILLATORS WITH NON-LINEAR STIFFNESS

The analysis will now be extended to deal with simple mechanical or structural systems, in which one element is non-linear. The single degree of freedom mass–spring–damper system, or oscillator, is the simplest type of system of engineering importance and will be studied in the remainder of this chapter.

Initially the non-linear element of the system is taken to be the spring component. Such a system may be represented schematically, as shown in Figure 5.3. Here y is the displacement of the system, measured from the static equilibrium position where it is assumed that one, and only one, such position exists. The restoring force, due to the non-linear spring element, is denoted $h(y)$. It is noted that, for the corresponding linear system, $h(y) = ky$ where k is the linear spring constant (see Figure 4.3). The equation of motion of such a system is given by

$$m\ddot{y} + c\dot{y} + h(y) = F(t) \tag{5.34}$$

where m is the mass of the oscillator, c the linear damping coefficient and $F(t)$ the external excitation force.

It will be convenient to decompose $h(y)$ into a linear component, ky, and a

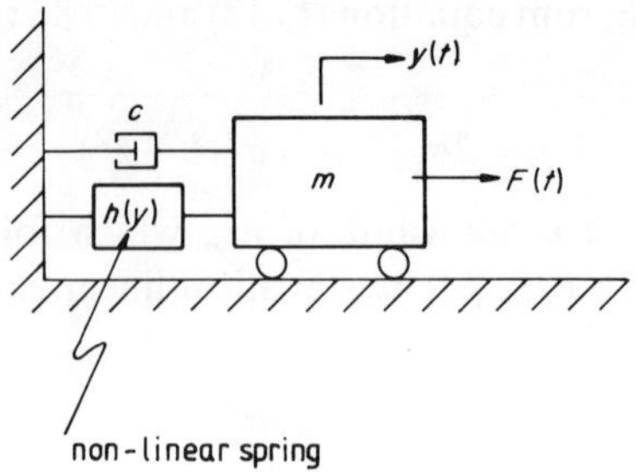

Figure 5.3
A single degree of freedom system, with linear damping and non-linear stiffness

non-linear component, $k\lambda G(y)$, where λ is a scaling factor. Thus

$$h(y) = k[y + \lambda G(y)] \tag{5.35}$$

For most real systems $h(y)$ can be linearized quite accurately when y is small, i.e. $G(y) \to 0$ as $y \to 0$. On combining equations (5.34) and (5.35), and dividing throughout by m, the equation of motion may be rewritten as

$$\ddot{y} + 2\zeta\omega_n\dot{y} + g(y) = f(t) \tag{5.36}$$

where

$$g(y) = \omega_n^2\{y + \lambda G(y)\} \quad f(t) = F(t)/m \tag{5.37}$$

Here ζ is the critical damping factor, and ω_n is the undamped natural frequency, for the linear system.

Equation (5.36) is a generalization of the standard equation of motion for a SDOF linear oscillator. It is useful in a variety of problems where only one 'mode' (usually the first) is significant, and where the principal non-linearity is in the restoring force. For example, as discussed by Lin (1967), equation (5.36) may be used to study the transverse vibrations of a plate excited by a random external pressure field. Here the non-linearity in the restoring force is due to the presence of membrane forces in the plate, which results in the effective stiffness of the plate increasing with increasing amplitude of transverse displacement (see Chapter 2).

To keep matters simple it will be assumed here, and throughout the remainder of this chapter, that the input process, $f(t)$, is Gaussian, with a power spectral density function denoted by $S_f(\omega)$. It is worth noting that, as in the linear case (see Section 4.3.4), the distribution of the response process, $y(t)$, will be relatively insensitive to the precise distribution of the input process, provided that the damping is light (e.g. see Roberts, 1981c). Thus the approximation of a real, somewhat non-Gaussian excitation process by an idealized Gaussian model is unlikely to lead to serious errors, in this situation.

5.3.1 The statistical linearization approximation

For the non-linear oscillator governed by equation (5.36) an exact solution is known only for the case where $f(t)$ is a white noise process (see Chapter 10). In the more general case, where $f(t)$ is non-white, the perturbation method may be used to investigate the effect of a small non-linearity in stiffness, when λ is small. However, this technique is ineffective when dealing with the case of large stiffness non-linearity. It is shown next how the technique of statistical linearization, as applied to a single non-linear element, may be extended to obtain an approximate solution to the problem of determining this non-linear oscillator response statistics.

Initially it will be assumed that $f(t)$ has a zero mean and that the function $G(y)$ is odd (i.e., antisymmetric). In this case the mean of the response process, $y(t)$, will also be zero. The analysis will then be generalized, in a later section, to deal with the more general case where $f(t)$ has a non-zero mean, and $G(y)$ is arbitrary.

Following the method described in Section 5.2.1, the non-linear element in equation (5.36) is replaced by a linear element. Specifically the term $g(y)$ is replaced by $\omega_{eq}^2 y$, where ω_{eq} is the natural frequency of the equivalent linear oscillator. Thus, as a substitute for equation (5.36) the equivalent linear system equation, given by

$$\ddot{y} + 2\zeta\omega_n\dot{y} + \omega_{eq}^2 y = f(t) \tag{5.38}$$

or

$$\ddot{y} + 2\zeta_{eq}\omega_{eq}\dot{y} + \omega_{eq}^2 y = f(t) \tag{5.39}$$

is considered, where

$$\zeta_{eq} = \frac{\omega_n}{\omega_{eq}}\zeta \tag{5.40}$$

is the critical damping factor of the equivalent linear system.

To find an expression for ω_{eq} it is necessary to minimize the expected value of the difference between equations (5.36) and (5.38), in a least square sense. Now this difference is simply the difference between the non-linear and the linear stiffness terms, i.e.

$$\varepsilon = g(y) - \omega_{eq}^2 y \tag{5.41}$$

Thus the value of ω_{eq} which minimizes $E\{\varepsilon^2\}$ is obtained from the equation

$$\frac{d}{d\omega_{eq}^2} E\{\varepsilon^2\} = 0 \tag{5.42}$$

On combining the above two equations, and performing the necessary differentiation, the following expression for ω_{eq} is obtained

$$\omega_{eq}^2 = \frac{E\{yg(y)\}}{E\{y^2\}} = \frac{E\{yg(y)\}}{\sigma_y^2} \tag{5.43}$$

where σ_y is the standard deviation of $y(t)$. On substituting the expression for $g(y)$, given by equation (5.37), into equation (5.43), the result

$$\omega_{eq}^2 = \omega_n^2\left(1 + \lambda\frac{E\{yG(y)\}}{\sigma_y^2}\right) \tag{5.44}$$

is obtained. This expression shows, very clearly, how the non-linear component of the stiffness element influences the value of ω_{eq}. The same result could, of course, have been obtained in a slightly more direct manner by replacing only the non-linear component of $g(y)$ with an equivalent linear component.

The above expression for ω_{eq}^2 is in agreement with the basic result found earlier for a single non-linear element, as given by equation (5.7). Indeed, equation (5.44) could have been obtained through a direct application of this earlier result (noting that here the input to the non-linear element is $y(t)$, rather than $x(t)$). However, the evaluation of ω_{eq}, according to equation (5.44), poses a problem which was not encountered when considering a single non-linear element, in isolation. For an isolated element the distribution of the input process is usually known, and hence the evaluation of the expected quantities in the expression for the linear parameter is straightforward. However, when the non-linear element is incorporated into an overall system, as in the present problem, the distribution of the input to the non-linear element is unknown. Thus, in equation (5.44), the exact evaluation of $E\{yG(y)\}/\sigma_y^2$ requires a knowledge of the first-order density function of the response process, $y(t)$, which is, of course, unknown.

Evidently, to evaluate ω_{eq}^2 it is necessary to adopt an approximation to the true, unknown distribution of $y(t)$. Two possible approaches are considered here. The first approach can be used if the original system is almost linear, so that x is small. In this case, it can be argued that the distribution of $y(t)$ is close to that of the linear system, obtained by setting $\lambda = 0$ in equation (5.37). Thus, the linear equation

$$\ddot{y} + 2\zeta\omega_n\dot{y} + \omega_n^2 y = f(t) \tag{5.45}$$

can be used to compute the distribution of $y(t)$. Here, since $f(t)$ is assumed to be Gaussian, it follows that the distribution of $y(t)$, found from equation (5.45), is also Gaussian.

The second, alternative approach is not limited to situations where λ is small. Specifically, the equivalent linear system, given by equation (5.39), is used to obtain an approximation to the distribution of $y(t)$. Thus, as in the first approach, the process $y(t)$ is assumed to be Gaussian, but now the standard deviation, σ_y, of $y(t)$ is found from equation (5.39), rather than equation (5.45). Since ω_{eq} depends on σ_y, when evaluated according to equation (5.44), with $y(t)$ taken to be Gaussian, and, in turn, σ_y evaluated from equation (5.39) will depend on ω_{eq}, it is evident that this second approach will generally lead to a non-linear algebraic equation for σ_y, which may be difficult to solve analytically. However, this disadvantage is heavily outweighted by the ability of the method to deal with quite severe non-linearities, with reasonable accuracy. Moreover, it can be shown that if λ is small it gives results which are identical with those found using the first approach, to the first order in λ. For this reason, the second approach has been generally preferred in the literature, and will be used throughout the remainder of this book.

With $y(t)$ assumed to be Gaussian, equation (5.43) can be simplified by using equation (3.56) again. Hence one finds that

$$\omega_{eq}^2 = E\left\{\frac{\partial g}{\partial y}\right\} \tag{5.46}$$

or

$$\omega_{eq}^2 = \omega_n^2\left(1 + \lambda E\left\{\frac{dG}{dy}\right\}\right) \tag{5.47}$$

In this context it is worth remarking that the value of the quantity $E\{dG/dy\}$, appearing in equation (5.47), will generally be rather insensitive to the precise shape of the density function of $y(t)$, provided that this function has the correct standard deviation. This is the principal reason why the second approach gives reasonably good estimates of σ_y, even when the non-linearity parameter is quite large (see Chapter 10). However, it must be remembered that the true distribution of $y(t)$ will tend to be markedly non-Gaussian when the non-linearity in the system is severe. The statistical linearization procedure obviously can not give information on the nature of such deviations from the Gaussian form.

In applying the second approach it is usual to implicitly assume that the resulting non-linear algebraic relationships do have a unique solution, over the whole parameter space. This does seem to be generally true, for systems with a unique, stable equilibrium position, and wide-band random excitation, although supporting mathematical arguments do not seem to be available. However, for systems with more then one equilibrium position, which may be stable or unstable, or with narrow band random excitation, it is possible to have either multiple solutions, or no solution, in some regions of the parameter space. Such solution characteristics can normally be associated with certain physical features of the response of the original non-linear system, as will be demonstrated later, by means of specific examples (see Sections 5.3.6 to 5.3.9).

5.3.2 Standard deviation of the response

To illustrate the procedure for finding the standard deviation, σ_y, of the response of the non-linear oscillator governed by equation (5.36), it will now be supposed that the non-linear stiffness function, $G(y)$, can be represented in a power series form, as follows

$$G(y) = \sum_{n\,\text{odd}} \alpha_n y^n \tag{5.48}$$

Here α_n are positive constants and only odd powers of n (i.e. $n = 3, 5, \ldots$) are included in the summation, to ensure that $G(y)$ (and hence $g(y)$) is an odd function.

As a first step, ω_{eq} is evaluated according to equation (5.47), taking the density function of $y(t)$ to be of the Gaussian form

$$f_y(y) = \frac{1}{(2\pi)^{1/2}\sigma_y} \exp\left(-\frac{y^2}{2\sigma_y^2}\right) \tag{5.49}$$

It is noted that σ_y, in this expression, is to be evaluated and is not known at this stage. Using the definition of the expectation operator (see Section 3.4) it is found from equation (5.47) that

$$\omega_{\text{eq}}^2 = \omega_{\text{n}}^2\left(1 + \lambda \sum_{n\,\text{odd}} n\alpha_n A_n\right) \tag{5.50}$$

where

$$A_n = E\{y^{n+1}\} = \int_{-\infty}^{\infty} y^{n+1} f_y(y)\,\mathrm{d}y \tag{5.51}$$

A combination of equations (5.49) and (5.51) enables A_n to be found in terms of the Gamma function, $\Gamma(x)$ (Abramovitch and Stegun, 1972). The result is (see also Appendix A, Table A1)

$$A_n = \frac{1}{\pi^{1/2}} (\sqrt{2}\sigma_y)^{n-1} \Gamma\left(\frac{n}{2}\right). \tag{5.52}$$

The next step is to evaluate σ_y, using the equivalent linear system equation, given by equation (5.39), for this purpose. The simplest procedure is to use the frequency domain, input–output formula derived in Chapter 4 (see equation (4.20)); i.e. the spectrum of $y(t)$ is determined by

$$S_y(\omega) = |\alpha(\omega)|^2 S_f(\omega) \tag{5.53}$$

where the appropriate frequency response function, $\alpha(\omega)$, is given by (see Section 4.5.1)

$$\alpha(\omega) = \frac{1}{(\omega_{\text{eq}}^2 - \omega^2 + 2\mathrm{i}\zeta_{\text{eq}}\omega\omega_{\text{eq}})} \tag{5.54}$$

Once $S_y(\omega)$ is determined, σ_y is found from the area under this spectrum (since the mean of $y(t)$ is here zero), i.e.

$$\sigma_y^2 = \int_{-\infty}^{\infty} S_y(\omega)\,\mathrm{d}\omega \tag{5.55}$$

Equations (5.53) to (5.55) yield a direct relationship between σ_y and ω_{eq}. Another independent relationship between these two quantities may be found from equations (5.50) to (5.52). Thus, two algebraic relationships are obtained for the two unknowns, σ_y and ω_{eq}, and straightforward manipulation enables algebraic expressions for σ_y and ω_{eq} to be formulated.

In the specific case where the excitation process is wide-band, and can be modelled as a white noise process, the evaluation of σ_y from equations (5.53) and (5.55) is considerably simplified. If $S_f(\omega) = S_0$, a constant, then it is found that (see Appendix B)

$$\sigma_y^2 = \frac{\pi S_0}{2\zeta_{eq}\omega_{eq}^3} = \frac{\pi S_0}{2\zeta\omega_n\omega_{eq}^2} \tag{5.56}$$

Eliminating ω_{eq} between equations (5.50) and (5.56) a single algebraic equation for σ_y emerges, i.e.

$$\sigma_y^2\left[1 + \frac{\lambda}{\pi^{1/2}} \sum_{n\,\mathrm{odd}} n\alpha_n(\sqrt{2}\sigma_y)^{n-1}\Gamma\left(\frac{n}{2}\right)\right] = \sigma_{yo}^2 \tag{5.57}$$

where σ_{yo} is the standard deviation of $y(t)$ in the linear case, for which $\lambda = 0$, i.e.

$$\sigma_{yo}^2 = \frac{\pi S_0}{2\zeta\omega_n^3} \tag{5.58}$$

In general a numerical algorithm will be required to compute σ_y, from equation (5.57).

In many applications, where the series expansion of equation (5.48) is appropriate, a good approximation is achieved by neglecting all but the first non-linear term in the expansion, i.e. by assuming that

$$G(y) = y^3 \tag{5.59}$$

The equation of motion for the oscillator then becomes

$$\ddot{y} + 2\zeta\omega_n\dot{y} + \omega_n^2(y + \lambda y^3) = f(t) \tag{5.60}$$

This type of system is generally referred to as a Duffing oscillator, and has been extensively studied (see Chapter 2).

For the Duffing oscillator, with white noise excitation, an explicit expression for σ_y can be found analytically. Thus, setting $\alpha_3 = 1$ and $\alpha_n = 0$ for $n \neq 3$, in equation (5.57) the quadratic expression

$$\sigma_y^2(1 + 3\lambda\sigma_y^2) = \sigma_{yo}^2 \tag{5.61}$$

for σ_y^2 is obtained. Only one root is positive and hence there is only one possible solution for σ_y^2. The result can be written in the form

$$\frac{\sigma_y^2}{\sigma_{yo}^2} = \frac{(1 + 12\Lambda)^{1/2} - 1}{6\Lambda}, \tag{5.62}$$

where Λ is the non-dimensional parameter

$$\Lambda = \lambda\sigma_{yo}^2 \tag{5.63}$$

In Chapter 10, this approximate solution for σ_y will be compared with the corresponding exact solution, obtained by the Fokker–Planck (FPK) method.

5.3.3 The case of small non-linearity

If the non-linear term in the equation of motion is small the solution of the σ_y equation can be considerably simplified. The procedure is to assume that the solution for σ_y^2 can be expanded in powers of the non-linearity parameter, λ. Thus

$$\sigma_y^2 = \sigma_{yo}^2 + \lambda\sigma_{y_1}^2 + \cdots \tag{5.64}$$

Clearly, for sufficiently small λ the terms of order λ^2, and higher powers, are negligible.

As a specific illustration of this approach, the case where the excitation is a white noise, and $G(y)$ is given by the power series of equation (5.48) will be reconsidered. The appropriate expression for σ_y is given by equation (5.57). If equation (5.64) is substituted into equation (5.57), and terms of order λ^2, and higher order, are neglected, it is found that the solution for σ_y^2, correct to order λ, is (Crandall, 1964)

$$\sigma_y^2 = \sigma_{yo}^2(1 - \lambda B) \tag{5.65}$$

where

$$B = \frac{1}{\pi^{1/2}} \sum_{n\,\text{odd}} n\alpha_n(\sqrt{2}\sigma_{yo})^{n-1}\Gamma\left(\frac{n}{2}\right) \tag{5.66}$$

The evaluation of σ_y, according to this expression, requires only a knowledge of σ_{yo}, which is given directly by equation (5.58).

In the specific case of a Duffing oscillator, equation (5.65) reduces to

$$\frac{\sigma_y^2}{\sigma_{yo}^2} = 1 - 3\Lambda \tag{5.67}$$

A comparison of this result with equation (5.62), for arbitrary Λ, reveals that the two expressions agree, to order Λ.

A simple expression for ω_{eq}, can also be found by neglecting terms of order higher than λ. Thus, from equations (5.50) and (5.52) it is found that, correct to order λ

$$\omega_{eq}^2 = \omega_n^2(1 + \lambda B) \tag{5.68}$$

It is noted that this result involves using σ_{yo} in place of σ_y; the calculation of

ω_{eq} is thus drastically simplified. For the Duffing oscillator $B = 3\sigma_{yo}^2$, and hence equation (5.68) reduces to

$$\omega_{eq}^2 = \omega_n^2(1 + 3\Lambda) \tag{5.69}$$

5.3.4 Power spectrum of the response

Exact expressions for the power spectrum of the oscillator response, $y(t)$, are unknown, even for the case of white noise excitation. However, a simple approximate expression for this spectrum is obtained directly by the statistical linearization method. A combination of equations (5.53) and (5.54) gives the following approximation for $S_y(\omega)$

$$S_y(\omega) = \frac{S_f(\omega)}{[(\omega_{eq}^2 - \omega^2)^2 + 4\zeta_{eq}^2\omega^2\omega_{eq}^2]} \tag{5.70}$$

where ω_{eq} is given by equation (5.50) and ζ_{eq} by equation (5.40).

If the non-linearity parameter, λ, is small a simplified result can be found by neglecting terms of order λ^2, and higher order (Crandall, 1964); see also Roy and Spanos (1990). Thus, for example, if $G(y)$ is given by the power series of equation (5.48) and $f(t)$ is a white noise, then equation (5.68) for ω_{eq} may be combined with equation (5.70). On expanding, and neglecting terms of order λ^2, and higher order, again, the following result is found

$$S_y(\omega) = \frac{S_0}{D} - \frac{4\lambda\zeta B}{\pi}\frac{\omega_n^3(\omega_n^2 - \omega^2)}{D^2} \tag{5.71}$$

where

$$D = (\omega_n^2 - \omega^2)^2 + 4\zeta^2\omega^2\omega_n^2 \tag{5.72}$$

Evidently the first term on the right-hand side of equation (5.71) is the linear oscillator response spectrum, for the case where $\lambda = 0$. The second term represents the effect of the non-linear stiffness contribution, to the first order in λ, the stiffness parameter.

5.3.5 Inputs with non-zero means

The foregoing analysis can be readily generalized to deal with the situation where the input process, $f(t)$, has a non-zero mean. Here $f(t)$ and $y(t)$ may be written as

$$f(t) = f_0(t) + m_f \tag{5.73}$$

and

$$y(t) = y_0(t) + m_y \tag{5.74}$$

where m_f and m_y are, respectively, the mean values of $f(t)$ and $y(t)$, and $f_0(t)$ and $y_0(t)$ are zero-mean processes.

The equation of motion of an oscillator with non-linear stiffness, i.e.

$$\ddot{y} + 2\zeta\omega_n\dot{y} + g(y) = f(t) \tag{5.75}$$

where $g(y)$ is an odd function, given by equation (5.37) will now be considered again. The problem is to obtain, by the statistical linearization procedure, approximate expressions for m_y and σ_y, the standard deviation of $y(t)$.

Following the approach described earlier, in Section 5.2.1, an appropriate equivalent linear system is

$$\ddot{y} + 2\zeta\omega_n\dot{y} + ay_0 + b = f(t) \tag{5.76}$$

Therefore, the function $g(y)$ has been replaced by a linear function, in the manner indicated by equation (5.3). Using the fact that $\dot{y} = \dot{y}_0$ and $\ddot{y} = \ddot{y}_0$, and taking expectations throughout in equation (5.76), it is evident that

$$b = m_f \tag{5.77}$$

Thus, the equivalent linear system equation may be written in the following simpler form

$$\ddot{y}_0 + 2\zeta\omega_n\dot{y}_0 + \omega_{eq}^2 y_0 = f_0(t) \tag{5.78}$$

where

$$a = \omega_{eq}^2 \tag{5.79}$$

and ω_{eq} is the natural frequency of the equivalent linear system, as before.

Expressions for a and b can be found by minimizing the difference between equations (5.75) and (5.76). This is the same as minimizing the difference between $g(y)$ and $ay_0 + b$. The analysis given in Section 5.2.1 shows that

$$a = \omega_{eq}^2 = \frac{E\{y_0 g(y)\}}{\sigma_y^2} = E\left\{\frac{dg}{dy}\right\} \tag{5.80}$$

and

$$b = m_f = E\{g(y)\} \tag{5.81}$$

These equations enable two algebraic relationships to be derived between the three unknowns, m_y, σ_y and ω_{eq}. A third relationship, between σ_y and ω_{eq}, may be found by computing σ_y from the equivalent linear system. This enables the three unknowns to be evaluated, from three non-linear algebraic equations. In general a numerical solution procedure will be necessary to obtain specific results.

As an illustration, the case of a Duffing oscillator excited by white noise, with a non-zero mean will be considered. Here $g(y) = \omega_n^2(y + \lambda y^3)$ and hence, from equations (5.80) and (5.81),

$$\omega_{eq}^2 = \omega_n^2[1 + 3\lambda E\{y^2\}] \tag{5.82}$$

and

$$m_f = \omega_n^2[m_y + \lambda E\{y^3\}] \tag{5.83}$$

The expectations appearing in these expressions have already been evaluated earlier in this chapter (example a in Section 5.2.3). Thus, employing equations (5.23) and (5.24), it is found that the above relationships lead to the following equations

$$\omega_{eq}^2 = \omega_n^2[1 + 3\lambda\sigma_y^2(1 + \eta^2)] \tag{5.84}$$

$$m_f = \omega_n^2[m_y + \lambda\sigma_y^3(3\eta + \eta^3)] \tag{5.85}$$

where

$$\eta = \frac{m_y}{\sigma_y} \tag{5.86}$$

In addition, since $f_0(t)$ is a white noise, equation (5.56) provides a simple relationship between σ_y and ω_{eq}.

Equations (5.56), (5.84) and (5.85) may be solved as three simultaneous, non-linear algebraic equations, to find m_y, σ_y and ω_{eq}. On eliminating ω_{eq}, and non-dimensionalizing, the following two equations emerge

$$1/r^2 = [1 + 3\Lambda r^2(1 + \eta^2)] \quad \eta = \rho/r - \Lambda r^2(3\eta + \eta^3) \tag{5.87}$$

where

$$r = \frac{\sigma_y}{\sigma_{yo}} \quad \rho = \frac{m_f}{\sigma_{yo}\omega_n^2} \quad \Lambda = \lambda\sigma_{yo}^2 \tag{5.88}$$

and σ_{yo} is the value of σ_y when $\lambda = 0$. A numerical solution of equation (5.87) will

Table 5.1

No. of iterations	r	η
0	1	0.2
1	0.625	−0.104
2	0.792	0.381
3	0.693	−0.124
4	0.759	0.378
5	0.709	−0.080
10	0.750	0.281
20	0.743	0.200
50	0.738	0.151
100	0.738	0.148

yield values of η and r, and hence values of m_y and σ_y, for particular values of ρ and Λ.

Equations (5.87) can be solved by a simple iterative scheme, if Λ is not too large. Thus, starting with the linear case, where $\Lambda = 0$, one has $r = 1$ and $\eta = \rho$. If these values are substituted into the right-hand sides of the relationships in equation (5.87) one can obtain updated values of r and η. These can be fed back into the right-hand sides of the equations, and so on, until convergence is achieved.

Table 5.1 shows a typical result, for $\Lambda = 0.5$ and $\rho = 0.2$. In this case the convergence is quite slow, but after 50 iterations the values of r and η have almost converged, and are correct to two decimal places. An alternative procedure for solving equation (5.85) is to eliminate η; this will result in a sixth-order polynomial equation for r^2. Standard numerical algorithms exist for obtaining all the roots of r^2. Numerical studics show that, over the whole $\Lambda - \rho$ parameter plane there is only one root of the polynomial equation which is real and positive. This means that there is a unique solution for r, and hence η.

5.3.6 Asymmetric non-linearities

If the non-linear stiffness function $g(y)$ is asymmetric with respect to the origin then the mean of $y(t)$ may be non-zero, even when the mean of the input is zero. In these circumstances the method used in the preceding section can be employed to find expressions for m_y and σ_y (see also Spanos, 1978, 1980).

The case of an oscillator with a quadratic non-linearity function, i.e. $G(y) = y^2$, provides an appropriate illustration. Here the equation of motion is as follows

$$\ddot{y} + 2\zeta\omega_n\dot{y} + \omega_n^2(y + \lambda y^2) = f(t) \tag{5.89}$$

For simplicity it will be assumed that $f(t)$ is a white noise with a non-zero mean, m_f. Here the mean of $y(t)$, m_y, will be non-zero even when $m_f = 0$.

An appropriate equivalent linear system is again given by equation (5.76), where $a = \omega_{eq}^2$ and $b = m_f$, with a and b given by equations (5.80) and (5.81). Application of these equations to the present example shows that

$$\omega_{eq}^2 = \omega_n^2(1 + 2\lambda E\{y\}) \tag{5.90}$$

and

$$m_f = \omega_n^2(m_y + \lambda E\{y^2\}) \tag{5.91}$$

Hence (see also example b in Section 5.2.5),

$$\omega_{eq}^2 = \omega_n^2(1 + 2\lambda\sigma_y\eta) \tag{5.92}$$

and

$$m_f = \omega_n^2[m_y + \lambda\sigma_y^2(1 + \eta^2)] \tag{5.93}$$

where $\eta = m_y/\sigma_y$, as before.

These expressions, together with equation (5.56), enable m_y and σ_y to be evaluated. On introducing the non-dimensional parameters r and ρ again, defined by equation (5.88), together with the non-dimensional parameter

$$\Theta = \lambda \sigma_{yo} \tag{5.94}$$

and eliminating ω_{eq}, the following expressions are obtained for η and r

$$\frac{1}{r^2} = 1 + 2\Theta r\eta \qquad \eta = \frac{\rho}{r} - \Theta r(1 + \eta^2) \tag{5.95}$$

These equations can be solved iteratively, in a similar way to that discussed in the previous section, in connection with equations (5.87). However, here a better method is to eliminate η and solve the resulting third-order polynomial equation for r^2 numerically, using a standard algorithm. This approach has the advantage that all possible solutions are readily obtained, for any choice of ρ and Θ. Figures 5.4a and 5.4b show, respectively, the variation of η and r with Θ, for several values of ρ. An interesting feature here is that there is only a finite region (Ω_1) in the Θ–ρ parameter plane, in which real solutions for r and η exist (see Figure 5.4c). In this region there are actually two solutions, for each choice of ρ and Θ. For the first of these solutions $r \to 1$ and $\eta \to \rho$ as $\Theta \to 0$, as one would expect from physical considerations. The second solution is such that $r \to \infty$ as $\Theta \to 0$. Clearly this does not relate to a physically realizable response. The two solutions do, however, converge as Θ increases, and coincide at the boundary of Ω_1.

A plausible explanation for the existence of only a finite Ω_1 region emerges if one considers the form of the potential energy function, corresponding to the complete stiffness term $\omega_n^2(y + \lambda y^2)$. Figures 5.5a and 5.5b show, respectively, the variation of stiffness and potential energy, with displacement. There is a 'potential well' centred at the position of stable, static equilibrium, $y = 0$. When the oscillator is randomly excited an 'escape' from this well, in the direction of positive displacement, is not possible, since the stiffness progressively increases with displacement. However, escape is possible in the negative direction and, in fact, will occur with some non-zero probability over any period of time, no matter how long. Thus, it is actually not possible to treat the response process as stationary, in the usual sense, since individual sample functions will eventually exceed the critical negative level, $-y_c = -1/\lambda$; after this they will usually grow without bound. This phenomenon is illustrated in Figure 5.5c.

However, if the level of the excitation is not too strong, and the mean of the excitation is not too negative, the probability that the response will escape from the potential well will be so small that, over long periods of time the response can, for practical purposes, be regarded as stationary. Since Θ is a measure of the excitation strength, and ρ is proportional to its mean level, there will be some region (Ω_2) in the plane where the response is approximately stationary, in the sense given above. The region Ω_2 is actually fairly well defined since the

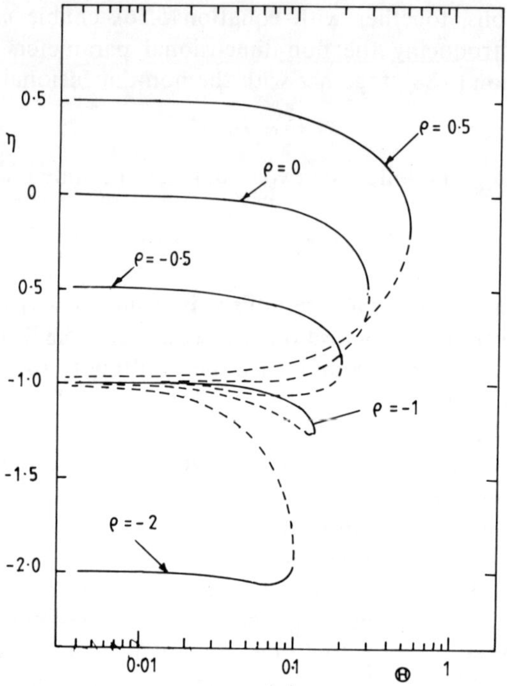
0.5
η
0
0.5
-1.0
-1.5
-2.0
ρ = 0.5
ρ = 0
ρ = -0.5
ρ = -1
ρ = -2
0.01
0.1
Θ
1

(a)

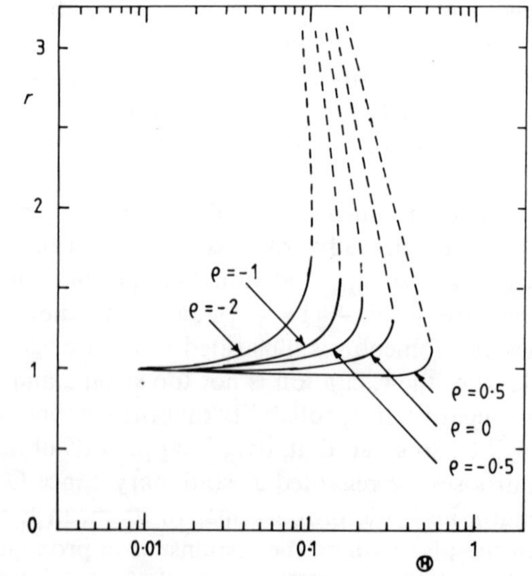
3
r
2
1
0
ρ = -1
ρ = -2
ρ = 0.5
ρ = 0
ρ = -0.5
0.01
0.1
Θ
1

(b)

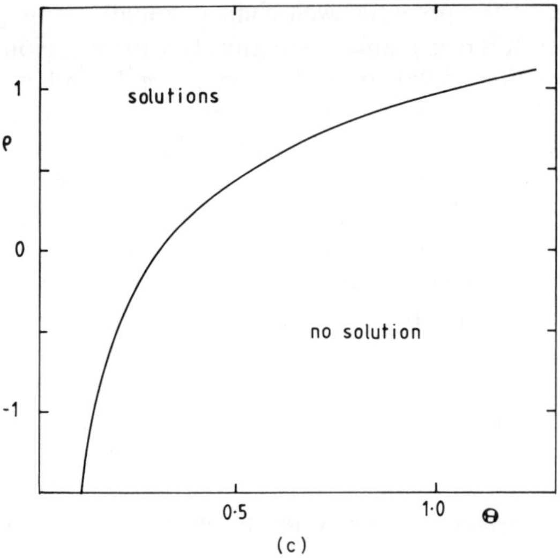

Figure 5.4
Statistical linearization results for an oscillator with an asymmetric stiffness. (a) Variation of η with Θ for various values of ρ. (b) Variation of r with Θ for various values of ρ. (———first solution, ---second solution) (c) Solution regions in the ρ–Θ plane

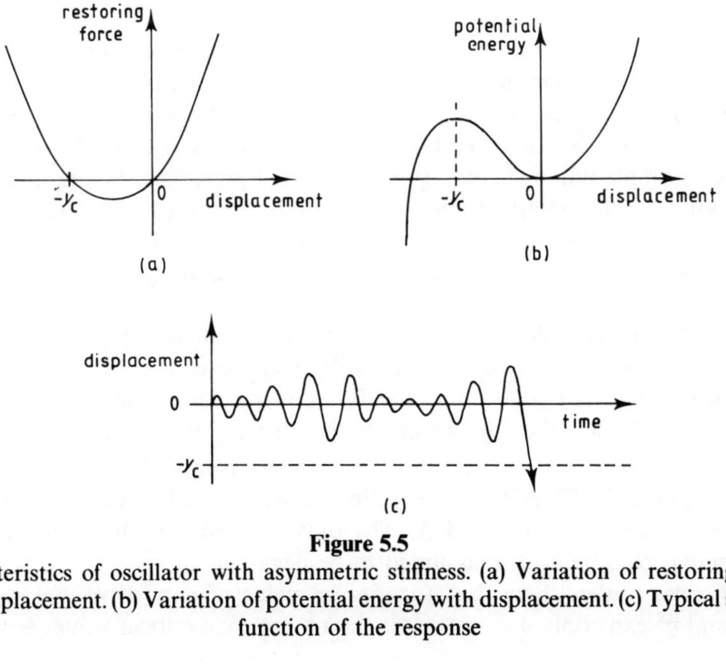

Figure 5.5
Characteristics of oscillator with asymmetric stiffness. (a) Variation of restoring force with displacement. (b) Variation of potential energy with displacement. (c) Typical sample function of the response

mean time spent in the potential well changes rapidly in magnitude near the region's perimeter. It is reasonable to identify Ω_2 with the region Ω_1 within which solutions to equation (5.95) exist. Certainly the boundary of Ω_1 depends qualitatively on Θ and ρ, in the manner one would expect for an Ω_2 region. Thus, the range of input strength for which a physically realistic solution exists decreases as the mean of the excitation, and hence the mean of the response, reduces. Further support for this conjecture can be obtained through a comparison with simulation results (see Chapter 10).

It may be concluded, therefore, that the existence, or otherwise, of an equivalent linear system with a stationary response provides an indication of the behaviour of sample functions of the response process of the non-linear system and can be used to delineate the region in parameter space within which the response may be treated as approximately stationary.

5.3.7 Systems with a softening restoring characteristic

In some applications the restoring term, $g(y)$, is of the softening kind. Examples occur in the equations of motion of a pendulum undergoing large amplitude oscillations and a ship executing large amplitude rolling motion (see Chapter 2 and Figures 2.3 and 2.5). In such cases $g(y)$ is antisymmetric and, as $|y|$ increases, $g(y)$ reaches an absolute maximum value and then falls progressively to zero, at some critical value, y_c. A typical softening restoring characteristic is shown in Figure 5.6a. The corresponding potential energy function is shown in Figure 5.6b. If the excitation is a stationary random process then it is evident that the response will eventually 'escape' from the potential well, as in the case discussed in the preceding section. However, in the present case, escape is possible in either direction, i.e. it can occur if y exceeds y_c, or falls below $-y_c$. A typical sample function for $y(t)$ is shown in Figure 5.6c. In the light of the discussion in the preceding section it can be expected that the statistical linearization method will yield a stationary solution, provided that the excitation level is below a critical, threshold level, despite the fact that, strictly, a stationary response process does not exist.

As a specific example, the case of a Duffing oscillator, excited by a white noise process, with a non-zero mean, will again be considered. The equation of motion is given by equation (5.60). By allowing the parameter λ to become *negative*, a suitable softening spring characteristic is obtained and the critical value, y_c, is here equal to $1/\sqrt{-\lambda}$. The analysis given earlier, in Section 5.3.5, is again applicable. Thus it is once again necessary to solve equations (5.87) for η and r, as functions of ρ and Λ. The only difference is that now the non-dimensional parameter Λ is assumed to be negative.

Numerical studies show that, for a given value of ρ, positive real solutions for r^2 and η^2 exist only if Λ is greater than a negative critical value, Λ_c. Above

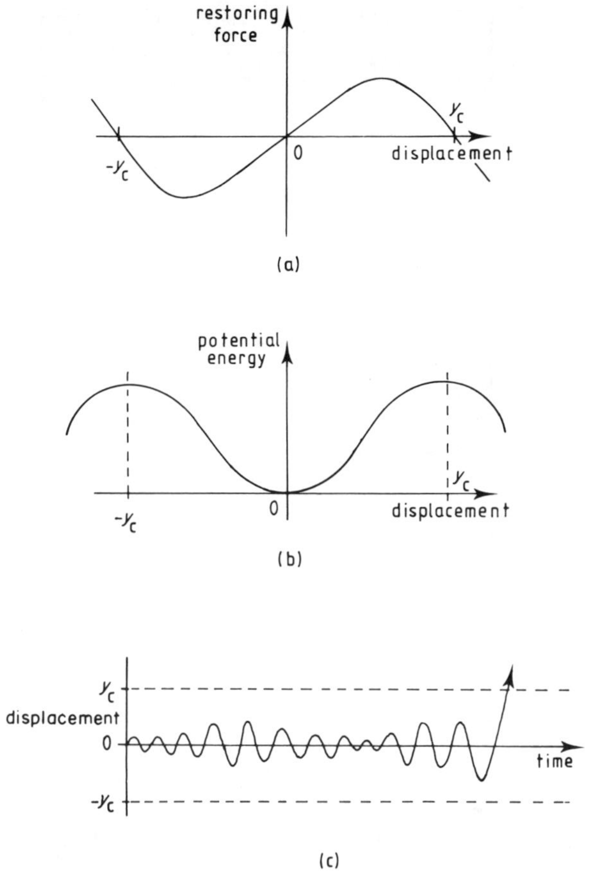

Figure 5.6
Characteristics of an oscillator with a softening spring. (a) Variation of restoring force with displacement. (b) Variation of potential energy with displacement. (c) Typical sample function of the response

this value there are two solutions, which converge as $\Lambda \to \Lambda_c$. Figures 5.7a and 5.7b show, respectively, the variation of η^2 and r^2 with Λ, for various values of ρ. The first solution converges to the expected asymptotic limiting values $r = 1$, $\eta = \rho$ as $\Lambda \to 0$ and thus is physically realistic. In contrast, the second solution is not physically meaningful since $r \to \infty$ as $\Lambda \to 0$.

Thus, as anticipated, the solution characteristics for the softening Duffing oscillators are qualitatively similar to those obtained in the preceding section. There is a finite region, Ω_1, in the Λ, ρ parameter plane inside which real, positive solutions for r and η exist. This corresponds to the fact that the response of the non-linear system can be regarded as approximately stationary. It is

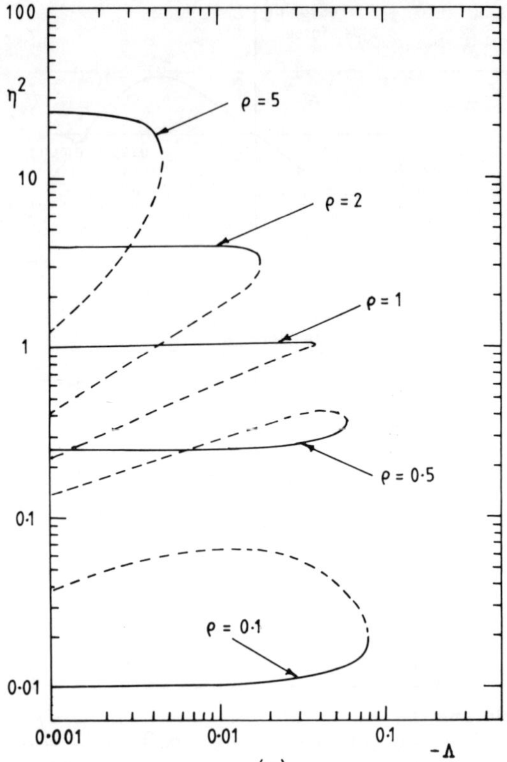
100
η^2
10
1
0·1
0·01
ρ = 5
ρ = 2
ρ = 1
ρ = 0·5
ρ = 0·1
0·001
0·01
0·1
-Λ

(a)

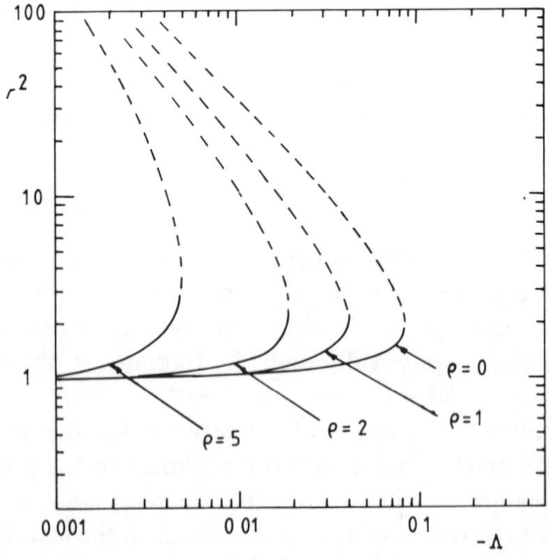
100
10
1
ρ = 0
ρ = 1
ρ = 2
ρ = 5
0 001
0 01
0 1
-Λ

(b)

stationary in the sense that the mean duration within the potential well is very long when the excitation is sufficiently weak, and its mean value is not too large.

For the special case where the mean of the excitation is zero the analysis is particularly simple. Thus, setting $\rho = 0$ in equation (5.87) one finds that a real, positive solution for r exists only if $\eta = 0$. $r^2 = \sigma_y^2/\sigma_{yo}^2$ is then given by equation (5.62) and it follows that real solutions for r^2 only exist if $\Lambda > \Lambda_c = -1/12$. Figures 5.7a and 5.7b show that, as one would expect, Λ_c progressively increases as the mean level of the excitation, as measured by ρ, increases.

5.3.8 Systems with multiple static equilibrium positions

It is possible for non-linear systems to have one or more positions of stable static equilibrium. A specific example is the case of a simple mass–spring model of a shallow arch, with transverse loading (see Chapter 2, Section 2.3.4, and Figure 2.14). Referring also to Figure 2.15, it is seen that there are two positions of stable static equilibrium, A and E, and one position of unstable static equilibrium, C. If y denotes the transverse displacement of the mass from the unstable equilibrium position ($\alpha = \lambda$), then the restoring characteristic appears as shown in Figure 5.8a. The corresponding potential energy function, shown in Figure 5.8b, has two potential wells, symmetrically disposed about $y = 0$, and with minima at $y = \pm y_c$.

If such a system is excited with a zero mean random process, of low intensity, then the response will tend to stay within one of the potential wells, for a long period of time. Eventually the response will 'escape' into the neighbouring well, where it will stay until another escape occurs. Thus, sample functions will tend to have an appearance similar to that shown in Figure 5.8c. As the intensity of the excitation is increased then the mean time spent in a particular well will decrease, and hence 'jumps' from one well to another will become progressively less well defined.

It is interesting to enquire if the statistical linearization method is capable of producing results which reflect the occurrence, or otherwise, of the jumping phenomenon described above. For this purpose equation (5.75) will again be adopted as an appropriate equation of motion, and $g(y)$ will be assumed to be of the form

$$g(y) = \omega_n^2(-y + \lambda y^3) \tag{5.96}$$

Figure 5.7

Statistical linearization results for an oscillator with a softening spring. (a) Variation of η^2 with Λ for various values of ρ. (b) Variation of r^2 with Λ for various values of ρ (——first solution, ---second solution)

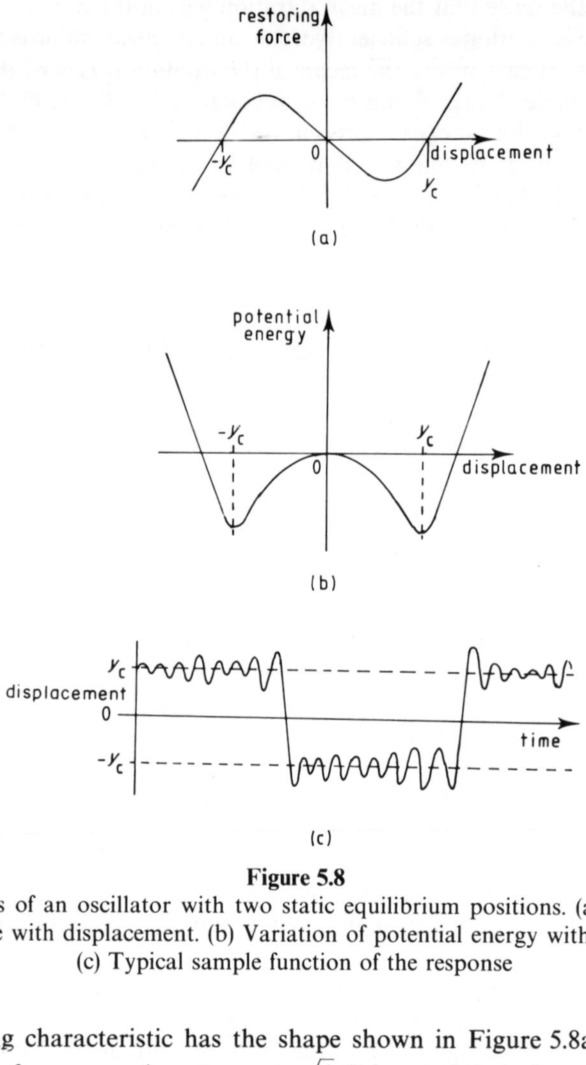

Figure 5.8
Characteristics of an oscillator with two static equilibrium positions. (a) Variation of restoring force with displacement. (b) Variation of potential energy with displacement. (c) Typical sample function of the response

This restoring characteristic has the shape shown in Figure 5.8a, the critical points of zero force occurring at $y_c = \pm\sqrt{\lambda}$. It is noted that, since the stiffness is negative at $y=0, \omega_n$ no longer has its usual meaning of a 'linear' natural frequency. For simplicity it will be assumed that $f(t)$ is a zero-mean white noise process.

Following an analysis similar to that given in Section 5.3.5, the statistical linearization method leads to equations for the non-dimensional mean response, $m = \lambda^{1/2} m_y$, and the non-dimensional variance of the response, $\sigma^2 = \lambda \sigma_y^2$. Specifically

$$\frac{\beta}{\sigma^2} = -1 + 3(\sigma^2 + m^2) \tag{5.97}$$

$$m[-1+3\sigma^2+m^2]=0 \tag{5.98}$$

where

$$\beta=\frac{\pi S_0\lambda}{2\zeta\omega_n^3}=\Lambda^2 \tag{5.99}$$

is a measure of the strength of the excitation. The symbol Λ is defined as before (see equation (5.88)) but it is noted that $\sigma_{y0}^2=\pi S_0/2\zeta\omega_n^3$ no longer has the physical significance of being the 'linear' variance of the response.

If $m \neq 0$ then the following quadratic equation for σ^2 can be obtained, by eliminating m^2 between equations (5.97) and (5.98); thus

$$6\sigma^4-2\sigma^2+\beta=0 \tag{5.100}$$

Hence one finds that, if $\beta<\frac{1}{6}$, there are two possible real, positive solutions for σ^2. These solutions, together with the corresponding expressions for m^2, appear in Table 5.2, as solutions 1 and 2.

A third solution also exists. Setting $m=0$ (which satisfies equation (5.98)) in equation (5.97), the following quadratic equation for σ^2 is obtained

$$3\sigma^4-\sigma^2-\beta=0 \tag{5.101}$$

There is only one positive real root of this equation. This result is also given in Table 5.2. Unlike the first two solutions, it is valid over the whole range of β.

Figures 5.9a and 5.9b show the variation of m^2 and σ^2 with β, according to these three solutions.

The physical significance of these solutions will now be examined. At very low levels of excitation (low β values) the mean duration time in a potential well will be very long. Thus one expects a sample function of the response to oscillate about a mean value in the region of unity (noting the scaling involved in m) with a variance which will increase as the level of excitation increases. This is precisely the behaviour indicated by the first solution. Thus, it is reasonable to associate the first solution with the 'quasi-stationary' behaviour of the response while it remains in one of the two potential wells. It is also

Table 5.2

Solution number	*Range of* validity	σ^2	m^2
1	$0<\beta<1/6$	$\dfrac{1+(1-6\beta)^{1/2}}{6}$	$\frac{1}{2}[1-(1-6\beta)^{1/2}]$
2	$0<\beta<1/6$	$\dfrac{1-(1-6\beta)^{1/2}}{6}$	$\frac{1}{2}[1+(1-6\beta)^{1/2}]$
3	$0<\beta<\infty$	$\dfrac{1+(1+12\beta)^{1/2}}{6}$	0

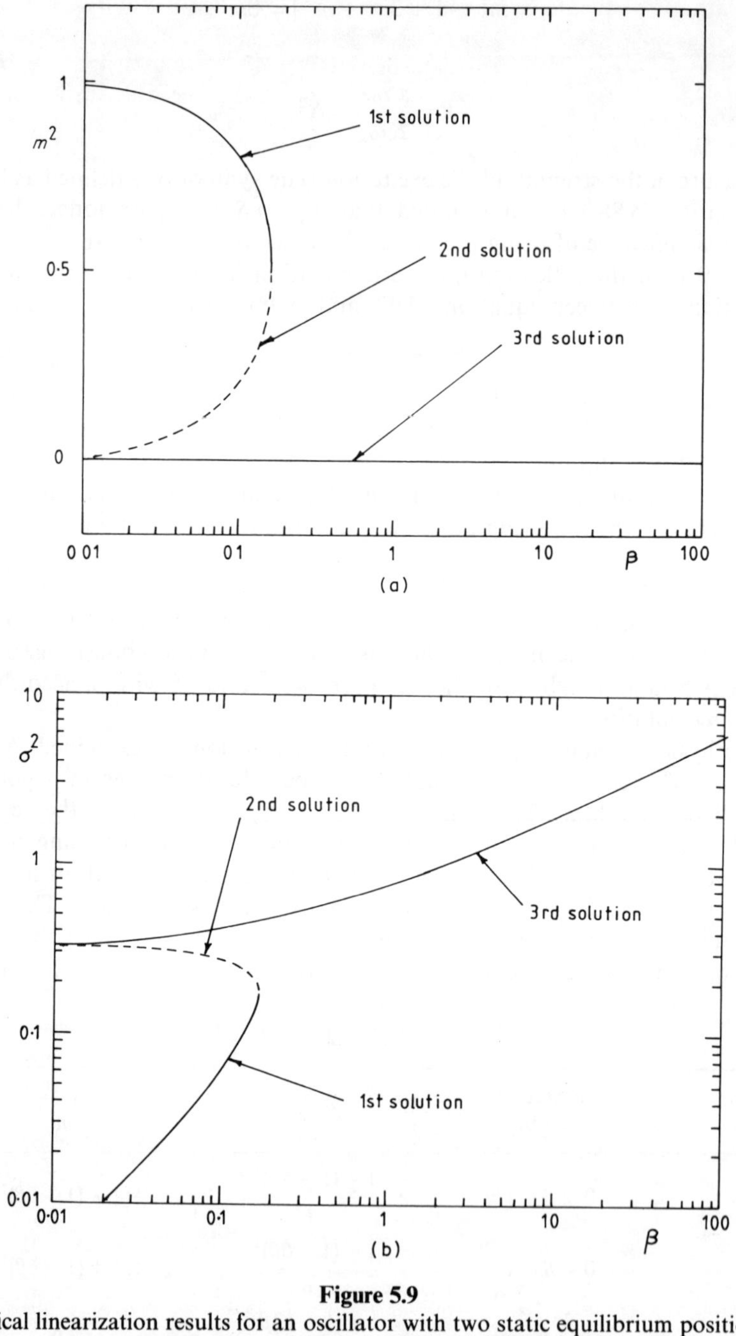

Figure 5.9
Statistical linearization results for an oscillator with two static equilibrium positions. (a) Variation of m^2 with β. (b) Variation of σ^2 with β

reasonable to associate the third solution with the 'true' stationary response statistics relating to sample functions of infinite duration. In the very long term the response will spend, on average, an equal time in each well, no matter what the level of excitation, and the variance will be correspondingly higher than that relating to oscillations within one of the wells. This is confirmed by the fact that the third solution is valid over the whole range of β and converges towards the normal statistical linearization result for an oscillator with cubic stiffness (i.e., $\sigma^2 \to (\beta/3)^{1/2}$) at high β values. The second solution appears to have much less physical significance. It converges to the first solution as $\beta \to \frac{1}{6}$, and to the third solution as $\beta \to 0$. However, in the intermediate range of β, σ^2 decreases as β increases, which is not physically realistic.

One can now infer that the existence of three solutions, for $\beta < \frac{1}{6}$, reflects the fact that jumping, of the type shown in Figure 5.8c, is clearly discernible in sample functions of the response. In these circumstances it is meaningful to consider both quasi-stationary statistics (motion within one well) and normal, stationary statistics (motion in both wells). The existence of a finite regime of triple solutions is a reflection of the fact that discernible jumping in sample functions will only occur if β is low enough. The statistical linearization method gives a convenient critical value of $\beta = 1/6$, below which jumps are well defined.

The above argument is, to some extent, conjectural in nature but is strongly supported by digital simulation studies (see Chapter 10). It is also supported by a comparison with the exact solution which is available for this case; this relates to the 'true', infinite duration statistics (see Chapter 10).

For systems with more than two stable static equilibrium positions then the analysis is, of course, more complicated, but similar conclusions relating to the physical meaning of multiple solutions can still be drawn. A detailed discussion of an oscillator with three potential wells has been given recently by Langley (1988b).

5.3.9 Response to narrow-band excitation

It is well known that, if an oscillator with a hardening non-linear stiffness is subjected to sinusoidal excitation, it is possible for the oscillator response to exhibit the phenomenon of sharp jumps in amplitude (Den Hartog, 1956). This jumping behaviour is associated with the fact that, over a range of excitation frequency, ω, the response amplitude, A, is triple-valued. Thus, a plot of response amplitude against frequency will appear as sketched in Figure 5.10a, with a triple-valued solution for A in the frequency range ω_1 to ω_2. It can be demonstrated that the section BC of the A against ω curve is associated with an unstable solution. This has the effect that, if the frequency of the excitation is gradually increased, the response follows the path shown in Figure 5.10b, with an abrupt jump in amplitude from B to D. For decreasing frequency the

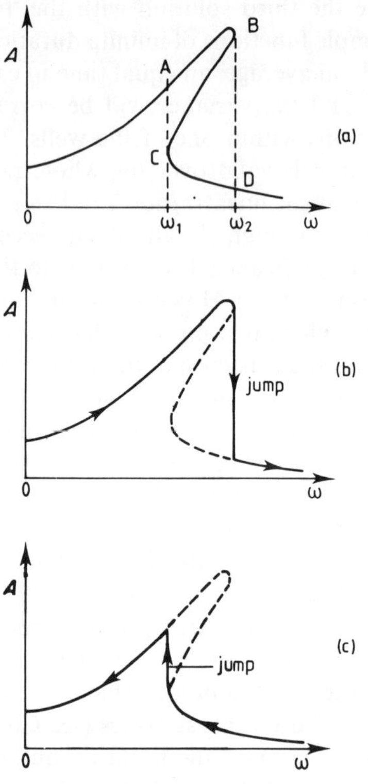

Figure 5.10
Amplitude response characteristic of an oscillator with a hardening non-linear stiffness. (a) Variation of response amplitude with frequency of excitation. (b) Effect of increasing the frequency of excitation. (c) Effect of decreasing the frequency of excitation

path followed is indicated in Figure 5.10c; there is an abrupt jump from C to A.

The same jumping phenomenon can occur if an oscillator with hardening non-linear stiffness is subjected to narrow-band random excitation, provided that the band-width of the excitation is sufficiently small. Jumps in response amplitude, with random excitation, were first observed experimentally by Lyon *et al.* (1961). They have been studied theoretically by a number of authors (Lyon *et al.*, 1961; Iyengar, 1975; Richard and Anand, 1983; Dimentberg, 1970; Davies and Nandlall, 1986; Lennox and Kuak, 1976). In particular, it has been demonstrated that the technique of statistical linearization may be used to investigate the conditions under which jumps may occur, when the excitation is random.

To demonstrate this application of statistical linearization the specific case of the Duffing oscillator, with the equation of motion given by equation (5.60) will be considered. Here the excitation process, $f(t)$, will be assumed to be generated by passing white noise through a linear, second-order filter. Thus $f(t)$ is governed by an equation of the form

$$\ddot{f} + 2\xi\Omega\dot{f} + \Omega^2 f = \Omega^2 n(t) \tag{5.102}$$

where $n(t)$ is a white noise with zero mean and a constant spectral level, S_0, i.e.

$$E\{n(t)n(t+\tau)\} = 2\pi S_0 \delta(\tau) \tag{5.103}$$

If the parameter ξ in equation (5.102) is small ($\xi \ll 1$) then the spectrum of $f(t)$ will exhibit a single sharp peak, at a frequency close to Ω. Thus, for small ξ, $f(t)$ as given by equation (5.102), is a suitable narrow-band excitation process, for the purposes of investigating the jump phenomenon. The scaling is such that, if $\Omega \gg \omega_\mathrm{n}$, then the spectrum of $f(t)$ is flat in the neighborhood of the frequency ω_n and $f(t)$ is effectively a white noise, with spectral level S_0. It follows that, for very large values of Ω, the results obtained here should correspond to those found earlier (in Section 5.3.2) for white noise excitation.

For the Duffing oscillator the natural frequency of the equivalent linear system, ω_eq (see equation (5.39)) may be found from equation (5.50), by setting $\alpha_3 = 1$ and $\alpha_n = 0$, for $n \neq 3$. Therefore

$$\omega_\mathrm{eq}^2 = \omega_\mathrm{n}^2(1 + 3\lambda\sigma_y^2) \tag{5.104}$$

Clearly this is a result is independent of the shape of the spectrum of $f(t)$. An expression for σ_y^2 can be found by using equations (5.53) to (5.55), where the appropriate expression for $\alpha(\omega)$ may be found from equations (5.38) and (5.102). Thus

$$\alpha(\omega) = \frac{\Omega^2}{(-\omega^2 + 2\mathrm{i}\xi\Omega\omega + \Omega^2)(-\omega^2 + 2\mathrm{i}\zeta\omega_\mathrm{n}\omega + \omega_\mathrm{eq}^2)} \tag{5.105}$$

Here the integral equation (5.55) may be evaluated by using the theory in Appendix B. The result may be written in the following form (Davies and Nandlall, 1986):

$$\frac{\sigma_y^2}{\sigma_f^2} = \frac{[\omega_\mathrm{eq}^2\beta + \gamma(\Omega^2 + \beta^2) + \gamma^2\beta]}{\omega_\mathrm{eq}^2\beta[(\Omega^2 - \omega_\mathrm{eq}^2)^2 + (\beta + \gamma)(\beta\Omega^2 + \gamma\omega_\mathrm{eq}^2)]} \tag{5.106}$$

Here σ_f is the standard deviation of $f(t)$, given by

$$\sigma_f^2 = \frac{\pi S_0 \Omega^2}{\gamma} \tag{5.107}$$

$$\beta = 2\zeta_\mathrm{eq}\omega_\mathrm{eq} = 2\zeta\omega_\mathrm{n} \tag{5.108}$$

and

$$\gamma = 2\zeta\Omega \tag{5.109}$$

There are two interesting limiting cases of equation (5.106). Firstly, if $\Omega \gg \omega_\mathrm{eq}$, then $f(t)$ is effectively a white noise and equation (5.106) approaches the limit

$$\sigma_y^2 = \frac{\pi S_0}{\beta \omega_{\mathrm{eq}}^2} \tag{5.110}$$

This is in agreement with the result for white noise excitation found earlier (see equation (5.56)). Secondly, if the excitation bandwidth is very small, so that $\gamma \to 0$, then equation (5.106) approaches the limit

$$\frac{\sigma_y}{\sigma_f} = \left(\frac{1}{(\Omega^2 - \omega_{\mathrm{eq}}^2)^2 + \beta^2 \Omega^2} \right)^{1/2} \tag{5.111}$$

Here the right-hand side of the equation is equal to the ratio of the output to input amplitudes, for the equivalent linear system driven by a *sinusoidal* input. It can be concluded that the results obtained from the present analysis will coincide with those found for sinusoidal excitation, as the input bandwidth approaches zero.

If ω_{eq} is eliminated between equations (5.104) and (5.106) then a quadratic equation for σ_y^2 is obtained. By examining the signs of the terms in this equation it is possible to deduce that, of the four roots,

(a) one must be real and negative
(b) one must be real and positive
(c) two may be real, or complex conjugates.

Since σ_y^2 must, of course, be real and positive, it follows that, for a given set of parameters, there is either a single value for σ_y^2, or three values. As in the case of sinusoidal excitation one can associate the triple-valued solution with the occurrence of jumps in the response process. However, it should be borne in mind that the response process can have only *one* value of σ_y^2. Thus, the occurrence of a triple valued solution for σ_y^2, when the statistical linearization technique is applied, must be viewed as only an *indication* that jumps will occur in sample functions of the response.

In fact the correspondence between the occurrence of multiple solutions from the statistical linearization procedure, and the occurrence of jumps in sample functions of the response, is similar to that discussed in the previous section, for the case of an oscillator with two static equilibrium positions. Thus, if jumps are clearly discernible in sample functions of the response then it is reasonable to consider quasi-stationary statistics, relating to short-term behaviour. The existence of multiple solutions mirrors the fact that such short-term statistics are physically meaningful, in circumstances where pronounced jumping occurs.

A numerical analysis is necessary to determine the parameter space, where $\lambda, \beta, \gamma, \Omega$ and ω_{n} are the relevant parameters for which triple-valued solutions for σ_y^2 occur. Studies have shown (Dimentberg, 1970) that the input bandwidth, governed by γ, is a critical parameter. Figure 5.11 sketches a typical variation

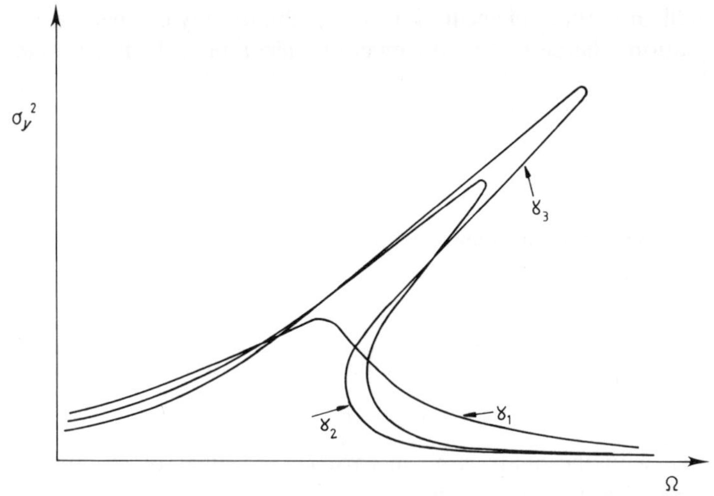

Figure 5.11
Variation of the mean square response with Ω for various input bandwidths.
$\gamma_3 < \gamma_2 < \gamma_1$

of σ_y^2 with Ω, for various values of γ, with ω_n, λ and β held constant. For very small γ values a triple-valued frequency range can exist; the magnitude of this range decreases as γ increases until a critical value, γ^*, is reached. For $\gamma > \gamma^*$ only single-valued solutions for σ_y^2 exist.

A detailed study of the influence of parameter variations on the magnitude of σ_y^2 has been undertaken by Dimentberg (1970), who has also compared the results from the statistical linearization analysis with the findings of an analogue simulation study. This comparison suggests that the method of statistical linearization yields an acceptable quantitative estimate for the boundary of the region, in parameter space, where jumping exists, provided that the excitation is not too severe.

5.4 OSCILLATORS WITH NON-LINEAR STIFFNESS AND DAMPING

For many mechanical oscillators non-linearity is present in both damping and stiffness. A suitable general form for the equation of motion of such a system is as follows

$$\ddot{y} + g(y, \dot{y}) = f(t) \tag{5.112}$$

For simplicity it will be assumed, initially, that $f(t)$ has zero mean and that $g(y, \dot{y})$ is an antisymmetric function, such that the mean of $y(t)$ is also zero.

For small amplitudes of oscillation, $g(y, \dot{y})$ can usually be linearised, to a good approximation. Therefore it is convenient to decompose the function $g(y, \dot{y})$ into linear damping and stiffness components, $\beta\dot{y}$ (where $\beta = 2\zeta\omega_n$) and $\omega_n^2 y$ respectively, together with a non-linear component, $\mu G(y, \dot{y})$, where μ is a scaling factor. Thus

$$g(y, \dot{y}) = \beta\dot{y} + \omega_n^2 y + \mu G(y, \dot{y}) \tag{5.113}$$

To apply the statistical linearization procedure the non-linear system (5.112) is replaced by an equivalent linear system, with an equation of motion

$$\ddot{y} + \beta_{eq}\dot{y} + \omega_{eq}^2 y = f(t) \tag{5.114}$$

where $\beta_{eq} = 2\zeta_{eq}\omega_{eq}$. The difference ε between equations (5.112) and (5.114) is

$$\varepsilon = g(y, \dot{y}) - \beta_{eq}\dot{y} - \omega_{eq}^2 y \tag{5.115}$$

and, in line with the analysis given earlier, the values of β_{eq} and ω_{eq} which minimize $E\{\varepsilon^2\}$ are required. Evidently these values must satisfy the following relationships

$$\frac{\partial}{\partial \beta_{eq}} E\{\varepsilon^2\} = 0 \quad \frac{\partial}{\partial \omega_{eq}^2} E\{\varepsilon^2\} = 0 \tag{5.116}$$

On substituting (5.115) into equations (5.116), and carrying out the differentiation, the following two simultaneous equations for β_{eq} and ω_{eq} emerge

$$E\{\dot{y}g(y, \dot{y})\} - \beta_{eq}E\{\dot{y}^2\} - \omega_{eq}^2 E\{y\dot{y}\} = 0 \tag{5.117}$$

$$E\{yg(y, \dot{y})\} - \beta_{eq}E\{y\dot{y}\} - \omega_{eq}^2 E\{y^2\} = 0 \tag{5.118}$$

Since $y(t)$ is a stationary process, with zero mean, it follows that (see Section 3.7)

$$E\{y\dot{y}\} = 0 \quad E\{y^2\} = \sigma_y^2 \quad E\{\dot{y}^2\} = \sigma_{\dot{y}}^2 \tag{5.119}$$

where σ_y and $\sigma_{\dot{y}}$ are, respectively, the standard deviations of $y(t)$ and $\dot{y}(t)$. With these substitutions, equations (5.117) and (5.118) may be easily solved, to yield the following expressions for β_{eq} and ω_{eq}^2

$$\beta_{eq} = \frac{E\{\dot{y}g(y, \dot{y})\}}{\sigma_{\dot{y}}^2} \quad \omega_{eq}^2 = \frac{E\{yg(y, \dot{y})\}}{\sigma_y^2} \tag{5.120}$$

If, following the approach described earlier, the assumption that $y(t)$ and $\dot{y}(t)$ are jointly Gaussian is made, alternative and simpler expressions for β_{eq} and ω_{eq}^2 can be found. Specifically using again the result given in Chapter 3 for Gaussian random variables (see equation (3.56)), it is easily found that

$$\beta_{eq} = E\left\{\frac{\partial g(y, \dot{y})}{\partial \dot{y}}\right\} \quad \omega_{eq}^2 = E\left\{\frac{\partial g(y, \dot{y})}{\partial y}\right\} \tag{5.121}$$

Thus β_{eq} and ω_{eq}^2 are simply equal to the expectations of the gradients of the function $g(y, \dot{y})$, with respect to $\dot{y}$ and y, respectively.

With $g(y, \dot{y})$ in the form of equation (5.113), application of equation (5.121) shows that

$$\beta_{\mathrm{eq}} = \beta + \mu E\left\{\frac{\partial G}{\partial \dot{y}}\right\} \tag{5.122}$$

and

$$\omega_{\mathrm{eq}}^2 = \omega_{\mathrm{n}}^2 + \mu E\left\{\frac{\partial G}{\partial y}\right\} \tag{5.123}$$

In the special case where $G(y, \dot{y})$ depends only on y it is found, from the above, that $\beta_{\mathrm{eq}} = \beta$ and that equation (5.123) is in agreement with the result for non-linear stiffness only, found earlier (see equation (5.47), where here $\lambda\omega_{\mathrm{n}}^2 = \mu$).

Frequently, in engineering applications, it is possible to represent the non-linear function $G(y, \dot{y})$ in a separable form, as follows

$$G(y, \dot{y}) = G_1(y) + G_2(\dot{y}) \tag{5.124}$$

In this case equations (5.122) and (5.123) reduce to

$$\beta_{\mathrm{eq}} = \beta + \mu E\left\{\frac{\mathrm{d}G_2}{\mathrm{d}\dot{y}}\right\} \tag{5.125}$$

$$\omega_{\mathrm{eq}}^2 = \omega_{\mathrm{n}}^2 + \mu E\left\{\frac{\mathrm{d}G_1}{\mathrm{d}y}\right\} \tag{5.126}$$

The evaluation of β_{eq} and ω_{eq}^2, according to the above formulae, depends on a knowledge of σ_y and $\sigma_{\dot{y}}$. Further relationships between σ_y and $\sigma_{\dot{y}}$, and ω_{eq} and β_{eq}, may be obtained by using the equivalent linear system to compute σ_y and $\sigma_{\dot{y}}$. Specifically, relying on spectral input–output relationships again, one has

$$\sigma_y^2 = \int_{-\infty}^{\infty} |\alpha(\omega)|^2 S_f(\omega)\,\mathrm{d}\omega \tag{5.127}$$

and (see equation (3.113))

$$\sigma_{\dot{y}}^2 = \int_{-\infty}^{\infty} |\alpha(\omega)|^2 \omega^2 S_f(\omega)\,\mathrm{d}\omega \tag{5.128}$$

where

$$\alpha(\omega) = \frac{1}{(\omega_{\mathrm{eq}}^2 - \omega^2 + \mathrm{i}\beta_{\mathrm{eq}}\omega)} \tag{5.129}$$

Therefore, four simultaneous equations can be found for the four unknowns, $\beta_{\mathrm{eq}}, \omega_{\mathrm{eq}}, \sigma_y$ and $\sigma_{\dot{y}}$. A numerical solution procedure will usually be necessary, in specific cases.

5.4.1 Standard deviation of the response

To enable specific results to be obtained, the case where the non-linearity function, $G(y, \dot{y})$, is of the separable form, and can be represented as the power series

$$G(y, \dot{y}) = \sum_{n \text{ odd}} \alpha_n y^n + \sum_{n \text{ odd}} \beta_n \dot{y}^n \tag{5.130}$$

will now be considered. Here the α_n and $\beta_n (n = 3, 5, \ldots)$ are constants and the restriction to odd integer values of n ensures that $G(y, \dot{y})$ is antisymmetric.

Assuming that both $y(t)$ and $\dot{y}(t)$ are Gaussian, with zero means, a substitution of equation (5.130) into equations (5.122) and (5.123) leads to the following expressions for β_{eq} and ω_{eq}^2 (see Appendix A, Table A1)

$$\beta_{\text{eq}} = \beta + \mu\kappa_1 \tag{5.131}$$

$$\omega_{\text{eq}}^2 = \omega_{\text{n}}^2 + \mu\kappa_2 \tag{5.132}$$

where

$$\kappa_1 = \frac{1}{\pi^{1/2}} \sum_{n \text{ odd}} n\beta_n (\sqrt{2}\sigma_{\dot{y}})^{n-1} \Gamma\left(\frac{n}{2}\right) \tag{5.133}$$

and

$$\kappa_2 = \frac{1}{\pi^{1/2}} \sum_{n \text{ odd}} n\alpha_n (\sqrt{2}\sigma_y)^{n-1} \Gamma\left(\frac{n}{2}\right) \tag{5.134}$$

These equations must be combined with expressions for σ_y and $\sigma_{\dot{y}}$, deduced from the equivalent linear system. To illustrate this, the case where $f(t)$ is modelled as a white noise, with spectral level S_0 will be considered. Here, using equations (5.127) to (5.129), it is found that

$$\sigma_y^2 = \frac{\pi S_0}{\beta_{\text{eq}} \omega_{\text{eq}}^2} \qquad \sigma_{\dot{y}}^2 = \frac{\pi S_0}{\beta_{\text{eq}}} \tag{5.135}$$

There are now four equations in the four unknowns, $\beta_{\text{eq}}, \omega_{\text{eq}}, \sigma_y$ and $\sigma_{\dot{y}}$, enabling these quantities to be evaluated. A combination of these four equations enables the following two relationships to be established

$$\sigma_{\dot{y}}^2 \left(1 + \frac{\mu}{\beta}\kappa_1\right) = \sigma_{\dot{y}\text{o}}^2 \tag{5.136}$$

and

$$\sigma_y^2 \left(1 + \frac{\mu}{\omega_{\text{n}}^2}\kappa_2\right) \frac{\beta_{\text{eq}}}{\beta} = \sigma_{y\text{o}}^2 \tag{5.137}$$

where $\sigma_{y\text{o}}$ and $\sigma_{\dot{y}\text{o}}$ are the values of σ_y and $\sigma_{\dot{y}}$, respectively, as $\mu \to 0$. It is noted that, since κ_1 is dependent only on $\sigma_{\dot{y}}$, it is possible to solve the first of these equations, independently of the second. This enables $\sigma_{\dot{y}}$, and hence β_{eq} to be found. The second equation can then be solved to find σ_y, and hence ω_{eq}.

Frequently, in applications, a good approximation may be achieved by retaining only the cubic terms in the separable form of $G(y, \dot{y})$, expressed as a series. One then has a Duffing type oscillator, with linear-plus-cubic damping. The corresponding the equation of motion is

$$\ddot{y} + \beta(\dot{y} + \eta\dot{y}^3) + \omega_n^2(y + \lambda y^3) = f(t) \tag{5.138}$$

Further, comparison with equation (5.130) shows that

$$\omega_n^2\lambda = \mu\alpha_3 \quad \beta\eta = \mu\beta_3 \tag{5.139}$$

When $\eta = 0$, equation (5.138) reduces to the equation of motion considered earlier (in Section 5.3.2).

For this particular oscillator it is found, from equations (5.133) and (5.134), that

$$\kappa_1 = 3\beta_3\sigma_{\dot{y}}^2 = 3\beta\eta\sigma_{\dot{y}}^2/\mu \tag{5.140}$$

and

$$\kappa_2 = 3\alpha_3\sigma_y^2 = 3\omega_n^2\lambda\sigma_y^2/\mu \tag{5.141}$$

Hence, from equations (5.136) and (5.137)

$$\sigma_{\dot{y}}^2(1 + 3\eta\sigma_{\dot{y}}^2) = \sigma_{\dot{y}o}^2 \tag{5.142}$$

and

$$\sigma_y^2(1 + 3\lambda\sigma_y^2)\frac{\beta_{eq}}{\beta} = \sigma_{yo}^2 \tag{5.143}$$

From the first of these two equations it follows that

$$\frac{\sigma_{\dot{y}}^2}{\sigma_{\dot{y}o}^2} = \frac{\beta}{\beta_{eq}} = \frac{(1 + 12H)^{1/2} - 1}{6H} \tag{5.144}$$

where

$$H = \eta\sigma_{\dot{y}o}^2 \tag{5.145}$$

With this result the second equation may be solved, to give

$$\frac{\sigma_y^2}{\sigma_{yo}^2} = \left(\frac{(1 + 12\Psi)^{1/2} - 1}{6\Psi}\right)\frac{\beta}{\beta_{eq}} \tag{5.146}$$

where

$$\Psi = \lambda\sigma_y^2\beta/\beta_{eq} \tag{5.147}$$

It is interesting to note that the result for $\sigma_{\dot{y}}^2$, and hence β_{eq}, is independent of the non-linear stiffness parameter, λ, whereas the result for σ_y^2, and hence ω_{eq}^2, does depend on the non-linear damping parameter, η. In the case where $\eta = 0, \beta_{eq} = \beta$ and the result for σ_y^2, and hence ω_{eq}^2, is the same as that found earlier for the Duffing oscillator with linear damping (see equations (5.62) and (5.63)).

5.4.2 The case of small non-linearity

For oscillators with combined non-linear stiffness and damping, the solution can be considerably simplified (as in the case of non-linear stiffness only), if the non-linearity parameter μ is sufficiently small for terms of order μ^2, and higher order, to be considered as negligible.

As an illustration, the case where $G(y, \dot{y})$ is of the separable, series form given by equation (5.130), and the excitation is a white noise may be considered. First σ_y^2 and $\sigma_{\dot{y}}^2$ are expressed as perturbation series in μ, i.e.

$$\sigma_y^2 = \sigma_{yo}^2 + \mu\sigma_{y_1}^2 + \cdots \tag{5.148}$$

and

$$\sigma_{\dot{y}}^2 = \sigma_{\dot{y}o}^2 + \mu\sigma_{\dot{y}_1}^2 + \cdots \tag{5.149}$$

Next, substituting into equations (5.136) and (5.137), and neglecting terms of order μ^2, and higher, it is found that

$$\sigma_{\dot{y}}^2 = \sigma_{\dot{y}o}^2\left(1 - \frac{\mu}{\beta}\kappa_{10}\right) \tag{5.150}$$

and

$$\sigma_y^2 = \sigma_{yo}^2\left(1 - \frac{\mu}{\beta}\kappa_{10} - \frac{\mu}{\omega_n^2}\kappa_{20}\right) \tag{5.151}$$

where

$$\kappa_{10} = \frac{1}{\pi^{1/2}}\sum_{n\,\text{odd}} n\beta_n(\sqrt{2}\sigma_{\dot{y}o})^{n-1}\Gamma\left(\frac{n}{2}\right) \tag{5.152}$$

$$\kappa_{20} = \frac{1}{\pi^{1/2}}\sum_{n\,\text{odd}} n\alpha_n(\sqrt{2}\sigma_{yo})^{n-1}\Gamma\left(\frac{n}{2}\right) \tag{5.153}$$

Evidently, from these expressions, σ_y and $\sigma_{\dot{y}}$ may be evaluated directly, without the need to solve non-linear algebraic equations. Similarly, expressions for ω_{eq} and β_{eq}, may be readily found, correct to order μ.

For the special case where the system is governed by equation (5.138), equations (5.150) and (5.151) reduce to

$$\frac{\sigma_{\dot{y}}^2}{\sigma_{\dot{y}o}^2} = 1 - 3H \quad \frac{\sigma_y^2}{\sigma_{yo}^2} = 1 - 3(H + \Psi) \tag{5.154}$$

where

$$\Psi = \lambda\sigma_{yo}^2 \tag{5.155}$$

These results are consistent with equations (5.144) and (5.146), for arbitrary values of H and Ψ, as may be readily shown.

Appropriate expressions for ω_{eq} and β_{eq} can be found by returning to equations (5.131) and (5.132). Thus, with $G(y, \dot{y})$ in the present example of the separable form given by equation (5.130), it is found, by neglecting terms of order higher than μ, that

$$\beta_{eq} = \beta + \mu\kappa_{10} \quad \omega_{eq}^2 = \omega_n^2\mu\kappa_{20} \tag{5.156}$$

In the special case of a Duffing oscillator with linear-plus-cubic damping, excited by white noise, it follows that

$$\beta = \beta(1 + 3H) \quad \omega_{eq}^2 = \omega_n^2(1 + 3\Psi) \tag{5.157}$$

correct to order μ.

5.4.3 Power spectrum of the response

From the equivalent linear system equation (5.114) the following expression for the power spectrum of the response process, $y(t)$, is readily obtained

$$S_y(\omega) = \frac{S_f(\omega)}{[(\omega_{eq}^2 - \omega^2)^2 + \beta_{eq}^2\omega^2]} \tag{5.158}$$

For the case of a small value of the non-linearity parameter, μ, an approximation for $S_y(\omega)$ may be derived, correct to order μ, without difficulty. In the case where $G(y, \dot{y})$ is given by the separable, series form of equation (5.130), it is found that

$$S_y(\omega) = \frac{S_f(\omega)}{D} - \frac{2\mu[(\omega_n^2 - \omega^2)\kappa_{20} + \omega^2\beta\kappa_{10}]S_f(\omega)}{D^2} \tag{5.159}$$

where

$$D = (\omega_n^2 - \omega^2)^2 + \beta^2\omega^2 \tag{5.160}$$

Here the first term on the right-hand side of equation (5.159) is the linear result (for $\mu = 0$), and the second term represents the adjustment due to non-linearity, correct to order μ. In the special case of white noise excitation, and linear damping, equation (5.159) reduces to the result found earlier (see equation (5.71)).

5.4.4 Input and output with non-zero means

The analysis given here for oscillators with combined non-linear damping and stiffness can be readily extended to the more general case where the mean of the input and the mean of the output process are both non-zero. As in the case of non-linear stiffness only (see Section 5.3.6) $f(t)$ and $y(t)$ are written in the form of equations (5.73) and (5.74), respectively. The appropriate linear equivalent for equation (5.112) is then

$$\ddot{y} + a_1 y_0 + a_2 \dot{y}_0 + b = f(t) \tag{5.161}$$

Taking expectations of both sides of this equation it becomes evident that

$$b = m_f \tag{5.162}$$

Hence, the linear equation may be written as

$$\ddot{y}_0 + a_1 y_0 + a_2 \dot{y}_0 = f_0(t) \tag{5.163}$$

A comparison with equation (5.114) shows that

$$a_1 = \omega_{\text{eq}}^2 \quad a_2 = \beta_{\text{eq}} \tag{5.164}$$

where β_{eq} and ω_{eq} have the same meaning as before.

On following through with an analysis similar to that given in Section 5.4, and assuming that y and $\dot{y}$ are Gaussian, it is found that the least square minimization criterion gives the following expressions for $\beta_{\text{eq}}, \omega_{\text{eq}}$, and b

$$\beta_{\text{eq}} = E\left\{\frac{\partial g}{\partial \dot{y}}\right\} \quad \omega_{\text{eq}}^2 = E\left\{\frac{\partial g}{\partial y}\right\} \tag{5.165}$$

and

$$b = m_f = E\{g(y, \dot{y})\} \tag{5.166}$$

These expressions, together with expressions for σ_y and $\sigma_{\dot{y}}$, deduced from the equivalent linear system, provide five equations for the five unknowns, $\beta_{\text{eq}}, \omega_{\text{eq}}, \sigma_y$ and $\sigma_{\dot{y}}$ and m_y (noting that the mean of $\dot{y}(t)$ is zero). It is observed that the expressions for β_{eq} and ω_{eq}^2, given above, do not involve m_y, explicitly. However m_y is involved implicitly, through the distribution of y and $\dot{y}$.

5.5 HIGHER ORDER LINEARIZATION

In the discussion so far, linearization of non-linear oscillators has been achieved by replacing the non-linear terms in the equation of motion by zero-memory linear terms. As pointed out in Section 5.2.2, this procedure actually corresponds to optimal linearization provided that the 'input' is Gaussian. However, in the case of a non-linear oscillator the relevant 'input' to the non-linear term, $g(y, \dot{y})$, in equation (5.112)) is the response process, $y(t)$, which is often distinctly non-Gaussian. Thus, when dealing with non-linear oscillators a conventional, zero-memory linear substitution for the non-linear terms is not necessarily optimal and increased accuracy may be obtained through a judicious introduction of memory into the linear substitution.

One simple approach to introducing memory has been given recently by Iyengar (1988). Suppose one starts with the general form of equation of motion given by equation (5.112), with $g(y, \dot{y})$ expressed by equation (5.113). In the case where the non-linearity is of separable form (see equation (5.124)) the governing equation may be written as

$$\ddot{y} + \beta\dot{y} + \omega_{\mathrm{n}}^2 y + \mu z = f(t) \tag{5.167}$$

where

$$z = G_1(y) + G_2(\dot{y}) \tag{5.168}$$

If equation (5.168) is differentiated with respect to time then one finds that

$$\dot{z} = \frac{\mathrm{d}G_1}{\mathrm{d}y}\dot{y} + \frac{\mathrm{d}G_2}{\mathrm{d}\dot{y}}\ddot{y} \tag{5.169}$$

Eliminating $\ddot{y}$ between equations (5.167) and (5.169) gives

$$\dot{z} + \mu\frac{\mathrm{d}G_2}{\mathrm{d}\dot{y}}z = \frac{\mathrm{d}G_1}{\mathrm{d}y}\dot{y} + \frac{\mathrm{d}G_2}{\mathrm{d}\dot{y}}(-\beta\dot{y} - \omega_{\mathrm{n}}^2 y + f(t)) \tag{5.170}$$

Equation (5.170) may be replaced by the following equivalent linear equation

$$\dot{z} + C_1 z = C_2 f(t) + C_3 y + C_4\dot{y} \tag{5.171}$$

where C_1 to C_4 are constants chosen to minimise the mean square error between equations (5.170) and (5.171). A general procedure for this purpose will be given in the next chapter. The combined equivalent linear system, represented by equations (5.167) and (5.171) is now of third order, i.e. of one order of greater than the original non-linear system.

Linearization of higher order can be obtained by an *ad hoc* extension of the above procedure. Iyengar (1988) has shown that, for the case of a Duffing oscillator driven by white noise, a fourth-order equivalent linear system leads to a significant improvement in accuracy, with regard to the estimation of the mean square of the response. Moreover, the power spectrum obtained from the fourth-order linear substitute equation yields an estimate of the power spectrum of the response which shows two peaks, reflecting the existence of sub-harmonics in the system. The secondary resonance occurs at about three times the primary resonance frequency, in reasonable agreement with digital simulation results.

A rational general procedure for developing equivalent linear systems of higher order than the original system does not seem to be available. This is an area which merits further study.

5.6 APPLICATIONS

5.6.1 Friction controlled slip of a structure on a foundation

One method of protecting a structure from earthquake ground motion, proposed by Williams (1973), is to allow the building to slip, on a frictional foundation. If μ is the coefficient of friction, and g is the acceleration due to gravity, then the structure will be protected from accelerations with a magnitude greater than μg. Clearly, when designing such an isolation system, it is necessary to predict the statistical features of the slip displacement for a given probabilistic earthquake excitation model. This problem has been addressed recently by a number of workers (Crandall *et al.*, 1974, Crandall and Lee, 1976; Ahmadi, 1983; Constantinou and Tadjbaksh, 1984; Noguchi, 1985).

Here a basic version of this problem is studied, using the statistical linearization method (Constantinou and Tadjbaksh, 1984; Noguchi 1985). A structure of mass m, assumed to be effectively rigid, rests on a frictional foundation, as shown in Figure 5.12. It will be supposed that random, earthquake excitation results in a time-varying horizontal displacement of the foundation, $x_g(t)$. If $y(t)$ is the horizontal, absolute displacement of the structure, and $F(t)$ is the frictional force between the foundation and the structure, then, the equation of motion of the structure is as follows

$$m\ddot{y} = -F(t) \tag{5.172}$$

The frictional force will have a Coulomb characteristic and can be modelled, approximately, as shown in Figure 5.13 (see Chapter 2). Thus

$$F(t) = \mu g \operatorname{sgn}|\dot{q}| \tag{5.173}$$

where q is the relative, slip displacement between the mass and the foundation; i.e.

$$q = y - x_g \tag{5.174}$$

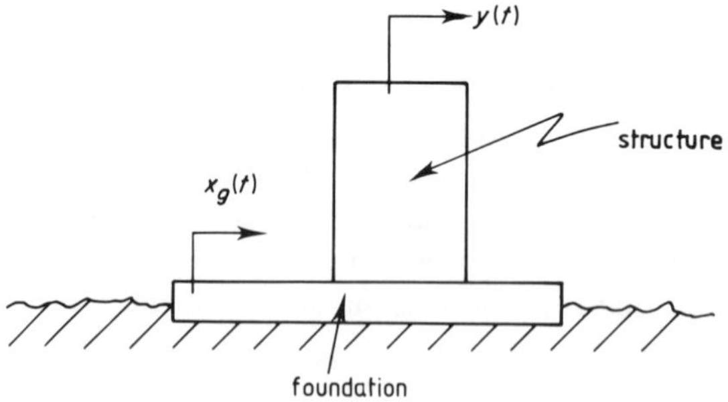

Figure 5.12
A structure resting on a frictional foundation

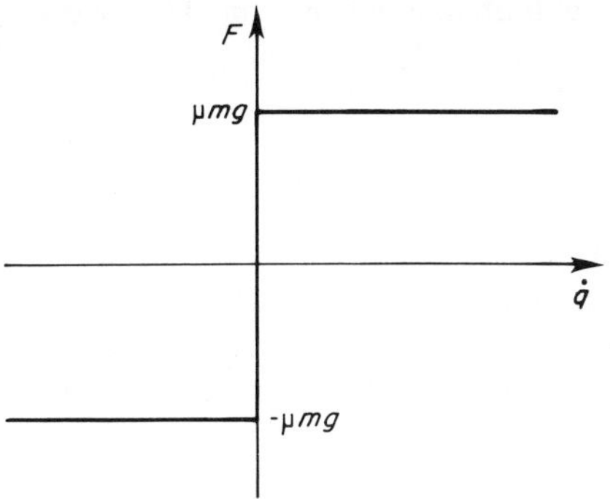

Figure 5.13
Idealized Coulomb frictional force against velocity characteristic

Combining equations (5.172) to (5.174) one has the following equation for the slip displacement

$$\ddot{q} + \mu g \,\mathrm{sgn}\,(\dot{q}) = -\ddot{x}_g \tag{5.175}$$

This equation is, of course, only valid during the intervals when slipping occurs. In what follows it will be assumed that the intensity of the random excitation, $\ddot{x}_g(t)$, is sufficiently high to ensure that the probability of sticking is negligible. In this circumstance, equation (5.175) can be assumed to be valid for all time. It will also be assumed, here, that $\ddot{x}_g(t)$ can be modelled as a stationary random process, with a zero mean. It will be shown in Chapter 7 how the analysis can be extended to incorporate a more realistic, non-stationary model of the input process.

Application of the statistical linearization to the present problem involves replacing the highly non-linear equation (5.175) with the linear equation

$$\ddot{q} + \beta_{\mathrm{eq}}\dot{q} = -\ddot{x}_g(t) \tag{5.176}$$

Thus the non-linear Coulomb friction term is replaced by a linear, viscous damping term. On minimizing the mean square of the error

$$\varepsilon = \mu g \,\mathrm{sgn}\,(\dot{q}) - \beta_{\mathrm{eq}}\dot{q} \tag{5.177}$$

with respect to β_{eq}, the following expression for β_{eq} is obtained

$$\beta_{\mathrm{eq}} = \frac{\mu g E\{\dot{q}\,\mathrm{sgn}\,(\dot{q})\}}{\sigma_{\dot{q}}^2} = \mu g E\left\{\frac{\mathrm{d}}{\mathrm{d}\dot{q}}(\mathrm{sgn}\,(\dot{q}))\right\} \tag{5.178}$$

where

$$\sigma_{\dot{q}}^2 = E\{\dot{q}^2\} \tag{5.179}$$

is the mean square of $\dot{q}$, and $\dot{q}$ is assumed to be Gaussian. The required expectation is readily evaluated, in terms of $\sigma_{\dot{q}}$. Hence one obtains (see Appendix A, Table A1)

$$\beta_{eq} = \left(\frac{2}{\pi}\right)^{1/2} \frac{\mu g}{\sigma_{\dot{q}}} \tag{5.180}$$

To proceed further it is necessary to use equation (5.176) to obtain another relationship between β_{eq} and $\sigma_{\dot{q}}$. For this purpose a specification of the spectrum, $S_g(\omega)$, of the excitation process is necessary. One can then use equations (5.53) and (5.55) to evaluate $\sigma_{\dot{q}}$, in terms of β_{eq}. Here the appropriate frequency response function is obtained by substituting $\ddot{x}_g = \exp(i\omega t), \dot{q} = \alpha(\omega)\exp(i\omega t)$ into equation (5.176). Hence

$$\alpha(\omega) = -\frac{1}{(i\omega + \beta_{eq})} \tag{5.181}$$

and it follows that

$$\sigma_{\dot{q}}^2 = \int_{-\infty}^{\infty} \frac{S_g(\omega)\,d\omega}{(\omega^2 + \beta_{eq}^2)} \tag{5.182}$$

The simplest case is where the excitation is modelled as a white noise. Setting $S_g(\omega) = S_0$, a constant, one finds, from equation (5.182), that

$$\sigma_{\dot{q}}^2 = \int_{-\infty}^{\infty} \frac{S_0\,d\omega}{(\omega^2 + \beta_{eq}^2)} = \frac{\pi S_0}{\beta_{eq}} \tag{5.183}$$

Hence, on combining equations (5.180) and (5.183), it is found that

$$\beta_{eq} = \frac{2(\mu g)^2}{\pi^2 S_0} \tag{5.184}$$

and

$$\sigma_{\dot{q}}^2 = \frac{\pi^3 S_0^2}{2(\mu g)^2} \tag{5.185}$$

For white noise excitation an exact solution is available, and the accuracy of the statistical linearization solution, given by equation (5.185) can thus be assessed, in this instance (see Chapter 10).

Filtered white noise, characterised by a Kanai–Tajimi spectrum (Kanai, 1961; Tajimi, 1960) provides a more realistic model of actual earthquake accelerations. This spectrum is given by

$$S_g(\omega) = G_0 \frac{[1 + 4\zeta_g^2(\omega/\omega_g)^2]}{\{[1 - (\omega/\omega_g)^2]^2 + 4\zeta_g^2(\omega/\omega_g)^2\}} \tag{5.186}$$

where G_0 is a scaling parameter, ζ_g is the 'ground damping' factor and ω_g is the 'predominant ground frequency'. It is evident that such a model relates to the output of a second-order linear filter, with parameters ζ_g and ω_g, driven by white noise.

Once again $\sigma_{\dot{q}}^2$ can be evaluated by using equation (5.182). The required integral can be cast into the general form given by equation (B1), in Appendix B. Thus here $m = 3$ and

$$\Xi_3(\omega) = G_0[1 + 4\zeta_g^2(\omega/\omega_g)^2] \tag{5.187}$$

$$\Lambda_3(\omega) = (-i\omega + \beta_{eq})[1 - (\omega/\omega_g)^2 + 2i\zeta_g(\omega/\omega_g)] \tag{5.188}$$

On comparing equations (5.187) and (5.188) with equations (B2) and (B3), respectively, one has

$$\xi_0 = G_0 \tag{5.189}$$

$$\xi_1 = 4G_0\zeta_g^2/\omega_g^2 \tag{5.190}$$

$$\xi_2 = 0 \tag{5.191}$$

and

$$\lambda_0 = \beta_{eq} \tag{5.192}$$

$$\lambda_1 = -1 + 2\beta_{eq}\zeta_g/\omega_g \tag{5.193}$$

$$\lambda_2 = -2\zeta_g/\omega_g + \beta_{eq}/\omega_g^2 \tag{5.194}$$

$$\lambda_3 = 1/\omega_g^2 \tag{5.195}$$

Hence, using equation (B26), equation (5.182) yields

$$\sigma_{\dot{q}}^2 = \frac{\pi}{\lambda_3} \frac{\begin{vmatrix} \xi_2 & \xi_1 & \xi_0 \\ -\lambda_3 & \lambda_1 & 0 \\ 0 & -\lambda_2 & \lambda_0 \end{vmatrix}}{\begin{vmatrix} \lambda_2 & -\lambda_0 & 0 \\ -\lambda_3 & \lambda_1 & 0 \\ 0 & -\lambda_2 & \lambda_0 \end{vmatrix}} \tag{5.196}$$

On combining equations (5.189) to (5.196) one obtains

$$\sigma_{\dot{q}}^2 = \frac{\pi G_0}{\omega_g K(\Omega, \zeta_g)} \tag{5.197}$$

where

$$K(\Omega, \zeta_g) = \frac{(2\zeta_g + \Omega + 4\zeta_g^2\Omega)}{2\Omega\zeta_g(1 + \Omega^2 + 2\Omega\zeta_g)} \tag{5.198}$$

and

$$\Omega = \beta/\omega_g \tag{5.199}$$

A further combination of equations (5.180) and (5.197) gives the following algebraic equation for β_{eq}

$$\beta_{eq} = \frac{\mu g}{\pi}\left(\frac{2\omega_g K(\Omega, \zeta_g)}{G_0}\right)^{1/2} \tag{5.200}$$

Equation (5.200) can be solved numerically, for β_{eq}, and hence $\sigma_{\dot{q}}^2$ can be found, from equation (5.197). Constantinou and Tadjbaksh (1984) have numerically obtained values of β_{eq}, for a wide spectrum of values of the parameters μ, G_0, ζ_g and ω_g, and have displayed their results in graphical form. They found that, below a certain critical value of ω_g, the solution procedure broke down; i.e., no real solution could be found.

The mean square of the slip displacement, σ_q^2, can also be determined, in principle, once β_{eq} is known, using the linearised equation of motion (5.176). Thus

$$\sigma_q^2 = \int_{-\infty}^{\infty} \frac{S_g(\omega)\,d\omega}{\omega^2(\omega^2 + \beta_{eq}^2)} \tag{5.201}$$

However, this integral will only be finite if $S_g(\omega) \to 0$ sufficiently rapidly, as $\omega \to 0$. In the case of white noise excitation, or excitation with the Kanai–Tajimi spectrum, the integral is infinite. This implies that the mean-square slip displacement continues to grow, indefinitely, as time progresses. Results for the variation of σ_q^2 with time have been given by Constantinou and Tadjbakhsh (1984).

5.6.2 Ship roll motion in irregular waves

For a ship rolling in irregular beam waves, the roll angle, Φ, shown in Figure 5.14, can be related to the zero-mean, roll excitation moment, $M(t)$, due to waves, through the following non-linear, SDOF equation of motion (e.g., see Roberts,

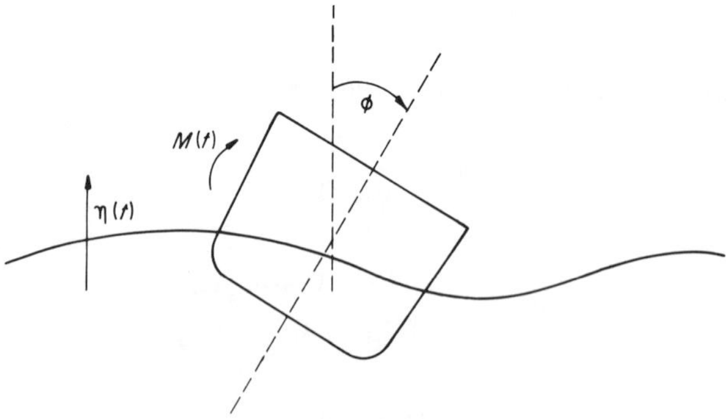

Figure 5.14
Sectional view of a ship rolling in irregular waves

1982; Roberts and Dacunha, 1985; Gawthrop *et al.*, 1988)

$$\ddot{\Phi} + (\beta + n_1|\dot{\Phi}|)\dot{\Phi} + (\omega_n^2 + n_2\Phi^2)\Phi = bM(t) \tag{5.202}$$

Here the parameters β and n_1 are the linear and quadratic damping factors, respectively (see Chapter 2). The symbol ω_n denotes the undamped natural frequency of roll and n_2 is a non-linear stiffness factor. b is the inverse of the total, effective roll inertia of the ship. Equation (5.202) is appropriate for small to moderate angle of roll ($\Phi \leqslant 35°$, say). It is noted that for large angles of roll the restoring moment reduces to zero at a critical roll angle. If this effect is incorporated into the model, there will be only a finite range of excitation level for which the statistical linearization method gives a stationary result (see Section 5.3.7).

Following the statistical linearization procedure (see also, Goodman and Sargent, 1961; Kaplan, 1966; Vassilopoulos, 1971), equation (5.202) can be replaced by the linear equation

$$\ddot{\Phi} + \beta_{eq}\dot{\Phi} + \omega_{eq}^2\Phi = bM(t) \tag{5.203}$$

It is noted that the present problem is of the general type discussed earlier, in Section 5.4, and that, furthermore, the non-linearity is of the 'separable' type (see equation (5.124)). Hence, applying equations (5.125) and (5.126) one obtains (assuming $\dot{\Phi}$ to be Gaussian)

$$\beta_{eq} = \beta + n_1 E\left\{\frac{d}{d\dot{\Phi}}(\dot{\Phi}|\dot{\Phi}|)\right\} \tag{5.204}$$

and

$$\omega_{eq}^2 = \omega_n^2 + n_2 E\left\{\frac{d}{d\Phi}(\Phi^3)\right\} \tag{5.205}$$

These expectations will depend on the mean squares of the roll displacement, and roll velocity, given, respectively, by

$$\sigma_\Phi^2 = E\{\Phi^2\} \quad \sigma_{\dot{\Phi}}^2 = E\{\dot{\Phi}^2\} \tag{5.206}$$

One finds from Appendix A, Table A1 that

$$\beta_{eq} = \beta + \left(\frac{8}{\pi}\right)^{1/2} n_1\sigma_{\dot{\Phi}} \tag{5.207}$$

$$\omega_{eq}^2 = \omega_n^2 + 3n_2\sigma_\Phi^2 \tag{5.208}$$

To obtain further relationships between $\sigma_\Phi, \sigma_{\dot{\Phi}}, \beta_{eq}$ and ω_{eq}^2 one can use equations (5.127) to (5.129). For this it is, of course, necessary to specify the power spectrum of the excitation process

$$f(t) = bM(t) \tag{5.209}$$

It has been demonstrated by Roberts and Dacunha (1985), and Gawthrop *et al.*

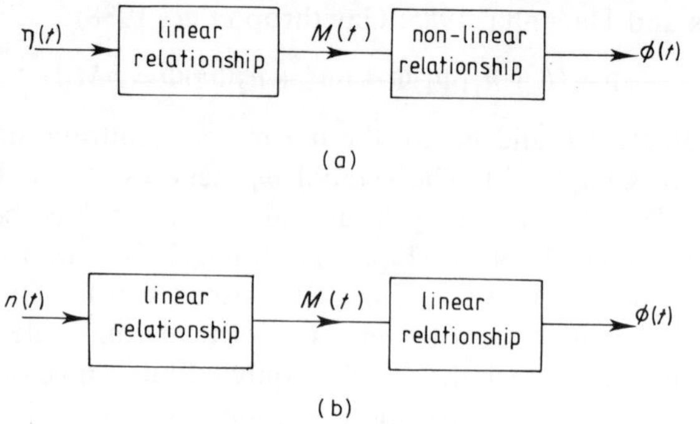

Figure 5.15
Block diagram relationships between the wave elevation process, the roll moment process and the roll displacement process. (a) Overall system non-linear. (b) Overall system linear

(1988), on the basis of a comparison between theory and experiment, that a linear theory is adequate, for the purpose of relating the roll excitation process, $f(t)$, to the wave elevation process, $\eta(t)$, say. Thus, a frequency response function, $\alpha^*(\omega)$, exists, relating $\eta(t)$ to $f(t)$, as illustrated in Figure 5.15. From equations (5.127) to (5.129) one can write

$$\sigma_{\Phi}^2 = \int_{-\infty}^{\infty} S_{\Phi}(\omega)\, d\omega \tag{5.210}$$

$$\sigma_{\dot{\Phi}}^2 = \int_{-\infty}^{\infty} \omega^2 S_{\Phi}(\omega)\, \omega \tag{5.211}$$

where $S_{\Phi}(\omega)$, the power spectrum of $\Phi(t)$, is given by

$$S_{\Phi}(\omega) = |\alpha(\omega)|^2 |\alpha^*(\omega)|^2 S_{\eta}(\omega) \tag{5.212}$$

and

$$\alpha(\omega) = \frac{1}{(\omega_{eq}^2 - \omega^2 + i\beta_{eq}\omega)} \tag{5.213}$$

is the frequency response function relating $f(t)$ to $\Phi(t)$.

Standard parametric forms of wave elevation spectra are frequently used in design studies. In particular, the JONSWAP spectrum is often used. In dimensionless form this is given by

$$S_{\eta}^*(\omega) = \frac{S_{\eta}(\omega)}{S_{\eta}(\omega_0)} = \exp(1.25)\hat{\omega}^{-5} \exp(-1.25\hat{\omega}^{-4})\gamma^{\mu-1} \tag{5.214}$$

where

$$\hat{\omega} = \omega/\omega_0 \tag{5.215}$$

and

$$\mu = \exp\left(-\frac{1}{2\lambda^2}(1-\hat{\omega})^2\right) \tag{5.216}$$

with

$$\lambda = 0.07 \quad \hat{\omega} \leqslant 1 \tag{5.217}$$

$$\lambda = 0.09 \quad \hat{\omega} > 1 \tag{5.218}$$

In equation (5.214) ω_0 is the peak spectral frequency, which depends on the wind velocity and the fetch. In addition, γ is a sharpness magnification factor which takes values in the range 1 to 7. For $\gamma = 1$ equation (5.214) reduces to the well known Pierson–Moskowitch spectrum (e.g. see Spanos, 1983, 1986).

For design purposes one also requires the magnitude of the frequency response function, $\alpha^*(\omega)$. This function can be computed by using sophisticated hydrodynamic, wave-diffraction theory (e.g. see Roberts and Dacunha, 1985). However, in many cases, a simple approximation, based on the use of the wave slope at the ship position (known as the so-called Froude–Krylov theory) is sufficiently accurate. Thus (e.g. see Vassilopoulos, 1971) one can write, approximately

$$\alpha^*(\omega) = \frac{b\omega^2}{g}\omega_{\text{eq}}^2 \tag{5.219}$$

A combination of equations (5.210) to (5.219) enables σ_Φ^2 and $\sigma_{\dot{\Phi}}^2$ to be calculated, for specified values of ω_{eq} and β_{eq}. This, together with equations (5.207) and (5.208), enables β_{eq} and ω_{eq} to be evaluated. Usually a simple iterative scheme will converge reasonably quickly. Thus, one starts with $\beta_{\text{eq}} = \beta$, $\omega_{\text{eq}}^2 = \omega_{\text{n}}^2$ and evaluates $\sigma_\Phi^2, \sigma_{\dot{\Phi}}^2$ from equations (5.210) to (5.219). These values are then used in equations (5.207) and (5.208) to update the values of β_{eq} and ω_{eq}^2; these are then fed back into equations (5.210) and (5.219), and so on, until convergence is achieved.

This iteration procedure can be made computationally more efficient by replacing equation (5.214) by a rational function approximation. Spanos (1986) has shown that the JONSWAP spectrum can be approximated, very closely, by using a model involving two cascaded, linear, second-order filters. Thus equation (5.214) can be replaced by

$$S_\eta^*(\omega) = \frac{G\hat{\omega}^4}{[(\hat{\omega}^2 - k_1)^2 + (c_1\hat{\omega})^2][(\hat{\omega} - k_2)^2 + (c_2\hat{\omega})^2]} \tag{5.220}$$

where the parameters G, k_1, k_2, c_1 and c_2 are determined by using a least-square algorithm, i.e. by minimizing the difference between equations (5.214) and (5.220), in a least-square sense. Figure 5.16 shows a typical comparison between equations (5.214) and (5.220), for the case where $\gamma = 3$. Using equation (5.220), together with equations (5.219) and (5.213), it is possible to express the integrals in equations (5.210) and (5.211) in the standard form given in Appendix B,

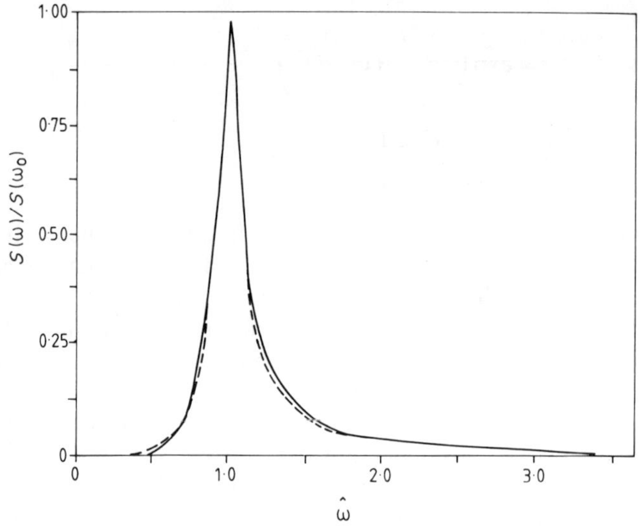

Figure 5.16
Typical comparison between the Jonswap spectrum (———) and an approximation (- - -) based on two, cascaded linear second-order filters (reproduced from Spanos (1986) by permission of Computational Mechanics Publications)

equation (B1). Thus, the result given by equation (B26) can be used to evaluate σ_Φ^2 and $\sigma_{\dot{\Phi}}^2$, with a consequent improvement in computational efficiency.

In the case where the roll damping is light the analysis can be very considerably simplified by using a white noise approximation for the excitation process, $f(t)$ (see Figure 4.5). Thus, replacing the spectrum of $f(t)$

$$S_f(\omega) = |\alpha^*(\omega)|^2 S_\eta(\omega) \tag{5.221}$$

by a constant spectral level, S_0 say, one finds from equations (5.210) and (5.211) (see also equation (5.135)) that

$$\sigma_\Phi^2 = \frac{\pi S_0}{\beta_{eq}\omega_{eq}^2} \quad \sigma_{\dot{\Phi}}^2 = \frac{\pi S_0}{\beta_{eq}} \tag{5.222}$$

One can then proceed to solve for β_{eq} and ω_{eq}^2, in a manner similar to that described in Section 5.4.1.

Another approximation, which does not depend on a white noise assumption, is possible if the degree of non-linearity is small, i.e. if η_1 and η_2 are small. The spectrum of $\Phi(t)$ can then be approximated, in the manner described in Section 5.4.3. In fact equation (5.159) is applicable, where here $S_y(\omega) = S_\Phi(\omega)$

$$\mu k_{10} = \left(\frac{8}{\pi}\right)^{1/2} n_1 (\sigma_\Phi^2)_0 \tag{5.223}$$

$$\mu k_{20} = 3n_2(\sigma_{\dot{\Phi}}^2)_0 \tag{5.224}$$

and $(\sigma_{\Phi}^2)_0$ and $(\sigma_{\dot{\Phi}}^2)_0$ are the values of σ_{Φ}^2 and $\sigma_{\dot{\Phi}}^2$ when n_1 and n_2 are zero. A substitution of equation (5.159) into equations (5.210) and (5.211) enables σ_{Φ}^2 and $\sigma_{\dot{\Phi}}^2$ to be evaluated, once only, thus avoiding the need for iteration. In the special case where $S_f(\omega)$ is replaced by S_0, according to the white noise approximation, then the integrals are readily evaluated, as shown in Appendix B, with the following result

$$\sigma_{\Phi}^2 = (\sigma_{\Phi}^2)_0\left[1 - \left(\frac{8}{\pi}\right)^{1/2}\frac{n_1}{\beta}(\sigma_{\dot{\Phi}}^2)_0 - \frac{3n_2}{\omega_n^2}(\sigma_{\Phi}^2)_0\right] \tag{5.225}$$

$$\sigma_{\dot{\Phi}}^2 = (\sigma_{\dot{\Phi}}^2)_0\left[1 - \left(\frac{8}{\pi}\right)^{1/2}\frac{n_1}{\beta}(\sigma_{\dot{\Phi}}^2)_0\right] \tag{5.226}$$

It is noted that these results follow directly from equations (5.150) and (5.151), where μk_{10} and μk_{20} are defined by equations (5.223) and (5.224), respectively.

On the basis of equation (5.225) one can conclude that, if the ships restoring moment characteristic is of the 'hardening' type, corresponding to n_2 positive, for small to moderate roll angles, then the non-linear roll mean square is less than that derived by neglecting the non-linear term in stiffness. Similarly, the non-linear damping term has the effect of lowering the mean square variance.

5.6.3 Flow induced vibration of cylindrical structures

The recent extensive development and use of offshore drilling platforms has stimulated considerable interest in the problem of predicting the dynamic response of such structures to random, flow induced forces. In this application the appropriate equations of motion are found to be non-linear, which renders the determination of exact solutions very difficult, if not impossible, in most cases. However, the technique of statistical linearization offers a useful approach to the formulation of approximate solutions (e.g. see Spanos and Chen, 1981).

Here a simplified version of the general problem, which illustrates the application of the statistical linearization procedure, is analysed. A single, elastically restrained, rigid cylindrical element is considered, which is submerged in a randomly fluctuating flow, as shown in Figure 5.17. Only 'in-line' vibration will be considered, i.e. vibration in the same direction as the randomly fluctuating flow velocity, which is assumed here to be unidirectional. For simplicity the elastic restraint on the cylinder is represented by a linear spring, of stiffness k, whereas the structural damping is represented by a linear viscous damper, with coefficient c. Hence if M_0 is the effective mass of the rigid cylinder, and $f(t)$ is the hydrodynamic force, due to the flow, on the structure, then the appropriate

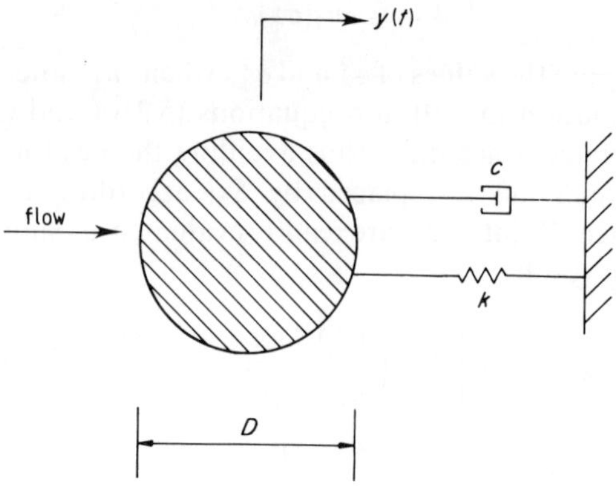

Figure 5.17
Simplified model of an elastically restrained cylindrical structure subjected to flow induced forces

equation of motion is of the usual, SDOF kind. Specifically

$$M_0\ddot{y} + \alpha\dot{y} + k\dot{y} = f(t) \tag{5.227}$$

where y is the absolute displacement of the cylinder.

The flow induced force, $f(t)$, may be related to the fluctuating flow by means of the Morison equation (see Chapter 2 and Blevins, 1977; Chen, 1977; Morison *et al.*, 1950). If $\dot{v}(t)$ is the flow velocity then the Morison equation for $f(t)$, modified to take into account the movement of the structure (see Chapter 2), is

$$f(t) = \rho A\ddot{v} + C_I\rho A(\ddot{v} - \ddot{y}) + \tfrac{1}{2}C_D\rho D|\dot{v} - \dot{y}|(\dot{v} - \dot{y}) \tag{5.228}$$

In this equation ρ, A and D represent the fluid density, the cross sectional area and the diameter of the cylinder, respectively; C_I and C_D are the added mass and drag coefficients, respectively.

The flow velocity $\dot{v}(t)$ can be decomposed into a mean component, V, and a zero-mean, fluctuating component, $\dot{u}(t)$, as follows

$$\dot{v}(t) = V + \dot{u}(t) \tag{5.229}$$

For simplicity it will be assumed that $u(t)$ is a stationary Gaussian process. It follows that $\dot{u}(t)$, and higher derivatives, are also stationary, zero-mean Gaussian processes.

For convenience the relative displacement coordinate

$$q = u - y \tag{5.230}$$

is now introduced. On substituting equation (5.230) into equation (5.228), and combining the result with equation (5.227), the following equation of motion is obtained

$$M\ddot{q} + c\dot{q} + \tfrac{1}{2}C_{\mathrm{D}}\rho D|V + \dot{q}|(V + \dot{q}) + kq = (M_0 - \rho A)\ddot{u} + c\dot{u} + ku \quad (5.231)$$

where

$$M = M_0 + C_{\mathrm{I}}\rho A \quad (5.232)$$

Dividing equation (5.231) throughout by M, an equation of the form

$$\ddot{q} + \beta\dot{q} + \omega_{\mathrm{n}}^2 q + \mu G(\dot{q}) = F(t) \quad (5.233)$$

is obtained, where

$$\beta = \frac{c}{M} \quad \omega_{\mathrm{n}}^2 = \frac{k}{M} \quad \mu = \frac{1}{2}\frac{C_{\mathrm{D}}\rho D}{M} \quad (5.234)$$

$$G(\dot{q}) = |V + \dot{q}|(V + \dot{q}) \quad (5.235)$$

and

$$F(t) = \frac{(M_0 - \rho A)}{M}\ddot{u} + \beta\dot{u} + \omega_{\mathrm{n}}^2 u \quad (5.236)$$

Equation (5.233) is of the general form discussed earlier (see equations (5.112) and (5.113)), where the non-linearity function $G(\dot{q})$ is here of the damping type, i.e. dependent on $\dot{q}$ only. Moreover, in the general case where V is non-zero, $G(\dot{q})$ is asymmetrical and hence the mean of the output variable $q(t)$ will be non-zero, even though the mean of $F(t)$ is zero.

The method outlined in Section 5.4.4 can be used to find the appropriate equivalent linear system. Referring to equations (5.162) and (5.163) it is evident that $b = 0$, since, in this application, $m_f = 0$. Moreover, since the non-linearity depends only on $\dot{q}$, $\omega_{\mathrm{eq}}^2 = \omega_{\mathrm{n}}^2$ and it remains to find the linear damping coefficient β_{eq} in the equivalent linear system

$$\ddot{q}_0 + \beta_{\mathrm{eq}}\dot{q}_0 + \omega_{\mathrm{n}}^2 q_0 = F(t) \quad (5.237)$$

where

$$q_0(t) = q(t) - m_q \quad (5.238)$$

and m_q is the mean of $q(t)$. Using equation (5.165) for β_{eq}, together with equation (5.235), on the assumption that $q(t)$ is Gaussian, it is found that

$$\beta_{\mathrm{eq}} = \beta + \mu E\left\{\frac{\mathrm{d}}{\mathrm{d}q}[|V + q|(V + q)]\right\} \quad (5.239)$$

Hence, using the fact that the mean of $\dot{q}(t)$ is zero, one obtains (see Appendix A, Table A1)

$$\beta_{\mathrm{eq}} = \beta + \mu\left[\left(\frac{8}{\pi}\right)^{1/2}\sigma_{\dot{q}}\exp(-\eta^2) + 2V\,\mathrm{erf}\,\eta\right] \quad (5.240)$$

where $\sigma_{\dot{q}}$ is the standard deviation of $\dot{q}(t)$ and

$$\eta = \frac{V}{\sqrt{2}\sigma_{\dot{q}}} \tag{5.241}$$

A second relationship between β_{eq} and $\sigma_{\dot{q}}$ can be found from the equivalent linear system (see equation (5.128)). This step requires a knowledge of the spectrum of $F(t), S_F(\omega)$, which can easily be related to the spectrum of $u(t), S_u(\omega)$ (Spanos and Chen, 1981).

Finally, it is noted that the mean value of $q(t), m_q$, may be found by employing equation (5.166). This shows that, in the present application,

$$\omega_{\mathrm{n}}^2 m_q + \mu E\{|V + \dot{q}|(V + \dot{q})\} = 0 \tag{5.242}$$

Assuming that $Q(t)$ is Gaussian again, the expectation in this equation may be evaluated in terms of $\sigma_{\dot{q}}$, the standard deviation of $\dot{q}(t)$. Hence it is found that (using a result in Appendix A, Table A1)

$$m_q = -\frac{\mu}{\omega_{\mathrm{n}}^2}\left[(\sigma_{\dot{q}}^2 + V^2)\operatorname{erf}\eta + \left(\frac{2}{\pi}\right)^{1/2} V\sigma_{\dot{q}}\exp(-\eta^2)\right] \tag{5.243}$$

When $V = 0$, the above equation gives $m_q = 0$, as expected.

Equations (5.240) and (5.243), together with the relationship between $\sigma_{\dot{y}}$ and β_{eq} deduced from the equivalent linear system, enable the unknowns, $\beta_{\mathrm{eq}}, \sigma_{\dot{q}}$ and m_q to be found. The equivalent linear system also yields estimates of $\sigma_{\dot{q}}$, together with statistics of the absolute response displacement, $y(t)$. Further details may be found in Spanos and Chen (1981), where the reliability of the method is assessed through a comparison with corresponding digital simulation estimates.

Chapter 6

Statistical linearization of multi-degree of freedom systems with stationary response

6.1 INTRODUCTION

So far the method of statistical linearization has been discussed in connection with the dynamic behaviour of oscillators, which possess a single degree of freedom (SDOF). Although SDOF systems may provide an adequate mathematical representation of a considerable number of physical problems, dynamic systems with multiple degrees of freedom (MDOF) must be used in most cases of engineering decision making. Thus, it is appropriate to address the issue of application of statistical linearization to lumped parameter MDOF systems subjected to random excitation. The transition from the SDOF to the MDOF problem has been gradual over a period of three decades (see Caughey, 1959, 1960b, 1963; Kazakov, 1965a, 1965b; Foster, 1968, 1970; Gelb and Warren 1972; Iwan and Yang, 1971, 1972; Iwan, 1973; Atalik and Utku, 1976; Spanos and Iwan, 1978; Spanos, 1976, 1978, 1981a, 1981b; Roberts, 1981b, 1984; Roberts and Dunne, 1988).

It will be assumed through this chapter that the excitation and response processes of concern are stationary.

6.2 THE NON-LINEAR SYSTEM

In this section the response to external load of a non-linear MDOF vibratory system will be considered. It is shown in Chapter 2 that the equations of motion of such a system may be written in the form

$$\mathbf{M}\ddot{\mathbf{q}} + \mathbf{C}\dot{\mathbf{q}} + \mathbf{K}\mathbf{q} + \mathbf{\Phi}(\mathbf{q}, \dot{\mathbf{q}}, \ddot{\mathbf{q}}) = \mathbf{Q}(t) \tag{6.1}$$

The symbols $\mathbf{M}, \mathbf{C}$ and $\mathbf{K}$ denote constant $n \times n$ matrices, defined as the inertia, damping and stiffness matrices, respectively. Further, $\mathbf{\Phi}(\mathbf{q}, \dot{\mathbf{q}}, \ddot{\mathbf{q}})$ is a non-linear n-vector function of the generalized coordinate vector $\mathbf{q}$ and its derivatives, and $\mathbf{Q}(t)$ is an n-vector random process of the independent variable t. In the linear case $\mathbf{\Phi}$ vanishes and equation (6.1) reduces to equation (4.21), the solution to which has been extensively discussed in Chapter 4.

Due to the scarcity of exact solutions of equation (6.1) when the function

$\mathbf{\Phi}(\mathbf{q}, \dot{\mathbf{q}}, \ddot{\mathbf{q}})$ is non-zero and non-linear, attention has been directed toward techniques of approximate analysis. For n small and with certain restrictions on $\mathbf{Q}(t)$ and $\mathbf{\Phi}(\mathbf{q}, \dot{\mathbf{q}}, \ddot{\mathbf{q}})$, a number of standard analytic techniques can be adapted to the problem of generating an approximate solution of equation (6.1). Among these are perturbation methods and the methods of energy balance and slowly varying parameters (see Chapter 1). However, for n large or for $\mathbf{Q}(t)$ and/or $\mathbf{\Phi}(\mathbf{q}, \dot{\mathbf{q}}, \ddot{\mathbf{q}})$ of a more general form, the standard techniques often break down and are at best quite difficult to apply. The concept of statistical linearization offers a systematic and readily mechanized method for generating an approximate solution of equation (6.1).

Similarly to the case of a SDOF problem, the objective of the method for a MDOF problem is the replacement of the non-linear dynamical system equation (6.1) by another auxiliary system for which the exact analytic formula for the solution is known. The replacement is made so as to be optimum with respect to some measure of the difference between the original and the auxiliary system. In general, the auxiliary system need not necessarily be linear (see Chapter 9). However, for the case of MDOF systems only the response of linear systems is readily available. Therefore, it is convenient here to deal exclusively with the case of a linear auxiliary system. Hereafter, this system will be called the equivalent linear system.

6.3 THE EQUIVALENT LINEAR SYSTEM

6.3.1 Formulation

The equivalent linear system of equation (6.1) is defined as

$$(\mathbf{M} + \mathbf{M}_e)\ddot{\mathbf{q}} + (\mathbf{C} + \mathbf{C}_e)\dot{\mathbf{q}} + (\mathbf{K} + \mathbf{K}_e)\mathbf{q} = \mathbf{Q}(t) \tag{6.2}$$

where $\mathbf{M}_e, \mathbf{C}_e$ and $\mathbf{K}_e$ are deterministic matrices. They are to be determined so that the n-vector difference $\boldsymbol{\varepsilon}$ between the actual and the equivalent linear system is minimized, in some statistical sense, for every $\mathbf{q}$ belonging to a certain class of functions of the independent variable t. The difference $\boldsymbol{\varepsilon}$ may be written as (Spanos, 1976; Spanos and Iwan, 1978)

$$\begin{aligned} \boldsymbol{\varepsilon} &= \mathbf{M}\ddot{\mathbf{q}} + \mathbf{C}\dot{\mathbf{q}} + \mathbf{K}\mathbf{q} + \mathbf{\Phi}(\mathbf{q}, \dot{\mathbf{q}}, \ddot{\mathbf{q}}) - (\mathbf{M} + \mathbf{M}_e)\ddot{\mathbf{q}} - (\mathbf{C} + \mathbf{C}_e)\dot{\mathbf{q}} - (\mathbf{K} + \mathbf{K}_e)\mathbf{q} \\ &= \mathbf{\Phi}(\mathbf{q}, \dot{\mathbf{q}}, \ddot{\mathbf{q}}) - \mathbf{M}_e\ddot{\mathbf{q}} - \mathbf{C}_e\dot{\mathbf{q}} - \mathbf{K}_e\mathbf{q} \end{aligned} \tag{6.3}$$

According to this formulation the matrices $\mathbf{M}_e, \mathbf{C}_e$ and $\mathbf{K}_e$ depend on $\mathbf{q}(t)$. If $\mathbf{M}_e, \mathbf{C}_e$ and $\mathbf{K}_e$ are determined for that $\mathbf{q}(t)$ which is the solution of the linear equation (6.2), it is reasonable to expect that this solution will also be a fairly good approximation to the solution of the non-linear problem governed by equation (6.1). Since the solution of the equivalent linear system depends on $\mathbf{M}_e$, $\mathbf{C}_e$ and $\mathbf{K}_e$, a cyclic relation may be established between $\mathbf{M}_e$, $\mathbf{C}_e$, $\mathbf{K}_e$ and

$\mathbf{q}(t)$, as in the SDOF case (see Chapter 5). Utilizing this cyclic scheme the system response can be determined. At this stage it is evident that the smaller the non-linear force $\mathbf{\Phi}(\mathbf{q}, \dot{\mathbf{q}}, \ddot{\mathbf{q}})$ is, the more an equivalent linear system is suitable for the description of the response of the non-linear system equation (6.1). The smallness of $\mathbf{\Phi}(\mathbf{q}, \dot{\mathbf{q}}, \ddot{\mathbf{q}})$ is usually emphasized by including a small coefficient in front of this term, in equation (6.1). However, the present formulation will be carried out irrespective of the smallness of $\mathbf{\Phi}(\mathbf{q}, \dot{\mathbf{q}}, \ddot{\mathbf{q}})$.

6.3.2 Minimization procedure

In order to determine the matrices $\mathbf{M}_e, \mathbf{C}_e$ and $\mathbf{K}_e$ of the equivalent linear system it is necessary to establish a criterion for the minimization of $\boldsymbol{\varepsilon}$, based on a suitable norm of this difference. Here the Euclidean norm $\|\boldsymbol{\varepsilon}\|_2$ will be used as a measure of $\boldsymbol{\varepsilon}$, where this is defined as

$$\|\boldsymbol{\varepsilon}\|_2^2 = \boldsymbol{\varepsilon}^T\boldsymbol{\varepsilon} \tag{6.4}$$

In this equation $\boldsymbol{\varepsilon}^T$ denotes the transpose of the vector $\boldsymbol{\varepsilon}$. The minimization of $\boldsymbol{\varepsilon}$ is performed according to the criterion

$$E\{\boldsymbol{\varepsilon}^T\boldsymbol{\varepsilon}\} = \text{minimum} \tag{6.5}$$

Minimization is with respect to the class of functions of t which are solutions of equation (6.2). In equation (6.5) $E\{\ \}$ denotes, as before, the operation of mathematical expectation which possesses several properties which assure certain characteristics of the equivalent linear system. Criterion (6.5) is a generalization of that used earlier, in Chapter 5, for SDOF systems.

6.3.3 Equations for the equivalent linear system parameters

The necessary conditions for equation (6.5) to be true are

$$\frac{\partial}{\partial m_{ij}^e} E\{\boldsymbol{\varepsilon}^T\boldsymbol{\varepsilon}\} = 0 \tag{6.6}$$

$$\frac{\partial}{\partial c_{ij}^e} E\{\boldsymbol{\varepsilon}^T\boldsymbol{\varepsilon}\} = 0 \tag{6.7}$$

and

$$\frac{\partial}{\partial k_{ij}^e} E\{\boldsymbol{\varepsilon}^T\boldsymbol{\varepsilon}\} = 0 \quad i, j = 1, 2, \ldots, n \tag{6.8}$$

where m_{ij}^e, c_{ij}^e and k_{ij}^e are the (i, j) elements of the matrices $\mathbf{M}_e, \mathbf{C}_e$, and $\mathbf{K}_e$, respectively. Equation (6.5) can be rewritten as

$$E\{\varepsilon_1^2 + \cdots + \varepsilon_n^2\} = \text{minimum} \tag{6.9}$$

where $\varepsilon_1, \varepsilon_2, \ldots, \varepsilon_n$ are the elements of $\boldsymbol{\varepsilon}$; i.e.

$$\boldsymbol{\varepsilon} = [\varepsilon_1, \ldots, \varepsilon_n]^{\mathrm{T}} \tag{6.10}$$

Upon using the linearity property of the expectation operator $E\{\ \}$, equation (6.9) can be put in the form

$$\sum_{i=1}^{n} D_i^2 = \text{minimum} \tag{6.11}$$

where D_i is defined by

$$D_i^2 = E\{\varepsilon_i^2\} \quad i = 1, 2, \ldots, n \tag{6.12}$$

Because of equation (6.3),

$$D_i^2 = E\left\{\left[\Phi_i - \sum_{j=1}^{n} (m_{ij}^{\mathrm{e}}\ddot{q}_j + c_{ij}^{\mathrm{e}}\dot{q}_j + k_{ij}^{\mathrm{e}}q_j)\right]^2\right\} \quad i = 1, 2, \ldots, n \tag{6.13}$$

where $\Phi_1, \Phi_2, \ldots, \Phi_n$ are the elements of $\boldsymbol{\Phi}$; i.e.

$$\boldsymbol{\Phi} = [\Phi_1, \ldots, \Phi_n]^{\mathrm{T}} \tag{6.14}$$

On examining equation (6.13) it is seen that D_i depends only on m_{ij}^{e}, c_{ij}^{e} and k_{ij}^{e}, where $j = 1, 2, \ldots, n$. Therefore, the minimization criterion given by equation (6.9) can be expressed as

$$D_i^2 = \text{minimum} \quad i = 1, 2, \ldots, n \tag{6.15}$$

The necessary conditions for equation (6.15) to be true are

$$\frac{\partial}{\partial m_{ij}^{\mathrm{e}}}(D_i^2) = 0 \quad j = 1, 2, \ldots, n \tag{6.16}$$

$$\frac{\partial}{\partial c_{ij}^{\mathrm{e}}}(D_i^2) = 0 \quad j = 1, 2, \ldots, n \tag{6.17}$$

and

$$\frac{\partial}{\partial k_{ij}^{\mathrm{e}}}(D_i^2) = 0 \quad j = 1, 2, \ldots, n \tag{6.18}$$

Expanding equations (6.16) to equation (6.18) and utilizing equation (6.13), gives

$$E\{\ddot{q}_j\Phi_i\} = \sum_{s=1}^{n} [m_{is}^{\mathrm{e}}E\{\ddot{q}_s\ddot{q}_j\} + c_{is}^{\mathrm{e}}E\{\dot{q}_s\ddot{q}_j\} + k_{is}^{\mathrm{e}}E\{q_s\ddot{q}_j\}] \tag{6.19}$$

$$E\{\dot{q}_j\Phi_i\} = \sum_{s=1}^{n} [m_{is}^{\mathrm{e}}E\{\ddot{q}_s\dot{q}_j\} + c_{is}^{\mathrm{e}}E\{\dot{q}_s\dot{q}_j\} + k_{is}^{\mathrm{e}}E\{q_s\dot{q}_j\}] \tag{6.20}$$

and

$$E\{q_j\Phi_i\} = \sum_{s=1}^{n} [m_{is}^{e} E\{\ddot{q}_s q_j\} + c_{is}^{e} E\{\dot{q}_s q_j\} + k_{is}^{e} E\{q_s q_j\}] \tag{6.21}$$

Equations (6.19) to (6.21) can be rewritten in the compact form

$$E\{\Phi_i \hat{\mathbf{q}}\} = E\{\hat{\mathbf{q}}\hat{\mathbf{q}}^{\mathrm{T}}\} \begin{bmatrix} \mathbf{k}_{i*}^{e\mathrm{T}} \\ \mathbf{c}_{i*}^{e\mathrm{T}} \\ \mathbf{m}_{i*}^{e\mathrm{T}} \end{bmatrix} \quad i = 1, 2, \ldots, n \tag{6.22}$$

where

$$\hat{\mathbf{q}} = [\mathbf{q}, \dot{\mathbf{q}}, \ddot{\mathbf{q}}]^{\mathrm{T}} \tag{6.23}$$

and $\mathbf{m}_{i*}^{e}$, $\mathbf{c}_{i*}^{e}$ and $\mathbf{k}_{i*}^{e}$ are the ith rows of the matrices $\mathbf{M}_e$, $\mathbf{C}_e$ and $\mathbf{K}_e$, respectively.

6.3.4 Examination of the minimum

From equation (6.13) it is clear that the quantity D_i^2 is simply a quadratic polynomial in the parameters m_{ij}^{e}, c_{ij}^{e} and k_{ij}^{e}. Therefore, its mixed partial derivatives with respect to m_{ij}^{e}, c_{ij}^{e} and k_{ij}^{e}, of order higher than two, vanish. Hence, if the value of D_i^2 which corresponds to another set of parameters (Spanos, 1976; Spanos and Iwan, 1978)

$$m_{ij}^{\prime e} = m_{ij}^{e} + \Delta m_{ij}^{e} \tag{6.24}$$

$$c_{ij}^{\prime e} = c_{ij}^{e} + \Delta c_{ij}^{e} \tag{6.25}$$

$$k_{ij}^{\prime e} = k_{ij}^{e} + \Delta k_{ij}^{e} \tag{6.26}$$

is considered, the following Taylor's expansion around m_{ij}^{e}, c_{ij}^{e} and k_{ij}^{e} can be made

$$D_i^2(m_{ij}^{\prime e}, c_{ij}^{\prime e}, k_{ij}^{\prime e}) = D_i^2(m_{ij}^{e}, c_{ij}^{e}, k_{ij}^{e}) + \sum_{j=1}^{n} \left(\frac{\partial D_i^2}{\partial m_{ij}^{e}} \Delta m_{ij}^{e} + \frac{\partial D_i^2}{\partial c_{ij}^{e}} \Delta c_{ij}^{e} + \frac{\partial D_i^2}{\partial k_{ij}^{e}} \Delta k_{ij}^{e} \right)$$
$$+ \frac{1}{2!} E\left\{ \left[\sum_{j=1}^{n} (\Delta m_{ij}^{e} \ddot{q}_j + \Delta c_{ij}^{e} \dot{q}_j + \Delta k_{ij}^{e} q_j) \right]^2 \right\} \quad i = 1, 2, \ldots, n \tag{6.27}$$

Conditions (6.16) to (6.18), mean that the first sum in equation (6.27) is zero. Defining

$$J_i = \sum_{j=1}^{n} (\Delta m_{ij}^{e} \ddot{q}_j + \Delta c_{ij}^{e} \dot{q}_j + \Delta k_{ij}^{e} q_j) \tag{6.28}$$

then

$$E\{J_i^2\} > 0 \quad i = 1, 2, \ldots, n \tag{6.29}$$

if

$$J_i \neq 0 \quad i=1,2,\ldots,n \tag{6.30}$$

and

$$E\{J_i^2\}=0 \quad i=1,2,\ldots,n \tag{6.31}$$

if

$$J_i=0 \quad i=1,2,\ldots,n \tag{6.32}$$

Hence, equation (6.27) yields

$$D_i^2(m_{ij}'^{e}, c_{ij}'^{e}, k_{ij}'^{e}) \geqslant D_i^2(m_{ij}^{e}, c_{ij}^{e}, k_{ij}^{e}) \quad i=,2,\ldots,n \tag{6.33}$$

where the equality holds if, and only if, equation (6.32) is true. Inequality (6.33) assures that if a linear system exists with mass matrix $\mathbf{M}_e$, damping matrix $\mathbf{C}_e$, and stiffness matrix $\mathbf{K}_e$ satisfying condition (6.22), then the value of $E\{\boldsymbol{\varepsilon}^T\boldsymbol{\varepsilon}\}$ which corresponds to $\mathbf{M}_e, \mathbf{C}_e$ and $\mathbf{K}_e$ is *not larger* than the value of $E\{\boldsymbol{\varepsilon}^T\boldsymbol{\varepsilon}\}$ which corresponds to any other set of mass, damping and stiffness matrices.

Obviously, relation (6.32) corresponds to the case that the member $\hat{\mathbf{q}}$ of the class of possible solutions of the equivalent linear system (6.2) has linearly dependent components for every value of the independent variable t. Hence, if the components are linearly independent it is certain that the value of $E\{\boldsymbol{\varepsilon}^T\boldsymbol{\varepsilon}\}$ which corresponds to $\mathbf{M}_e, \mathbf{C}_e$ and $\mathbf{K}_e$ is an absolute minimum.

6.3.5 Existence and uniqueness of the equivalent linear system

In this section it will be proved that the linear independence of the components of $\hat{\mathbf{q}}$ is also important for the existence and uniqueness of the equations governing the components of the equivalent linear system.

Equation (6.22) must be solved so that the matrices $\mathbf{M}_e, \mathbf{C}_e$ and $\mathbf{K}_e$ can be expressed in terms of $\hat{\mathbf{q}}$. It is readily recognized that this equation is merely a system of $3n$ equations, linear in m_{ij}^e, c_{ij}^e and k_{ij}^e. Therefore, equation (6.22) leads to a unique set of equations for $\mathbf{M}_e, \mathbf{C}_e$ and $\mathbf{K}_e$ if, and only if, the matrix $E\{\hat{\mathbf{q}}\hat{\mathbf{q}}^T\}$ is non-singular. It can be shown that the matrix $E\{\hat{\mathbf{q}}\hat{\mathbf{q}}^T\}$ is singular if, and only if, at least one of the components of $\hat{\mathbf{q}}$ can be expressed as a linear combination of the remaining elements. To prove the sufficiency of this condition, it is noted that linear dependence is equivalent to the existence of $3n$ real numbers $a_j, b_j, c_j; j=1,2,\ldots,n$, not all zero, such that

$$\sum_{j=1}^{n} (a_j q_j + b_j \dot{q}_j + c_j \ddot{q}_j) = 0 \tag{6.34}$$

Using vector notation, equation (6.34) can be rewritten as

$$\hat{\mathbf{q}}^T\mathbf{u} = \mathbf{0} \tag{6.35}$$

where

$$\mathbf{u} = [a_1, \ldots, a_n, b_1, \ldots, b_n, c_1, \ldots, c_n]^{\mathrm{T}} \neq \mathbf{0} \tag{6.36}$$

Pre-multiplying equation (6.35) by $\hat{\mathbf{q}}$ gives

$$\hat{\mathbf{q}}\hat{\mathbf{q}}^{\mathrm{T}}\mathbf{u} = \mathbf{0} \tag{6.37}$$

Applying the operator $E\{\ \}$ to both sides of equation (6.37), and taking into consideration the linear properties of this operator, one finds that

$$E\{\hat{\mathbf{q}}\hat{\mathbf{q}}^{\mathrm{T}}\}\mathbf{u} = \mathbf{0} \tag{6.38}$$

Since $\mathbf{u} \neq \mathbf{0}$, the matrix $E\{\hat{\mathbf{q}}\hat{\mathbf{q}}^{\mathrm{T}}\}$ is singular.

To prove the necessity condition, it is assumed that the matrix $E\{\hat{\mathbf{q}}\hat{\mathbf{q}}^{\mathrm{T}}\}$ is singular. Hence, there exists a vector $\mathbf{u} \neq \mathbf{0}$ such that equation (6.38) is satisfied. Pre-multiplying both sides of equation (6.38) by $\mathbf{u}^{\mathrm{T}}$ yields

$$\mathbf{u}^{\mathrm{T}}E\{\hat{\mathbf{q}}\hat{\mathbf{q}}^{\mathrm{T}}\}\mathbf{u} = E\{\mathbf{u}^{\mathrm{T}}\hat{\mathbf{q}}\hat{\mathbf{q}}^{\mathrm{T}}\mathbf{u}\} = E\{(\hat{\mathbf{q}}^{\mathrm{T}}\mathbf{u})^2\} = 0 \tag{6.39}$$

Since $E\{0\} = 0$, equation (6.39) leads to equation (6.35). Therefore, the $3n$ elements $q_i, \dot{q}_i, \ddot{q}_i$ $(i = 1, \ldots, n)$ are linearly dependent.

Clearly, for a dynamical system to have linearly dependent $q_1, \ldots, q_n, \dot{q}_1, \ldots, \dot{q}_n, \ddot{q}_1, \ldots, \ddot{q}_n$ requires that the number of degrees of freedom is less than n. That is, the system of equations of motion includes a redundant equation. However, it is always possible to remove such equations to reduce the number of equations of motion to the number of the degrees of freedom of the system (e.g. see Bishop *et al.*, 1965). Thus, without loss of generality, in the ensuing development, it will be assumed that $q_1, \ldots, q_n$, $\dot{q}_1, \ldots, \dot{q}_n, \ddot{q}_1, \ldots, \ddot{q}_n$ are linearly independent. It follows that the matrix $E\{\hat{\mathbf{q}}\hat{\mathbf{q}}^{\mathrm{T}}\}$ is non-singular, and that a unique set of equations for the matrices $\mathbf{M}_e$, $\mathbf{C}_e$ and $\mathbf{K}_e$ exists, corresponding to the minimization of the error defined by equation (6.3).

6.4 MECHANIZATION OF THE METHOD

In this section a brief outline of the manner in which the presented method may be applied is given. The details for specific categories of problems will be discussed in subsequent sections.

Given the non-linear system (6.1) it is first necessary to specify an acceptable class of approximate solutions $\mathbf{q}$. The selection of the class of approximate solutions depends on the excitation $\mathbf{Q}(t)$ and the system (6.1). For the class of acceptable approximate solution functions the equivalent linear system (6.2) must be determined. That is, the matrices $\mathbf{M}_e$, $\mathbf{C}_e$ and $\mathbf{K}_e$ need to be found by solving equation (6.22). Clearly, these matrices depend on the 'identification parameters' of $\hat{\mathbf{q}}$, such as statistical moments up to a certain order. The last step in this procedure is to solve the equivalent linear system equation (6.2) and derive equations for the identification parameters of $\hat{\mathbf{q}}$ in terms of the matrices $\mathbf{M}_e, \mathbf{C}_e$ and $\mathbf{K}_e$. Evidently, the method provides equations, either algebraic or differential, for the determination of important parameters of the approximate solution of the non-linear system.

Due to the complexity of the outlined procedure, one can expect that it will be necessary to resort to numerical methods for solving the appropriate algebraic or differential equations.

6.5 DETERMINATION OF THE ELEMENTS OF THE EQUIVALENT LINEAR SYSTEM

6.5.1 Gaussian approximation

So far in this chapter the formulation of the statistical linearization method for MDOF systems has been presented irrespective of the particular probability density which is used in computing the expectations appearing in equation (6.22). In this section the elements of the equivalent linear system when $\hat{\mathbf{q}}$ is a jointly Gaussian random vector will be determined. Clearly, if the excitation of the original non-linear system is Gaussian then the response of a linear system to this excitation will also be Gaussian. Therefore, a Gaussian approximation of the response is logical, in view of the incurred statistical linearization. It turns out that for a large class of non-linear functions, the Gaussian approximation yields quite convenient formulae for the determination of the elements of the equivalent linear system. This is accomplished by using a formula given earlier, in Chapter 3 (equation (3.56)).

Utilizing equation (3.56), the left-hand side of equation (6.22) can be rewritten as

$$E[\Phi_i \hat{\mathbf{q}}] = E\{\hat{\mathbf{q}}\hat{\mathbf{q}}^{\mathrm{T}}\} E \left\{ \begin{bmatrix} \dfrac{\partial \Phi_i}{\partial \mathbf{q}} \\ \dfrac{\partial \Phi_i}{\partial \dot{\mathbf{q}}} \\ \dfrac{\partial \Phi_i}{\partial \ddot{\mathbf{q}}} \end{bmatrix} \right\} \quad i = 1, 2, \ldots, n \tag{6.40}$$

where

$$\frac{\partial \Phi_i}{\partial \mathbf{q}} = \left[\frac{\partial \Phi_i}{\partial \mathbf{q}_1}, \frac{\partial \Phi_i}{\partial \mathbf{q}_2}, \ldots, \frac{\partial \Phi_i}{\partial \mathbf{q}_n} \right]^{\mathrm{T}},$$

and similarly for $(\partial \Phi_i / \partial \dot{\mathbf{q}}), (\partial \Phi_i / \partial \ddot{\mathbf{q}})$. Then, combining equations (6.22) and

(6.40), and using the fact that the matrix $E\{\hat{\mathbf{q}}\hat{\mathbf{q}}^{\mathrm{T}}\}$ is non-singular, yields

$$\mathbf{m}_{i*}^{\mathrm{e}^{\mathrm{T}}} = E\left\{\frac{\partial \Phi_i}{\partial \ddot{\mathbf{q}}}\right\} \tag{6.41}$$

$$\mathbf{c}_{i*}^{\mathrm{e}^{\mathrm{T}}} = E\left\{\frac{\partial \Phi_i}{\partial \dot{\mathbf{q}}}\right\} \tag{6.42}$$

and

$$\mathbf{k}_{i*}^{\mathrm{e}^{\mathrm{T}}} = E\left\{\frac{\partial \Phi_i}{\partial \mathbf{q}}\right\} \tag{6.43}$$

Thus, the elements of the matrices $\mathbf{M}_{\mathrm{e}}$, $\mathbf{C}_{\mathrm{e}}$ and $\mathbf{K}_{\mathrm{e}}$ are given by the simple expressions

$$m_{ij}^{\mathrm{e}} = E\left\{\frac{\partial \Phi_i}{\partial \ddot{q}_j}\right\} \tag{6.44}$$

$$c_{ij}^{\mathrm{e}} = E\left\{\frac{\partial \Phi_i}{\partial \dot{q}_j}\right\} \tag{6.45}$$

and

$$k_{ij}^{\mathrm{e}} = E\left\{\frac{\partial \Phi_i}{\partial q_j}\right\} \tag{6.46}$$

These formulae are due to Kazakov (1965a). They were first used for stationary random vibration analysis by Atalik and Utku (1976). Subsequently, Spanos (1978, 1980a) pointed out that they are also applicable for non-stationary random vibration analysis (see Chapter 7).

6.5.2 Chain-like systems

In many practical problems the system of concern consists of a number of lumped masses, or nodal points, interconnected by non-linear elements whose behaviour depends only upon the relative coordinates between adjacent points. In such situations, where one has a chain-like system, it is convenient to equate the generalized coordinate vector $\mathbf{q}$ to the vector of the relative displacements between the nodal points. The non-linear function $\mathbf{\Phi}$ appearing in equation (6.1) then has the following specific form (Spanos and Iwan, 1978)

$$\mathbf{\Phi}(\mathbf{q}, \dot{\mathbf{q}}, \ddot{\mathbf{q}}) = \sum_{j=1}^{n} \mathbf{h}_j(q_j, \dot{q}_j, \ddot{q}_j) \tag{6.47}$$

The elements h_{ij} $(i = 1, 2, \ldots, n)$ in the vector

$$\mathbf{h}_j = [h_{1j}, h_{2j}, \ldots, h_{nj}]^{\mathrm{T}} \tag{6.48}$$

represent the actual non-linear elements, or parallel combinations of elementary elements, in the system. As an example, the non-linear function described by equation (6.47), in the absence of acceleration terms, could represent the non-linear stiffness and energy dissipation effects of the structural elements in multi-storey buildings. It is logical to assume that these effects depend only on the relative displacement and velocities between adjacent floors.

If equation (6.47) is substituted into equations (6.44) to (6.46) one obtains the following simple expressions for the parameters in the equivalent linear system given by equation (6.2)

$$m_{ij}^{\mathrm{e}} = E\left\{\frac{\partial h_{ij}}{\partial \ddot{q}_j}\right\} \tag{6.49}$$

$$c_{ij}^{\mathrm{e}} = E\left\{\frac{\partial h_{ij}}{\partial \dot{q}_j}\right\} \tag{6.50}$$

$$k_{ij}^{\mathrm{e}} = E\left\{\frac{\partial h_{ij}}{\partial q_j}\right\} \tag{6.51}$$

This result has the simple physical interpretation that each non-linear element, h_{ij}, in the system is linearized *individually*.

6.5.3 Treatment of asymmetric non-linearities

In the preceding developments it has been tacitly assumed that the non-linearity of the system is antisymmetrical, i.e.

$$\mathbf{\Phi}(\mathbf{q}, \dot{\mathbf{q}}, \ddot{\mathbf{q}}) = -\mathbf{\Phi}(-\mathbf{q}, -\dot{\mathbf{q}}, -\ddot{\mathbf{q}}) \tag{6.52}$$

for an arbitrary combination of $\mathbf{q}, \dot{\mathbf{q}}$ and $\ddot{\mathbf{q}}$. In this case, with the additional provision that the excitation process has a zero mean, the probability density function of the response may be approximated by an even function of $\mathbf{q}, \dot{\mathbf{q}}$ and $\ddot{\mathbf{q}}$, and it may be assumed that the response is a zero-mean process. However, in many practical cases the condition expressed by equation (6.52) does not hold and the mean of the excitation is non-zero. In these cirmustances the mean of the response may be non-zero, and some modification to the preceding analysis is required. (Spanos, 1978, 1980a)

This can be achieved by writing the excitation as

$$\mathbf{Q}(t) = \mathbf{Q}_0(t) + \mathbf{m}_Q \tag{6.53}$$

where $\mathbf{m}_Q = E\{\mathbf{Q}(t)\}$, and $\mathbf{Q}_0(t)$ is a zero-mean process. Similarly, for the response one can write (Spanos, 1981a)

$$\mathbf{q}(t) = \mathbf{q}_0(t) + \mathbf{m}_q \tag{6.54}$$

where $\mathbf{m}_q = E\{\mathbf{q}(t)\}$ and $\mathbf{q}_0(t)$ is a zero-mean process. Since it is assumed here that both $\mathbf{Q}(t)$ and $\mathbf{q}(t)$ are stationary processes, the means $\mathbf{m}_Q$ and $\mathbf{m}_q$ are constant vectors.

Using equations (6.53) and (6.54) the equation of motion (6.1) can be written as

$$\mathbf{M}\ddot{\mathbf{q}}_0 + \mathbf{C}\dot{\mathbf{q}}_0 + \mathbf{K}\mathbf{q}_0 + \mathbf{K}\mathbf{m}_q + \mathbf{\Phi}(\mathbf{q}_0 + \mathbf{m}_q, \dot{\mathbf{q}}_0, \ddot{\mathbf{q}}_0) = \mathbf{Q}_0(t) + \mathbf{m}_Q \tag{6.55}$$

Taking expectations of this last equation yields

$$\mathbf{K}\mathbf{m}_q + E\{\mathbf{\Phi}(\mathbf{q}_0 + \mathbf{m}_q, \dot{\mathbf{q}}_0, \ddot{\mathbf{q}}_0)\} = \mathbf{m}_Q \tag{6.56}$$

Subtracting equation (6.56) from equation (6.55) yields

$$\mathbf{M}\ddot{\mathbf{q}}_0 + \mathbf{C}\dot{\mathbf{q}}_0 + \mathbf{K}\mathbf{q}_0 + \mathbf{h}(\mathbf{q}_0, \dot{\mathbf{q}}_0, \ddot{\mathbf{q}}_0) = \mathbf{Q}_0(t) \tag{6.57}$$

where

$$\mathbf{h}(\mathbf{q}_0, \dot{\mathbf{q}}_0, \ddot{\mathbf{q}}_0) = \mathbf{\Phi}(\mathbf{q}_0 + \mathbf{m}_q, \dot{\mathbf{q}}_0, \ddot{\mathbf{q}}_0) - E\{\mathbf{\Phi}(\mathbf{q}_0 + \mathbf{m}_q, \dot{\mathbf{q}}_0, \ddot{\mathbf{q}}_0)\} \tag{6.58}$$

It may now be observed that equation (6.57) is identical in form to equation (6.1). Thus the formulae given earlier for the determination of the equivalent linear system (equations (6.44) to (6.46)) are also applicable in the present case, provided that the modifications given by equation (6.53), (6.54) and (6.58) are incorporated into the analysis.

6.6 SOLUTION PROCEDURES

6.6.1 General remarks

Upon determining the elements of the equivalent linear system, several options are available to the analyst for calculating pertinent response statistics. The choice of a particular method depends on the nature of the information which is being sought and the background of the analyst. Nevertheless, one aspect is crucial in the selection of the analysis method; i.e., whether or not the excitation process is modelled as a white noise.

In the general case of non-white excitation a relatively simple procedure is to work in the frequency domain using spectral input–output relationships (see Section 4.3). If the excitation is a white noise then the alternative state variable time domain analysis (see Section 4.4), is often simpler. One can note here that, even if the excitation is non-white, the problem can be usually reduced to that of a dynamic system with white noise excitation, provided that the dimension of the system is properly augmented to account for the non-white character of the input. However, if the dimension of the original system has to be increased very significantly then the relative advantage of the state-variable approach diminishes considerably and the frequency domain approach may be preferable.

If the excitation is non-white, but of a simple character (e.g., the output of a first-order, or second-order filter, driven by white noise) then an analysis based on the use of complex modes (see Chapter 4) offers an interesting alternative computational approach. This can be regarded as a variant of the usual state-variable approach.

In the remainder of this chapter these three approaches will be discussed within the context of MDOF systems, and illustrated through an analysis of a specific two degree of freedom system.

6.6.2 Spectral matrix solution procedure

For simplicity, a rather less general form of equation (6.1) will now be considered, where the non-linearity depends only on displacements and velocities; thus

$$\mathbf{M}\ddot{\mathbf{q}} + \mathbf{C}\dot{\mathbf{q}} + \mathbf{K}\mathbf{q} + \mathbf{g}(\mathbf{q}, \dot{\mathbf{q}}) = \mathbf{Q}(t) \tag{6.59}$$

where $\mathbf{g}(\mathbf{q}, \dot{\mathbf{q}})$ will be assumed to be an antisymmetric non-linearity and $\mathbf{Q}(t)$ is a zero-mean stationary random vector process with the spectral density matrix (see Section 3.11)

$$\mathbf{S}_Q(\omega) = \begin{bmatrix} S_{Q_1Q_1}(\omega) & S_{Q_1Q_2}(\omega) & \cdots & S_{Q_1Q_n}(\omega) \\ \vdots & & & \\ S_{Q_nQ_1}(\omega) & S_{Q_nQ_2}(\omega) & \cdots & S_{Q_nQ_n}(\omega) \end{bmatrix} \tag{6.60}$$

Here $S_{Q_iQ_j}(\omega)$ is the spectral density of the components $Q_i(t), Q_j(t)$ of $\mathbf{Q}(t)$, as defined by equation (3.121).

The response determination problem may be defined as that of seeking the corresponding spectral density matrix of the response process, $\mathbf{q}(t)$, which is of the form

$$\mathbf{S}_q(\omega) = \begin{bmatrix} S_{q_1q_1}(\omega) & S_{q_1q_2}(\omega) & \cdots & S_{q_1q_n}(\omega) \\ \vdots & & & \\ S_{q_nq_1}(\omega) & S_{q_nq_2}(\omega) & \cdots & S_{q_nq_n}(\omega) \end{bmatrix} \tag{6.61}$$

The first step is to compute the elements in the equivalent linear system, according to equations (6.44) to (6.46). Here the terms m_{ij}^{e} are zero and one has only the equivalent damping and stiffness elements, as follows

$$c_{ij}^{\mathrm{e}} = E\left\{\frac{\partial g_i}{\partial \dot{q}_j}\right\} \quad k_{ij}^{\mathrm{e}} = E\left\{\frac{\partial g_i}{\partial q_j}\right\} \tag{6.62}$$

The evaluation of the expectations here requires the solution to the equivalent linear system equation

$$\mathbf{M}\ddot{\mathbf{q}} + (\mathbf{C} + \mathbf{C}_{\mathrm{e}})\dot{\mathbf{q}} + (\mathbf{K} + \mathbf{K}_{\mathrm{e}})\mathbf{q} = \mathbf{Q}(t) \tag{6.63}$$

Using the matrix spectral input–output relationship given in Chapter 4 (see equation (4.19)) one can write

$$\mathbf{S}_q(\omega) = \boldsymbol{\alpha}(\omega)\mathbf{S}_Q(\omega)\boldsymbol{\alpha}^{\mathrm{T}*}(\omega) \tag{6.64}$$

where $\boldsymbol{\alpha}(\omega)$ is the matrix of frequency response functions. It is given by (see equation (4.22))

$$\boldsymbol{\alpha}(\omega) = [-\omega^2\mathbf{M} + \mathrm{i}\omega(\mathbf{C} + \mathbf{C}_{\mathrm{e}}) + (\mathbf{K} + \mathbf{K}_{\mathrm{e}})]^{-1} \tag{6.65}$$

Moreover, the cross-variance of the response can be calculated using the equation

$$E\{q_i(t)q_j(t)\} = \int_{-\infty}^{\infty} S_{q_iq_j}(\omega)\,\mathrm{d}\omega \tag{6.66}$$

where $S_{q_iq_j}(\omega)$ is the (i, j)th element of $\mathbf{S}_q(\omega)$.

Equations (6.62) to (6.66) establish a set of non-linear algebraic equations which must be solved iteratively. For instance, this procedure can be initialized by neglecting $\mathbf{g}(\mathbf{q}, \dot{\mathbf{q}})$ in equation (6.59) and calculating the elements of the matrix $E\{q_i(t)q_j(t)\}$. Then the elements of this matrix can be used in equation (6.62) to determine a first estimate of the damping and stiffness matrices of the equivalent linear system. Next the frequency response function matrix of the equivalent linear system is substituted into equation (6.64). This provides an update of $\mathbf{S}_q(\omega)$ which leads from equation (6.66) to an update of $E\{q_i(t)q_j(t)\}$. This procedure can be repeated until a satisfactory accuracy is achieved.

Example

Consider the two degree of freedom (TDOF) system shown in Figure 6.1. This system is dynamically equivalent to a wide variety of TDOF systems, including the two-storey building structure discussed earlier in Section 4.5 (see Figure 4.6a).

It will be assumed that an external force, $F(t)$, acts on the mass m_1. The absolute displacements of m_1 and m_2, measured from the static equilibrium position, are denoted by y_1 and y_2, respectively. A linear damper, with coefficient c_1 connects m_1 to the foundation, whereas a linear spring, of stiffness k_2 connects

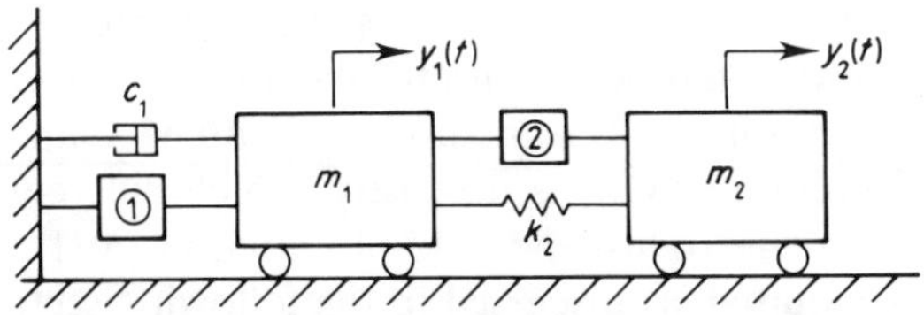

Figure 6.1
Two degree of freedom system. Element 1 non-linear spring, element 2 non-linear damper

m_1 and m_2. In addition, it is assumed that a non-linear spring, of the linear-plus-cubic type connects m_1 to the foundation, whereas m_1 and m_2 are connected by a non-linear damper of the linear-plus-quadratic type. Thus, the force in the non-linear spring is given by $k_1 y_1(1 + \varepsilon_1 y_1^2)$, whereas the force in the non-linear damper is given by $c_2(\dot{y}_2 - \dot{y}_1)(1 + \varepsilon_2 |\dot{y}_2 - \dot{y}_1|)$.

It is noted that in the special case where $\varepsilon_1 = \varepsilon_2 = 0$ the system is linear, and corresponds to that discussed in some detail in Section 4.5.2.

The equations of motion, in terms of the coordinates y_1 and y_2, may be written as

$$
\begin{aligned}
m_1\ddot{y}_1 + k_1 y_1(1 + \varepsilon_1 y_1^2) + c_1\dot{y}_1 - k_2(y_2 - y_1) \\
- c_2(\dot{y}_2 - \dot{y}_1)[1 + \varepsilon_2|\dot{y}_2 - \dot{y}_1|] = F(t) \\
m_2\ddot{y}_2 + k_2(y_2 - y_1) + c_2(\dot{y}_2 - \dot{y}_1)[1 + \varepsilon_2|\dot{y}_2 - \dot{y}_1|] = 0
\end{aligned}
\tag{6.67}
$$

Since the non-linear damping element force and the linear spring force $k_2(y_2 - y_1)$ depend on the relative velocity $\dot{y}_2 - \dot{y}_1$ and relative displacement, $y_2 - y_1$, respectively, it is convenient to introduce the transformation

$$
q_1 = y_1 \quad q_2 = y_2 - y_1 \tag{6.68}
$$

In terms of q_1, q_2 the equations can be written as

$$
\begin{aligned}
m_1\ddot{q}_1 + k_1 q_1(1 + \varepsilon_1 q_1^2) + c_1\dot{q}_1 - k_2 q_2 - c_2\dot{q}_2(1 + \varepsilon_2|\dot{q}_2|) = F(t) \\
m_2(\ddot{q}_1 + \ddot{q}_2) + k_2 q_2 + c_2\dot{q}_2(1 + \varepsilon_2|\dot{q}_2|) = 0
\end{aligned}
\tag{6.69}
$$

These equations are of the standard form given by equation (6.59). Here

$$
\mathbf{M} = \begin{bmatrix} m_1 & 0 \\ m_2 & m_2 \end{bmatrix} \quad \mathbf{C} = \begin{bmatrix} c_1 & -c_2 \\ 0 & c_2 \end{bmatrix} \tag{6.70}
$$

$$
\mathbf{K} = \begin{bmatrix} k_1 & -k_2 \\ 0 & k_2 \end{bmatrix} \quad \mathbf{Q}(t) = \begin{bmatrix} F(t) \\ 0 \end{bmatrix} \tag{6.71}
$$

and

$$
\mathbf{g}(\mathbf{q}, \dot{\mathbf{q}}) = \begin{bmatrix} \varepsilon_1 k_1 q_1^3 \\ \varepsilon_2 c_2 \dot{q}_2 |\dot{q}_2| \end{bmatrix} \tag{6.72}
$$

It is noted that the matrices **M** and **K** are not symmetric, in this case. This is because the force vector $\mathbf{Q}(t)$, as defined by equation (6.71), is not the generalized force vector corresponding to $\mathbf{q}$. For the present purposes this is not important, and we can proceed with the equations of motion as they stand. To render them into a completely standard form, with **M**, **C** and **K** all symmetric, a more careful definition of $\mathbf{Q}(t)$ is necessary, which differentiates it from the vector appearing on the right-hand side of equation (6.71). In fact, a linear transformation is required, as discussed in the following section.

The first step in the equivalent linearization procedure is to replace equation (6.69) by the linear equations given by (6.63). Here the non-linearity function in $\mathbf{g}(\mathbf{q}, \dot{\mathbf{q}})$ is of the special form discussed in Section 6.5.2, i.e. the system is 'chain-like' and so the non-linear elements can be linearized individually. A comparison of equation (6.72) with equations (6.47) and (6.48) shows that

$$\mathbf{g}(\mathbf{q}, \dot{\mathbf{q}}) = \begin{bmatrix} h_{11}(q_1) \\ h_{22}(\dot{q}_2) \end{bmatrix} \tag{6.73}$$

where

$$h_{11}(q_1) = \varepsilon_1 k_1 q_1^3 \quad h_{22}(\dot{q}_2) = \varepsilon_2 c_2 \dot{q}_2 |\dot{q}_2| \tag{6.74}$$

Hence, using equations (6.50) and (6.51) one has

$$k_{11}^{\mathrm{e}} = E\left\{\frac{\partial h_{11}}{\partial q_1}\right\} \quad c_{22}^{\mathrm{e}} = E\left\{\frac{\partial h_{22}}{\partial \dot{q}_2}\right\} \tag{6.75}$$

and the other elements of the matrices $\mathbf{C}_{\mathrm{e}}$ and $\mathbf{K}_{\mathrm{e}}$ are all zero.

From equations (6.74) and (6.75) one easily finds k_{11}^{e} to be given by

$$k_{\mathrm{e}} \equiv k_{11}^{\mathrm{e}} = 3\varepsilon_1 k_1 \sigma_{q_1}^2 \tag{6.76}$$

where σ_{q_1} is the standard deviation of q_1, i.e.

$$\sigma_{q_2}^2 = E\{q_1^2\} \tag{6.77}$$

For c_{22}^{e} one finds that

$$c_{22}^{\mathrm{e}} = 2\varepsilon_2 c_2 E\{|\dot{q}_2|\} = 2\varepsilon_2 c_2 \int_{-\infty}^{\infty} |\dot{q}_2| f(\dot{q}_2)\,\mathrm{d}\dot{q}_2 \tag{6.78}$$

where $f(\dot{q}_2)$ is the probability density function for $\dot{q}_2$. Taking this density to be of the Gaussian form (see equation (3.51)) one finds that

$$c_{\mathrm{e}} \equiv c_{22}^{\mathrm{e}} = \frac{4\varepsilon_2 c_2}{(2\pi)^{1/2}} \sigma_{\dot{q}_2} \tag{6.79}$$

where $\sigma_{\dot{q}_2}$ is the standard deviation of $\dot{q}_2$, i.e.

$$\sigma_{\dot{q}_2}^2 = E\{\dot{q}_2^2\} \tag{6.80}$$

With the equivalent linear parameters established it is possible to write the resulting linearized equations in the following standard matrix form

$$\begin{bmatrix} m_1 & 0 \\ m_2 & m_2 \end{bmatrix}\begin{bmatrix} \ddot{q}_1 \\ \ddot{q}_2 \end{bmatrix} + \begin{bmatrix} c_1 & -c_{2\mathrm{e}} \\ 0 & c_{2\mathrm{e}} \end{bmatrix}\begin{bmatrix} \dot{q}_1 \\ \dot{q}_2 \end{bmatrix} + \begin{bmatrix} k_{1\mathrm{e}} & -k_2 \\ 0 & k_2 \end{bmatrix}\begin{bmatrix} q_1 \\ q_2 \end{bmatrix} = \begin{bmatrix} F(t) \\ 0 \end{bmatrix} \tag{6.81}$$

where

$$k_{1\mathrm{e}} = k_1 + k_{\mathrm{e}} \tag{6.82}$$

$$c_{2\mathrm{e}} = c_2 + c_{\mathrm{e}} \tag{6.83}$$

Alternatively, they can be written in the following normalized form on dividing the first equation throughout by m_1, and the second equation throughout by m_2

$$\begin{bmatrix}1 & 0\\ 1 & 1\end{bmatrix}\begin{bmatrix}\ddot{q}_1\\ \ddot{q}_2\end{bmatrix}+\begin{bmatrix}2\zeta_{1e}\omega_e & -2\mu\zeta_{2e}\omega_2\\ 0 & 2\zeta_{2e}\omega_2\end{bmatrix}\begin{bmatrix}\dot{q}_1\\ \dot{q}_2\end{bmatrix}+\begin{bmatrix}\omega_e^2 & -\mu\omega_2^2\\ 0 & \omega_2^2\end{bmatrix}\begin{bmatrix}q_1\\ q_2\end{bmatrix}=\begin{bmatrix}p(t)\\ 0\end{bmatrix} \tag{6.84}$$

where

$$\omega_e^2=\frac{k_{1e}}{m_1}\quad \omega_2^2=\frac{k_2}{m_2} \tag{6.85}$$

$$\zeta_{1e}=\frac{c_1}{2(k_{1e}m_1)^{1/2}}\quad \zeta_{2e}=\frac{c_{2e}}{2(k_2m_2)^{1/2}} \tag{6.86}$$

$$\mu=\frac{m_2}{m_1}\quad p(t)=\frac{F(t)}{m_1} \tag{6.87}$$

The next step is to determine the appropriate elements of the frequency response function matrix, $\boldsymbol{\alpha}(\omega)$. Since the second element of $\mathbf{Q}(t)$ is here zero, only the first column of $\boldsymbol{\alpha}(\omega)$ is required. Defining

$$H_1(\omega)\equiv\alpha_{11}(\omega)\quad H_2(\omega)\equiv\alpha_{21}(\omega) \tag{6.88}$$

then $H_1(\omega)$ and $H_2(\omega)$ may be found by considering the steady-state response to an harmonic input. Thus, if

$$p(t)=\exp(\mathrm{i}\omega t) \tag{6.89}$$

then

$$q_1=H_1(\omega)\exp(\mathrm{i}\omega t)\quad q_2=H_2(\omega)\exp(\mathrm{i}\omega t) \tag{6.90}$$

Substituting equation (6.90) into equation (6.81) yields

$$\begin{aligned}(\omega_e^2-\omega^2+2\mathrm{i}\zeta_{1e}\omega_e\omega)H_1(\omega)-(\mu\omega_2^2+2\mathrm{i}\mu\zeta_{2e}\omega_2\omega)H_2(\omega)&=1\\ -\omega^2H_1(\omega)+(\omega_2^2-\omega^2+2\mathrm{i}\zeta_{2e}\omega_2\omega)H_2(\omega)&=0\end{aligned} \tag{6.91}$$

Solving for $H_1(\omega)$ and $H_2(\omega)$ one obtains

$$H_1(\omega)=\frac{\begin{vmatrix}0 & (\omega_2^2-\omega^2+2\mathrm{i}\zeta_{2e}\omega_2\omega)\\ 1 & -(\mu\omega_2^2+2\mathrm{i}\mu\zeta_{2e}\omega_2\omega\end{vmatrix}}{\Delta} \tag{6.92}$$

$$H_2(\omega)=\frac{\begin{vmatrix}-\omega^2 & 0\\ (\omega_e^2-\omega^2+2\mathrm{i}\zeta_{1e}\omega_e\omega) & 1\end{vmatrix}}{\Delta} \tag{6.93}$$

where

$$\Delta=\begin{vmatrix}-\omega^2 & (\omega_2^2-\omega^2+2\mathrm{i}\zeta_{2e}\omega_2\omega)\\ (\omega_e^2-\omega^2+2\mathrm{i}\zeta_{1e}\omega_e\omega) & (-\mu\omega_2^2-2\mathrm{i}\mu\zeta_{2e}\omega_2\omega)\end{vmatrix} \tag{6.94}$$

Now that $H_1(\omega)$ and $H_2(\omega)$ are determined, the variances $\sigma_{q_1}^2$ and $\sigma_{\dot{q}_2}^2$ can be expressed in terms of appropriate integrals. Specifically

$$\sigma_{q_1}^2 = \int_{-\infty}^{\infty} |H_1(\omega)|^2 S_p(\omega)\,\mathrm{d}\omega \tag{6.95}$$

and (see equation (3.113))

$$\sigma_{\dot{q}_2}^2 = \int_{-\infty}^{\infty} \omega^2 |H_2(\omega)|^2 S_p(\omega)\,\mathrm{d}\omega \tag{6.96}$$

where $S_p(\omega)$ is the spectral density of $p(t)$.

To proceed further a specific form for $S_p(\omega)$ will be assumed, i.e. a 'first-order' spectrum of the form

$$S_p(\omega) = \frac{S_0}{\alpha^2 + \omega^2} \tag{6.97}$$

A process with such a spectrum may be obtained by passing white noise through a linear, first-order filter.

On substituting equation (6.97) into equation (6.95) one obtains

$$\sigma_{q_1}^2 = S_0 \int_{-\infty}^{\infty} \frac{|H_1(\omega)|^2}{\alpha^2 + \omega^2}\,\mathrm{d}\omega = S_0 \int_{-\infty}^{\infty} \frac{\Xi_5^{(1)}(\omega)\,\mathrm{d}\omega}{\Lambda_5^{(1)}(-\mathrm{i}\omega)\Lambda_5^{(1)}(\mathrm{i}\omega)} \tag{6.98}$$

where

$$\Xi_5^{(1)}(\omega) = \xi_2\omega^4 + \xi_1\omega^2 + \xi_0 \tag{6.99}$$

and

$$\Lambda_5^{(1)}(\mathrm{i}\omega) = \lambda_5(\mathrm{i}\omega)^5 + \lambda_4(\mathrm{i}\omega)^4 + \lambda_3(\mathrm{i}\omega)^3 + \lambda_2(\mathrm{i}\omega)^2 + \lambda_1(\mathrm{i}\omega) + \lambda_0 \tag{6.100}$$

The parameters of the polynomials $\Xi(\omega)$ and $\Lambda(\mathrm{i}\omega)$ are given by the equations

$$\xi_2 = 1 \tag{6.101}$$

$$\xi_1 = \omega_2^2(4\zeta_{2e}^2 - 2) \tag{6.102}$$

$$\xi_0 = \omega_2^4 \tag{6.103}$$

$$\lambda_5 = 1 \tag{6.104}$$

$$\lambda_4 = 2\mu\zeta_{2e}\omega_2 + 2\zeta_{2e}\omega_2 + 2\zeta_{1e}\omega_e + \alpha \tag{6.105}$$

$$\lambda_3 = \mu\omega_2^2 + \omega_e^2 + \omega_2^2 + 4\zeta_{1e}\zeta_{2e}\omega_e\omega_2 + 2\mu\zeta_{2e}\omega_2\alpha + 2\zeta_{2e}\omega_2\alpha + 2\zeta_{1e}\omega_e\alpha \tag{6.106}$$

$$\lambda_2 = 2\zeta_{2e}\omega_2\omega_e^2 + 2\zeta_{1e}\omega_e\omega_2^2 + \mu\omega_2^2\alpha + \omega_e^2\alpha + \omega_2^2\alpha + 4\zeta_{1e}\zeta_{2e}\alpha\omega_e\omega_2 \tag{6.107}$$

$$\lambda_1 = 2\zeta_{2e}\alpha\omega_2\omega_e^2 + 2\zeta_{1e}\alpha\omega_e\omega_2^2 + \omega_e^2\omega_2^2 \tag{6.108}$$

$$\lambda_0 = \alpha\omega_e^2\omega_2^2 \tag{6.109}$$

The integral appearing in equation (6.98) can be calculated as the ratio of two determinants (see Appendix B)

$$\sigma_{q_1}^2 = \frac{\pi S_0}{\lambda_5} \frac{\begin{vmatrix} 0 & 0 & \xi_2 & \xi_1 & \xi_0 \\ -\lambda_5 & \lambda_3 & -\lambda_1 & 0 & 0 \\ 0 & -\lambda_4 & \lambda_2 & -\lambda_0 & 0 \\ 0 & \lambda_5 & -\lambda_3 & \lambda_1 & 0 \\ 0 & 0 & \lambda_4 & -\lambda_2 & \lambda_0 \end{vmatrix}}{\begin{vmatrix} \lambda_4 & -\lambda_2 & \lambda_0 & 0 & 0 \\ -\lambda_5 & \lambda_3 & -\lambda_1 & 0 & 0 \\ 0 & -\lambda_4 & \lambda_2 & -\lambda_0 & 0 \\ 0 & \lambda_5 & -\lambda_3 & \lambda_1 & 0 \\ 0 & 0 & \lambda_4 & -\lambda_2 & \lambda_0 \end{vmatrix}} \tag{6.110}$$

Similarly to the calculation of σ_{q_1} the variance of $\dot{q}_2$ can be determined by the equation

$$\sigma_{\dot{q}_2}^2 = S_0 \int_{-\infty}^{\infty} \frac{\omega^2 |H_2(\omega)|^2}{\alpha^2 + \omega^2} \mathrm{d}\omega = S_0 \int_{-\infty}^{\infty} \frac{\Xi^{(2)}(\omega)\,\mathrm{d}\omega}{\Lambda_5^{(1)}(-\mathrm{i}\omega)\Lambda_5^{(1)}(\mathrm{i}\omega)} \mathrm{d}\omega \tag{6.111}$$

where

$$\Xi^{(2)}(\omega) = \omega^6 \tag{6.112}$$

Thus,

$$\sigma_{\dot{q}_2}^2 = \frac{\pi S_0}{\lambda_5} \frac{\begin{vmatrix} 0 & 1 & 0 & 0 & 0 \\ -\lambda_5 & \lambda_3 & -\lambda_1 & 0 & 0 \\ 0 & -\lambda_4 & \lambda_2 & -\lambda_0 & 0 \\ 0 & \lambda_5 & -\lambda_3 & \lambda_1 & 0 \\ 0 & 0 & \lambda_4 & -\lambda_2 & \lambda_0 \end{vmatrix}}{\begin{vmatrix} \lambda_4 & -\lambda_2 & \lambda_0 & 0 & 0 \\ -\lambda_5 & \lambda_3 & -\lambda_1 & 0 & 0 \\ 0 & -\lambda_4 & \lambda_2 & -\lambda_0 & 0 \\ 0 & \lambda_5 & -\lambda_3 & \lambda_1 & 0 \\ 0 & 0 & \lambda_4 & -\lambda_2 & \lambda_0 \end{vmatrix}} \tag{6.113}$$

The preceding procedure establishes a cyclic relationship between the element of the equivalent linear system, and σ_{q_1} and $\sigma_{\dot{q}_2}$. Thus, an iterative scheme must be used to solve the set of non-linear equations. For this, initial estimates of σ_{q_1} and $\sigma_{\dot{q}_2}$ can be obtained by neglecting the non-linear elements of the system, that is $k_e = c_e = 0$. These values can be used in equations (6.76) and (6.79) to update the values of k_e and c_e. Then, the new values of c_e and k_e can be substituted into equations (6.110) and (6.113) to determine new values of σ_{q_1} and $\sigma_{\dot{q}_2}$. This procedure may be repeated several times to yield sufficiently accurate solutions. Several elements of the determinants appearing in equations (6.110) and (6.113) are equal to zero, a fact which facilitates the evaluation of these determinants.

To illustrate the rate of convergence which is achieved with this computation scheme, some typical results are given in Tables 6.1 and 6.2. Here, in both cases,

Table 6.1

Number of iterations	k_{1e} ($N m^{-1}$)	c_{2e} ($Ns m^{-1}$)	$\sigma^2_{q_1}$ (m^2)	$\sigma^2_{\dot{q}_2}$ ($m^2 s^{-2}$)
1	0.373	0.0778	2.485	0.950
2	0.163	0.0760	1.084	0.908
3	0.249	0.0759	1.660	0.904
4	0.207	0.0760	1.382	0.908
5	0.226	0.0760	1.507	0.907
6	0.217	0.0760	1.449	0.908
—	—	—	—	—
10	0.220	0.0760	1.467	0.907
—	—	—	—	—
20	0.220	0.0760	1.467	0.907

$\omega_1 = 1$ rad s^{-1}, $\omega_2 = 1$ rad s^{-1},
$\zeta_1 = 0.05$, $\zeta_2 = 0.2$, $\mu = 1$
$S_0 = 1\ N^2 s\, kg^{-2}$, $\alpha = 2$ rad s^{-1}, $\varepsilon_1 k_1 = 0.05\ N m^{-3}$,
$\varepsilon_2 c_2 = 0.05\ Ns^2 m^{-2}$.

Table 6.2

Number of iterations	k_{1e} ($N m^{-1}$)	c_{2e} ($Ns m^{-1}$)	$\sigma^2_{q_1}$ (m^2)	$\sigma^2_{\dot{q}_2}$ ($m^2 s^{-2}$)
1	1.491	0.311	2.485	0.950
5	0.760	0.280	1.266	0.770
10	0.416	0.278	0.693	0.757
15	0.529	0.279	0.881	0.764
20	0.483	0.279	0.805	0.762
—	—	—	—	—
30	0.493	0.279	0.822	0.762
40	0.495	0.279	0.825	0.762
50	0.495	0.279	0.826	0.762

$\omega_1 = 1$ rad s^{-1}, $\omega_2 = 1$ rad s^{-1}
$\zeta_1 = 0.05$, $\zeta_2 = 0.2$, $\mu = 1$
$S_0 = 1\ Ns^2 kg^{-2}$, $\alpha = 2$ rad s^{-1}, $\varepsilon_1 k_1 = 0.2\ N m^{-3}$, $\varepsilon_2 c_2 = 0.2\ Ns^2 m^{-2}$

$\omega_1 = \omega_2 = 1\ \text{rad s}^{-1}$, $\zeta_1 = 0.05$, $\zeta_2 = 0.2$, $\mu = 1$, $S_0 = 1\ \text{Ns}^2\ \text{kg}^{-2}$ and $\alpha = 2\ \text{rad s}^{-1}$, where

$$\omega_1 = \left(\frac{k_1}{m_1}\right)^{1/2} \quad \omega_2 = \left(\frac{k_2}{m_2}\right)^{1/2} \tag{6.114}$$

$$\zeta_1 = \frac{c_1}{2(k_1 m_1)^{1/2}}, \quad \zeta_2 = \frac{c_2}{2(k_2 m_2)^{1/2}} \tag{6.115}$$

For the case of a relatively weak non-linearity ($\varepsilon_1 k = 0.05\ \text{N m}^{-3}$, $\varepsilon_2 c = 0.05\ \text{Ns}^2\ \text{m}^{-2}$) it is seen from Table 6.1 that the convergence rate is reasonably fast, and that stabilization of the estimates of $k_{1e}, c_{2e}, \sigma^2_{q_1}$ and $\sigma^2_{q_2}$ is achieved after ten iterations. Thus a further ten iterations can be seen to have little affect on these estimates. As the degree of non-linearity increases, the rate of convergence will tend to slow down. Thus, for example, in Table 6.2, where $\varepsilon_1 k_1 = 0.2\ \text{N m}^{-3}$ and $\varepsilon_2 c_2 = 0.2\ \text{Ns}^2\ \text{m}^{-2}$, it is seen that about fifty iterations are required to achieve stable estimates.

6.6.3 Modal analysis

Clearly it is possible, in some cases, to calculate the integral in equation (6.66) exactly, using the expressions given in Appendix B, (as the preceding example has demonstrated), for complex systems this approach can lead to excessive algebra. However, if the damping is light, one can usually simplify the analysis, with little loss of accuracy, by employing a modal expansion of the matrix, $\boldsymbol{\alpha}(\omega)$, of frequency response functions. It is shown in Chapter 4 that $\boldsymbol{\alpha}(\omega)$ can be expanded in the following series

$$\boldsymbol{\alpha}(\omega) = \sum_{r=1}^{n} \mathbf{k}^{(r)} \frac{1}{(\omega_r^2 - \omega^2 + 2\mathrm{i}\zeta_r \omega \omega_r)} \tag{6.116}$$

where

$$\mathbf{k}^{(r)} = \frac{\boldsymbol{\lambda}^{(r)} \boldsymbol{\lambda}^{\mathrm{T}(r)}}{l_r} \tag{6.117}$$

ω_r is the rth natural frequency of the undamped, linearized system, ζ_r the rth modal damping factor, $\boldsymbol{\lambda}^{(r)}$ is the rth normalized eigenvector and l_r is the rth modal mass (see Section 4.4). The elements of $\mathbf{k}^{(r)}$ are dependent on the mode number, r, and are usually called 'modal participation factors'. The parameters ω_r and $\mathbf{k}^{(r)}$ can be deduced from an eigenvalue analysis of the free, undamped linearized system equations, given by

$$\mathbf{M}\ddot{\mathbf{q}} + (\mathbf{K} + \mathbf{K}_e)\mathbf{q} = \mathbf{0} \tag{6.118}$$

The use of a modal expansion for $\boldsymbol{\alpha}(\omega)$ has two distinct advantages. Firstly, the integrals for the cross-covariance, $E\{q_i(t)q_j(t)\}$ (see equation (6.66)) can be written in the form

$$\int_{-\infty}^{\infty} \frac{S(\omega)}{(\omega_r^2 - \omega^2 + 2\mathrm{i}\zeta_r\omega\omega_r)(\omega_s^2 - \omega^2 + 2\mathrm{i}\zeta_s\omega\omega_s)}\,\mathrm{d}\omega \tag{6.119}$$

where $S(\omega)$ is a component of the matrix $\mathbf{S}_Q(\omega)$. If $S(\omega)$ is a rational function of ω then this integral can be evaluated fairly easily by the method given in Appendix B. It is noted that the order of the polynomial in $(\mathrm{i}\omega)$, in the denominator of the integral in equation (6.119), is four, irrespective of the number of degrees of freedom, n. In the exact approach the order of this polynomial is $2n$. The second advantage of the modal approximation is that, in many applications, only a few modes contribute significantly to the response. Thus, it is usually necessary to consider only a few terms in the expansion (6.116). A disadvantage of the modal approximation is that, during the iteration process, the new natural modes and frequencies of the system must be calculated as the equivalent linear stiffnesses are being updated. However, if the non-linearity of the system involves only damping terms, such as wave induced forces on offshore structures (see equation (2.109)), this disadvantage does not apply.

Example

To illustrate the application of modal analysis the system shown in Figure 6.1 will again be considered.

The linearized equations of motion, in terms of q_1 and q_2, are given by equation (6.81). As previously noted, the mass and stiffness matrices are not symmetrical, here, due to the fact that the elements on the right-hand side are not the generalized forces corresponding to q_1 and q_2.

Consider the case where forces F_1 and F_2 are applied to m_1 and m_2, respectively. During a small virtual displacement, δy_1, for m_1 and δy_2 for m_2, the virtual work done, δW is given by

$$\delta W = F_1\delta y_1 + F_2\delta y_2 \tag{6.120}$$

By definition, however, one has

$$\delta W = Q_1\delta q_1 + Q_2\delta q_2 \tag{6.121}$$

where $\delta q_1, \delta q_2$ are displacements corresponding to $\delta y_1, \delta y_2$, and Q_1, Q_2 are the generalized forces corresponding to q_1, q_2. Using equation (6.68) it follows that

$$\delta y_1 = \delta q_1 \quad \delta y_2 = \delta q_1 + \delta q_2 \tag{6.122}$$

Hence, from equations (6.120) to (6.122),

$$\mathbf{F} = \mathbf{TQ} \tag{6.123}$$

where

$$\mathbf{F} = \begin{bmatrix} F_1 \\ F_2 \end{bmatrix} \quad \mathbf{Q} = \begin{bmatrix} Q_1 \\ Q_2 \end{bmatrix} \quad \mathbf{T} = \begin{bmatrix} 1 & -1 \\ 0 & 1 \end{bmatrix} \tag{6.124}$$

If $\mathbf{F}$ on the right-hand side of equation (6.81) is replaced by the product $\mathbf{TQ}$, then on pre-multiplying equation (6.81) throughout by the inverse of $\mathbf{T}$, one obtains the equation

$$\begin{bmatrix} (m_1+m_2) & m_2 \\ m_2 & m_2 \end{bmatrix}\begin{bmatrix} \ddot{q}_1 \\ \ddot{q}_2 \end{bmatrix}+\begin{bmatrix} c_1 & 0 \\ 0 & c_{2e} \end{bmatrix}\begin{bmatrix} \dot{q}_1 \\ \dot{q}_2 \end{bmatrix}+\begin{bmatrix} k_{1e} & 0 \\ 0 & k_2 \end{bmatrix}\begin{bmatrix} q_1 \\ q_2 \end{bmatrix}=\begin{bmatrix} Q_1 \\ Q_2 \end{bmatrix} \tag{6.125}$$

The equations are now in the completely standard form, with **M**, **C** and **K** all symmetric. Furthermore

$$\begin{bmatrix} Q_1 \\ Q_2 \end{bmatrix}=\begin{bmatrix} 1 & 1 \\ 0 & 1 \end{bmatrix}\begin{bmatrix} F_1 \\ F_2 \end{bmatrix}=\begin{bmatrix} F(t) \\ 0 \end{bmatrix} \tag{6.126}$$

since, in the specific case under discussion, $F_1=F(t)$ and $F_2=0$.

If equation (6.125) is divided throughout by m_1, and the definitions (6.85) to (6.87) are used, together with equation (6.126), then the equations of motion can be written as

$$\begin{bmatrix} (1+\mu) & \mu \\ \mu & \mu \end{bmatrix}\begin{bmatrix} \ddot{q}_1 \\ \ddot{q}_2 \end{bmatrix}+\begin{bmatrix} 2\zeta_{1e}\omega_{1e} & 0 \\ 0 & 2\zeta_{2e}\mu\omega_2 \end{bmatrix}\begin{bmatrix} \dot{q}_1 \\ \dot{q}_2 \end{bmatrix}$$
$$+\begin{bmatrix} \omega_{1e}^2 & 0 \\ 0 & \mu\omega_2^2 \end{bmatrix}\begin{bmatrix} q_1 \\ q_2 \end{bmatrix}=\begin{bmatrix} p(t) \\ 0 \end{bmatrix} \tag{6.127}$$

As a first step in a modal analysis, the reduced case of free, undamped vibration is considered (see Section 4.4.2). The equation for the natural frequencies is given by equation (4.28), where here

$$\mathbf{M}=\begin{bmatrix} (1+\mu) & \mu \\ \mu & \mu \end{bmatrix} \quad \mathbf{K}=\begin{bmatrix} \omega_{1e}^2 & 0 \\ 0 & \mu\omega_2^2 \end{bmatrix} \tag{6.128}$$

Hence one obtains

$$\begin{vmatrix} -\omega^2(1+\mu)+\omega_{1e}^2 & -\mu\omega^2 \\ -\mu\omega^2 & \mu(\omega_2^2-\omega^2) \end{vmatrix}=0 \tag{6.129}$$

Solving this equation the following expression for the two natural frequencies, ω_1 and ω_2, is obtained

$$2\omega_{1,2}^2=[\omega_{1e}^2+\omega_2^2(1+\mu)]$$
$$\pm\{[\omega_{1e}^2+\omega_2^2(1+\mu)]^2-4\omega_{1e}^2\omega_2^2\}^{1/2} \tag{6.130}$$

Thus, there are two modes of free undamped vibration, as follows

$$\mathbf{q}_0^{(1)}\exp(\mathrm{i}\omega_1 t) \quad \mathbf{q}_0^{(2)}\exp(\mathrm{i}\omega_2 t) \tag{6.131}$$

Further, the modal matrix, $\boldsymbol{\lambda}$ (see Section 4.4.2)), is given by

$$\boldsymbol{\lambda}=[\mathbf{q}_0^{(1)} \quad \mathbf{q}_0^{(2)}] \tag{6.132}$$

The two modal columns, or eigenvectors, $\mathbf{q}_0^{(1)}, \mathbf{q}_0^{(2)}$, are easily found by substituting equations (6.131) into the reduced equation of motion (6.127), corresponding to no damping and no excitation. On normalizing $\mathbf{q}_0^{(1)}, \mathbf{q}_0^{(2)}$ so that their first elements are unity it is found, after some algebra, that

$$\boldsymbol{\lambda} = \begin{bmatrix} 1 & 1 \\ r_1 & r_2 \end{bmatrix} \tag{6.133}$$

where

$$r_i = \frac{\omega_{1e}^2 - (1+\mu)\omega_i^2}{\mu\omega_i^2} \quad i = 1, 2 \tag{6.134}$$

Using $\boldsymbol{\lambda}$ one can rewrite the equation of motion (6.127) in terms of normal coordinates, $\boldsymbol{\Phi}$, where

$$\mathbf{q} = \boldsymbol{\lambda}\boldsymbol{\Phi} \tag{6.135}$$

as described in Section 4.4.3. Thus the equation of motion can be written in the form of equation (4.39), where **L** and **N** are defined by equation (4.35) and **D** and **P** are defined by equation (4.40). In the present case

$$\mathbf{L} = \begin{bmatrix} 1 & r_1 \\ 1 & r_2 \end{bmatrix} \begin{bmatrix} (1+\mu) & \mu \\ \mu & \mu \end{bmatrix} \begin{bmatrix} 1 & 1 \\ r_1 & r_2 \end{bmatrix} = \begin{bmatrix} l_1 & 0 \\ 0 & l_2 \end{bmatrix} \tag{6.136}$$

where

$$l_i = 1 + \mu(1 + r_i)^2 \quad i = 1, 2 \tag{6.137}$$

Further

$$\mathbf{N} = \begin{bmatrix} 1 & r_1 \\ 1 & r_2 \end{bmatrix} \begin{bmatrix} \omega_{1e}^2 & 0 \\ 0 & \mu\omega_2^2 \end{bmatrix} \begin{bmatrix} 1 & 1 \\ r_1 & r_2 \end{bmatrix} = \begin{bmatrix} n_1 & 0 \\ 0 & n_2 \end{bmatrix} \tag{6.138}$$

where

$$n_i = \omega_{1e}^2 + \mu\omega_2^2 r_i^2 \quad i = 1, 2 \tag{6.139}$$

The transformed damping matrix

$$\mathbf{D} = \begin{bmatrix} 1 & r_1 \\ 1 & r_2 \end{bmatrix} \begin{bmatrix} 2\zeta_{1e} & 0 \\ 0 & 2\zeta_{2e}\mu\omega_2 \end{bmatrix} \begin{bmatrix} 1 & 1 \\ r_1 & r_2 \end{bmatrix} = \begin{bmatrix} d_{11} & d_{12} \\ d_{21} & d_{22} \end{bmatrix} \tag{6.140}$$

is no longer diagonal as, in general, one can expect. However, as explained in Section 4.4, if the damping is light, a very good approximation is to neglect the off-diagonal terms d_{12}, d_{21}. Then one has

$$\mathbf{D} = \begin{bmatrix} d_1 & 0 \\ 0 & d_2 \end{bmatrix} \tag{6.141}$$

where

$$d_i \equiv d_{ii} = 2\zeta_{1e}\omega_{1e} + 2\zeta_{2e}\mu\omega_2 r_i^2 \quad i = 1, 2 \tag{6.142}$$

The frequency response function matrix, $\mathbf{H}(\omega)$, can now be expressed, approximately, by the modal expansion (4.43), where here $n = 2$ and the two modal damping factors ζ_1 and ζ_2 are given by equation (4.47). In particular

$$H_1(\omega) \equiv \alpha_{11}(\omega) = \sum_{j=1}^{2} k_{j1}\alpha_j(\omega) \tag{6.143}$$

and

$$H_2(\omega) \equiv \alpha_{21}(\omega) = \sum_{j=1}^{2} k_{j2}\alpha_j(\omega) \qquad (6.144)$$

where

$$\alpha_j(\omega) = \frac{1}{(\omega_j^2 - \omega^2 + 2\mathrm{i}\zeta_j\omega_j\omega)} \qquad (6.145)$$

and the 'modal participation factors', k_{j1}, k_{j2} are given by

$$k_{j1} = 1/l_j \quad k_{j2} = r_j^2/l_j \quad j = 1, 2 \qquad (6.146)$$

These expressions may be compared with the corresponding exact expressions given by equations (6.92) and (6.93).

On substituting expressions (6.143) and (6.144) into equations (6.95) and (6.96), one can evaluate the variances $\sigma_{q_1}^2, \sigma_{\dot{q}_2}^2$ and, as before, establish a cyclic relationship between the elements of the equivalent linear system and σ_{q_1} and $\sigma_{\dot{q}_2}$. In this process it is necessary to evaluate an integral of the form

$$I = \int_{-\infty}^{\infty} \frac{\alpha^2}{(\alpha^2 + \omega^2)} \alpha_i(\omega)\alpha_j^*(\omega)\,\mathrm{d}\omega \qquad (6.147)$$

This integral can be evaluated by using the result given in Appendix B. Note that of the integral discussed earlier in Chapter 4 (see equation (4.122)) is a special case of the integral in equation (6.147). In this context it is necessary to multiply both the numerator and the denominator of the integrand in equation (6.147) by a suitable common factor. Here this is as follows

$$P(\mathrm{i}\omega) \equiv (\omega_i^2 - \omega^2 - 2\mathrm{i}\zeta_i\omega_i\omega)(\omega_j^2 - \omega^2 + 2\mathrm{i}\zeta_j\omega_j\omega) \qquad (6.148)$$

This operation converts the required integral into the standard form given by equation (B1). Thus

$$I = \int_{-\infty}^{\infty} \frac{\alpha^2 P(\mathrm{i}\omega)}{|Q(\mathrm{i}\omega)|^2}\,\mathrm{d}\omega \qquad (6.149)$$

where

$$Q(\mathrm{i}\omega) = (\alpha + \mathrm{i}\omega)(\omega_i^2 - \omega^2 + 2\mathrm{i}\zeta_i\omega_i\omega)(\omega_j^2 - \omega^2 + 2\mathrm{i}\zeta_j\omega_j\omega) \qquad (6.150)$$

and, on comparing equation (6.149) to equation (B1), it is evident that $I = I_5$, where, in equation (B1),

$$\Xi_5(\omega) = \alpha^2\,\mathrm{Re}\{P(\mathrm{i}\omega)\} \quad \Lambda_5(\mathrm{i}\omega) = Q(\mathrm{i}\omega) \qquad (6.151)$$

From equation (B26), I_5 is proportional to the ratio of two 5×5 determinants, as in equations (6.110) and (6.113). In the present case, from a comparison of equations (6.150) and (6.151) with equations (B2) and (B3) one finds that

$$\xi_0 = \alpha^2\omega_i^2\omega_j^2 \qquad (6.152)$$

$$\xi_1 = -\alpha^2(\omega_i^2 + \omega_j^2 - 4\zeta_i\zeta_j\omega_i\omega_j) \tag{6.153}$$

$$\xi_2 = \alpha^2 \tag{6.154}$$

$$\xi_3 = \xi_4 = 0 \tag{6.155}$$

Also,

$$\lambda_0 = \omega_i^2\omega_j^2 \tag{6.156}$$

$$\lambda_1 = \alpha(2\zeta_i\omega_i\omega_j^2 + 2\zeta_j\omega_i^2\omega_j) + \omega_i^2\omega_j^2 \tag{6.157}$$

$$\lambda_2 = \alpha(\omega_i^2 + \omega_j^2 + 4\zeta_i\zeta_j\omega_i\omega_j) + 2\zeta_i\omega_i\omega_j^2 + 2\zeta_j\omega_i^2\omega_j \tag{6.158}$$

$$\lambda_3 = \alpha(2\zeta_i\omega_i + 2\zeta_j\omega_j) + \omega_i^2 + \omega_j^2 + 4\zeta_i\zeta_j\omega_i\omega_j \tag{6.159}$$

$$\lambda_4 = \alpha + 2\zeta_i\omega_i + 2\zeta_j\omega_j \tag{6.160}$$

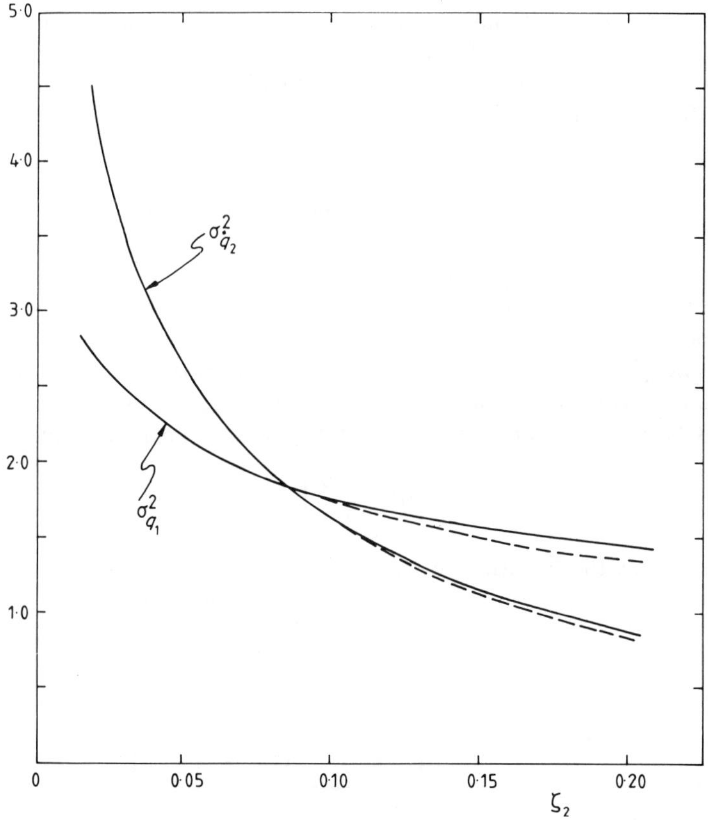

Figure 6.2
Variation of the mean square displacement response, $\sigma^2_{q_1}$, and the mean square velocity response, $\sigma^2_{q_2}$, with ζ_2, obtained by using the exact frequency response functions (———), and the modal series approximations (– – –) for these functions

It is noted that, in the limiting case where $\alpha \to \infty$, corresponding to a white noise input, I is similar to the integral considered earlier, in Section (4.5.2). The appropriate result here is given by equation (4.132).

Figure 6.2 shows a comparison between estimates of $\sigma^2_{q_1}$ and $\sigma^2_{\dot{q}_2}$, obtained by the equivalent linearization procedure, using (a) the exact expressions for the frequency response functions $H_1(\omega)$ and $H_2(\omega)$ of the equivalent linear system and (b) the modal series approximation for these frequency response functions. For the numerical results the following parameter values were selected

$$\omega_1 = \omega_2 = 1 \text{ rad s}^{-1} \quad \zeta_1 = 0.05$$
$$\mu = 1 \quad S_0 = 1 \text{ N}^2 \text{ s kg}^{-2} \quad \alpha = 2 \text{ rad s}^{-1}$$
$$\varepsilon_1 k_1 = 0.05 \text{ N m}^{-3} \quad \varepsilon_2 c = 0.05 \text{ Ns}^2 \text{ m}^{-2}$$

The values of $\sigma^2_{q_1}$ and $\sigma^2_{\dot{q}_2}$ are plotted in Figure 6.2, against ζ_2. It is observed that, as expected, the modal expansion gives good agreement with the more exact method when ζ_2 (and ζ_1) are small. As ζ_2 increases, the accuracy of the modal approximation clearly tends to diminish but it is still reasonably good for ζ_2 values as high as 0.20.

6.6.4 State variable solution procedure

In this section the state variable approach (see Section 4.5.4) to dealing with an equivalent linear MDOF system will be exemplified by considering, once again, the case of a system with a non-linearity which is a function of the velocity and displacement vectors only. Such a system is governed by equation (6.59), and the equivalent linear system is given by equation (6.63).

As in Chapter 4 (see Section 4.4.4) a state vector $\mathbf{z}(t)$ can be defined as

$$\mathbf{z}(t) = \begin{bmatrix} \mathbf{q} \\ \dot{\mathbf{q}} \end{bmatrix} \tag{6.161}$$

In terms of $\mathbf{z}(t)$, the linearized equation of motion (6.63) can be written as a first-order matrix equation, as follows (see also equations (4.53) and (4.54)),

$$\dot{\mathbf{z}} = \mathbf{G}\mathbf{z} + \mathbf{f} \tag{6.162}$$

where

$$\mathbf{G} = \begin{bmatrix} \mathbf{0} & \mathbf{I} \\ -\mathbf{M}^{-1}(\mathbf{K}+\mathbf{K}_e) & -\mathbf{M}^{-1}(\mathbf{C}+\mathbf{C}_e) \end{bmatrix} \tag{6.163}$$

and

$$\mathbf{f} = \begin{bmatrix} \mathbf{0} \\ \mathbf{M}^{-1}\mathbf{Q} \end{bmatrix} = \begin{bmatrix} \mathbf{0} \\ \mathbf{p}(t) \end{bmatrix} \tag{6.164}$$

If, as before, it is assumed that the mean of $\mathbf{Q}(t)$ is zero, and that $\boldsymbol{\phi}(\mathbf{q}, \dot{\mathbf{q}})$ is antisymmetric, then it follows that

$$\mathbf{m}_z = E\{\mathbf{z}(t)\} = \mathbf{0} \tag{6.165}$$

and that the covariance matrix $\mathbf{V}$ may be defined here as (see equation (3.19))

$$\mathbf{V} = E\{\mathbf{z}(t)\mathbf{z}^{\mathrm{T}}(t)\} \tag{6.166}$$

A differential equation for $\mathbf{V}$ can be written in terms of the covariance matrix of $\mathbf{f}(t)$

$$\mathbf{w}_f(t, \tau) = E\{\mathbf{f}(t)\mathbf{f}^{\mathrm{T}}(\tau)\} \tag{6.167}$$

as shown in Section 4.5.4. Here attention is focused on the stationary response, and hence it can be assumed that the elements of $\mathbf{V}$ are constants. Then $\mathrm{d}\mathbf{V}/\mathrm{d}t = \mathbf{0}$ in equation (4.153) and, since $\mathbf{w}_f(t, \tau) = \mathbf{w}_f(t - \tau)$ here, the following equation governs $\mathbf{V}$

$$\mathbf{G}\mathbf{V}^{\mathrm{T}} + \mathbf{V}\mathbf{G}^{\mathrm{T}} + \int_0^{\infty} \exp(\mathbf{G}\mathbf{u})[\mathbf{w}_f(\mathbf{u}) + \mathbf{w}_f^{\mathrm{T}}(\mathbf{u})]\, \mathrm{d}\mathbf{u} = \mathbf{0} \tag{6.168}$$

This equation can be solved numerically to yield the elements of $\mathbf{V}$, in terms of the elements $\mathbf{K}_{\mathrm{e}}$ and $\mathbf{C}_{\mathrm{e}}$. The latter are given by equations (6.45) and (6.46) and hence, as before, a cyclic relationship is established, which can be solved iteratively.

The calculations are much simplified if the elements of the input, $\mathbf{Q}(t)$, can be modelled as white noises. Then

$$\mathbf{w}_f(t - \tau) = \mathbf{R}\delta(t - \tau) \tag{6.169}$$

where $\mathbf{R}$ is a matrix of constants, and equation (6.168) can be reduced to

$$\mathbf{G}\mathbf{V} + \mathbf{V}\mathbf{G}^{\mathrm{T}} + \mathbf{R} = \mathbf{0} \tag{6.170}$$

In those cases where $\mathbf{Q}(t)$ is not a white noise vector it is possible to augment the order of the overall system by introducing a 'shaping filter'. A specific example of this approach, for a SDOF system, was given earlier, in Section 5.3.5. Thus the input, $\mathbf{p}(t) = \mathbf{M}^{-1}\mathbf{Q}$ can be considered to arise as the response of a linear filter to a white noise vector, $\mathbf{n}(t)$. If the filter is of mth order, one can write

$$\mathbf{v}_{m-1}\boldsymbol{\phi}^{(m-1)} + \mathbf{v}_{m-2}\boldsymbol{\phi}^{(m-2)} + \cdots + \mathbf{v}_0\boldsymbol{\phi}^{(0)} = \mathbf{p} \tag{6.171}$$

$$\boldsymbol{\phi}^{(m)} + \boldsymbol{\lambda}_{m-1}\boldsymbol{\phi}^{(m-1)} + \cdots + \boldsymbol{\lambda}_0\boldsymbol{\phi}^{(0)} = \mathbf{n}(t) \tag{6.172}$$

where $\boldsymbol{\phi}$ is an n-vector and the superscript (r) denotes the rth time derivative. $\boldsymbol{\lambda}_0, \boldsymbol{\lambda}_1, \ldots, \lambda_{m-1}$ and $\mathbf{v}_0, \mathbf{v}_1, \ldots, \mathbf{v}_{m-1}$ are $n \times n$ matrices, the elements of which are constants.

The filter equation (6.171) may be written in the following state variable form

$$\dot{\boldsymbol{\theta}} = \mathbf{F}\boldsymbol{\theta} + \mathbf{w} \tag{6.173}$$

where

$$\boldsymbol{\theta} = \begin{bmatrix} \boldsymbol{\phi}^{(0)} \\ \boldsymbol{\phi}^{(1)} \\ \vdots \\ \boldsymbol{\phi}^{(m-1)} \end{bmatrix} \qquad \mathbf{w} = \begin{bmatrix} \mathbf{0} \\ \mathbf{0} \\ \vdots \\ \mathbf{n}(t) \end{bmatrix} \tag{6.174}$$

and

$$\mathbf{F} = \begin{bmatrix} \mathbf{0} & \mathbf{I} & \cdots & \mathbf{0} \\ \mathbf{0} & \mathbf{0} & \cdots & \mathbf{0} \\ \vdots & & & \\ -\boldsymbol{\lambda}_0 & -\boldsymbol{\lambda}_1 & & -\boldsymbol{\lambda}_{m-1} \end{bmatrix} \tag{6.175}$$

The complete system equation, in state variable form, is, from equations (6.162) and (6.173),

$$\dot{\mathbf{x}} = \mathbf{H}\mathbf{x} + \mathbf{W} \tag{6.176}$$

where

$$\mathbf{x} = \begin{bmatrix} \mathbf{z} \\ \boldsymbol{\theta} \end{bmatrix} \qquad \mathbf{W} = \begin{bmatrix} \mathbf{0} \\ \mathbf{w} \end{bmatrix} \tag{6.177}$$

$$\mathbf{H} = \begin{bmatrix} \mathbf{G} & \mathbf{D} \\ \mathbf{0} & \mathbf{F} \end{bmatrix} \qquad \mathbf{D} = \begin{bmatrix} \mathbf{0} \\ \boldsymbol{\Lambda}^{\mathrm{T}} \end{bmatrix} \tag{6.178}$$

and

$$\boldsymbol{\Lambda} = [\mathbf{v}_0, \mathbf{v}_1, \ldots, \mathbf{v}_{m-1}]^{\mathrm{T}} \tag{6.179}$$

Equation (6.170) is now applicable, for non-white excitation, where $\mathbf{V}$ is now the covariance of the augmented vector $\mathbf{x}$, $\mathbf{G}$ is replaced by the augmented matrix $\mathbf{H}$, and $\mathbf{R}$ is the matrix of constants relating to the white noise vector $\mathbf{W}$.

Example

To illustrate the application of the state variable method to a MDOF system with non-white excitation, the system shown in Figure 6.1 is once again considered. In this case $n = 2$ and a convenient form of the equations of motion is given by equation (6.127). The process $p(t)$ has a spectrum given by equation (6.97), and can thus be regarded as the output of a first-order filter driven by white noise, i.e.

$$\dot{p} + \alpha p = n(t) \tag{6.180}$$

If the spectrum of $n(t)$ is of constant level S_0, then it follows from equation (6.173) that the spectrum of p is given by equation (6.97).

In the present example

$$\mathbf{M}^{-1} = \begin{bmatrix} (1+\mu) & \mu \\ \mu & \mu \end{bmatrix}^{-1} = \begin{bmatrix} \mu & -\mu \\ -\mu & (1+\mu) \end{bmatrix} \tag{6.181}$$

and hence **p**, in equation (6.164), is here given by

$$\mathbf{p} = \mathbf{M}^{-1}\begin{bmatrix} p \\ 0 \end{bmatrix} = \begin{bmatrix} \mu p(t) \\ -\mu p(t) \end{bmatrix} \tag{6.182}$$

On combining equations (6.180) and (6.182) one can write the following filter equations for $\boldsymbol{\phi}$

$$\mathbf{v}_0 \boldsymbol{\phi}^{(0)} = \mathbf{p} \tag{6.183}$$

$$\boldsymbol{\phi}^{(1)} + \boldsymbol{\lambda}_0 \boldsymbol{\phi}^{(0)} = \mathbf{n}(t) \tag{6.184}$$

where

$$\mathbf{n}(t) = \begin{bmatrix} n(t) \\ n(t) \end{bmatrix} \quad \boldsymbol{\lambda}_0 = \begin{bmatrix} \alpha & 0 \\ 0 & \alpha \end{bmatrix} \quad \mathbf{v}_0 = \begin{bmatrix} \mu & 0 \\ 0 & -\mu \end{bmatrix} \tag{6.185}$$

and it is evident that $\phi_1 = \phi_2 = p$. It follows, from the analysis in the preceding section, that the combined equations of motion ((6.127), (6.183) and (6.184)) can be written in the form of equation (6.176), where **H** is given by equation (6.178). Specifically

$$\mathbf{F} = -\lambda_0 \quad \mathbf{D} = \begin{bmatrix} \mathbf{0} \\ \mathbf{v}_0 \end{bmatrix} \tag{6.186}$$

and **H** is a 6×6 matrix. From equation (6.170) the covariance matrix of the augmented vector **x** is given by

$$\mathbf{HV} + \mathbf{VH}^{\mathrm{T}} + \mathbf{R} = \mathbf{0} \tag{6.187}$$

where here

$$\mathbf{R} = \begin{bmatrix} 0 & 0 & \cdots & & 0 \\ 0 & 0 & \cdots & & 0 \\ \vdots & & & & \\ 0 & 0 & \cdots & 1 & 1 \\ 0 & 0 & \cdots & 1 & 1 \end{bmatrix} \tag{6.188}$$

is also a 6×6 matrix. Equation (6.187) can be solved numerically, once $\mathbf{C}_\mathrm{e}$ and $\mathbf{K}_\mathrm{e}$ are specified. Thus, as before, a cyclic relationship is established, which can be solved iteratively.

6.6.5 Complex modal analysis

The method of complex modes, described earlier (in Section 4.5.5) can also be used within the context of the statistical linearization scheme, for MDOF systems. Again, the basis of the method is the state variable formulation of the equations of motion. However, an advantage of the complex mode method is that, if the excitation has a relatively simple spectrum, it is possible to obtain results fairly easily, without resorting to the introduction of a shaping filter.

The method is best illustrated through a specific example. Once again, the two degree of freedom system shown in Figure 6.1 will be considered, for this purpose.

Example

As a starting point, the equations of motion, as given by equation (6.127) will be taken. On premultiplying this matrix equation throughout by the inverse of **M** (see equation (6.181)) one obtains the equation

$$\ddot{\mathbf{q}} + \boldsymbol{\alpha}\dot{\mathbf{q}} + \boldsymbol{\beta}\mathbf{q} = \mathbf{p}(t) \tag{6.189}$$

where $\mathbf{p}(t)$ has been defined previously, by equation (6.182) and $\boldsymbol{\alpha}$ and $\boldsymbol{\beta}$ are given by

$$\boldsymbol{\alpha} = \begin{bmatrix} \mu & -\mu \\ -\mu & (1+\mu) \end{bmatrix} \begin{bmatrix} 2\zeta_{1e} & 0 \\ 0 & 2\zeta_{1e}\mu\omega_2 \end{bmatrix} = \begin{bmatrix} 2\zeta_{1e}\mu\omega_{1e} & -2\zeta_{2e}\mu^2\omega_2 \\ -2\zeta_{1e}\mu\omega_{1e} & 2\zeta_{2e}\mu(1+\mu)\omega_2 \end{bmatrix} \tag{6.190}$$

and

$$\boldsymbol{\beta} = \begin{bmatrix} \mu & -\mu \\ -\mu & (1+\mu) \end{bmatrix} \begin{bmatrix} \omega_{1e}^2 & 0 \\ 0 & \mu\omega_2^2 \end{bmatrix} = \begin{bmatrix} \omega_{1e}^2\mu & -\omega_2^2\mu^2 \\ \omega_{1e}^2\mu & \omega_2^2\mu(1+\mu) \end{bmatrix} \tag{6.191}$$

Furthermore

$$\mathbf{q} = [q_1, q_2]^{\mathrm{T}} \tag{6.192}$$

It is now possible to recast the equations of motion directly into the standard state variable form (see also equation (4.53))

$$\dot{\mathbf{z}} = \mathbf{G}\mathbf{z} + \mathbf{f} \tag{6.193}$$

where

$$\mathbf{z} = [\mathbf{q}, \dot{\mathbf{q}}]^{\mathrm{T}} \tag{6.194}$$

$$\mathbf{G} = \begin{bmatrix} \mathbf{0} & \mathbf{I} \\ -\boldsymbol{\beta} & -\boldsymbol{\alpha} \end{bmatrix} \tag{6.195}$$

and

$$\mathbf{f}(t) = [\mathbf{0}, \mathbf{p}(t)]^{\mathrm{T}} \tag{6.196}$$

The analysis can now proceed along the lines given in Section 4.5.5. The first step is to obtain the four complex eigenvalues, λ_i, $(i = 1, \ldots, 4)$ of the matrix **G**, and the associated four complex eigenvectors, $\mathbf{d}_i$ $(i = 1, \ldots, 4)$. The complex modal matrix, **T** (here 4×4), as defined by equation (4.64), can then be assembled.

The next stage is to transform the covariance matrix, $\mathbf{w}_f(\tau)$, of the excitation vector $\mathbf{f}(t)$. In the present example, $p(t)$ has the first order spectrum given by equation (6.97). It follows that the correlation function of $p(t)$ is given by

$$w_p(\tau) = \int_{-\infty}^{\infty} S_p(\omega)\exp(\mathrm{i}\omega\tau)\,\mathrm{d}\omega = \int_{-\infty}^{\infty} \frac{S_0}{(\alpha^2+\omega^2)}\cos\omega\tau\,\mathrm{d}\omega \tag{6.197}$$

On evaluating this integral one finds that

$$w_p(\tau) = \frac{\pi S_0}{\alpha}\exp(-\alpha|\tau|) \tag{6.198}$$

From the definitions of $\mathbf{p}$ (see equation (6.182) and $\mathbf{f}(t)$ (see equation (6.196)) the covariance matrix of $\mathbf{f}(t)$ can now be evaluated by combining results. Hence one obtains

$$\mathbf{w}_f(\tau) = \frac{\mu^2\pi S_0}{\alpha}\exp(-\alpha|\tau|)\mathbf{J} \tag{6.199}$$

where

$$\mathbf{J} = \begin{bmatrix} 0 & 0 & 0 & 0 \\ 0 & 0 & 0 & 0 \\ 0 & 0 & 1 & -1 \\ 0 & 0 & -1 & 1 \end{bmatrix} \tag{6.200}$$

This covariance matrix is of the general form given by equation (4.189), where here

$$\mathbf{D} = \frac{\mu^2\pi S_0}{\alpha}\mathbf{J} \tag{6.201}$$

and

$$\rho(\tau) = \exp(-\alpha|\tau|) \tag{6.202}$$

The covariance matrix $\mathbf{w}_f(\tau)$ is readily converted to the covariance matrix, $\mathbf{w}_g(\tau)$, of the transformed state variable vector $\mathbf{v}$ (see equation (4.70)), using equation (4.181). Hence, in the present case, $\mathbf{w}_g(\tau)$ is of the general form given by equation (4.190), where $\mathbf{R}$ is given by equation (4.183).

Once the elements of $\mathbf{R}$, R_{ij}, are computed then, for the present example, one can take advantage of the result given by equation (4.193), since here $\rho(\tau)$ is exponential. Thus, from a knowledge of the eigenvalues, λ_i, and the elements of $\mathbf{R}$, the elements of the covariance matrix $\mathbf{w}_v(\tau)$ for $\mathbf{v}(t)$, at zero time lag, can be computed easily, according to equation (4.193). Then $\mathbf{w}_v(\tau)$ can be transformed to $\mathbf{w}_z(0)$, the covariance matrix of $\mathbf{z}(t)$, at zero time lag, according to equation (4.180).

For the present purpose, only the mean square quantities $\sigma^2_{q_1}$ and $\sigma^2_{\dot{q}_2}$ are required, in the iteration scheme. From the definition of $\mathbf{z}(t)$, it follows that

$$\sigma_{q_1}^2 = w_z^{11}(0) \tag{6.203}$$

$$\sigma_{\dot{q}_2}^2 = w_z^{44}(0) \tag{6.204}$$

The analysis outlined above can be readily incorporated into the iterative scheme, previously described. It is noted that the calculations are very cumbersome to perform manually, due to the complex arithmetic involved. However, with the use of the complex arithmetic facilities available on most digital computers, the algorithm is easily implemented. The necessary complex eigenvalue analysis of $\mathbf{G}$ can be carried out using a standard computer library routine. Routines for inverting, and multiplying together, complex matrices are also required.

Some specific numerical results are presented in Table 6.3, for the same set of parameters that relate to Table 6.2. Here, due to the relatively light damping, the four eigenvalues occur in two complex conjugate pairs. Thus $\lambda_2 = \lambda_1^*$ and $\lambda_4 = \lambda_3^*$. In Table 6.3 the real and imaginary values of λ_1 and λ_3 are given, for various numbers of iterations (the zeroth iteration referring to the initial 'guess', where c_e and k_e are zero). As expected for a lightly damped system, the real parts of the eigenvalues are small, compared with the imaginary parts. The values of k_{1e}, c_{2e}, $\sigma_{q_1}^2$ and $\sigma_{\dot{q}_2}^2$ which are obtained are, of course, identical to those obtained by the spectral method (see Table 6.2).

Table 6.3

Number of iterations	λ_1 *real*	λ_1 *imaginary*	λ_3 *real*	λ_3 *imaginary*
0	−0.415	1.56	−0.035	0.620
1	−0.469	1.86	−0.105	0.817
5	−0.490	1.60	−0.072	0.750
10	−0.504	1.61	−0.056	0.702
15	−0.500	1.64	−0.062	0.719
20	−0.501	1.63	−0.060	0.712
30	−0.501	1.63	−0.060	0.714
40	−0.501	1.63	−0.060	0.714
50	−0.501	1.63	−0.060	0.714

$\omega_1 = 1\ \mathrm{rad\,s^{-1}}$, $\omega_2 = 1\ \mathrm{rad\,s^{-1}}$
$\zeta_1 = 0.05$, $\zeta_2 = 0.2$, $\mu = 1$
$S_0 = 1$, $\alpha = 2\ \mathrm{rad\,s^{-1}}$, $\varepsilon_1 k_1 = 0.2\ \mathrm{N\,m^{-3}}$, $\varepsilon_2 c_2 = 0.2\ \mathrm{N\,s^2\,m^{-2}}$

6.7 MODE-BY-MODE LINEARIZATION

The method of complex modes opens up the possibility of an alternative method of linearization, for MDOF systems, as pointed out by Fang and Wang (1986a).

Suppose, for simplicity, that the non-linearity depends only on $\mathbf{q}$ and $\dot{\mathbf{q}}$. The appropriate general form of the equations of motion is then equation (6.59). A complex modal analysis can be performed on the linear part of (6.59); i.e. on the equation

$$\mathbf{M}\ddot{\mathbf{q}} + \mathbf{C}\dot{\mathbf{q}} + \mathbf{K}\mathbf{q} = \mathbf{Q}(t) \tag{6.205}$$

according to the method given in Section 4.5.5. Thus the state vector, $\mathbf{z}$, defined by equation (4.52) can be transformed to another state vector $\mathbf{v}$, according to

$$\mathbf{z} = \mathbf{T}\mathbf{v} \tag{6.206}$$

where $\mathbf{T}$ is the complex modal matrix, formed from the eigenvectors of $\mathbf{G}$, where this matrix is defined by equation (4.54).

In terms of $\mathbf{z}$ the original, non-linear equation of motion can be written as

$$\dot{\mathbf{z}} = \mathbf{G}\mathbf{z} + \mathbf{N}(\mathbf{z}) + \mathbf{f} \tag{6.207}$$

where

$$\mathbf{N}(z) = \begin{bmatrix} \mathbf{0} \\ \mathbf{g}(\mathbf{q}, \dot{\mathbf{q}}) \end{bmatrix} = \begin{bmatrix} \mathbf{0} \\ \mathbf{g}(\mathbf{z}) \end{bmatrix} \tag{6.208}$$

This equation can be transformed into an equation of motion in terms of $\mathbf{v}$, by using equation (6.206). Thus, on combining equations (6.206) and (6.207), and using equations (4.72) and (4.74), one obtains

$$\dot{\mathbf{v}} = \boldsymbol{\eta}\mathbf{v} + \mathbf{n}(\mathbf{v}) + \mathbf{q} \tag{6.209}$$

where

$$\mathbf{n}(\mathbf{v}) = \mathbf{T}^{-1}\mathbf{N}\mathbf{T}\mathbf{v} \tag{6.210}$$

Here the matrix $\boldsymbol{\eta}$ is diagonal, and contains the $2n$ eigenvalues, λ_i $(i = 1, \ldots, 2n)$ (see Section 4.4.5). Hence, equation (6.209) can be written as the set of equations

$$\dot{v}_i = \lambda_i v_i + n_i(\mathbf{v}) + q_i \tag{6.211}$$

where n_i is the ith element of $\mathbf{n}$ and, as before, v_i and q_i are, respectively, the ith elements of $\mathbf{v}$ and $\mathbf{q}$. In the absence of the non-linearity function $n_i(\mathbf{v})$, equation (6.211) represents a set of *uncoupled* equations, as pointed out previously.

The statistical linearization method may now be applied to the transformed equations (6.211), rather than to the original equation in terms of $\mathbf{z}$. Thus, one can replace the set (6.211) by an equivalent linear set

$$\dot{v}_i = \lambda_{ie} v_i + q_i \tag{6.212}$$

where λ_{ie} $(i = 1, \ldots, 2n)$ are the equivalent eigenvalues of the system. The difference between equations (6.211) and (6.212) can be minimized on a mode-by-mode basis. This means that minimization can be carried out for each value of i (corresponding to the ith mode). Thus, for the ith mode, the difference

$$\varepsilon_i = (\lambda_i - \lambda_{ie})v_i + n_i(\mathbf{v}) \tag{6.213}$$

can be minimized. It is important to note that, in general, the ε_i quantities will be complex.

Minimising the expected mean square error, as before, it is evident that, in the present case, one wishes to minimize the quantities

$$D_i \equiv E\{\varepsilon_i \varepsilon_i^*\} \tag{6.214}$$

for every i ($i = 1, \ldots, 2n$). Writing

$$\lambda_i - \lambda_{ie} = \alpha + \mathrm{i}\beta \tag{6.215}$$

then it is clear that α and β must be determined such that

$$\frac{\partial D_i}{\partial \alpha} = \frac{\partial D_i}{\partial \beta} = 0 \tag{6.216}$$

After some algebra one finds, on combining equation (6.213) to (6.215), that the equivalent eigenvalues, λ_{ie}, are given by (Fang and Wang, 1986a)

$$\lambda_{ie} = \lambda_i - \frac{E\{n_i(\mathbf{v})v_i^*\}}{E\{v_i v_i^*\}} \tag{6.217}$$

These equations form the basis for an alternative cyclic, iterative procedure for finding an equivalent linear system. As with the standard method (see Section 6.6.5) one first sets the non-linearity function to zero, and finds the complex modal matrix **T**, for the linear part of the equations of motion. On the basis of this linear equation one can then compute the expectations appearing in equation (6.217), to arrive at a new set of eigenvalues, which specify the equivalent linear system. One can then recompute the expectations in equation (6.217), to update λ_{ie}, continuing this process until convergence is achieved. A final step is to transform the equations back into the original z state-vector form. If only the covariance matrix of $\mathbf{z}$, at zero lag, is required then one can use equation (4.180) to transform $\mathbf{w}_v(0)$ into $\mathbf{w}_z(0)$.

It is noted that this linearization method will *not* give results identical to those obtained with the standard procedure, described earlier in this chapter. The difference arises from the different minimum error criteria adopted in the two techniques. However, the accuracy of the mode-by-mode method of linearization appears to be comparable with that of the standard method, as a particular numerical result for a two degree of freedom system, obtained by Wang and Fang (1986a), indicates. An advantage of the method under discussion is that it is not necessary to compute **T**, and its inverse, at every iteration. On the other hand, $\mathbf{n}(\mathbf{v})$ can be complicated, even if the original non-linearity, expressed in terms of $\mathbf{z}$, is simple. Thus the evaluation of $E\{n_i(\mathbf{v})v_i^*\}$ in equation (6.217) can be cumbersome, despite the apparent simplicity of this equation.

The mode-by-mode analysis described here can be regarded as an extension of Caughey's normal mode method (Caughey, 1963), where a classical modal analysis was employed, under certain restricted conditions.

Chapter 7

Non-stationary problems

7.1 INTRODUCTION

In many cases, such as in earthquake engineering, the non-stationary characteristics of the system response are quite important, from an analysis and design perspective. The general methodology for statistical linearization, developed so far in this book, can be adapted for the case of non-stationary excitation. However, a major complicating factor does arise in the non-stationary situation; the elements of the equivalent linear system are no longer time invariant. Thus it is necessary to extend the theory to cope with time-varying equivalent linear systems.

In this chapter suitable extensions of the linearization procedure will be described and illustrated by means of simple examples. It is noted that theoretical developments in this area are of fairly recent origin (Spanos, 1978, 1980a, 1981a; Iwan and Mason, 1980; Ahmadi, 1980; Sakata and Kimura, 1980; Kimura and Sakata, 1981). Applications have been discussed by Ahmadi (1981, 1983), Constantinou and Tadjibash (1984), Noguchi (1985), Sakata *et al.* (1984) and Harrison and Hammond (1986).

7.2 GENERAL THEORY

Once again an n degree of freedom system will be considered, where the non-linearities depend on the instantaneous values of the displacements and velocities. Non-linearities dependent on acceleration will be excluded here, since they rarely occur in practice. The general form of the equations of motion can be written, as before, in matrix form (see also equation (6.59)) as follows

$$\mathbf{M}\ddot{\mathbf{q}} + \mathbf{C}\dot{\mathbf{q}} + \mathbf{K}\mathbf{q} + \mathbf{g}(\mathbf{q}, \dot{\mathbf{q}}) = \mathbf{Q}(t) \tag{7.1}$$

where the symbols have the same meaning as before (see Chapter 6).

In general $\mathbf{Q}(t)$ will be a non-stationary n vector random process, with a mean vector function

$$\mathbf{m}_Q(t) = E\{\mathbf{Q}(t)\} \tag{7.2}$$

and a covariance matrix function

$$\mathbf{w}_Q(t, s) = E\{[\mathbf{Q}(t) - \mathbf{m}_Q(t)][\mathbf{Q}(s) - \mathbf{m}_Q(s)]^{\mathrm{T}}\} \tag{7.3}$$

which, unlike the stationary case, does not depend solely on $t-s$. Also, in general, the non-linearity function $\mathbf{g}(\mathbf{q}, \dot{\mathbf{q}})$ will *not* be antisymmetric i.e.

$$\mathbf{g}(\mathbf{q}, \dot{\mathbf{q}}) \neq -\mathbf{g}(-\mathbf{q}, -\dot{\mathbf{q}}) \tag{7.4}$$

These two factors of the problem guarantee that, in general, the vector mean of the response, $\mathbf{q}(t)$, will also be non-zero. One can denote this mean variation with time as

$$\mathbf{m}_q(t) = E\{\mathbf{q}(t)\} \tag{7.5}$$

and can define a vector zero-mean response process as

$$\mathbf{q}_0(t) = \mathbf{q}(t) - \mathbf{m}_q(t) \tag{7.6}$$

Similarly, a vector zero-mean excitation process may be defined by

$$\mathbf{Q}_0(t) = \mathbf{Q}(t) - \mathbf{m}_Q(t) \tag{7.7}$$

If equations (7.6) and (7.7) are substituted into the equations of motion, as given by equation (7.1), then an alternative expression for these equations is obtained, as follows (Spanos, 1978, 1980a)

$$\mathbf{M}\ddot{\mathbf{q}}_0 + \mathbf{C}\dot{\mathbf{q}}_0 + \mathbf{K}\mathbf{q}_0 + \mathbf{M}\ddot{\mathbf{m}}_q + \mathbf{C}\dot{\mathbf{m}}_q + \mathbf{K}\mathbf{m}_q + \mathbf{g}(\mathbf{q}_0 + \mathbf{m}_q, \dot{\mathbf{q}}_0 + \dot{\mathbf{m}}_q) = \mathbf{Q}_0(\mathrm{t}) + \mathbf{m}_Q(\mathrm{t}) \tag{7.8}$$

If expectations are taken of all the terms in the above equation, then one finds that

$$\mathbf{M}\ddot{\mathbf{m}}_q + \mathbf{C}\dot{\mathbf{m}}_q + \mathbf{K}\mathbf{m}_q + E\{\mathbf{g}(\mathbf{q}_0 + \mathbf{m}_q, \dot{\mathbf{q}} + \dot{\mathbf{m}}_q)\} = \mathbf{m}_Q(t) \tag{7.9}$$

If this is subtracted from equation (7.8) it is found that

$$\mathbf{M}\ddot{\mathbf{q}}_0 + \mathbf{C}\dot{\mathbf{q}}_0 + \mathbf{K}\mathbf{q}_0 + \mathbf{h}(\mathbf{q}_0, \dot{\mathbf{q}}_0) = \mathbf{Q}_0(t) \tag{7.10}$$

where

$$\mathbf{h}(\mathbf{q}_0, \dot{\mathbf{q}}_0) = \mathbf{g}(\mathbf{q}_0 + \mathbf{m}_q, \dot{\mathbf{q}}_0 + \dot{\mathbf{m}}_q) - E\{\mathbf{g}(\mathbf{q}_0 + \mathbf{m}_q, \dot{\mathbf{q}}_0 + \dot{\mathbf{m}}_q)\} \tag{7.11}$$

It is noted that, if $\mathbf{m}_q$ and $\mathbf{m}_Q$ are constants, independent of time, as in the special case of stationary excitation and response, the above equations reduce to those given earlier, in Section 6.5.3.

Equation (7.10) can now be linearized by using the procedure developed in Chapter 6. Thus, the equivalent linear system may be defined by

$$\mathbf{M}\ddot{\mathbf{q}}_0 + (\mathbf{C} + \mathbf{C}_e)\dot{\mathbf{q}}_0 + (\mathbf{K} + \mathbf{K}_e)\mathbf{q}_0 = \mathbf{Q}_0(t) \tag{7.12}$$

where $\mathbf{C}_e$ and $\mathbf{K}_e$ are $n \times n$ matrices, selected, as in the case of stationary response, to satisfy the criterion described by equation (6.5). Thus, from equation (6.22), the ith rows, $\mathbf{c}^e_{i*}$ and $\mathbf{k}^e_{i*}$ of the matrices $\mathbf{C}_e$ and $\mathbf{K}_e$, respectively, satisfy

$$E\{h_i\mathbf{z}\} = \mathbf{V}(t)\begin{bmatrix} \mathbf{k}^{e\mathrm{T}}_{i*} \\ \mathbf{c}^{e\mathrm{T}}_{i*} \end{bmatrix} \tag{7.13}$$

where h_i is the ith element of $\mathbf{h}(\mathbf{q}_0, \dot{\mathbf{q}}_0)$

$$\mathbf{z} = \begin{bmatrix} \mathbf{q}_0 \\ \dot{\mathbf{q}}_0 \end{bmatrix} \tag{7.14}$$

and $\mathbf{V}$ is the covariance matrix of $\mathbf{z}(t)$, defined by

$$\mathbf{V}(t) = E\{\mathbf{z}(t)\mathbf{z}^{\mathrm{T}}(t)\} \tag{7.15}$$

For the special case of a Gaussian approximation of the response then, as in the stationary case, equation (7.13) reduces to the solution given by equations (6.45) and (6.46). Thus

$$c_{ij}^{\mathrm{e}} = E\left\{\frac{\partial h_i}{\partial \dot{q}_{0j}}\right\} \quad i = 1, 2, \ldots n \tag{7.16}$$

and

$$k_{ij}^{\mathrm{e}} = E\left\{\frac{\partial h_i}{\partial q_{0j}}\right\} \quad i = 1, 2, \ldots, n \tag{7.17}$$

where q_{0j} is the jth element of $\mathbf{q}_0$ and c_{ij}^{e} and k_{ij}^{e} are the elements of $\mathbf{C}_{\mathrm{e}}$ and $\mathbf{K}_{\mathrm{e}}$, respectively. Although equations (7.16) and (7.17) are, interestingly enough, valid irrespective of the stationarity, or otherwise of the response (Spanos, 1978, 1981a), it should be noted that, in the non-stationary case, the elements of $\mathbf{C}_{\mathrm{e}}$ and $\mathbf{K}_{\mathrm{e}}$ are *time-dependent*. Thus a set of non-linear differential equations, rather than algebraic equations, must be solved to determine the response statistics.

If, as in the stationary case, the response is assumed to be Gaussian, with $\mathbf{V}$ determined from the equivalent linear system of equation (7.12), then it is possible to evaluate the expectations of equation (7.9), (7.16) and (7.17) in terms of the elements of $\mathbf{V}$. These elements will, of course, depend in turn on the coefficients in $\mathbf{C}_{\mathrm{e}}$ and $\mathbf{K}_{\mathrm{e}}$. The evolution of $\mathbf{V}$ with time can be computed using the state variable method outlined earlier, in Section 6.6.4. Thus $\mathbf{z}$, defined by equation (7.14), is governed by

$$\dot{\mathbf{z}} = \mathbf{G}(t)\mathbf{z} + \mathbf{f} \tag{7.18}$$

where $\mathbf{G}$ is given by equation (6.163) and $\mathbf{f}$ by equation (6.164) and here $\mathbf{Q} = \mathbf{Q}_0$. However, as indicated in equation (7.18), the elements of $\mathbf{G}$ are now time-dependent so some modification of the analysis presented earlier, in Chapter 4, is required to accommodate this feature of the analysis.

Let $\mathbf{Y}(t)$ be the solution of the homogeneous equation

$$\dot{\mathbf{Y}} = \mathbf{G}(t)\mathbf{Y} \tag{7.19}$$

with initial conditions

$$\mathbf{Y}(0) = \mathbf{I} \tag{7.20}$$

Then, from standard linear theory (see Reid, 1983) the general solution of equation (7.18) can be expressed as

$$\mathbf{z}(t) = \mathbf{Y}(t)\mathbf{z}(0) + \mathbf{Y}(t)\int_0^t \mathbf{Y}^{-1}(s)\mathbf{f}(s)\,\mathrm{d}s \tag{7.21}$$

In the special case where the elements of $\mathbf{G}$ are time-invariant, one has (see equation (4.56))

$$\mathbf{Y}(t) = \exp(\mathbf{G}t) \tag{7.22}$$

and since

$$[\exp \mathbf{G}s]^{-1} = \exp(-\mathbf{G}s) \tag{7.23}$$

it follows that equation (7.21) reduces to the result given earlier, in Chapter 4 (see equation (4.58)). In the more general case $\mathbf{Y}(t)$ may be written, formally, as

$$\mathbf{Y}(t) = \exp\left(\int_0^t \mathbf{G}(\tau)\,\mathrm{d}\tau\right) \tag{7.24}$$

To obtain the required differential equation for $\mathbf{V}$ one may differentiate both sides of equation (7.15). Therefore

$$\dot{\mathbf{V}} = E\{\dot{\mathbf{z}}\mathbf{z}^{\mathrm{T}}\} + E\{\mathbf{z}\dot{\mathbf{z}}^{\mathrm{T}}\} \tag{7.25}$$

If expression (7.18) for $\dot{\mathbf{z}}$ is substituted into equation (7.25) then one obtains

$$\dot{\mathbf{V}} = \mathbf{G}(t)\mathbf{V}^{\mathrm{T}} + \mathbf{V}\mathbf{G}^{\mathrm{T}}(t) + \mathbf{U}(t) + \mathbf{U}^{\mathrm{T}}(t) \tag{7.26}$$

where (similar to equation (4.147)).

$$\mathbf{U}(t) = E\{\mathbf{z}\mathbf{f}^{\mathrm{T}}\} \tag{7.27}$$

Using equation (7.21) for $\mathbf{z}(t)$ and assuming, as in Section 4.5.4, that the initial conditions are deterministic, or at least uncorrelated with the input process, $\mathbf{f}(t)$, then the $2n \times 2n$ matrix $\mathbf{U}(t)$ may be expressed as

$$\mathbf{U}(t) = \mathbf{Y}(t)\int_0^t \mathbf{Y}^{-1}(s)\mathbf{w}_f(t,s)\,\mathrm{d}s \tag{7.28}$$

In this equation $\mathbf{w}_f(t,s)$ is the covariance matrix for $\mathbf{f}$, defined by

$$\mathbf{w}_f(t,s) = E\{\mathbf{f}(t)\mathbf{f}^{\mathrm{T}}(s)\} \tag{7.29}$$

Since

$$\mathbf{f} = \begin{bmatrix} \mathbf{0} \\ \mathbf{M}^{-1}\mathbf{Q}_0 \end{bmatrix} \tag{7.30}$$

it follows, from equations (7.3), (7.7), (7.29) and (7.30), that the following relation exists between $\mathbf{w}_f$ and $\mathbf{w}_Q$

$$\mathbf{w}_f = \begin{bmatrix} \mathbf{0} & \mathbf{0} \\ \mathbf{0} & \mathbf{M}^{-1}\mathbf{w}_Q\mathbf{M} \end{bmatrix} \tag{7.31}$$

The above expressions provide a sufficient number of equations to determine the evolution of the covariance matrix with time.

In many cases the non-stationary excitation has a zero mean, so that $\mathbf{m}_Q = \mathbf{0}$, and can be modelled, with satisfactory accuracy, as a modulated stationary vector process. Thus, one may write

$$\mathbf{Q}(t) = \mathbf{a}(t)\mathbf{Q}_s(t) \tag{7.32}$$

where $\mathbf{a}(t)$ is an $n \times m$ matrix of deterministic modulating functions and $\mathbf{Q}_s(t)$ is an m vector of stationary random processes. Then the covariance matrix for $\mathbf{Q}(t)$ is of the simpler form

$$\mathbf{w}_Q(t, s) = \mathbf{a}(t)\mathbf{w}_{Q_s}(t - s)\mathbf{a}^{\mathrm{T}}(s) \tag{7.33}$$

where

$$\mathbf{w}_{Q_s}(t - s) = E\{\mathbf{Q}_s(t)\mathbf{Q}_s^{\mathrm{T}}(s)\} \tag{7.34}$$

is the covariance matrix of $\mathbf{Q}_s(t)$; as indicated this depends only on $t - s$.

In solving the non-linear differential equations arising from the statistical linearization procedure it is, of course, necessary to specify initial conditions for the covariance matrix, $\mathbf{V}$. Normally, the system is initially at rest. It is then appropriate to set $\mathbf{V}(0) = \mathbf{0}$. When dealing with pre-filters, as a means of modelling non-white excitations, it is, however, necessary to modify this initial condition, as explained later (in Section 7.4).

7.3 WHITE NOISE EXCITATION

In the special case where the excitation consists of modulated, white noise processes, the analysis simplifies considerably.

If the elements of $\mathbf{Q}_s(t)$, in equation (7.32) are approximated as ideal white noise processes then the covariance of $\mathbf{Q}_s(t)$, defined by equation (7.34), reduces to

$$\mathbf{w}_{Q_s}(t - s) = \mathbf{D}\delta(t - s) \tag{7.35}$$

where $\mathbf{D}$ is real, symmetric, non-negative matrix of constants. It follows, from equation (7.33), that

$$\mathbf{w}_Q(t, s) = \mathbf{\Lambda}(t)\delta(t - s) \tag{7.36}$$

where

$$\mathbf{\Lambda}(t) = \mathbf{a}(t)\mathbf{D}\mathbf{a}^{\mathrm{T}}(t) \tag{7.37}$$

For this special form of covariance matrix function the evaluation of $\mathbf{U}(t)$, as defined by equation (7.28), is very simple. Thus, from equations (7.28), (7.31) and (7.36), using the properties of the delta function,

$$\mathbf{U}(t) = \tfrac{1}{2}\mathbf{\Theta}(t) \tag{7.38}$$

where

$$\mathbf{\Theta}(t) = \begin{bmatrix} \mathbf{0} & \mathbf{0} \\ \mathbf{0} & \mathbf{M}^{-1}\mathbf{\Lambda}(t)\mathbf{M} \end{bmatrix} \tag{7.39}$$

Hence, from equation (7.26),

$$\dot{\mathbf{V}} = \mathbf{G}(t)\mathbf{V}^{\mathrm{T}} + \mathbf{V}\mathbf{G}^{\mathrm{T}}(t) + \mathbf{\Theta}(t) \tag{7.40}$$

In the special case of no modulation ($\boldsymbol{\alpha} = \mathbf{I}$) then this last result reduces to that given earlier, in Chapter 4 (see equation (4.155)).

7.3.1 Friction controlled slip of a structure on a foundation

As an illustration of the statistical linearization method, when dealing with non-stationary white noise excitation, the system shown in Figure 5.12 will once again be considered. As before, q will be used to denote the relative slip between the mass and the foundation (see equation (5.174)) and the equation of motion will be assumed to be that given by equation (5.175).

To keep matters simple, it will be assumed here that the excitation process

$$Q(t) \equiv -\ddot{x}_g(t) \tag{7.41}$$

has a zero mean and can be modelled as

$$Q(t) = \alpha(t)Q_s(t) \tag{7.42}$$

Further, it will be assumed that $\alpha(t)$ is a deterministic modulating function and that $Q_s(t)$ is a stationary white noise process, with covariance function

$$w_{Q_s}(t-s) = 2\pi S_0 \delta(t-s) \tag{7.43}$$

Thus, from equations (7.36) and (7.37), the covariance function for $Q(t)$ is

$$w_Q(t,s) = \Lambda(t)\delta(t-s) \tag{7.44}$$

where

$$\Lambda(t) = 2\pi S_0 \alpha^2(t) \tag{7.45}$$

Since the non-linearity function, $\mu g \operatorname{sgn}(\dot{q})$ is antisymmetric, it follows that the mean of $\mathbf{q}$ will be zero.

A comparison of equation (5.175) with the standard form given by equation (7.10) makes it evident that here $\mathbf{M} = 1$, $\mathbf{C} = \mathbf{K} = 0$ and

$$\ddot{q} + h(q, \dot{q}) = Q(t) \tag{7.46}$$

where

$$h(q, \dot{q}) = \mu g \operatorname{sgn}(\dot{q}) \tag{7.47}$$

and the subscript ‘o’ has been dropped. The equivalent linear system is, from equation (7.12),

$$\ddot{q} + c_e \dot{q} = Q(t) \tag{7.48}$$

and, from equation (7.16),

$$c_e = E\left\{\frac{\partial h}{\partial \dot{q}}\right\}, \tag{7.49}$$

Now

$$\frac{\partial h}{\partial \dot{q}} = 2\mu g \delta(\dot{q}) \tag{7.50}$$

and using the Gaussian density function for $\dot{q}$ (see equation (3.51)) and the properties of the delta function it is easily seen that, if the response is assumed to be Gaussian, one has

$$c_e = \left(\frac{2}{\pi}\right)^{1/2} \frac{\mu g}{\sigma_{\dot{q}}} \tag{7.51}$$

This result agrees with that found earlier, in Chapter 5 (see equation (5.180)), for the special case of stationary excitation.

In the case of non-stationary excitation, $\sigma_{\dot{q}}$ will vary with time, and hence c_e will also be time varying. To obtain another relationship between c_e and $\sigma_{\dot{q}}$, one can employ equation (7.40). Firstly, it is observed that the original, linearized equation of motion can be written in state space form, as

$$\frac{d}{dt}\begin{bmatrix} q \\ \dot{q} \end{bmatrix} = \begin{bmatrix} 0 & 1 \\ 0 & -c_e \end{bmatrix}\begin{bmatrix} q \\ \dot{q} \end{bmatrix} + \begin{bmatrix} 0 \\ Q(t) \end{bmatrix}, \tag{7.52}$$

On comparing this with the standard form given by equation (7.18), one has

$$\mathbf{G}(t) = \begin{bmatrix} 0 & 1 \\ 0 & -c_e \end{bmatrix} \tag{7.53}$$

and

$$\mathbf{f}(t) = \begin{bmatrix} 0 \\ Q(t) \end{bmatrix} \tag{7.54}$$

Hence, from equation (7.40),

$$\frac{d}{dt}\begin{bmatrix} v_{11} & v_{12} \\ v_{21} & v_{22} \end{bmatrix} = \begin{bmatrix} 0 & 1 \\ 0 & -c_e \end{bmatrix}\begin{bmatrix} v_{11} & v_{12} \\ v_{21} & v_{22} \end{bmatrix} + \begin{bmatrix} v_{11} & v_{12} \\ v_{21} & v_{22} \end{bmatrix}\begin{bmatrix} 0 & 0 \\ 1 & -c_e \end{bmatrix}$$

$$+ 2\pi S_0 \alpha^2(t)\begin{bmatrix} 0 & 0 \\ 0 & 1 \end{bmatrix} \tag{7.55}$$

where

$$v_{11} = E\{q^2\} = \sigma_q^2 \tag{7.56}$$

$$v_{12} = v_{21} = E\{q\dot{q}\} \tag{7.57}$$

$$v_{22} = E\{\dot{q}^2\} = \sigma_{\dot{q}}^2 \tag{7.58}$$

Equation (7.55) is equivalent to the following three differential equations

$$\dot{v}_{11} = 2v_{12} \tag{7.59}$$

$$\dot{v}_{12} = v_{22} - c_e v_{12} \tag{7.60}$$

$$\dot{v}_{22} = -2c_e v_{22} + 2\pi S_0 \alpha^2(t) \tag{7.61}$$

Clearly equation (7.61) can be solved for $v_{22} = \sigma_{\dot{q}}^2 \equiv \sigma_2^2$, independently of v_{11} and v_{12}. Hence, on combining equation (7.51) with equation (7.61) one has

$$\sigma_2 \frac{d\sigma_2}{dt} = -\left(\frac{2}{\pi}\right)^{1/2} \mu d\sigma_2 + \pi S_0 \alpha^2(t) \tag{7.62}$$

as a non-linear differential equation for $\sigma_2(t)$. Once $\sigma_2(t)$ is determined then $v_{11} = \sigma_q^2$ and v_{12} can be found by solving equations (7.59) and (7.60).

A case of special interest is where $\alpha(t)$ is in the form of a unit step function. This corresponds to a stationary excitation, suddenly 'switched on' at $t = 0$. The response will be non-stationary, during the transient response phase. Equation (7.62) then reduces to

$$\sigma_2 \frac{d\sigma_2}{dt} = -\left(\frac{2}{\pi}\right)^{1/2} \mu g \sigma_2 + \pi S_0 \quad t > 0 \tag{7.63}$$

As Ahmadi (1983) has shown, this equation can be solved analytically. Thus, first introducing the non-dimensional variables

$$t^* = \left(\frac{2\mu^2 g^2}{\pi^2 S_0}\right) t \tag{7.64}$$

and

$$\sigma_2^* = \left(\frac{2\mu g}{\pi(2\pi)^{1/2} S_0}\right) \sigma_2 \tag{7.65}$$

one finds that equation (7.65) can be expressed as

$$\sigma_2^* \frac{d\sigma_2^*}{dt^*} = -\sigma_2^* + 1 \tag{7.66}$$

The solution to this equation, which satisfies the initial condition

$$\sigma_2(0) = 0 \tag{7.67}$$

is as follows

$$t^* = -\sigma_2^* - \ln(1 - \sigma_2^*) \tag{7.68}$$

or

$$\sigma_2^* = 1 - \exp(-t^* - \sigma_2^*) \tag{7.69}$$

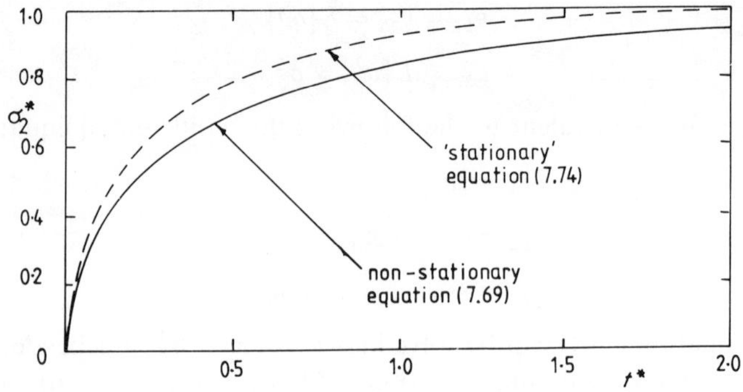

Figure 7.1
Friction controlled slip of a structure on a foundation. Variation of the non-dimensionalized standard deviation of the velocity response with time (reproduced from Ahmadi (1983) by permission of Pergamon Press)

Figure 7.1 shows the variation of σ_2^* with t^*. It is observed that σ_2^* approaches a 'stationary' limit

$$\sigma_2^* \to 1 \quad \text{as } t^* \to \infty \tag{7.70}$$

Hence, from equation (7.65),

$$\sigma_2 \to \frac{\pi^{3/2} S_0}{\sqrt{2\mu g}} \quad \text{as } t \to \infty \tag{7.71}$$

This agrees with the result found earlier, in Chapter 5, using the spectral method of calculation (see equation (5.185)). The corresponding 'stationary' value of c_e, c_e^s is, from equations (7.51) and (7.71)

$$c_e^s = \frac{2}{\pi^2} \frac{(\mu g)^2}{S_0} \tag{7.72}$$

If, for switched or stationary excitation, one approximates by setting c_e to the constant value given by equation (7.72), throughout the transient response, (as in Crandall *et al.*, (1974), so that the variation of c_e with time is not taken into account, then, from equation (7.63)

$$\sigma_2^2 = \frac{\pi S_0}{c_e^s} [1 - \exp(-2c_e^s t)] \tag{7.73}$$

Using the same non-dimensional variables as before, this can be written as

$$\sigma_2^* = [1 - \exp(-2t^*)]^{1/2} \tag{7.74}$$

In Figure 7.1 this result, shown by the broken line, is seen to overestimate the standard deviation of the velocity response, by up to about 15%.

Regarding the mean square of the slip displacement, for switched or stationary excitation, then if v_{11} and v_{12} are to reach 'stationary' values, as $t \to \infty$, $\dot{v}_{11}$ and $\dot{v}_{12}$ should tend to zero, and, from equations (7.59) and (7.60)

$$\sigma_2^2 = v_{22} \to 0 \quad \text{as } t \to \infty \tag{7.75}$$

Clearly equation (7.75) is incompatible with equation (7.71). This apparent anomaly is easily explained by the fact that v_{11} does *not* asymptote to a stationary value, as time elapses. A frequency domain explanation of this phenomenon is given earlier, in Chapter 5.

An equation for $m_{22} = \sigma_q^2 \equiv \sigma_1^2$ is easily obtained by eliminating v_{12} between equations (7.59) and (7.60); hence

$$\frac{\mathrm{d}\sigma_1}{\mathrm{d}t} + c_\mathrm{e}\frac{\mathrm{d}\sigma_1}{\mathrm{d}t} = 2\sigma_2^2 \tag{7.76}$$

If the non-dimensional variance

$$\sigma_1^{*2} = \left(\frac{2\mu^6 g^6}{\pi^7 S_0^4}\right)\sigma_1^2 \tag{7.77}$$

is introduced, and t is replaced by t^*, as defined by equation (7.64), then equation (7.76) can be expressed in non-dimensional form as

$$\frac{\mathrm{d}\sigma_1^*}{\mathrm{d}t} + \frac{1}{\sigma_2^*}\frac{\mathrm{d}\sigma_1^*}{\mathrm{d}t^*} = \tfrac{1}{2}\sigma_2^* \tag{7.78}$$

where σ_2^* is given by equation (7.65).

The solution to equation (7.78) may be expressed as a quadrature. Specifically

$$\sigma_1^* = \frac{1}{2}\int_0^{t^*}\int_0^{t^*}\left(\frac{1-\sigma_2^*(u)}{1-\sigma_2^*(u)}\right)\sigma_2^{*2}(v)\,\mathrm{d}u\,\mathrm{d}v \tag{7.79}$$

Using the previous result for σ_2^* (see equation (7.69)) an analytical solution for σ_1^* may be found from equation (7.79), by performing the necessary integration. The result is (Ahmadi, 1983),

$$\sigma_1^{*2}(t) = \tfrac{1}{2}(3\sigma_2^{*2} - 5)\ln(1-\sigma_2^*) + \tfrac{1}{4}(-10\sigma_2^* - 5\sigma_2^{*2} + \tfrac{8}{3}\sigma_2^{*3} + \tfrac{1}{2}\sigma_2^{*4}) \tag{7.80}$$

Figure 7.2 shows the variation of σ_1^* with t^*, according to equations (7.80) and (7.69). This is compared with the corresponding result obtained by approximating c_e by its stationary value, c_2^s (see equation (7.72)). In non-dimensional form this result (see also Crandall *et al*, 1974) may be written as

$$\sigma_1^{*2} = t^* - \tfrac{1}{2}[1-\exp(-t^*)][3-\exp(-t^*)] \tag{7.81}$$

Once again, the 'stationary' solution, obtained by ignoring the variation of c_e with time, overestimates the standard deviation of the response, the error inherent in the approximation being of order 15%.

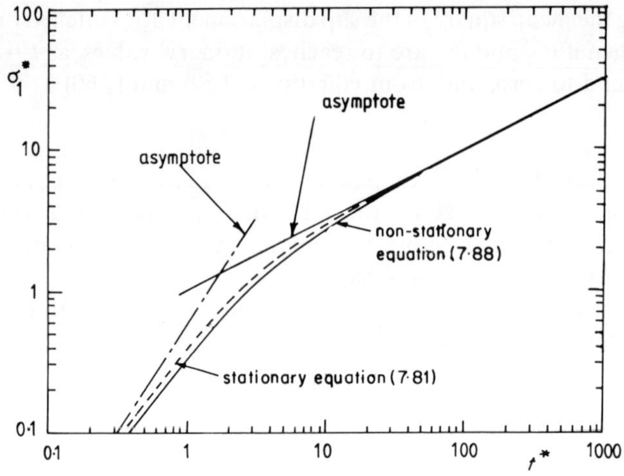

Figure 7.2

Friction controlled slip of a structure on a foundation. Variation of the non-dimensionalized standard deviation of the displacement response with time (reproduced from Ahmadi (1983) by permission of Pergamon Press)

It is interesting to note that, as t^* becomes large, both equation (7.80) and equation (7.81) yield the same asymptotic result, i.e.

$$\sigma_1^* \to (t^*)^{1/2} \quad \text{as } t^* \to \infty \tag{7.82}$$

This is to be expected since c_e approaches the constant value c_e^s, as time elapses. More surprisingly, for small times equations (7.80) and (7.81) also give the same asymptote, i.e.

$$\sigma_1^* \to \left(\frac{t^{*3}}{3}\right)^{1/2} \quad \text{as } t^* \to 0 \tag{7.83}$$

These asymptotes are shown in Figure 7.2.

7.3.2 Oscillator with asymmetric non-linearity

As an example of the application of the theory in a situation where the mean of the response is non-zero, due to an asymmetry in the non-linearity in the system, an oscillator governed by the following differential equation of motion will be considered (see also Spanos, 1978, 1980a, 1981b)

$$\ddot{q} + 2\zeta\dot{q} + q(1 + 3\varepsilon + 3\varepsilon q + \varepsilon q^2) = Q(t) \tag{7.84}$$

where

$$Q(t) = \lambda\alpha(t)Q_s(t) \tag{7.85}$$

As in the preceding example it will be assumed that $Q_s(t)$ is a zero-mean, stationary white noise process, with a covariance function given by equation (7.43). The symbol $\alpha(t)$, as before, is taken to represent a deterministic modulating function. Here it is convenient to choose the scaling factor, λ', as

$$\lambda' = \frac{2\zeta}{\pi S_0} \tag{7.86}$$

This ensures that, when $\varepsilon = 0$ and $\alpha(t) = 1$, the stationary variance of q is equal to unity.

Equation (7.84) can be derived by considering the time dependent deviation of a randomly excited Duffing oscillator from its position of static equilibrium under gravity. Thus, consider the mass–spring–damper system shown in Figure 7.3a. The mass is subjected to gravity and to an external force, $F(t)$, the non-linear spring characteristic is assumed to be that shown in Figure 7.3*b*. The symbol x denotes the displacement of the spring from its unstrained position. If the damping is taken to be linear, and viscous, with coefficient c, the equation of motion of the system is

$$m\ddot{x} + c\dot{x} + kx(1 + \lambda x^2) - mg = F(t) \tag{7.87}$$

Setting $F(t) = 0$, the static equilibrium position, $x + x_0$, is given by

$$kx_0(1 + \lambda x_0^2) = mg \tag{7.88}$$

and deviations

$$y = x - x_0 \tag{7.89}$$

from this position are governed by (from equations (7.87) and (7.89))

$$m\ddot{y} + c\dot{y} + k(y + 3\lambda x_0^2 y + 3\lambda x_0 y^2 + \lambda y^3) = F(t) \tag{7.90}$$

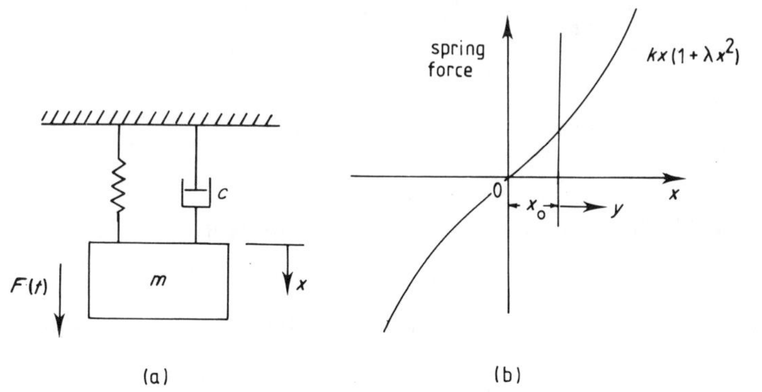

Figure 7.3
Mass–Spring–damper system under the influence of gravity. (a) System configuration. (b) Spring force variation with displacement

Defining

$$q = \frac{y}{x_0} \tag{7.91}$$

and replacing t by τ, where $\tau = \omega_n t$ and $\omega_n = (k/m)^{1/2}$, equation (7.90) can be cast into the dimensionless form of equation (7.84), where

$$2\zeta = c/(m\omega_n) \quad \varepsilon = x_0^2 \lambda \tag{7.92}$$

$$Q(\tau) = F(t)/m\omega_2^2 x_0 \tag{7.93}$$

and differentiation is now with respect to τ. Henceforth, t, in place of τ, will be used to denote non-dimensional time.

Comparing equation (7.84) with the standard forms of equations (7.10) and (7.11) it is clear that here $\mathbf{M} = 1$, $\mathbf{C} = \beta$, $\mathbf{K} = 1 + 3\varepsilon$; therefore, equation (7.84) can be written as

$$\ddot{q}_0 + 2\zeta \dot{q}_0 + (1 + 3\varepsilon) q_0 + h(q_0, \dot{q}_0) = Q_0(t) \tag{7.94}$$

where

$$h(q_0, \dot{q}_0) = \varepsilon(q_0 + m_q)^2[3 + (q_0 + m_q)] - E\{\varepsilon(q_0 + m_q)^2[3 + (q_0 + m_q)]\} \tag{7.95}$$

and $Q_0 \equiv Q$. Moreover, the mean, m_q of $q(t)$ is governed by (from equation (7.9))

$$\ddot{m}_q + 2\zeta \dot{m}_q + (1 + 3\varepsilon) m_q + E\{\varepsilon(q_0 + m)^2[3 + (q_0 + m_q)]\} = 0 \tag{7.96}$$

The equivalent linear system representation for equation (7.94) is (from equation (7.12))

$$\ddot{q}_0 + (2\zeta + c_e)\dot{q}_0 + (1 + 3\varepsilon + k_e) q_0 = Q_0(t) \tag{7.97}$$

Applying equations (7.16) and (7.17) one finds that

$$\begin{aligned} k_e &= E\left\{\frac{\partial}{\partial q_0} h(q_0, \dot{q}_0)\right\} = E\{6\varepsilon(q_0 + m_q) + 3\varepsilon(q_0 + m_q)^2\} \\ &= 6\varepsilon m_q + 3\varepsilon(\sigma_{q0}^2 + m_q^2) \end{aligned} \tag{7.98}$$

and

$$c_e = 0 \tag{7.99}$$

Equation (7.97) can be cast into the standard state form of (7.18), where here

$$\mathbf{z} = \begin{bmatrix} q_0 \\ \dot{q}_0 \end{bmatrix}, \quad \mathbf{f} = \begin{bmatrix} 0 \\ Q_0 \end{bmatrix} \tag{7.100}$$

and

$$\mathbf{G} = \begin{bmatrix} 0 & 1 \\ -(1 + 3\varepsilon + k_e) & -2\zeta \end{bmatrix} \tag{7.101}$$

Hence, applying equation (7.40) again, differential equations for the element of

$\mathbf{V} = [v_{ij}]$ are obtained. These may be supplemented with the differential equations for m_q, derived from equation (7.97). Denoting $X_1 = m_q$, $X_2 = \dot{m}_q$, $X_3 = v_{11}$, $X_4 = v_{12}$, $X_5 = v_{22}$, the complete set of equations for these unknowns can be written as

$$\dot{X}_1 = X_2 \tag{7.102}$$

$$\dot{X}_2 = -2\zeta X_2 - X_1(1 + 3\varepsilon + 3\varepsilon X_1 + 3\varepsilon X_3 + \varepsilon X_1^2 - 3\varepsilon X_3) \tag{7.103}$$

$$\dot{X}_3 = 2X_4 \tag{7.104}$$

$$X_4 = -(1 + 3\varepsilon + k_e)X_3 - 2\zeta X_4 + X_5 \tag{7.105}$$

$$\dot{X}_5 = -2(1 + 3\varepsilon + k_e)X_4 - 4\zeta X_5 + 4\zeta\alpha^2(t) \tag{7.106}$$

This set of equations, in conjunction with equation (7.98), can be easily solved numerically, using standard methods such as the fourth-order Runge–Kutta algorithm. In Chapter 10, results so obtained will be compared with corresponding digital simulation estimates, for X_1 and X_3.

7.4 NON-WHITE EXCITATION

As previously mentioned, the calculations become more complicated when the excitation cannot be modelled adequately, in terms of modulated white noises. Here two computationally efficient methods of analysis will be described, and illustrated through an application to a specific case. Both methods are suitable when the excitation can be modelled as a modulated, stationary process, according to equation (7.32). To keep matters simple, only the reduced case where $\mathbf{Q}(t)$ contains a single, non-zero element will be considered i.e. the situation where the input can be represented in terms of a single element of $\mathbf{Q}(t)$. Taking this element to be the last, one can write

$$\mathbf{Q}(t) = \mathbf{v}a(t)Q_s(t) \tag{7.107}$$

where $a(t)$ is a scalar modulating function, $Q_s(t)$ is a single, stationary, zero-mean random process and $\mathbf{v}$ is defined by

$$\mathbf{v} = [0, 0, \ldots, 1]^{\mathrm{T}} \tag{7.108}$$

It follows from equations (7.30), (7.31) and (7.107) that

$$\mathbf{w}_f(t, s) = \mathbf{R}a(t)a(s)w_s(t - s) \tag{7.109}$$

where

$$\mathbf{R} = \begin{bmatrix} \mathbf{0} & \mathbf{0} \\ \mathbf{0} & \mathbf{M}^{-1}\mathbf{P}\mathbf{M}^{-1\mathrm{T}} \end{bmatrix} \tag{7.110}$$

$$\mathbf{P} = \mathbf{v}\mathbf{v}^{\mathrm{T}} = \begin{bmatrix} 0 & 0 & \cdots & 0 \\ 0 & 0 & \cdots & 0 \\ \vdots & & & \\ 0 & 0 & \cdots & 1 \end{bmatrix} \tag{7.111}$$

and

$$w_s(t-s) = E\{Q_s(t)Q_s(s)\} \tag{7.112}$$

is the covariance function for $Q_s(t)$. It is noted that the elements of $\mathbf{R}$ are constants.

7.4.1 Decomposition method

Suppose that $a(t)$ can be decomposed into a sum of exponential terms, as follows

$$\begin{aligned} a(t) &= \sum_{i=1}^{M} A_i \exp(-\lambda_i t) \quad & t \geqslant 0 \\ &= 0 & t < 0 \end{aligned} \tag{7.113}$$

where A_i and λ_i are constants. Similarly, suppose that the covariance function of $Q_s(t)$ can be decomposed into exponential terms, as follows

$$w_s(t-s) = \sum_{k=1}^{N} \alpha_k \exp(-\beta_k |t-s|) \tag{7.114}$$

where α_k and β_k are constants.

A substitution of equation (7.109), (7.113) and (7.114) into equation (7.28) gives

$$\mathbf{U}(t) = \sum_{i=1}^{M} \sum_{j=1}^{M} \sum_{k=1}^{N} A_i A_j \alpha_k \mathbf{U}^{(ijk)}(t) \tag{7.115}$$

where

$$\mathbf{U}^{(ijk)}(t) = \exp[-(\lambda_i + \beta_k)t]\mathbf{Y}(t) \int_0^t \mathbf{W}(s) \exp[(\beta_k - \lambda_j)s]\, \mathrm{d}s \tag{7.116}$$

and

$$\mathbf{W}(s) = \mathbf{Y}^{-1}(s)\mathbf{R} \tag{7.117}$$

Differentiating the right-hand side of equation (7.116) with respect to time, and using equation (7.19), gives

$$\dot{U}^{(ijk)} = -(\lambda_i + \beta_k)\mathbf{U}^{(ijk)} + \mathbf{G}\mathbf{U}^{(ijk)} + \exp[-(\lambda_i + \lambda_j)t]\mathbf{R} \tag{7.118}$$

Equations (7.115) to (7.118) can be combined with equations (7.16), (7.17) and (7.26) to give a set of non-linear differential equations, which can be solved by standard methods.

The success, or otherwise, of this approach depends, to some extent, on the ability to express $a(t)$ and $w_s(t-s)$, according to equation (7.113) and (7.114), respectively, with a relatively small number of terms, in each case. Arbitrary forms for $a(t)$ and $w_s(t-s)$ may be decomposed by using a straightforward Gram–Schmidt orthogonalization procedure, and a Fourier series expansion (e.g. see Davis, 1963). It is noted that the coefficients in the expansions may be complex, in some cases.

7.4.2 Use of pre-filters

As discussed earlier, in Chapter 6 (see Section 6.6.4), in the stationary case one can deal with non-white excitations by introducing a 'shaping filter', or 'pre-filter', the input of which is white noise and the output is the non-white excitation process of the original system. The complete, augmented system, consisting of the pre-filter and the original system in series, can easily be analysed by the state variable method, using the basic result for systems driven by white noise, as given by equation (6.170).

Pre-filters can also be used when dealing with systems excited by modulated stationary processes. However, a modificaton is required, to avoid introducing the effect arising from the transient response of the pre-filter. Essentially, the output of the pre-filter must be allowed to reach stationarity, before it is multiplied by the modulating function, $a(t)$, to form the input to the original system. This point is illustrated schematically in Figure 7.4. If the switch is closed after the pre-filter output has reached stationarity then it is clear that the non-stationary response of the original system will not be influenced by the transient response of the pre-filter.

The effect of introducing a switch into the complete, augmented system can be realized by selecting suitable initial conditions for the covariance matrix of the overall state variable vector, $\mathbf{x}$. Thus, let

$$\mathbf{x} = \begin{bmatrix} \mathbf{z} \\ \mathbf{\Phi} \end{bmatrix} \tag{7.119}$$

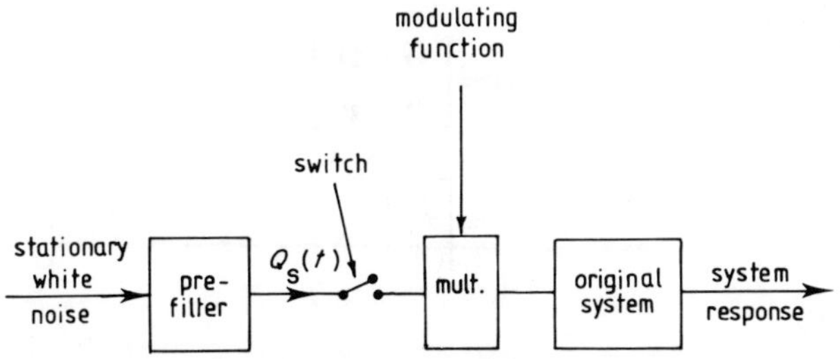

Figure 7.4
Block diagram representation of the use of pre-filters to determine system response in the non-stationary case

where $\mathbf{z}$ is the state variable vector of the original system (see equation (7.14)) and $\mathbf{\Phi}$ is the state variable vector associated with the pre-filter output. In the reduced case under discussion, where $\mathbf{Q}(t)$ is of the form of equation (7.107), the process $Q_s(t)$ is governed by pre-filter equations of the general form

$$\nu_{m-1}\Phi^{(m-1)} + \nu_{m-2}\Phi^{(m-2)} + \cdots + \nu_0\Phi^{(0)} = Q_s(t) \tag{7.120}$$

$$\Phi^{(m)} + \lambda_{m-1}\Phi^{(m-1)} + \cdots + \lambda_0\Phi^{(0)} = n(t) \tag{7.121}$$

where $\lambda_0, \lambda_1, \ldots, \lambda_{m-1}$ and $\nu_0, \nu_1, \ldots, \nu_{m-1}$ are filter constants and $n(t)$ is a stationary white noise process, with unit strength. As earlier, the superscript (n) denotes the nth derivative. The n vector, $\mathbf{\Phi}$, in equation (7.119) may be defined by

$$\mathbf{\Phi} = [\mathbf{\Phi}^{(0)}, \mathbf{\Phi}^{(1)}, \ldots, \mathbf{\Phi}^{(m-1)}]^{\mathrm{T}} \tag{7.122}$$

With this definition, equation (7.121) may be written in the standard state variable form i.e.

$$\dot{\mathbf{\Phi}} = \mathbf{F}\mathbf{\Phi} + \mathbf{w} \tag{7.123}$$

where

$$\mathbf{F} = \begin{bmatrix} 0 & 1 & \cdots & 0 \\ 0 & 0 & \cdots & 0 \\ \vdots & & & \\ -\lambda_0 & -\lambda_1 & \cdots & -\lambda_{m-1} \end{bmatrix} \tag{7.124}$$

and

$$\mathbf{w} = \begin{bmatrix} 0 \\ 0 \\ \vdots \\ n(t) \end{bmatrix} \tag{7.125}$$

For the case where the switch is closed it is easy to demonstrate (through an analysis similar to that in Section 6.6.4) that the complete system is governed by the usual state variable equation, of the form

$$\dot{\mathbf{x}} = \mathbf{H}(t)\mathbf{x} + \mathbf{W} \tag{7.126}$$

Here

$$\mathbf{H}(t) = \begin{bmatrix} \mathbf{G} & \mathbf{D} \\ \mathbf{0} & \mathbf{F} \end{bmatrix} \tag{7.127}$$

and

$$\mathbf{W} = \begin{bmatrix} \mathbf{0} \\ \mathbf{w} \end{bmatrix} \tag{7.128}$$

$\mathbf{D}$ is defined by

$$\mathbf{D} = a(t)\begin{bmatrix} \mathbf{0} \\ \mathbf{M}^{-1}\mathbf{v} \end{bmatrix}\mathbf{v}^{\mathrm{T}} \tag{7.129}$$

where

$$\mathbf{v} = [v_0, v_1, \ldots, v_{m-1}]^{\mathrm{T}} \tag{7.130}$$

As indicated, the augmented system matrix **H** is time dependent, partly through the time dependency of the elements in $\mathbf{C}_e$ and $\mathbf{K}_e$ (and hence in **G**) and partly through the introduction of the modulating function, $a(t)$, which appears in **D**.

The covariance matrix, **V**, for **x** is governed by the usual differential matrix equation for a system drive by white noise (e.g. see equation (7.40)). In the present notation this equation may be written as

$$\dot{\mathbf{V}} = \mathbf{H}(t)\mathbf{V} + \mathbf{V}\mathbf{H}^{\mathrm{T}}(t) + \mathbf{P} \tag{7.131}$$

where

$$\mathbf{P} = \begin{bmatrix} 0 & 0 & \cdots & 0 \\ 0 & 0 & \cdots & 0 \\ \vdots & & & \\ 0 & & \cdots & 1 \end{bmatrix} \tag{7.132}$$

In solving equation (7.131) it is essential to start the integration procedure at the instant the switch is closed. At this time ($t = 0$, say), some elements in **V**, which relate directly to $\mathbf{\Phi}$, will be non-zero. To demonstrate this, it is convenient to partition **V**, according to equation (7.119). Thus

$$\mathbf{V} = \begin{bmatrix} \mathbf{V}_{zz} & \mathbf{V}_{z\Phi} \\ \mathbf{V}_{z\Phi}^{\mathrm{T}} & \mathbf{V}_{\Phi\Phi} \end{bmatrix} \tag{7.133}$$

where

$$\mathbf{V}_{zz} = E\{\mathbf{z}\mathbf{z}^{\mathrm{T}}\} \tag{7.134}$$

$$\mathbf{V}_{z\Phi} = E\{\mathbf{z}\mathbf{\Phi}^{\mathrm{T}}\} \tag{7.135}$$

$$\mathbf{V}_{\Phi\Phi} = E\{\mathbf{\Phi}\mathbf{\Phi}^{\mathrm{T}}\} \tag{7.136}$$

If the original system is at rest, at $t = 0$, then clearly the response will be zero and all the elements of $\mathbf{V}_{zz}$ and $\mathbf{V}_{z\Phi}$ will be zero, at this time. However, the elements of $\mathbf{V}_{\Phi\Phi}$, at $t = 0$, will correspond to the stationary response of the pre-filter. Writing the stationary covariance matrix of $\mathbf{\Phi}$ as $\mathbf{V}^{s}_{\Phi\Phi}$, then it follows from equation (7.125) and (7.128) that $\mathbf{V}^{s}_{\Phi\Phi}$ is governed by

$$\mathbf{F}\mathbf{V}^{s}_{\Phi\Phi} + \mathbf{V}^{s}_{\Phi\Phi}\mathbf{F}^{\mathrm{T}} + \mathbf{P}' = \mathbf{0} \tag{7.137}$$

where $\mathbf{P}'$ has the same form as **P** (but is of lower dimension).

One can conclude, therefore, that a fairly small modification to the normal state variable analysis is required here. There are two basic steps, as follows:

(1) Equation (7.137) is solved to determine the stationary covariance matrix, $\mathbf{V}^{s}_{\Phi\Phi}$.
(2) Equation (7.131) is then integrated numerically with the following initial

condition for **V**

$$\mathbf{V} = \begin{bmatrix} \mathbf{0} & \mathbf{0} \\ \mathbf{0} & \mathbf{V}^{s}_{\Phi\Phi} \end{bmatrix} \tag{7.138}$$

7.4.3 An example

To illustrate the above methods of analysing the response of systems to non-white excitation, the case of a Duffing oscillator will once again be considered. An appropriate equation of motion is (see also equation (5.60))

$$\ddot{q} + 2\zeta\dot{q} + q + \lambda q^3 = Q(t) \tag{7.139}$$

where the symbols have the same meaning as before, and, similar to equation (7.84), time has been non-dimensionalized. It will be supposed that $Q(t)$ can be modelled as

$$Q(t) = a(t)Q_s(t) \tag{7.140}$$

where $a(t)$ is a deterministic modulating function and $Q_s(t)$ is a stationary response process, with zero mean, and a correlation function of the form

$$w_s(t-s) = \exp(-\gamma|t-s|)\cos\rho(t-s) \tag{7.141}$$

Here $a(t)$ will be taken to be of the form

$$\begin{aligned} a(t) &= A\exp(-\mu t) \quad & t \geqslant 0 \\ &= 0 & t < 0 \end{aligned} \tag{7.142}$$

Comparing equation (7.139) with the standard form of equation (7.10) one sees that here $\mathbf{M} = 1$, $\mathbf{C} = 2\zeta$, $\mathbf{K} = 1$ and

$$h(q, \dot{q}) = \lambda q^3 \tag{7.143}$$

Thus, from equation (7.12), the equivalent linear system is

$$\ddot{q} + 2\zeta\dot{q} + (1 + k_e)q = Q(t) \tag{7.144}$$

and, from equation (7.17)

$$k_e = E\left\{\frac{\partial h}{\partial q}\right\} = E\{3\lambda q^2\} = 3\lambda\sigma_q^2 \tag{7.145}$$

This agrees with the result given earlier, in Chapter 5 (see equation (5.82)), using a somewhat less direct method.

To apply the statistical linearization procedure it is first necessary to write the linearized equation of motion in state variable form (see equation (7.18)). One finds here that

$$\mathbf{G} = \begin{bmatrix} 0 & 1 \\ -(1 + 3\lambda\sigma_q^2) & -2\zeta \end{bmatrix} \tag{7.146}$$

and that $\mathbf{z}$ and $\mathbf{f}$ are given by

$$\mathbf{z} = \begin{bmatrix} q \\ \dot{q} \end{bmatrix} \quad \mathbf{f} = \begin{bmatrix} 0 \\ Q \end{bmatrix} \tag{7.147}$$

Denote

$$\begin{aligned} &X_1 = v_{11} \quad X_2 = v_{12} = v_{21} \quad X_3 = v_{22} \\ &X_4 = U_{11} \quad X_5 = U_{12} \quad X_6 = U_{21} \quad X_7 = U_{22} \end{aligned} \tag{7.148}$$

where v_{ij} are the elements of the covariance matrix, $\mathbf{V}$, for $\mathbf{z}$, then, from equation (7.26), one obtains the following set of equations

$$\dot{X}_1 = 2X_2 + 2X_4 \tag{7.149}$$

$$\dot{X}_2 = X_3 - (1 + 3\lambda X_1)X_1 - 2\zeta X_2 + X_5 + X_6 \tag{7.150}$$

and

$$\dot{X}_3 = -2(1 + 3\lambda X_1)X_2 - 4\zeta X_3 + 2X_7 \tag{7.151}$$

In applying the decomposition method, described in Section 7.4.1, it is necessary to express $w_s(t - s)$ in the form of equation (7.114). It is easy to see that equations (7.141) and (7.114) are equivalent if $N = 2$ and

$$\alpha_1 = \alpha_2 = 1/2 \tag{7.152}$$

$$\beta_1 = \gamma - \mathrm{i}\rho \tag{7.153}$$

$$\beta_2 = \gamma + \mathrm{i}\rho \tag{7.154}$$

Thus, referring to equation (7.115) to (7.118), it is seen that, in the present problem, with $M = 1$ ($A_1 = A$, $\lambda_1 = \mu$) and $N = 2$

$$\mathbf{U}(t) = \frac{A^2}{2}(\mathbf{U}^{(1)} + \mathbf{U}^{(2)}) \tag{7.155}$$

where $U^{(1)}$ and $U^{(2)}$ are governed by

$$\dot{\mathbf{U}}^{(1)} = -(\mu + \beta_1)\mathbf{U}^{(1)} + \mathbf{G}\mathbf{U}^{(1)} + \exp(-2\mu t)\mathbf{R} \tag{7.156}$$

and

$$\dot{\mathbf{U}}^{(2)} = -(\mu + \beta_2)\mathbf{U}^{(2)} + \mathbf{G}\mathbf{U}^{(2)} + \exp(-2\mu t)\mathbf{R} \tag{7.157}$$

Here $\mathbf{R}$ is defined by equation (7.110). In the present case this reduces to

$$\mathbf{R} = \begin{bmatrix} 0 & 0 \\ 0 & 1 \end{bmatrix} \tag{7.158}$$

A combination of equations (7.149) to (7.151) with equations (7.155) to (7.157) yields a set of seven coupled non-linear differential equations in the seven unknowns, X_1 to X_7. These can be solved by standard numerical integration routines, incorporating complex arithmetic facilities.

As an alternative, the pre-filer method, described in Section 7.4.2, may be employed. Here a pre-filter must be chosen to generate a stationary process, $Q_s(t)$, with the correct spectrum. By Fourier transforming equation (7.141), the required spectrum is found to be

$$S(\omega) = \frac{\gamma}{\pi}\left(\frac{(\gamma^2 + \rho^2) + \omega^2}{(\gamma^2 + \rho^2)^2 + 2\omega^2(\gamma^2 - \rho^2) + \omega^4}\right) \tag{7.159}$$

This may be compared with the output spectrum of a second-order filter. Thus, setting $m = 2$ in equations (7.120) and (7.121), the appropriate filter equations are

$$v_1\dot{\Phi} + v_0\Phi = Q_s(t) \tag{7.160}$$

$$\ddot{\Phi} + \lambda_1\dot{\Phi} + \lambda_0\Phi = n(t) \tag{7.161}$$

The frequency response of such a filter is

$$\alpha(\omega) = \frac{v_0 + v_1(\mathrm{i}\omega)}{\lambda_0 + \lambda_1(\mathrm{i}\omega) + (\mathrm{i}\omega)^2} \tag{7.162}$$

In this case $n(t)$ has unit strength, so that the constant spectral level of this process, S_0, is equal to $1/2\pi$. Hence, using the standard spectral input–output formula (see equation (4.20)),

$$S(\omega) = \frac{1}{2\pi}\left|\frac{v_0 + v_1(\mathrm{i}\omega)}{\lambda_0 + \lambda_1(\mathrm{i}\omega) + (\mathrm{i}\omega)^2}\right|^2 \tag{7.163}$$

Through a direct comparison between equation (7.159) and (7.163), the following expressions for the filter parameters may be derived

$$v_0 = [2\gamma(\gamma^2 + \rho^2)]^{1/2} \tag{7.164}$$

$$v_1 = (2\gamma)^{1/2} \tag{7.165}$$

$$\lambda_0 = \gamma^2 + \rho^2 \tag{7.166}$$

$$\lambda_1 = 2\gamma \tag{7.167}$$

One can now combine the equation of motion (7.139) with the filter equations (7.160) and (7.161) to obtain the overall system equation. The state variable form is given by equation (7.126), where here

$$\mathbf{x} = \begin{bmatrix} q \\ \dot{q} \\ \Phi \\ \dot{\Phi} \end{bmatrix} \qquad \mathbf{W} = \begin{bmatrix} 0 \\ 0 \\ 0 \\ n(t) \end{bmatrix} \tag{7.168}$$

and the augmented system matrix is, from equations (7.124), (7.127), (7.129) and (7.146)

$$\mathbf{H} = \begin{bmatrix} 0 & 1 & 0 & 0 \\ -(1+3\lambda\sigma_q^2) & -2\zeta & a(t)v_0 & a(t)v_1 \\ 0 & 0 & 0 & 1 \\ 0 & 0 & -\lambda_0 & -\lambda_1 \end{bmatrix} \tag{7.169}$$

Equation (7.126) must be integrated with the covariance matrix $\mathbf{V}$ set to the initial condition specified by equation (7.138). The 2×2 stationary covariance matrix $\mathbf{V}^s_{\Phi\Phi}$ must satisfy equation (7.137). Solving the latter equation gives the result

$$\mathbf{V}^s_{\Phi\Phi} = \begin{bmatrix} 1/(2\lambda_0\lambda_1) & 0 \\ 0 & 1/(2\lambda_1) \end{bmatrix} \tag{7.170}$$

Figure 7.5 shows some typical results obtained for this particular case. Here the mean square response, $\sigma_q^2 = v_{11}$, is plotted against the non-dimensional time, for the following set of parameters: $\gamma = 0.1$, $\rho = 1.99$, $\zeta = 0.1$. The value of $\mu = 0$ is chosen, which means that the modulating function is simply a unit step

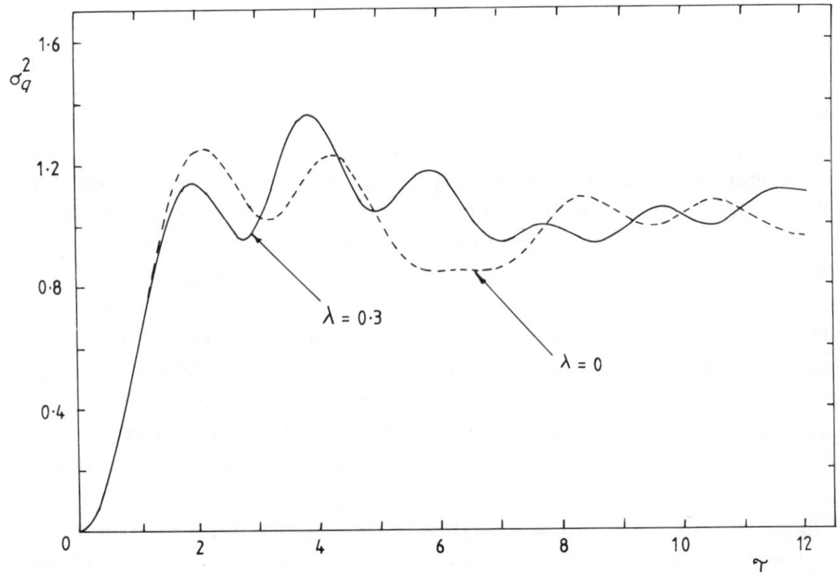

Figure 7.5
Variation of the mean square of the displacement response with time for $\lambda = 0$ and $\lambda = 0.3$. Step modulated excitation

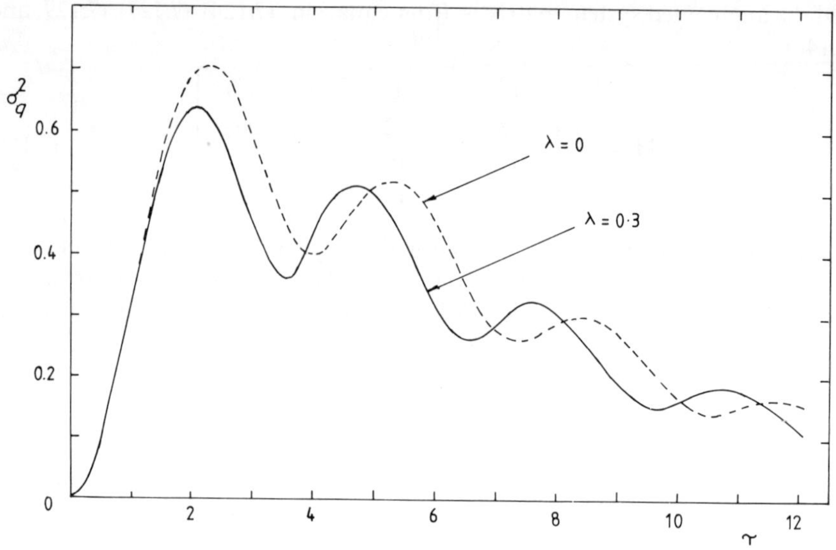

Figure 7.6
Variation of the mean square of the displacement response with time for $\lambda = 0$ and $\lambda = 0.3$. Exponentially modulated excitation

function. Hence, the results relate to the transient reponse to a suddenly applied, non-white, stationary excitation. In Figure 7.5 the mean square response is normalized by dividing by its limiting stationary value in the linear case ($\lambda = 0$). Thus this normalized response measure will asymptote towards a value of unity, as time elapses, in the linear case. Two transient responses are shown, one for $\lambda = 0$ (linear) and one for $\lambda = 0.3$ (non-linear). It is interesting to note that, in both cases, the instantaneous mean square response exceeds its limiting, stationary value, during some time intervals during the transient response. Moreover, the non-linear mean-square response exceeds the linear mean-square response, for some intervals of time.

The particular parameters relating to the results in Figure 7.5 were chosen to coincide with those adopted by Sakata and Kimura (1980) (see Figure 3b of their paper). Their numerical results, obtained by a different, and less general technique, agree completely with the present results. A comparison of their simulation estimates of the mean-square response with these statistical linearization results will be given later (in Chapter 10).

Finally, Figure 7.6 shows some typical mean square response histories for a case where μ is non-zero, so that the modulating function decays exponentially to zero. Here $A = 1$, $\mu = 0.5$, $\gamma = 0.1$, $\rho = 0.95$ and $\zeta = 0.2$. Two responses are shown, one for $\lambda = 0$ (linear) and one for $\lambda = 0.3$ (non-linear).

Chapter 8

Systems with hysteretic non-linearity

8.1 INTRODUCTION

As discussed earlier, in Chapter 2, many structures exhibit non-linear behaviour, in the form of a hysteretic restoring force–displacement characteristic, when subjected to severe dynamic excitations, such as those due to earthquakes. Hysteresis is also encountered in structures where relative sliding occurs between substructures when the amplitude of motion reaches a critical value (e.g., steel framed buildings with masonary walls can exhibit such behaviour).

In this chapter two approaches to applying the statistical linearization method to randomly excited hysteretic structures are discussed. The first, involving temporal averaging (Caughey, 1960a, 1963) is applicable to oscillators with a hysteresis loop of an arbitrary shape. It does, however, involve the assumption that the displacement response is narrow-band, in nature. This limits the usefulness of the analysis to situations where the energy dissipated in a cycle of oscillation, due to hysteresis, is a relatively small fraction of the total energy level associated with that cycle.

The second approach is to employ the fact that, in many cases, the hysteresis loops which occur can be modelled satisfactorily in terms of a differential relationship. Thus, for example, the differential model proposed by Bouc (1967) and Wen (1980) is capable of representing a wide variety of shapes of hysteresis loop, through a variation of the governing parameters, as shown in Chapter 2. When a differential model of the hysteretic force is incorporated into the equations of motion one has an extended set of differential equations, which can be written in the standard state-space form (see Chapter 4). The normal statistical linearization procedure, as described earlier, in Chapter 6, can then be applied.

In the first part of this chapter it will be assumed throughout, for simplicity, that the excitation is a stationary random process and that the statistics of the steady-state, stationary response are of interest. Extensions of the analysis to allow for non-stationarity in the response, will be briefly discussed in the last section of this chapter (Section 8.4).

8.2 AVERAGING METHOD

A single degree of freedom system will be considered here, with the following equation of motion

$$m\ddot{y} + c\dot{y} + h(t) = F(t) \tag{8.1}$$

This can represent, for example, a one-storey building structure (see Section 4.5.1), where m is the mass of the roof, c a linear damping coefficient, $h(t)$ the restoring force, due to the flexible walls, and $F(t)$ an excitation process, assumed here to be a random process with zero mean. It will also be assumed that $h(t)$ is antisymmetrical, such that the mean of $y(t)$ is also zero. On dividing throughout by m, equation (8.1) becomes

$$\varepsilon = z(t) + (\beta - \beta_{eq})\dot{y} - \omega_{eq}^2 y \tag{8.5}$$

where

$$\beta = \frac{c}{m} \quad z(t) = \frac{h(t)}{m} \quad f(t) = \frac{F(t)}{m} \tag{8.3}$$

The situation where $z(t)$ depends only on the instantaneous values of y and $\dot{y}$ has already been extensively discussed, in Chapter 5. For this 'non-hereditary', or 'zero-memory' type of restoring force, the statistical linearization procedure involves replacing equation (8.2) with the equivalent linear equation

$$\ddot{y} + \beta_{eq}\dot{y} + \omega_{eq}^2 y = f(t) \tag{8.4}$$

On minimizing the mean square of the error

$$\varepsilon = z(t) + (\beta - \beta_{eq})\dot{y} - \omega_{eq}^2 y \tag{8.5}$$

between equations (8.2) and (8.4) one obtains the following expressions for β_{eq} and ω_{eq}^2 (see equations (5.120))

$$\beta_{eq} = \beta + \frac{E\{\dot{y}z(t)\}}{\sigma_{\dot{y}}^2} \tag{8.6}$$

$$\omega_{eq}^2 = \frac{E\{yz(t)\}}{\sigma_y^2} \tag{8.7}$$

Equations (8.6) and (8.7) are still valid when the restoring force is hysteretic. However, in this situation, $z(t)$ depends on the entire history of the oscillator response, up to time t, and the appropriate technique for evaluating the expectations in equations (8.6) and (8.7) is not obvious.

One approach to the problem of evaluating β_{eq} and ω_{eq}^2, first proposed by Caughey (1960a), is useful when the damping, as measured by β_{eq}, is small and the excitation process, $f(t)$, is a wide-band process. In these circumstances the solution to equation (8.4) is a narrow-band process and one can write

$$y(t) = A(t)\cos(\omega_{eq}t + \theta(t)) \tag{8.8}$$

$$\dot{y}(t) = -\omega_{eq}A(t)\sin(\omega_{eq}t + \theta(t)) \tag{8.9}$$

This can be regarded as a transformation (usually known as the Van der Pol transformation) between the original response variables, $y(t)$, $\dot{y}(t)$, and new response variables, $A(t)$, $\theta(t)$, where $A(t)$ is an amplitude envelope process and $\theta(t)$ is a corresponding phase process. When β_{eq} is small, $A(t)$ and $\theta(t)$ will be

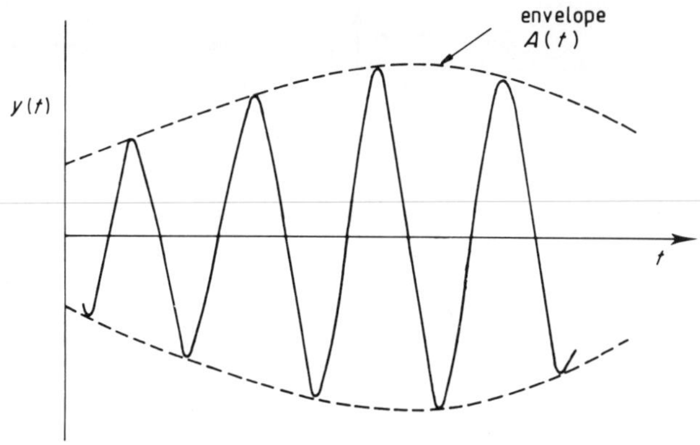

Figure 8.1
Relationship between the amplitude envelope process, $A(t)$, and the displacement response process, $y(t)$, for a narrow-band process

slowly-varying functions of time and, over one 'cycle' of the response, of period

$$T_c = \frac{2\pi}{\omega_{eq}} \tag{8.10}$$

can be treated as approximately constant. Figure 8.1 shows a typical variation of $y(t)$ with time, compared with the envelope process, $A(t)$.

Now, considering one cycle of the response, if $A(t)$ and $\theta(t)$ are approximated as constants, A and θ, then

$$y(t) = A\cos(\omega_{eq}t + \theta) \tag{8.11}$$

$$\dot{y}(t) = -\omega_{eq}A\sin(\omega_{eq}t + \theta) \tag{8.12}$$

and the variation of $z(t)$ with t, over the period T_c, is fixed. The steady-state variation of $z(t)$ with time, corresponding to a harmonic variation of $y(t)$ with time, will be denoted $z(A, t)$. A typical hysteresis loop, $z(A, t)$, is shown in Figure 8.2.

The essence of this method of evaluating β_{eq} and ω_{eq}^2 is to use $z(A, t)$ in place of $z(t)$, in equation (8.6) and (8.7), and to average the expectations over one cycle, of period T_c, treating A and θ constants. This procedure is an approximation, the error associated with which tends to zero as the energy dissipation per cycle tends to zero. Thus, substituting $z(A, t)$ for $z(t)$, and taking a time average, one has

$$\beta_{eq} = \beta + \frac{E\left\{\oint -\omega_{eq}A\sin\chi\, z(A,t)\,dt\right\}}{E\left\{\oint \omega_{eq}^2 A^2 \sin^2\chi\, dt\right\}} \tag{8.13}$$

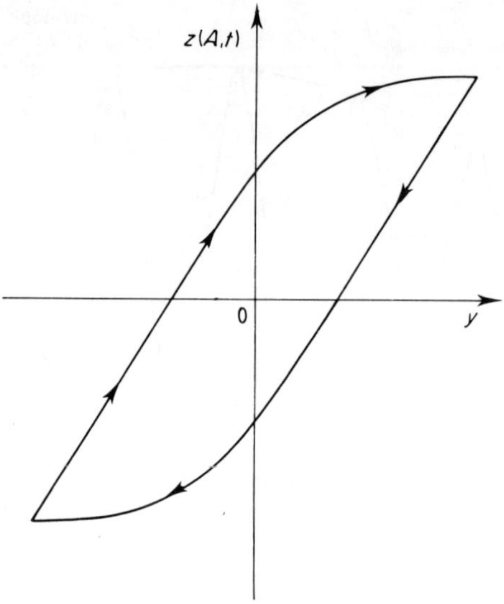

Figure 8.2
Typical hysteresis loop for an harmonic variation of displacement with time

and

$$\omega_{eq}^2 = \frac{E\left\{\oint A\cos\chi\, z(A,t)\,dt\right\}}{E\left\{\oint A^2\cos^2\chi\,dt\right\}} \tag{8.14}$$

where

$$\chi = \omega_{eq}t + \theta \tag{8.15}$$

Here the operations of expectation and averaging have been interchanged. This is permissible if integration is interpreted in mean-square sense (see Section 3.8). Using the fact that

$$\oint \sin^2\chi\,dt = \oint \cos^2\chi\,dt = \frac{1}{2\omega_{eq}} \tag{8.16}$$

and defining

$$C(A) = \frac{1}{\pi}\int_0^{2\pi} \cos\chi\, z(A,t)\,d\chi \tag{8.17}$$

$$S(A) = -\frac{1}{\pi}\int_0^{2\pi} \sin\chi\, z(A,t)\,d\chi \tag{8.18}$$

it is possible to express equations (8.13) and (8.14) in the following concise manner

$$\beta_{\mathrm{eq}} = \beta + \frac{E\{AS(A)\}}{\omega_{\mathrm{eq}} E\{A^2\}} \tag{8.19}$$

$$\omega_{\mathrm{eq}}^2 = \frac{E\{AC(A)\}}{E\{A^2\}} \tag{8.20}$$

If $f(A)$ is the probability density function of the envelope process $A(t)$ then one has

$$E\{AS(A)\} = \int_0^\infty AS(A) f(A)\,\mathrm{d}A \tag{8.21}$$

$$E\{AC(A)\} = \int_0^\infty AC(A) f(A)\,\mathrm{d}A \tag{8.22}$$

and

$$E\{A^2\} = \int_0^\infty A^2 f(A)\,\mathrm{d}A \tag{8.23}$$

8.2.1 An alternative approach

Instead of equation (8.4) one can write the following equivalent 'quasi-linear' equation

$$\ddot{y} + \beta(A)\dot{y} + \omega^2(A)y = f(t) \tag{8.24}$$

where now the equivalent damping factor, and natural frequency, are assumed to be functions of the amplitude, A. The error between equations (8.1) and (8.24) is given by

$$\varepsilon = z(t) + [\beta - \beta(A)]\dot{y} - \omega^2(A)y \tag{8.25}$$

This error can be minimized, in a mean-square sense, by minimizing the average value of ε^2, over one cycle of oscillation, i.e. minimize

$$\bar{\varepsilon} = \frac{\omega(A)}{2\pi} \oint \varepsilon^2\,\mathrm{d}t \tag{8.26}$$

and treat $A(t)$ as a constant over one cycle, as before. Thus it is required that

$$\frac{\partial \bar{\varepsilon}}{\partial \beta(A)} = \frac{\partial \bar{\varepsilon}}{\partial \omega^2(A)} = 0 \tag{8.27}$$

On substituting equation (8.25) into equation (8.26), and carrying out the differentiations indicated in equation (8.27), one obtains the following expressions for $\beta(A)$ and $\omega^2(A)$

$$\beta(A) = \beta + \frac{\oint \dot{y}z\,dt}{\oint \dot{y}^2\,dt} \tag{8.28}$$

$$\omega^2(A) = \frac{\oint yz\,dt}{\oint y^2\,dt} \tag{8.29}$$

These expressions are, of course, comparable with equations (8.6) and (8.7); here, effectively, the expectation operator has been replaced by an 'average-over-a-cycle' operator, resulting in amplitude dependent damping and stiffness factors.

The integrals in equations (8.28) and (8.29) can be evaluated by using a Van der Pol transformation similar to that expressed by equations (8.8) and (8.9), but with ω_{eq} replaced by $\omega(A)$; thus

$$y(t) = A(t)[\cos(\omega(A)t + \theta(t))] \tag{8.30}$$

$$\dot{y}(t) = -\omega(A)A(t)[\sin(\omega(A)t + \theta(t))] \tag{8.31}$$

Substituting equations (8.30) and (8.31) into equations (8.28) and (8.29), and assuming that $A(t)$ and $\theta(t)$ are constant, over one cycle, it is found that

$$\beta(A) = \beta + \frac{S(A)}{A\omega(A)} \tag{8.32}$$

$$\omega^2(A) = \frac{C(A)}{A} \tag{8.33}$$

where $S(A)$ and $C(A)$ are defined by equations (8.17) and (8.18), respectively.

For random excitation one can simply take expectations of the right-hand sides of equations (8.32) and (8.33), to obtain β_{eq} and ω_{eq}^2. Thus,

$$\beta_{eq} = \beta + E\left\{\frac{S(A)}{A\omega(A)}\right\} \tag{8.34}$$

$$\omega_{eq}^2 = E\left\{\frac{C(A)}{A}\right\} \tag{8.35}$$

This is the approach followed by Goto and Iemura (1973) and will lead to results which are different from those obtained using equations (8.19) and (8.20). The evaluation of the expectations in equations (8.34) and (8.35) requires a knowledge of $f(A)$. Thus, similar to equations (8.21) to (8.23),

$$E\left\{\frac{S(A)}{A\omega(A)}\right\} = \int_0^\infty \frac{S(A)f(A)}{A\omega(A)}\,\mathrm{d}A \tag{8.36}$$

$$E\left\{\frac{C(A)}{A}\right\} = \int_0^\infty \frac{C(A)f(A)}{A}\,\mathrm{d}A \tag{8.37}$$

Other methods of calculating β_{eq} and ω_{eq} from equations (8.32) and (8.33) are possible. For example, multiplying the top and bottom of the last term in equation (8.33) by A, and taking expectations of the top and bottom separately, the result given by equation (8.20) is recovered. However, a similar operation on equation (8.32) does not recover equation (8.19). Thus, there is no unique way of computing β_{eq} and ω_{eq}, using the slowly varying amplitude approach. It is only possible to assess the relative accuracy of the various possibilities by considering specific examples.

It is noted that the expression for $\beta(A)$, given by equation (8.32), can be derived from a rather different standpoint, by using the 'energy balance' method, first proposed by Jacobsen (1930) (see Goto and Iemura, 1973). In this method $\beta(A)$ is determined so as to equate the energy dissipated by the hysteretic oscillator to that dissipated by the linear oscillator. Thus

$$E = \oint (\beta\dot{y} + z)\,\mathrm{d}y = \oint \beta(A)\dot{y}\,\mathrm{d}y \tag{8.38}$$

where E is the energy dissipated in one cycle of the response. From equation (8.38) one obtains

$$\beta(A) = \beta + \frac{\oint z\,\mathrm{d}y}{\oint \dot{y}\,\mathrm{d}y} \tag{8.39}$$

If the substitution $\mathrm{d}y = \dot{y}\,\mathrm{d}t$ is made, in equation (8.39), it is evident that this result is identical to equation (8.28). It is well known that the energy balance method gives good results, for lightly damped non-linear oscillators (e.g. see Den Hartog, 1956).

8.2.2 Evaluation of the expectations

To evaluate the quantities such as $E\{AS(A)\}$ (see equation (8.21)), it is necessary to specify a probability density function, $f(A)$, for $A(t)$.

As for the standard statistical linearization method, described in Chapter 5, one can use the equivalent linear equation to find the appropriate response distribution. If $f(t)$ is Gaussian, in equation (8.4), then the joint process $\{y(t), \dot{y}(t)\}$ has a two-dimensional Gaussian distribution. Moreover, $y(t)$ and $\dot{y}(t)$ are statistically independent, since $E\{y\dot{y}\}=0$ (see equation (5.119)). Hence, from equation (3.49), with $n=2$,

$$f(y,\dot{y})=\frac{1}{2\pi\sigma_y\sigma_{\dot{y}}}\exp\left[-\frac{1}{2}\left(\frac{y^2}{\sigma_y^2}+\frac{\dot{y}^2}{\sigma_{\dot{y}}^2}\right)\right] \tag{8.40}$$

This result can be used to compute the density function for $A(t)$. Thus, from equations (8.8) and (8.9),

$$A(t)=\left(y^2+\frac{\dot{y}^2}{\omega_{\text{eq}}^2}\right)^{1/2} \tag{8.41}$$

and

$$f(A)=\iint_{\mathrm{d}A} f(y,\dot{y})\,\mathrm{d}y\,\mathrm{d}\dot{y} \tag{8.42}$$

where $\mathrm{d}A$ is an annular ring, in the y, $\dot{y}/\omega_{\text{eq}}$ plane, as shown in Figure 8.3. Since, from equations (8.8) and (8.9) one has, approximately,

$$\sigma_y^2=\frac{\sigma_{\dot{y}}^2}{\omega_{\text{eq}}^2} \tag{8.43}$$

if A and θ are assumed to be constant over one cycle, it follows, from equations (8.40) to (8.43), that

$$f(A)=\frac{A}{\sigma_y^2}\exp\left(-\frac{A^2}{2\sigma_y^2}\right) \tag{8.44}$$

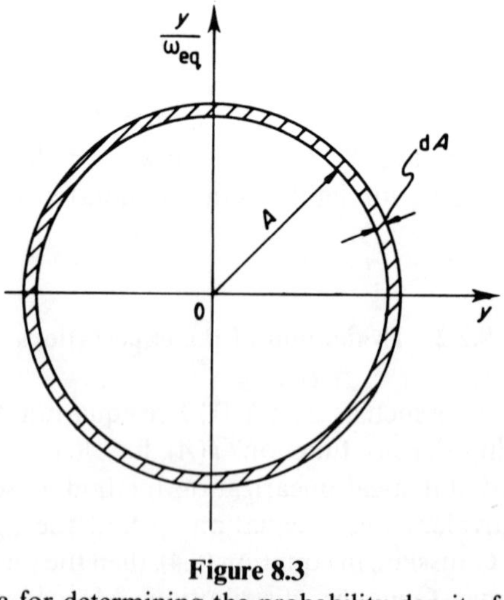

Figure 8.3
Integration area for determining the probability density function of $A(t)$

This density function corresponds to the well known Rayleigh distribution. It also applies if one adopts the alternative approach given in the preceding section, and replaces ω_{eq} by $\omega(A)$.

From equation (8.44) the following expression for the moments of A is readily obtained

$$E\{A^n\} = (2\sigma^2)^{(n-1)/2}\Gamma\left(\frac{n+1}{2}\right) \tag{8.45}$$

In particular,

$$E\{A^2\} = 2\sigma_y^2 \tag{8.46}$$

$$E\{A^4\} = 8\sigma_y^4 \tag{8.47}$$

8.2.3 Application to non-hysteretic oscillators

The relationships derived in the preceding sections of this chapter are valid for any type of lightly damped oscillator. They do not depend on whether $z(t)$ is hysteretic. Before proceeding with applications to specific hysteretic oscillators it is instructive to compare some results for a non-hysteretic oscillator with those derived previously, in Chapter 5, using the standard relationships (see equations (5.122) and (5.123)).

For this purpose an oscillator with cubic non-linearity in both the damping and stiffness terms is appropriate. The equation of motion of such an oscillator is given by equation (5.138). Comparing this with equation (8.2) one has

$$z(t) = \beta\eta\dot{y}^3 + \omega_n^2(y + \lambda y^3) \tag{8.48}$$

as a non-hysteretic force term. If the damping as measured by β is small, the averaging method should give good results, for this type of oscillator.

Initially the approach where β_{eq} and ω_{eq}^2 are treated as constants, at the outset, will be followed. Using equations (8.8), (8.9), (8.17) and (8.18), the functions $C(A)$ and $S(A)$ are readily evaluated, for the present example. Thus,

$$C(A) = \omega_n^2 A + \frac{3\omega_n^2\lambda}{4} A^3 \tag{8.49}$$

$$S(A) = \frac{3\beta\eta\omega_{eq}^3}{4} A^3 \tag{8.50}$$

On substituting these expressions into equations (8.19) and 8.20), and using equations (8.46) and (8.47), it is found that

$$\beta_{eq} = \beta + 3\beta\eta\omega_{eq}^2\sigma_y^2 \tag{8.51}$$

$$\omega_{eq}^2 = \omega_n^2 + 3\omega_n^2\lambda\sigma_y^2 \tag{8.52}$$

These results may be compared with those obtained earlier, in Chapter 5. From equations (5.131), (5.132), (5.140) and (5.141) one finds that

$$\beta_{eq} = \beta + 3\beta\eta\sigma_{\dot{y}}^2 \tag{8.53}$$

$$\omega_{eq}^2 = \omega_n^2 + 3\omega_n^2\lambda\sigma_y^2 \tag{8.54}$$

where these results are independent of the shape of the input spectrum, and the magnitude of the non-linearities.

If the excitation is a white noise then, from equation (5.135),

$$\sigma_{\dot{y}}^2 = \omega_{eq}^2\sigma_y^2 \tag{8.55}$$

and equations (8.53) and (8.54) agree exactly with equations (8.51) and (8.52), respectively. It is interesting to note that this agreement occurs independently of the magnitude of the damping. If the excitation is non-white then complete agreement is not obtained, in general. However, if the damping is light then, from equations (8.8) and (8.9) it follows that equation (8.55) again holds. This is due to the fact that $A(t)$ and $\theta(t)$ can be treated as approximately constant over one cycle, as previously noted; the results from the two methods thus coincide again.

If now the alternative approach outlined in Section 8.2.1 is followed then $C(A)$ is given by equation (8.49) again, but

$$S(A) = \tfrac{3}{4}\beta\eta\omega^3(A)A^3 \tag{8.56}$$

Substituting these expressions into equations (8.32) and (8.33) gives

$$\beta(A) = \beta + \tfrac{3}{4}\beta\eta\omega^2(A)A^2 \tag{8.57}$$

$$\omega^2(A) = \omega_n^2 + \tfrac{3}{4}\omega_n^2\lambda A^2 \tag{8.58}$$

On taking expectations of $\beta(A)$ and $\omega^2(A)$, according to equations (8.34) and (8.35), it is found that

$$\beta_{eq} = \beta + \tfrac{3}{2}\beta\eta\omega_n^2(1 + 3\lambda\sigma_y^2)\sigma_y^2 \tag{8.59}$$

$$\omega_{eq}^2 = \omega_n^2 + \tfrac{3}{2}\omega_n^2\lambda\sigma_y^2 \tag{8.60}$$

On comparing these results with equations (8.51) and (8.52) it is seen that there is a substantial difference. In fact the non-linear contribution to both β_{eq} and ω_{eq}^2 differs by a factor of two.

Since the standard method of Chapter 5, for non-hysteretic oscillators, is known to give good results (see Chapter 10), it can be concluded, on the basis of the limited evidence provided by this example, that the method of calculating β_{eq} and ω_{eq}^2, based on equations (8.19) and (8.20), respectively, is superior to the alternative approach of averaging $\beta(A)$ and $\omega^2(A)$, in terms of accuracy. Hence, in the following discussion of hysteretic oscillators, equations (8.19) and (8.20) will be used as the basis of the analysis, in preference to equations (8.34) and (8.35).

8.2.4 Inputs with non-zero means

The averaging method can be readily adapted to deal with situations where the mean of the excitation process, $f(t)$, is non-zero.

If $f(t)$ is represented as a sum of its mean value, m_f, and a zero-mean process $f_0(t)$, and similarly for $y(t)$ (see equations (5.73) and (5.74)), then the equation of motion can be rewritten as

$$\ddot{y}_0 + \beta\dot{y}_0 + z(t) = f_0(t) + m_f \tag{8.61}$$

Taking expectations of this last equation gives

$$E\{z(t)\} = m_f \tag{8.62}$$

and subtracting this relationship from equation (8.61) results in

$$\ddot{y}_0 + \beta\dot{y}_0 + z^*(t) = f_0(t) \tag{8.63}$$

where

$$z^*(t) = z(t) - E\{z(t)\} = z(t) - m_f \tag{8.64}$$

Equation (8.63) can be linearized by the technique described in Section 8.2, if the appropriate modifications are incorporated into the analysis. Thus $y(t)$ is replaced by $y_0(t)$ in equations (8.11) and (8.12). Therefore

$$y(t) = A\cos(\omega_{eq}t + \theta) + m_y \tag{8.65}$$

With $m_y = 0$ the symmetric loop corresponding to this harmonic fluctuation will be centred at the origin, as previously noted. Let this loop be denoted $z^*(A, t)$. Now for a wide class of hysteretic oscillator (including the differential hysteresis models, to be discussed in some detail later) the effect of including a

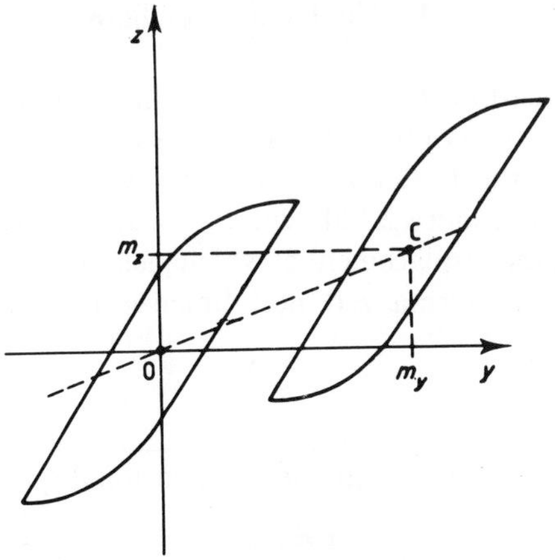

Figure 8.4
Effect of a non-zero mean on the position of the hysteresis loop

non-zero mean in equation (8.65) is simply to shift the loop to a new position, as shown in Figure 8.4. The coordinates of the centre of the loop are (m_y, m_z), where m_z is the mean value of the shifted loop amplitude. The quantity m_z is generally a function of m_y i.e.

$$m_z = F(m_y) \tag{8.66}$$

where this function describes the locus of C, as m_y varies (usually this locus is a straight line, through the origin). If $z(A,t)$ denotes the shifted loop then

$$z(A,t) = z^*(A,t) + m_z \tag{8.67}$$

An essential feature of the averaging method is the replacement of $z(t)$ by $z(A,t)$. Making this substitution in equation (8.62) one has

$$m_f = E\{z^*(A,t) + m_z\} = m_z \tag{8.68}$$

since the mean of $z^*(A,t)$ is zero. Thus, replacing m_z *by* m_f in equation (8.67) shows that

$$z(A,t) = z^*(A,t) + m_f \tag{8.69}$$

A comparison of equation (8.69) with equation (8.64) shows that $z^*(A,t)$ is the appropriate replacement for $z^*(t)$. It follows that β_{eq} and ω_{eq}^2 are unaffected by the presence of a non-zero mean in the excitation, if equation (8.67) is valid.

Further m_y can be evaluated by combining equations (8.66) and (8.68). Thus

$$m_y = F^{-1}(m_f) \tag{8.70}$$

where $F^{-1}(\ )$ is the inverse of the function in equation (8.66).

8.2.5 The bilinear oscillator

To illustrate the application of the averaging method to hysteretic oscillators, the case where the hysteresis is of the bilinear type (see Chapter 2) will be discussed in some detail. Thus the hysteretic force, $h(t)$, will be assumed to have the characteristic shown in Figure 2.32d, where $x(=q)$ is the displacement, and x^* is the critical value of the displacement at which yield first occurs. It is convenient here to replace x by the normalized, non-dimensional displacement

$$y = \frac{x}{x^*} \tag{8.71}$$

and to measure time in terms of the non-dimensional quantity

$$\tau = \lambda^{1/2} t \tag{8.72}$$

where λ is the tangent of the primary, elastic slope. The equation of motion for an oscillator with viscous damping and bilinear hysteresis is then given by

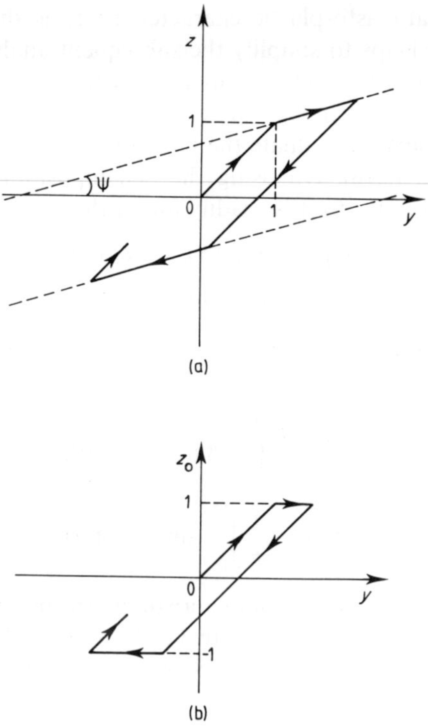

Figure 8.5
The bilinear hysteresis characteristic. (a) Normalized non-dimensional characteristic, $\alpha = \tan\psi$ (b) Non-dimensional elasto-plastic hysteresis

equation (8.2), where now differentiation is with respect to τ, and all quantities are non-dimensional. It will be assumed, initially, that $f(t)$ has zero mean. The non-dimensional bilinear hysteretic force, $z(\tau)$, in this equation is shown in Figure 8.5a. The tangent of the primary, elastic slope is unity. Further, it is noted that the tangent, α, of the secondary slope of $z(t)$, is related to the tangent, θ, of the secondary slope of $h(t)$ through the relationship

$$\alpha = \frac{\theta}{\lambda} \tag{8.73}$$

As pointed out by Suzuki and Minai (1987) (see also Chapter 2) the bilinear characteristic $z(t)$ can be expressed as a combination of a linear, non-hysteretic elastic contribution and a purely elasto–plastic combination, z_0 with a secondary slope, $\alpha = 0$. Thus

$$z = \alpha y + (1 - \alpha)z_0 \tag{8.74}$$

The non-dimensional elasto-plastic characteristic z_0 is shown in Figure 8.5b. This decomposition helps to simplify the subsequent analysis.

In applying the averaging method, equation (8.2) is replaced by equation (8.4), where β_{eq} and ω_{eq}^2 are given by equations (8.19) and (8.20). As a first step, therefore, it is necessary to evaluate the functions $S(A)$ and $C(A)$, as defined by equations (8.17) and (8.18). On using the decomposition of equation (8.74), together with equation (8.10) it is readily found that

$$C(A) = \alpha A + (1 - \alpha)C_0(A) \tag{8.75}$$

$$S(A) = (1 - \alpha)S_0(A) \tag{8.76}$$

where $C_0(A)$, $S_0(A)$ are the appropriate functions for the elasto–plastic characteristic $z_0(\tau)$. Thus

$$C_0(A) = \frac{1}{\pi}\int_0^{2\pi} \cos\chi\, z_0(A, \tau)\,\mathrm{d}\chi \tag{8.77}$$

$$S_0(A) = -\frac{1}{\pi}\int_0^{2\pi} \sin\chi\, z_0(A, \tau)\,\mathrm{d}\chi \tag{8.78}$$

where $z_0(A, \tau)$ is the symmetric characteristic (corresponding to an harmonic variation of $y(\tau)$ with τ) shown in Figure 8.6. A simple technique for evaluating the integrals in Equations (8.77) and (8.78) is described by Caughey (1960a). The results are as follows

$$\begin{aligned} C_0(A) &= \frac{A}{\pi}(\Lambda - \tfrac{1}{2}\sin 2\Lambda) \quad & A > 1 \\ &= A & A < 1 \end{aligned} \tag{8.79}$$

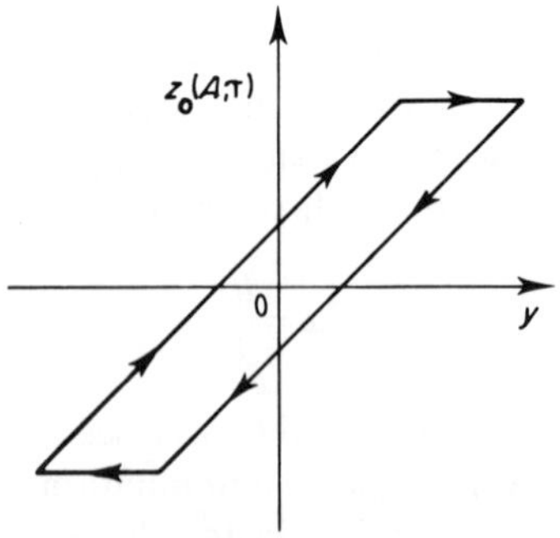

Figure 8.6
The elasto-plastic hysteresis loop corresponding to an harmonic variation of displacement with time

and

$$S_0(A) = \frac{A}{\pi}\sin^2\Lambda \quad A > 1$$
$$= 0 \qquad A < 1 \tag{8.80}$$

where

$$\cos\Lambda = 1 - \frac{2}{A} \tag{8.81}$$

It follows from equation (8.81) that an alternative form of equation (8.80) is as follows

$$S_0(A) = \frac{4}{\pi}\left(1 - \frac{1}{A}\right) \quad A > 1$$
$$= 0 \qquad A < 1 \tag{8.82}$$

The variation of $C_0(A)$ and $S_0(A)$ with A is shown graphically in Figure 8.7. It

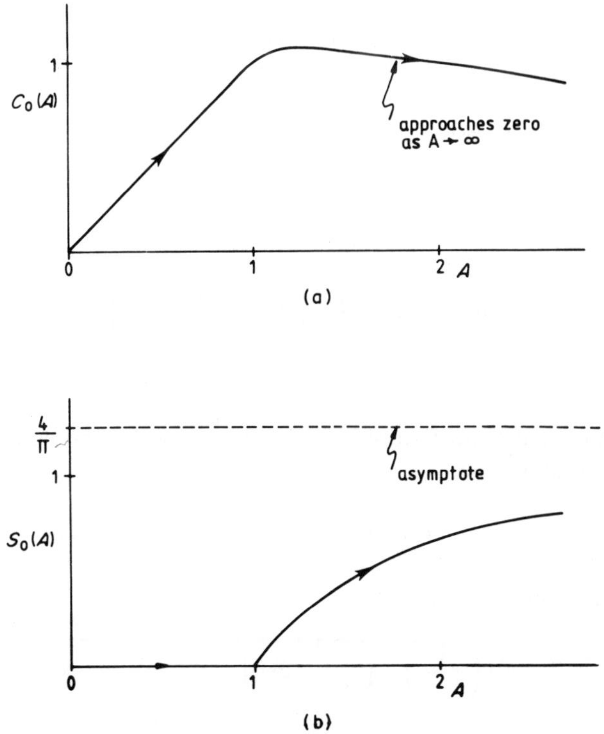

Figure 8.7
The variation of (a) $C_0(A)$ and (*b*) $S_0(A)$ with A

is noted that, as $A \to \infty$, $\Lambda \to \pi$ and the following asymptotes are approached

$$C_0(A) \to 0 \tag{8.83}$$

$$S_0(A) \to \frac{4}{\pi} \tag{8.84}$$

Equations (8.75), (8.76), (8.79) and (8.80) may now be combined with equations (8.19) and (8.20), to evaluate β_{eq} and ω_{eq}^2. Considering β_{eq}, initially, and using the assumption that $A(t)$ is Rayleigh distributed, so that $E\{A^2\} = 2\sigma_y^2$, it is found that

$$\beta_{eq} = \beta + \frac{(1-\alpha)}{2\omega_{eq}\sigma_y^2} E\{AS_0(A)\} \tag{8.85}$$

where

$$E\{AS_0(A)\} = \frac{4}{\pi}\int_1^\infty (A-1)\frac{A}{\sigma_y^2}\exp\left(-\frac{A^2}{2\sigma_y^2}\right)\mathrm{d}A \tag{8.86}$$

The integral here can be evaluated in terms of the error function, erf ( ) (see Appendix A). Hence one obtains the result

$$\beta_{eq} = \beta + \left(\frac{2}{\pi}\right)^{1/2}\frac{(1-\alpha)}{\omega_{eq}\sigma_y}\left[1 - \exp\left(\frac{1}{\sqrt{2}\sigma_y}\right)\right] \tag{8.87}$$

This result can be expressed as

$$\frac{(\beta_{eq} - \beta)\omega_{eq}}{(1-\alpha)} = g_1(\lambda) \tag{8.88}$$

where

$$g_1(\lambda) = \left(\frac{4}{\pi\lambda}\right)^{1/2}\left[1 - \exp\left(\frac{1}{\lambda^{1/2}}\right)\right] \tag{8.89}$$

and

$$\lambda = 2\sigma_y^2 \tag{8.90}$$

The function $g_1(\lambda)$ is shown graphically in Figure 8.8. It is noted that, as $\lambda \to \infty$, $\exp(1/\lambda^{1/2}) \to 0$ and hence

$$g_1(\lambda) \to \left(\frac{4}{\pi\lambda}\right)^{1/2} \qquad \lambda \to \infty \tag{8.91}$$

On the other hand, as $\lambda \to 0$,

$$1 - \operatorname{erf}\left(\frac{1}{\lambda^{1/2}}\right) \to \left(\frac{\lambda}{\pi}\right)^{1/2}\exp\left(-\frac{1}{\lambda^2}\right) \qquad \lambda \to 0 \tag{8.92}$$

and hence

$$g_1(\lambda) \to \frac{2}{\pi}\exp\left(-\frac{1}{\lambda^2}\right) \qquad \lambda \to 0 \tag{8.93}$$

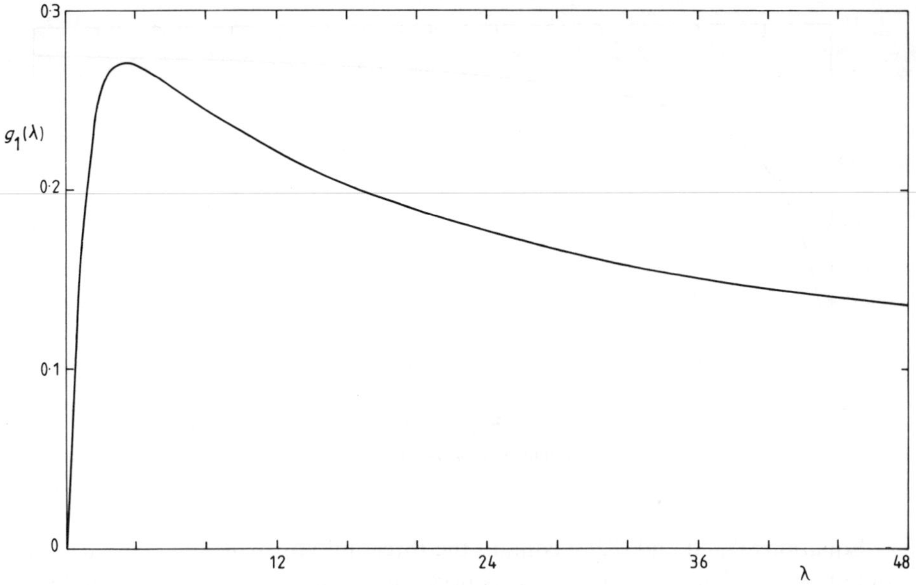

Figure 8.8
Variation of $g_1(\lambda)$ with λ

The evaluation of ω_{eq}^2 follows similar lines, except that here, unfortunately, the result cannot be expressed in terms of simple functions. It is found that (Caughey, 1960a)

$$\frac{1-\omega_{eq}^2}{(1-\alpha)}=g_2(\lambda) \tag{8.94}$$

where

$$g_2(\lambda)=\frac{2}{\pi\lambda^2}\int_1^{\infty} A^3(\pi-\Lambda+\tfrac{1}{2}\sin 2\Lambda)\exp\left(-\frac{A}{\lambda}\right)\mathrm{d}\alpha \tag{8.95}$$

The function $g_2(\lambda)$ is shown in Figure 8.9. It is noted that, if $\lambda\to\infty$, $g_2(\lambda)\to 1$ and hence

$$\omega_{eq}\to\alpha^{1/2} \quad \lambda\to\infty \tag{8.96}$$

This result is to be expected, since if σ_y is very large, most of the displacement is in the plastic regime, and the associated natural frequency of oscillation should correspond to the square root of the tangent of the secondary, plastic slope. At the other extreme, if $\sigma_y\to 0$, then $g_2(\sigma_y)\to 0$ and

$$\omega_{eq}^2\to 1 \quad \sigma_y\to 0 \tag{8.97}$$

Again, this is natural, since at low levels of excitation, the displacement is mostly in the elastic range, where the primary slope is (here) unity.

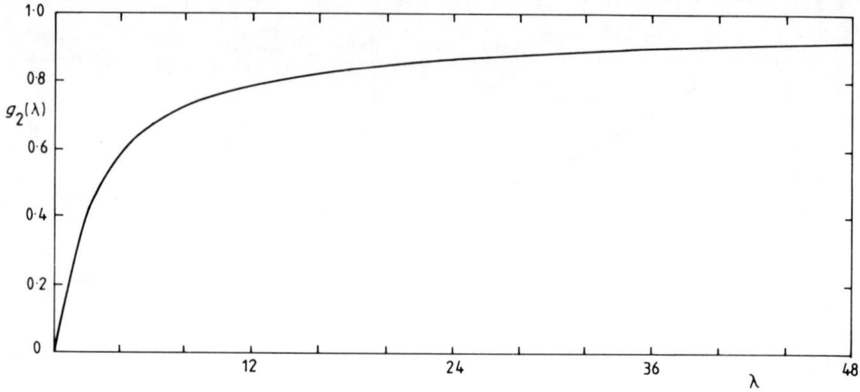

Figure 8.9
Variation of $g_2(\lambda)$ with λ

Equations (8.88) and (8.94) contain the three unknowns β_{eq}, ω_{eq}^2 and σ_y^2. To obtain a closed set of equations it is necessary to use the linearized equation, (8.4), to obtain a further relationship between the unknowns. As before, σ_y^2 can be computed from equation (5.127), where $\alpha(\omega)$ is given by equation (5.129). In the special case where $S_f(\omega)$ is approximated as a constant level, S_0, corresponding to the white noise approximation, one has, as before (see equation (5.135))

$$\sigma_y^2 = \frac{\pi S_0}{\beta_{eq}\omega_{eq}^2} = \frac{\sigma_{yo}^2 \beta}{\beta_{eq}} \tag{8.98}$$

In this equation

$$\sigma_{yo}^2 = \frac{\pi S_0}{\beta} \tag{8.99}$$

is the mean square response in the linear, non-hysteretic case ($\beta_{eq} = \beta, \omega_{eq}^2 = 1$).

In the case of white noise excitation, these equations can be solved by the following procedure. First it is necessary to specify a value of σ_y. Then, one can proceed to determine $g_1(\lambda)$ and $g_2(\lambda)$, from equations (8.89) and (8.95). From this ω_{eq} and β_{eq} can be found, from equations (8.94) and (8.88), respectively. Further, using ω_{eq} and β_{eq} one can find S_0 from equation (8.98). This procedure can be repeated to plot σ_y against S_0. Then, one can determine σ_y for a specified S_0 value, from the graph of σ_y against S_0.

Figure 8.10 shows some typical results obtained by the method. Here $\sigma_y/S_0^{1/2}$ is plotted against S_0, for $\alpha = 0.02$ and $\beta = 0$ and $\beta = 1$. Figure 8.11 shows similar results for the case $\alpha = 0.5$. It is noted that, as $S_0 \to 0$, $\beta_{eq} \to \beta$ and $\omega_{eq} \to 1$; thus

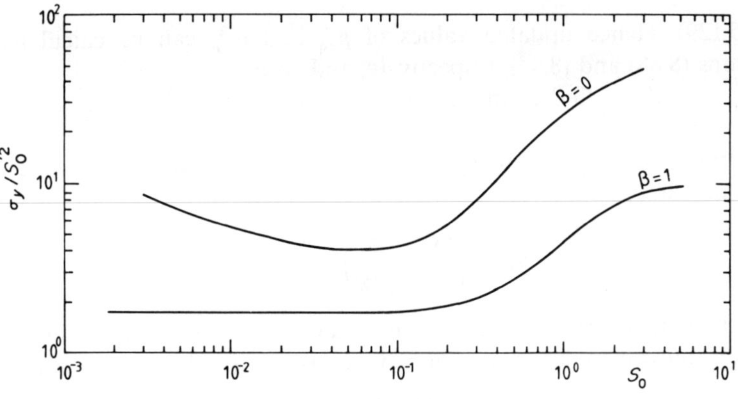

Figure 8.10
Variation of the standard deviation of the displacement response with input level, for $\beta = 0$ and $\beta = 1$. $\alpha = 0.02$. Bilinear hysteresis

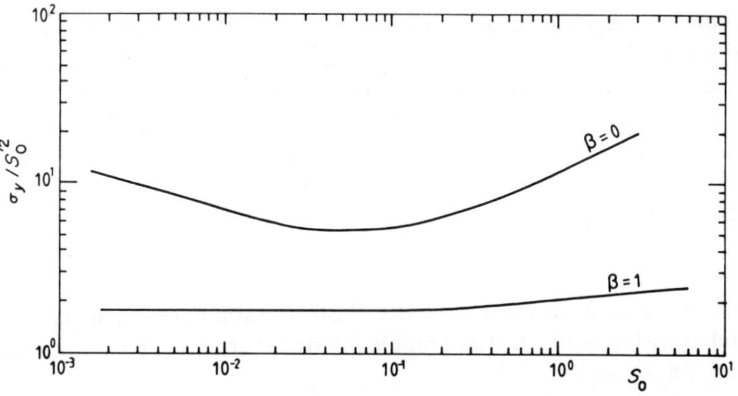

Figure 8.11
Variation of the standard deviation of the displacement response with input level, for $\beta = 0$ and $\beta = 1$. $\alpha = 0.5$. Bilinear hysteresis

$\sigma_y \to \sigma_{yo}$ and, from equation (8.98)

$$\frac{\sigma_y}{S_0^{1/2}} \to \left(\frac{\pi}{\beta}\right)^{1/2} \quad \text{as } S_0 \to 0 \qquad (8.100)$$

This asymptotic behaviour is clearly evident in Figures 8.10 and 8.11. Therefore, for $\beta = 1$, $\sigma_y/S_0^{1/2} \to \pi^{1/2}$ as $S_0 \to 0$.

For non-white excitation an iterative procedure will usually be necessary. Thus, with an initial guess of β_{eq} and ω_{eq}^2, σ_y^2 can be found from equations (5.127) and (5.129). Hence updated values of β_{eq} and ω_{eq}^2 can be calculated from equations (8.88) and (8.94), respectively, and so on.

It is possible to find a simple asymptotic result for σ_y^2, in the case of white noise excitation, valid as $\sigma_y \to \infty$, by using the asymptotes given by equations (8.91) and (8.96), together with equation (8.98). In the special case where $\beta = 0$ it is found that

$$\sigma_y \to \frac{\pi S_0}{(1-\alpha)} \left(\frac{\pi}{2\alpha} \right)^{1/2} \tag{8.101}$$

Similarly, if σ_y is very small, a combination of equations (8.92), (8.97) and (8.98) yields the following relationship for σ_y, in terms of S_0 (in the case where $\beta = 0$)

$$S_0 = \frac{2(1-\alpha)}{\pi^2 \sigma_y^2} \exp\left(-\frac{1}{2\sigma_y^2} \right) \tag{8.102}$$

Clearly, from this result, $\sigma_y \to 0$ as $S_0 \to 0$. It is worth pointing out, however, that this last result is actually an artifice of the statistical linearization approximation. A more exact analysis, based on the stochastic averagaing method, shows that $\sigma_y \to 0.5$, as $S_0 \to 0$, for the case where $\beta = 0$ (Roberts, 1978a).

For the case where the mean of $f(t)$ is non-zero then, as pointed out in Section 8.2.4, the effect of m_f is simply to shift the symmetric loop, $z(A, t)$, to a new position. Therefore, the expressions derived earlier for σ_y are still applicable. The mean of the response can be determined from equation (8.70). Using the decomposition of equation (8.74) again, it is evident that

$$m_z = \alpha m_y = m_f \tag{8.103}$$

since the m_z value associated with the elasto–plastic component is zero. The above relationship is illustrated graphically in Figure 8.12. Hence, in equation (8.70), $F^{-1}(m_f) = m_f/\alpha$.

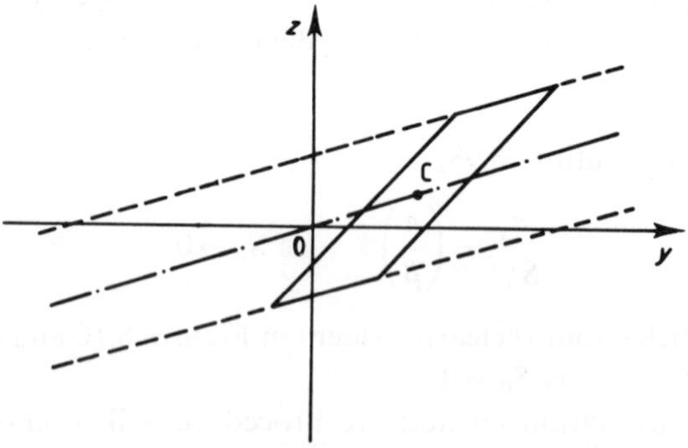

Figure 8.12
The effect of a non-zero mean on the position of the symmetrical bilinear hysteresis loop

8.2.6 Allowance for drift motion

The statistical linearization procedure given in the preceding section, for the bilinear oscillator, is based on the assumption that the response process, $y(t)$, is narrow-band in nature. This condition is satisfied if the response displacement is mostly in the elastic range (low excitation levels) or mostly in the plastic range (high excitation levels). When the excitation level is in the intermediate range, and the secondary slope is small, simulation studies reveal that the response process has a broader bandwidth (Roberts, 1978d). Moreover, in these circumstances, it is found that, in addition to the roughly cyclic nature of the response, there is an associated low frequency 'drift' motion. Thus the 'centre' of the hysteresis loops is found to fluctuate randomly, with a very low frequency. This is akin to a slow variation in the mean of the response, resulting in a slow variation of the loop centre coordinates, (m_y, m_z), as shown in Figure 8.4. The power spectrum of the response, in this situation, has the shape sketched in Figure 8.13.

Kobori *et al.* (1973) have proposed a generalization of the original averaging method (Caughey, 1960a) which allows for slowly varying random fluctuations in the 'frequency' of the response and for slow drift motion. Their approach is applicable to a variety of hysteretic oscillators, but is illustrated here, in a simplified form, by reference to the bilinear oscillator, with a zero-mean random input.

Figure 8.14 shows a bilinear, symmetrical hysteresis loop, whose centre, C, has drifted to some non-zero position, at some time τ. If $\delta_c(\tau)$ is the y coordinate of the point C then an appropriate generalization of equations (8.8) and (8.9) is as follows

$$y(\tau) = \delta_c(\tau) + A(\tau)\cos\chi \tag{8.104}$$

$$\dot{y}(\tau) = -\,\omega(A)A(\tau)\sin\chi \tag{8.105}$$

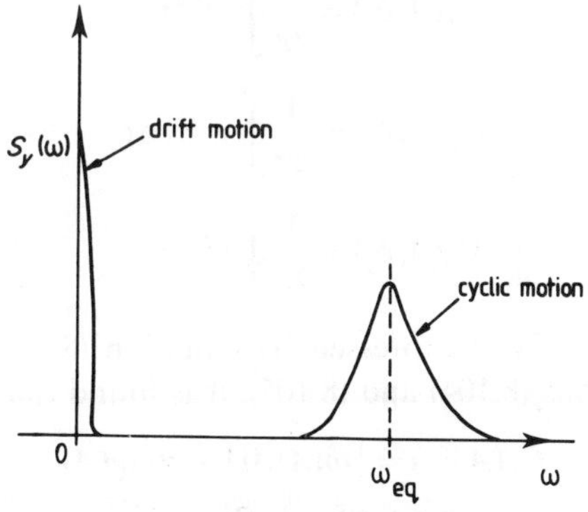

Figure 8.13
Power spectrum of the displacement response, for wide-band response

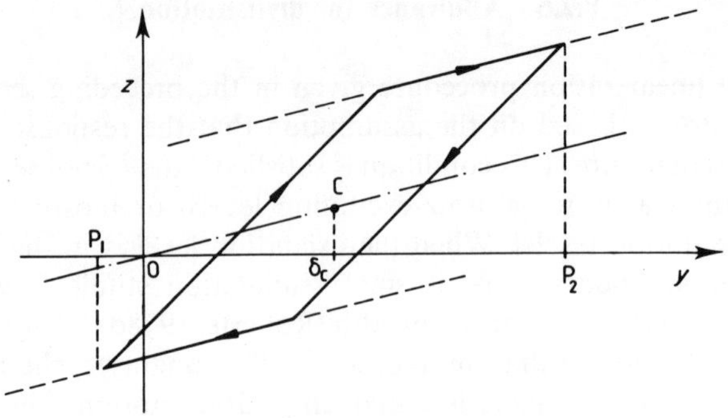

Figure 8.14
Effect of drift on the position of the bilinear symmetrical bilinear hysteresis loop

where

$$\chi = \omega(A)\tau + \theta(t) \tag{8.106}$$

These expressions can be used to evaluate β_{eq} and ω_{eq}^2, by using equations (8.6) and (8.7), and averaging again over one cycle of oscillation. Thus

$$\beta_{eq} = \beta + \frac{E\{C_\beta(A, \delta_c)\}}{E\{B_\beta(A, \delta_c)\}} \tag{8.107}$$

$$\omega_{eq}^2 = \frac{E\{C_\omega(A, \delta_c)\}}{E\{B_\omega(A, \delta_c)\}} \tag{8.108}$$

where

$$C_\beta(A, \delta_c) = \frac{1}{2\pi} \oint \dot{y} z(\tau)\, d\tau \tag{8.109}$$

$$B_\beta(A, \delta_c) = \frac{1}{2\pi} \oint \dot{y}^2\, d\tau \tag{8.110}$$

$$C_\omega(A, \delta_c) = \frac{1}{2\pi} \oint y z(\tau)\, d\tau \tag{8.111}$$

$$B_\omega(A, \delta_c) = \frac{1}{2\pi} \oint y^2\, d\tau \tag{8.112}$$

If the decomposition of $z(\tau)$ expressed by equation (8.74) is employed again, together with equations (8.104) and (8.105), it is found that

$$C_\beta(A, \delta_c) = \tfrac{1}{2}\omega(A)A(1 - \alpha)S_0(A) \tag{8.113}$$

$$B_\beta(A, \delta_c) = \tfrac{1}{2}\omega^2(A)A^2 \tag{8.114}$$

$$C_\omega(A, \delta_c) = \alpha\delta_c^2 + \tfrac{1}{2}A^2 + (1 - \alpha)C_0(A) \tag{8.115}$$

$$B_\omega(A, \delta_c) = \alpha\delta_c^2 + \tfrac{1}{2}A^2 \tag{8.116}$$

To evaluate the expectations of the above quantities one requires the joint density function, $p(A, \delta_c)$ of $A(t)$ and $\delta_c(t)$. Referring to Figure 8.14 again, it is noted that

$$\delta_c(t) = (P_1 + P_2)/2 \tag{8.117}$$

where P_1 and P_2 are the values of $y(t)$ at the extremes of the loop. If these extremes occur at times t_1 and t_2, respectively, then

$$\delta_c(t) = (A(t_1) - A(t_2))/2 \tag{8.118}$$

Since $(t_1 - t_2) \sim \pi/\omega(A)$, it is evident that the expectations in equations (8.107) and (8.108) can be found from a knowledge of the joint probability density function $p[A(t_1), A(t_2)]$. A method of estimating this density function, from the equivalent linear system, is given by Kobori *et al.* (1973). In fact, they also allow for random variations in the time difference between the successive peaks, $t_1 - t_2$, for a given value of $A(t_1)$. Unfortunately, however, the analysis is very complicated and does not lead to simple analytical results.

In the remainder of this chapter an alternative statistical linearization method for randomly excited hysteretic systems is discussed, which can, to some extent, account for the drift phenomenon, and has the advantage that it is relatively straightforward to implement.

8.3 USE OF DIFFERENTIAL MODELS OF HYSTERESIS

It was shown, in Chapter 2, that hysteresis loops can often be modelled in terms of a differential equation. If such an equation is incorporated into the normal equations of motion of the system, then an 'extended' set of differential equations is obtained. These can be linearized by using the technique for multi-degree of freedom systems, described in Chapter 6.

To illustrate this approach, the case of a SDOF system, such as an oscillator, with a hysteretic restoring force, will first be considered.

8.3.1 Oscillators with hysteresis

Consider an oscillator with viscous damping and a hysteretic restoring force, having the following equation of motion

$$\ddot{y} + \beta\dot{y} + \alpha y + (1 - \alpha)z = f(t) \tag{8.119}$$

Here y is a non-dimensional displacement, obtained by dividing the actual displacement $x(t)$ by a characteristic yield displacement value, x^* say, and time is measured in the non-dimensional units $\tau = \omega_n t$, where ω_n is a characteristic frequency, usually associated with small, non-yielding displacements. The parameter α is a measure of the relative contribution of the hysteretic component, z, of the restoring force. In the particular case of a bilinear oscillator, x^* is the response displacement at which yielding first occurs, ω_n is the frequency of oscillation corresponding to the primary, elastic slope and α is the 'rigidity ratio', giving the ratio of plastic to elastic stiffness (see Section 8.2.5). Also, for the bilinear oscillator, z is the hysteretic force corresponding to the elasto–plastic characteristic, denoted z_0 previously. As before, it will be assumed initially that $f(t)$ is a zero-mean stationary random process, and that $z(t)$ is antisymmetric, so that the mean of $y(t)$, in the steady state, is zero.

As indicated earlier, in Chapter 2, the non-dimensional hysteretic force, $z(t)$, can often be represented in terms of a first-order differential equation. A general form for this is as follows

$$\dot{z} = G(\dot{y}, z) \tag{8.120}$$

Thus, for example, elasto–plastic hysteresis is given by (Suzuki and Minai, 1987)

$$G(\dot{y}, z) = \dot{y}[1 - U(\dot{y})U(z-1) - U(-\dot{y})U(-z-1)] \tag{8.121}$$

where $U(\ \ )$ denotes the unit step function. As another example, a curvilinear hysteretic characteristic is obtained if (see also Chapter 2, equation (2.102))

$$G(\dot{y}, z) = -\gamma|\dot{y}|z|z|^{n-1} - \nu\dot{y}|z|^n + A\dot{y} \tag{8.122}$$

where γ, ν, A and n are 'loop parameters', which control the shape, and magnitude, of the hysteretic loop.

If the vector $\mathbf{q}$ is introduced, where

$$\mathbf{q} = \begin{bmatrix} q_1 \\ q_2 \end{bmatrix} = \begin{bmatrix} y \\ z \end{bmatrix} \tag{8.123}$$

then equations (8.119) and (8.120) can be combined into the following matrix equation

$$\mathbf{M}\ddot{\mathbf{q}} + \mathbf{C}\dot{\mathbf{q}} + \mathbf{K}\mathbf{q} + \mathbf{\Phi}(\mathbf{q}, \dot{\mathbf{q}}, \ddot{\mathbf{q}}) = \mathbf{Q}(t) \tag{8.124}$$

where

$$\mathbf{M} = \begin{bmatrix} 1 & 0 \\ 0 & 0 \end{bmatrix} \tag{8.125}$$

$$\mathbf{C} = \begin{bmatrix} \beta & 0 \\ 0 & 1 \end{bmatrix} \tag{8.126}$$

$$\mathbf{K} = \begin{bmatrix} \alpha & (1-\alpha) \\ 0 & 0 \end{bmatrix} \tag{8.127}$$

$$\mathbf{\Phi} = \begin{bmatrix} \Phi_1 \\ \Phi_2 \end{bmatrix} = \begin{bmatrix} 0 \\ -G(\dot{q}_1, q_2) \end{bmatrix} \tag{8.128}$$

$$\mathbf{Q}(t) = \begin{bmatrix} f(t) \\ 0 \end{bmatrix} \tag{8.129}$$

Equation (8.124) is recognized as that discussed earlier, is some detail, in Chapter 6. The methodology given there, for replacing it with the equivalent linear equation (see also equation (6.2))

$$(\mathbf{M} + \mathbf{M}_e)\ddot{\mathbf{q}} + (\mathbf{C} + \mathbf{C}_e)\dot{\mathbf{q}} + (\mathbf{K} + \mathbf{K}_e)\mathbf{q} = \mathbf{Q}(t) \tag{8.130}$$

is thus applicable. This exemplifies the very significant advantage of adopting a differential model of $z(t)$. Through this means the extended set of differential equations, here given by equation (8.124), can be linearized by the standard method, developed earlier in the context of non-hysteretic, non-linear systems.

The elements of the matrices $\mathbf{M}_e$, $\mathbf{C}_e$ and $\mathbf{K}_e$ can be found by employing the results given in Section 6.5.1. Specifically, using equations (6.44) to (6.46) one finds that

$$\mathbf{M}_e = \begin{bmatrix} 0 & 0 \\ 0 & 0 \end{bmatrix} \tag{8.131}$$

$$\mathbf{C}_e = \begin{bmatrix} 0 & 0 \\ c_{21}^e & 0 \end{bmatrix} \tag{8.132}$$

$$\mathbf{K}_e = \begin{bmatrix} 0 & 0 \\ 0 & k_{22}^e \end{bmatrix} \tag{8.133}$$

where the only two non-zero elements, c_{21}^e and k_{22}^e, are given by

$$c \equiv c_{21}^e = E\left\{\frac{\partial \Phi_2}{\partial \dot{q}_1}\right\} = -E\left\{\frac{\partial G}{\partial \dot{q}_1}\right\} \tag{8.134}$$

$$k \equiv k_{22}^e = E\left\{\frac{\partial \Phi_2}{\partial q_2}\right\} = -E\left\{\frac{\partial G}{\partial q_2}\right\} \tag{8.135}$$

It can be seen that the statistical linearization achieved here effectively consists of replacing equation (8.120) by the equivalent linear relationship

$$\dot{q}_2 + c\dot{q}_1 + kq_2 = 0 \tag{8.136}$$

Thus, equations (8.119) and (8.136) can be combined as equation (8.130).

Using a joint Gaussian probability density function, $f(\dot{q}_1, q_2)$, for $\dot{q}_1$ and q_2, the expectations in equations (8.134) and (8.135) can be evaluated, for specific functions, $G(\dot{q}_1, q_2)$. This density function can be written as (see equation (3.49), where $n = 2$)

$$f(\dot{q}_1 q_2) = \frac{1}{2\pi\sigma_1\sigma_2(1-\rho^2)^{1/2}} \exp\left(-\frac{(\sigma_1^2 q_2^2 + \sigma_2^2 \dot{q}_1^2 - 2\sigma_1\sigma_2\rho\dot{q}_1 q_2)}{2\sigma_1^2\sigma_2^2(1-\rho^2)}\right) \tag{8.137}$$

where

$$\sigma_1^2 = E\{\dot{q}_1^2\} \tag{8.138}$$

$$\sigma_2^2 = E\{q_2^2\} \tag{8.139}$$

and

$$\rho = \frac{E\{\dot{q}_1 q_2\}}{\sigma_1\sigma_2} \tag{8.140}$$

In terms of $f(\dot{q}_1 q_2)$ equations (8.134) and (8.135) can be written as

$$c = -\int_{-\infty}^{\infty}\int_{-\infty}^{\infty} \frac{\partial G}{\partial \dot{q}_1} f(\dot{q}_1, q_2)\,\mathrm{d}\dot{q}_1\,\mathrm{d}q_2 \tag{8.141}$$

$$k = -\int_{-\infty}^{\infty}\int_{-\infty}^{\infty} \frac{\partial G}{\partial q_2} f(\dot{q}_1, q_2)\,\mathrm{d}\dot{q}_1\,\mathrm{d}q_2 \tag{8.142}$$

As in the non-hysteretic case, the linearized equations of motion, (8.130), can be used to establish a further relationship between c, k and σ_1, σ_2 and ρ. Here, as described earlier, one has a choice between the spectral matrix solution procedure (possibly including classical modal analysis) and the state variable solution method (possibly incorporating complex modal analysis). Here the state variable solution method will be chosen since it has the advantage that it can be readily extended to deal with the case of non-stationary excitation.

If a state vector $\mathbf{x}$ is defined as

$$\mathbf{x} = \begin{bmatrix} y \\ \dot{y} \\ z \end{bmatrix} = \begin{bmatrix} q_1 \\ \dot{q}_1 \\ q_2 \end{bmatrix} \tag{8.143}$$

then the linearized equations of motion (8.130) can be written as the following first-order matrix equation (see equation (6.162))

$$\dot{\mathbf{x}} = \mathbf{G}\mathbf{x} + \mathbf{f} \tag{8.144}$$

where

$$\mathbf{G} = \begin{bmatrix} 0 & 1 & 0 \\ -\alpha & -\beta & -(1-\alpha) \\ 0 & -c & -k \end{bmatrix} \tag{8.145}$$

and

$$\mathbf{f} = \begin{bmatrix} 0 \\ f(t) \\ 0 \end{bmatrix} \tag{8.146}$$

As shown earlier, in Chapter 4, the covariance matrix of the response

$$\mathbf{V} = E\{\mathbf{x}(t)\mathbf{x}^{\mathrm{T}}(t)\} = [v_{ij}] \tag{8.147}$$

is governed by a differential equation of first order, involving the covariance matrix of $f(t)$, defined by

$$\mathbf{w}_f(u) = E\{\mathbf{f}(t)\mathbf{f}^{\mathrm{T}}(t+u)\} \tag{8.148}$$

(see equation (6.168)). In the special case where $f(t)$ is a white noise, such that

$$E\{\mathrm{f}(t)f(t+\tau)\} = 2\pi S_0\delta(u) \tag{8.149}$$

where S_0 is the constant spectral level, then

$$\mathbf{w}_f(u) = \mathbf{D}\delta(u) \tag{8.150}$$

where

$$\mathbf{D} = \begin{bmatrix} 0 & 0 & 0 \\ 0 & 2\pi S_0 & 0 \\ 0 & 0 & 0 \end{bmatrix} \tag{8.151}$$

and the equation for **V** becomes

$$\dot{\mathbf{V}} = \mathbf{G}\mathbf{V}^{\mathrm{T}} + \mathbf{V}\mathbf{G}^{\mathrm{T}} + \mathbf{D} \tag{8.152}$$

In the steady state, stationary response case, $\dot{\mathbf{V}} = \mathbf{0}$ and hence equation (8.152) reduces to

$$\mathbf{G}\mathbf{V}^{\mathrm{T}} + \mathbf{V}\mathbf{G}^{\mathrm{T}} + \mathbf{D} = \mathbf{0} \tag{8.153}$$

Efficient algorithms exist (e.g. see Bartels and Stewart, 1972, Beavers and Denman, 1975; Pace and Barnett, 1972; Rothschild and Jameson, 1968) for solving the Liapunov matrix equation given by equation (8.153). However, in the present case, one can proceed in a straightforward way, without difficulty, by recasting equation (8.153) as a standard matrix inversion problem. Thus, writing,

$$\mathbf{V} = [v_{ij}] \quad \mathbf{G} = [g_{ij}] \tag{8.154}$$

and using the fact that **V** is symmetric, one can write a standard matrix equation for the unknowns, v_{11}, v_{12}, v_{13}, v_{22}, v_{23}, v_{33}. In fact,

$$v_{12} = E\{y(t)\dot{y}(t)\} = 0 \tag{8.155}$$

(see Section 3.7) so that the number of unknowns reduces to five. After some algebra one obtains an equation of the form

$$\mathbf{Aw} = \mathbf{B} \tag{8.156}$$

where

$$\mathbf{w} = \begin{bmatrix} v_{11} \\ v_{13} \\ v_{22} \\ v_{23} \\ v_{33} \end{bmatrix} \qquad \mathbf{B} = \begin{bmatrix} 0 \\ -\pi S_0 \\ 0 \\ 0 \\ 0 \end{bmatrix} \tag{8.157}$$

and

$$\mathbf{A} = \begin{bmatrix} -\alpha & (\alpha-1) & 1 & 0 & 0 \\ 0 & 0 & -\beta & (\alpha-1) & 0 \\ 0 & -\alpha & -c & (-\beta-k) & (\alpha-1) \\ 0 & -k & 0 & 1 & 0 \\ 0 & 0 & 0 & -c & -k \end{bmatrix} \tag{8.158}$$

These equations can be solved fairly easily, to obtain the following

$$v_{11} = \frac{\pi S_0}{\alpha\beta} + v_{23}\frac{(\alpha-1)}{\alpha}\left(\frac{1}{k} + \frac{1}{\beta}\right) \tag{8.159}$$

$$v_{13} = \frac{v_{23}}{k} \tag{8.160}$$

$$v_{22} = \frac{\pi S_0}{\beta} + \frac{(\alpha-1)}{\beta}v_{23} \tag{8.161}$$

$$v_{33} = \frac{-cv_{23}}{k} \tag{8.162}$$

and

$$v_{23} = \frac{ck\pi S_0}{[c(1-\alpha) - k\beta](\beta+k) - \alpha\beta} \tag{8.163}$$

Noting that

$$v_{22} = E\{\dot{y}^2\} = E\{\dot{q}_1^2\} = \sigma_1^2 \tag{8.164}$$

$$v_{33} = E\{z^2\} = E\{q_2^2\} = \sigma_2^2 \tag{8.165}$$

$$v_{23} = E\{\dot{y}z\} = E\{\dot{q}_1 q_2\} = \rho\sigma_1\sigma_2 \tag{8.166}$$

it is evident that the above equations can be solved in an iterative manner, to obtain σ_1, σ_2 and ρ. For the case of non-zero linear damping ($\beta \neq 0$) a convenient start to the iteration procedure is to set $z=0$ in equation (8.119). Then

$$v_{13} = v_{33} = v_{23} = 0 \tag{8.167}$$

and it follows, from equations (8.159) and (8.161), that

$$v_{11} = \frac{\pi S_0}{\alpha\beta} \quad v_{22} = \frac{\pi S_0}{\beta} \tag{8.168}$$

These are the standard linear results. These values can be used to form the first estimates of c and k, using equations (8.141) and (8.142). If these estimates are incorporated into equations (8.159) to (8.163), new values of v_{22}, v_{33} and v_{23} are found. Hence new values of c and k are found, from equations (8.141) and (8.142), and so on. It should be noted that numerical difficulties can sometimes be experienced with these initial conditions, in specific cases (e.g. see the next section) and some adaptation may then be required.

If the excitation is non-white then a 'pre-filter', or 'shaping filter', as discussed in Section 6.6.4, can be used. The system equation can still be written in the form of equation (8.144), but the order of the system is increased.

To illustrate this point, the case where the excitation of the oscillator has a spectrum of the Kanai–Tajimi form, as given by equation (5.186). This spectrum is a fairly realistic model for earthquake excitation, as mentioned earlier, in Section 5.5.1.

If $n(t)$ is a white noise process, with a constant spectral level $G_0 = 2\pi S_0$, and $f(t)$ is a process with a spectrum given by equation (5.186), then it is easy to show, using the basic theory given in Chapter 4, that the appropriate differential relationships between $y(t)$ and $n(t)$ are as follows

$$f(t) = \omega_g^2 u + 2\zeta_g \omega_g \dot{u} \tag{8.169}$$

where $u(t)$ is governed by

$$\ddot{u} + 2\zeta_g \omega_g \dot{u} + \omega_g u = n(t) \tag{8.170}$$

The above equations can be combined with equations (8.119) and (8.136) to give a complete system equation, in matrix form. Specifically, if

$$\boldsymbol{\chi} = \begin{bmatrix} y \\ \dot{y} \\ z \\ u \\ \dot{u} \end{bmatrix} \quad \mathbf{f}(t) = \begin{bmatrix} 0 \\ 0 \\ 0 \\ 0 \\ n(t) \end{bmatrix} \tag{8.171}$$

then equation (8.144) is again applicable, where now

$$\mathbf{G} = \begin{bmatrix} 0 & 1 & 0 & 0 & 0 \\ -\alpha & -\beta & (\alpha-1) & \omega_g^2 & 2\zeta\omega_g \\ 0 & -c & -k & 0 & 0 \\ 0 & 0 & 0 & 0 & 1 \\ 0 & 0 & 0 & -\omega_g^2 & -2\zeta\omega_g \end{bmatrix} \tag{8.172}$$

Therefore, the method outlined earlier, for the case of white noise excitation, can also be applied to this situation, where the excitation is non-white. In particular, equation (8.153) can still be used, where now all the elements D_{ij} of $\mathbf{D}$ are zero, except $D_{55} = 2\pi S_0 = G_0$.

8.3.2 The bilinear oscillator

The specific case of an oscillator with a bilinear hysteresis characteristic will now be considered in some detail. Here the hysteretic non-linearity function $G(\dot{y}, z)$, is given by equation (8.121) and c and k may be evaluated through the use of equations (8.141) and (8.142).

Considering the evaluation of k, initially, one has, from equation (8.121) with $q_1 = y$, $q_2 = z$,

$$\frac{\partial G}{\partial z} = \dot{q}_1[-U(\dot{q}_1)\delta(q_2 - 1) + U(-\dot{q}_1)\delta(-q_2 - 1)] \tag{8.173}$$

where $\delta(\;)$ is the Dirac delta function. Substituting this result into equation (8.142) gives

$$k = -\int_{-\infty}^{\infty}\int_{-\infty}^{\infty} \dot{q}_1[-U(\dot{q}_1)\delta(q_2 - 1) + U(-\dot{q}_1)\delta(-q_2 - 1)]\, f(\dot{q}_1, q_2)\,\mathrm{d}\dot{q}_1\,\mathrm{d}q_2 \tag{8.174}$$

This expression can be considerably simplified by using the properties of the step and delta functions. Thus the double integral can be expressed as the sum of two single integrals, as follows

$$k = \int_0^{\infty} \dot{q}_1 f(\dot{q}_1, 1)\,\mathrm{d}\dot{q}_1 - \int_{-\infty}^{0} \dot{q}_1 f(\dot{q}_1, -1)\,\mathrm{d}\dot{q}_1 \tag{8.175}$$

Now the joint Gaussian density function $f(\dot{q}_1, q_2)$ is even, as expressed by equation (8.137) then, $f(\dot{q}_1, q_2) = f(-\dot{q}_1, -\dot{q}_2)$. It follows that the two terms in equation (8.175) are equal. Thus

$$k = 2J \tag{8.176}$$

where

$$J = \int_0^\infty \dot{q}_1 f(\dot{q}_1, 1)\,\mathrm{d}\dot{q}_1 \tag{8.177}$$

On substituting equation (8.137) into equation (8.177) and carrying out the integration it is found that

$$J = \frac{k\exp(\mu^2 - \beta)}{\alpha}\left[\tfrac{1}{2}\exp(-\mu^2) + \rho\left(\frac{\pi}{2}\right)^{1/2}(1 + \exp\mu)\right] \tag{8.178}$$

where

$$k = \frac{1}{2\pi\Delta^{1/2}} \tag{8.179}$$

$$\alpha = \frac{\sigma_2^2}{2\Delta} \tag{8.180}$$

$$\beta = \frac{\sigma_1^2}{2\Delta} \tag{8.181}$$

$$\gamma = \frac{\rho\sigma_1\sigma_2}{2\Delta} \tag{8.182}$$

$$\mu = \frac{\gamma}{\alpha^{1/2}} \tag{8.183}$$

and

$$\Delta = \sigma_1^2\sigma_2^2(1 - \rho^2) \tag{8.184}$$

Here erf ( ) is the error function.

The evaluation of the coefficient c turns out to be rather more complicated. Differentiating $G(\dot{q}_1, q_2)$ with respect to $\dot{q}_1$ gives (see equation (8.121))

$$\begin{aligned}\frac{\partial G}{\partial \dot{q}_1} &= [1 - U(\dot{q}_1)U(q_2 - 1) - U(-\dot{q}_1)U(-q_2 - 1)] \\ &\quad + \dot{q}_1[-\delta(\dot{q}_1)U(q_2 - 1) + \delta(\dot{q}_1)U(-q_2 - 1)]\end{aligned} \tag{8.185}$$

This may be substituted into the integral of equation (8.141), resulting in a double integral expression for c. The second part of the expression for $\partial G/\partial \dot{q}_1$, given by equation (8.185), can be seen to have zero contribution to c, since the delta function is such that

$$\int_{-\infty}^{\infty} u\delta(u)\,\mathrm{d}u = 0 \tag{8.186}$$

With this simplification, one obtains, therefore,

$$c = -\int_{-\infty}^{\infty}\int_{-\infty}^{\infty}[1 - U(\dot{q}_1)U(q_2 - 1) - U(-\dot{q}_1)U(-q_2 - 1)]f(\dot{q}_1, q_2)\,\mathrm{d}\dot{q}_1\,\mathrm{d}q_2 \tag{8.187}$$

Using the property of the step function, this expression can be further simplified as follows

$$c = -1 + \int_{1}^{\infty}\int_{0}^{\infty} f(\dot{q}_1, q_2)\,\mathrm{d}\dot{q}_1\,\mathrm{d}q_2 + \int_{-\infty}^{-1}\int_{-\infty}^{0} f(\dot{q}_1, q_2)\,\mathrm{d}\dot{q}_1\,\mathrm{d}q_2 \tag{8.188}$$

From the fact that $f(\dot{q}_1 q_2)$ is even, it is easy to see that the two integrals in equation (8.188) are, in fact, equal in value. Thus one can write

$$c = 2I - 1 \tag{8.189}$$

where

$$I = \int_{1}^{\infty}\int_{0}^{\infty} f(\dot{q}_1 q_2)\,\mathrm{d}\dot{q}_1\,\mathrm{d}q_2 \tag{8.190}$$

An analytical, closed form expression for the integral I seems difficult to obtain. However, it is relatively easy to reduce the double integral to a single integral, using the definition of the error function again. One finds that

$$I = k\left(\frac{\pi}{2\alpha\beta}\right)^{1/2}\int_{\beta^{1/2}}^{\infty} \exp[-(1+\rho)u^2][1 + \operatorname{erf}(\rho u)]\,\mathrm{d}u \tag{8.191}$$

where k, α and β are defined by equations (8.179), (8.180) and (8.181), respectively. The above single integral is readily evaluated numerically.

Once c and k are evaluated in terms of u_{22}, u_{23} and u_{33}, the iterative scheme discussed in the preceding section can be implemented. In the following, some numerical results will be presented for the special case of white noise excitation. Here there are only three non-dimensional parameters which govern the response. They are α, the 'rigidity ratio' (see equation (8.119)), β, the linear damping coefficient (see equation (8.119)) and S_0, the spectral level of the input process.

In starting with the initial conditions given by equation (8.167) and equation (8.168) one is faced immediately with the difficulty that $\Delta = 0$ (see equation (8.184)). Thus the expressions given earlier involve singularities. This difficulty is overcome by noting that, if $\sigma_2^2 = u_{33} = 0$ then, from equations (8.175) and (8.187)

$$k = 0 \quad c = -1 \tag{8.192}$$

These values of k and c can be used to *start* the iteration process. Thus, from equations (8.159) to (8.163), values such as u_{11}, can be calculated. The new

Table 8.1

Number of iterations	c	k	σ_y
0	-1	0	—
1	-0.499	0.318	1447
2	-0.540	0.605	6.90
3	-0.509	0.665	5.77
4	-0.509	0.774	5.92
...	...	...	...
10	-0.505	0.821	5.96
20	-0.505	0.830	5.97
30	-0.505	0.830	5.97

$\alpha = 0.5$, $\beta = 0$, $S_0 = 1$.

values of c and k may be found, from equations (8.176), (8.178), (8.189) and (8.191), and so on, until convergence is achieved.

It is noted that expressions for v_{11} etc. must be treated carefully when $k = 0$; in fact the proper limit as $k \rightarrow 0$ should be taken analytically. A simple alternative approach is to set k to a very small, but non-zero value at the beginning of the iteration process. Similarly, care must be exercised in the special case where $\beta = 0$, since terms involving $1/\beta$ occur in equations (8.159) to (8.163). Again this difficulty can be overcome by setting β to a very small, but non-zero value.

Table 8.1 shows a typical set of results for the case $\alpha = 0.5$, $\beta = 0$ and $S_0 = 1$. Actually, to avoid singularities, β is set to 10^{-6}. After a large, initial fluctuation in the value of σ_y, it is observed that the iteration scheme converges quite

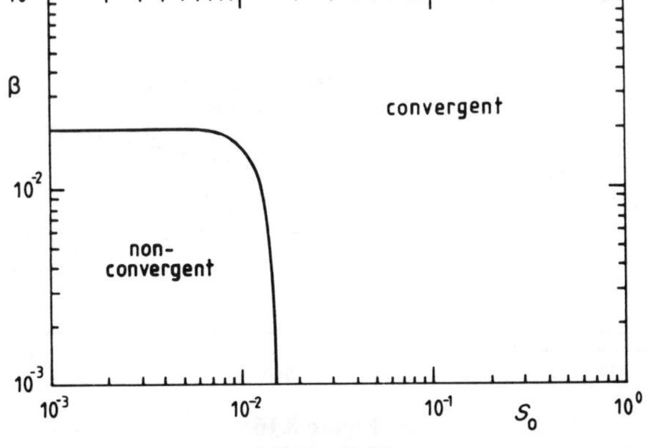

Figure 8.15
Solution regions in the β, S_0 plane, for the bilinear oscillator

Table 8.2

Number of iterations	c	k	σ_y
0	-1	0	—
1	-0.505	0.318	64.7
2	-0.999	0	0.307
3	-0.531	0.319	79.3
4	-0.999	0	0.299
5	-0.531	0.319	79.3
...	...	...	...
10	-0.991	0	0.299
11	-0.531	0.319	79.3

$\alpha=0.5$, $\beta=0$, $S_0=0.002$

rapidly, with results correct to three significant figures being obtained after about 20 iterations.

The rate of convergence is found to depend primarily on the values of β and S_0 and to slow down as those values are reduced. For sufficiently small values of S_0 and β, the iteration scheme fails to converge. Figure 8.15 indicates the region in the β, S_0 plane where convergence is not achieved. This region is virtually independent of the value of α. A typical set of results obtained within this region is given in Table 8.2. After a few iterations a cyclic behaviour is observed, the values of c, k and σ_y alternating between two sets of values. The values in these two sets are very sensitive to the precise value of β chosen; $\beta=10^{-6}$ in Table 8.2.

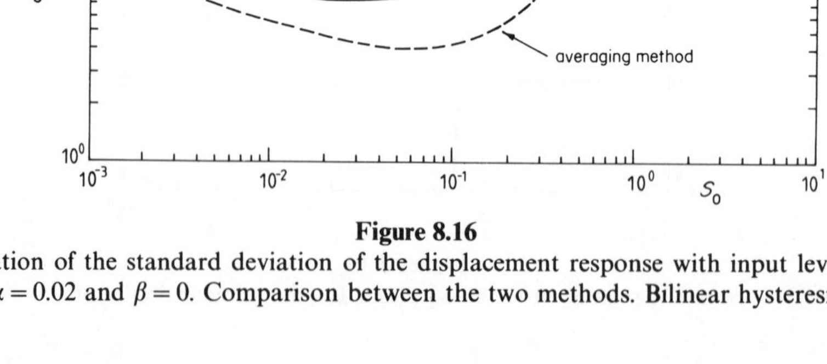

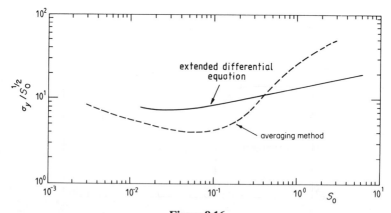

Figure 8.16
Variation of the standard deviation of the displacement response with input level, for $\alpha=0.02$ and $\beta=0$. Comparison between the two methods. Bilinear hysteresis

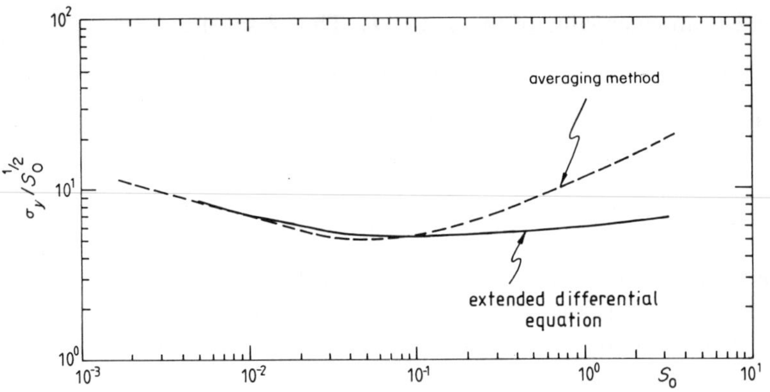

Figure 8.17
Variation of the standard deviation of the displacement response with input level, for $\alpha = 0.5$ and $\beta = 0$. Comparison between the two methods. Bilinear hysteresis

Figures 8.16 and 8.17 show typical comparisons between results for σ_y, computed by the present extended differential equation method and corresponding results for σ_y obtained by the averaging method (see Section 8.2.5). Here $\sigma_y/S_0^{1/2}$ is plotted against S_0, for $\beta = 0$ and $\alpha = 0.02$ (Figure 8.16) and $\alpha = 0.5$ (Figure 8.17). The present method does not give results at low values of β, for the reason discussed above. It is observed that for $\alpha = 0.5$, the two methods give results in reasonable agreement, at low to moderate values of S_0. There is a divergence, however, at high S_0 values. At $\alpha = 0.02$, the agreement between the two methods is generally poor over the whole range of S_0.

A proper assessment of the relative accuracy of the results from the two methods, where these diverge, can be obtained through a comparison with simulation results. This matter will be discussed later, in Chapter 10. However, here it is worth noting again (see also Section 8.2.6) that the response process will not be narrow-band when α is small and S_0 is in the intermediate range. Thus one can expect the results from the extended differential equation method in Figure 8.16, in the middle range of S_0, to be more accurate. This accounts, at least partly, for the divergence in the results from the two methods.

The present method does not require the response to be narrow-band, and can, in fact, also allow for the effect of low frequency drift. This fact is evident if one considers the unforced, free-decay response of the bilinear oscillator. Then, from equation (8.144) one has

$$\dot{\mathbf{x}} = \mathbf{G}\mathbf{x} \tag{8.193}$$

where $\mathbf{G}$ is given by equation (8.145). The solutions are of the form

$$\mathbf{x} = \mathbf{X}\exp(\lambda t) \tag{8.194}$$

where λ is an eigenvalue of the matrix $\mathbf{G}$. Thus λ satisfies

$$\mathbf{G} - \mathbf{I}\lambda = 0 \tag{8.195}$$

(see also Section 4.4.5). Since here $\mathbf{G}$ is a 3×3 matrix there are three, possibly complex, eigenvalues, and the solution for the displacement response takes the form

$$y(t) = y_1 \exp(\lambda_1 t) + y_2 \exp(\lambda_2 t) + y_3 \exp(\lambda_3 t) \tag{8.196}$$

where the amplitudes y_1, y_2 and y_3 may also be complex.

In contrast to this, there are only two eigenvalues associated with the averaging method; thus, from equation (8.4) with $f(t) = 0$, the equation corresponding to equation (8.193) is

$$\ddot{y} + \beta_{\mathrm{eq}}\dot{y} + \omega_{\mathrm{eq}}^2 y = 0 \tag{8.197}$$

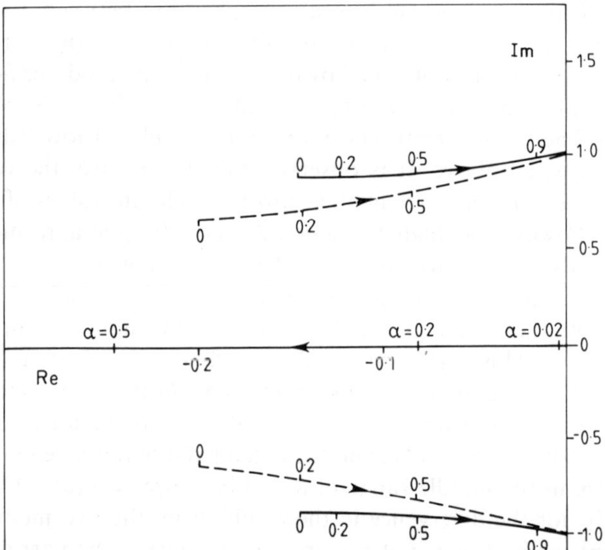

Figure 8.18
Eigenvalue locii in the complex plane for the case $S_0 = 0.1$, $\beta = 0$. Comparison between the two methods. Bilinear hysteresis. (——— extended differential equation, – – – averaging method)

The free-decay solution is of the form

$$y = y_1 \exp(\lambda_1 t) + y_2 \exp(\lambda_2 t) \tag{8.198}$$

where λ satisfies

$$\lambda^2 + \beta_{eq}\lambda + \omega_{eq}^2 = 0 \tag{8.199}$$

Figure 8.18 shows some typical numerical results for the eigenvalues, derived from the two methods. Here $S_0 = 0.1$ and $\beta = 0$ and the locii of the various eigenvalues, as α varies from 0 to 1, is depicted in the complex plane. For the extended differential equation method two of the eigenvalues (λ_1 and λ_2) form a complex conjugate pair, whilst the other eigenvalue (λ_3) is real. For values of α close to unity λ_1 and λ_2 are close to the imaginary axis, indicating a lightly damped response of the narrow-band type, whilst λ_3 has a large real negative value, associated with a rapid exponential decay. As α is decreased towards zero λ_1 and λ_2 move away from the imaginary axis, indicating an increase in effective damping. More interestingly, λ_3 moves towards the imaginary axis. When $\alpha = 0.02$, for example, λ_3 is very small in magnitude, and the term $y_3 \exp(\lambda_3 t)$ represents a very slowly decaying response. It is evident that it is this eigenvalue which accounts for the slowly varying, drift component of the response in the case of random forcing.

Figure 8.18 also shows the corresponding eigenvalue locii obtained from the averaging method. Here the two eigenvalues form a complex conjugate pair and, as $\alpha \to 1$, the values of these eigenvalues converge to the first two, complex eigenvalues obtained from equation (8.195). Clearly the use of the averaging method precludes the possibility of accommodating the effect of low frequency drift motion, in addition to cyclic response, unless one is prepared to undertake the very complex extension outlined in Section 8.2.6.

8.3.3 The curvilinear model

In the curvilinear model the hysteretic non-linearity function $G(\dot{y}, z)$ is given by equation (8.122). As with the bilinear oscillator, the parameters c and k may be evaluated through the use of equations (8.141) and (8.142).

The appropriate calculations have been carried out by Wen (1980). For the special case $n = 1$ the result is relatively simple. It can be written as follows

$$c = \left(\frac{2}{\pi}\right)^{1/2} \left(\frac{\gamma v_{23}}{(v_{22})^{1/2}} + \nu (v_{33})^{1/2}\right) - A \tag{8.200}$$

$$k = \left(\frac{2}{\pi}\right)^{1/2} \left(\gamma (v_{22})^{1/2} + \nu \frac{v_{23}}{(v_{33})^{1/2}}\right) \tag{8.201}$$

For $n \neq 1$ the result can be considerably more complicated (see Wen, 1980).

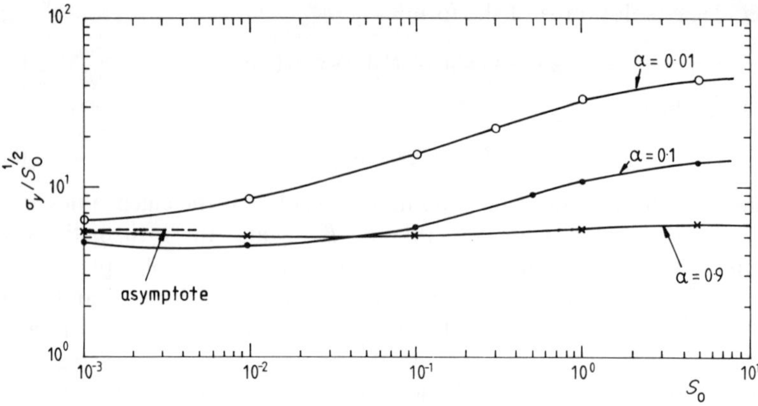

Figure 8.19
Variation of the standard deviation of the displacement response with input level, for the curvilinear hysteresis model. $\gamma = \nu = 0.5$, $A = n = 1$, $\beta = 0.1$. $\alpha = 0.01$, 0.1 and 0.9

The expressions for c and k can be incorporated into the iteration scheme, described in Section 8.3.1, in exactly the same manner as for the bilinear oscillator.

The standard deviation of the response here depends on the rigidity ratio, α, the loop parameters γ, ν, n and A, the linear damping coefficient β and the spectral level of the input process, S_0. Figure 8.19 shows a typical set of results, for the case where $A = 1$, $\gamma = \nu = 0.5$, $n = 1$ and $\beta = 0.1$. As in Figures 8.10 and 8.11, the quantity $\sigma_y/S_0^{1/2}$ is plotted against S_0, here for various values of α. Similar to the bilinear case, the relative effect of the hysteretic damping, as $S_0 \to 0$ diminishes and stiffness associated with z approaches unity; thus equation (8.119) approaches the linear equation

$$\ddot{y} + \beta\dot{y} + y = f(t) \tag{8.202}$$

and the asymptotic result given by equation (8.100) is approached. Hence, with $\beta = 0.1$, $\sigma_y/S_0^{1/2} \to (10\pi)^{1/2} \sim 5.60$.

Numerical studies, over a wide range of parameter values, do not reveal any regions in parameter space where the iteration procedure fails to converge. Care must, however, be exercised in some regions, regarding the choice of initial values for c and k, if the iteration is not to become numerically unstable. Generally, the initial values used previously for the bilinear oscillator, $c = -1$, $k = 0$, work well unless S_0 is large ($S_0 > 1$, say). For large S_0 it is advisable to start the iteration with a large value of k ($k = 10$, say).

In Chapter 10 the accuracy of results for the curvilinear model, such as those shown in Figure 8.19, will be assessed through a comparison with digital simulation estimates.

8.3.4 Inputs with non-zero means

If the input has a non-zero mean value then the analysis given earlier, in Section 8.3.1 can be readily generalized by using the procedure outlined in Chapter 6 (see Section 6.5.3).

Consider once again an oscillator, with an equation of motion of the form of equation (8.119), where z is governed by equation (8.120). Then the equation of motion can be written in matrix form, as in equations (8.123) to (8.129). Let $f(t)$ be written as

$$f(t) = f_0(t) + m_f \tag{8.203}$$

where m_f is the mean of $f(t)$, and $f_0(t)$ has a zero mean. Then, one can write

$$\mathbf{Q}(t) = \mathbf{Q}_0(t) + \mathbf{m}_Q \tag{8.204}$$

where here

$$\mathbf{Q}_0(t) = \begin{bmatrix} f_0(t) \\ 0 \end{bmatrix} \quad \mathbf{m}_Q = \begin{bmatrix} m_f \\ 0 \end{bmatrix} \tag{8.205}$$

Moreover, following the method of Section 6.5.3, one can define

$$\mathbf{q}_0 = \mathbf{q}(t) - \mathbf{m}_q \tag{8.206}$$

or

$$\mathbf{q}_0 = \begin{bmatrix} q_{01} \\ q_{02} \end{bmatrix} = \begin{bmatrix} y - m_y \\ z - m_z \end{bmatrix} \tag{8.207}$$

where

$$m_y = E\{y(t)\} \quad m_z = E\{z(t)\} \tag{8.208}$$

The governing equations can now be written as equations (6.56) and (6.57), where $\mathbf{h}(\;)$ is defined by equation (6.58). Linearizing, in the normal way, the following set of equations is obtained

$$\ddot{q}_{01} + \beta\dot{q}_{01} + \alpha q_{01} + (1-\alpha)q_{02} = f_0(t) \tag{8.209}$$

$$\dot{q}_{02} + c\dot{q}_{01} + kq_{02} = 0 \tag{8.210}$$

$$c = -E\left\{\frac{\partial G_0}{\partial \dot{q}_{01}}\right\} \tag{8.211}$$

$$k = -E\left\{\frac{\partial G_0}{\partial q_{02}}\right\} \tag{8.212}$$

$$G_0 = G(\dot{q}_{01}, q_{02} + m_z) - E\{G(\dot{q}_{01}, q_{02} + m_z)\} \tag{8.213}$$

Also, from equation (6.56), one obtains

$$\alpha m_y + (1-\alpha)m_z = m_f \tag{8.214}$$

and

$$E\{G(\dot{q}_{01}, q_{02} + m_z)\} = 0 \tag{8.215}$$

It follows from this last result that G is identical to G_0 (see equation (8.213)).

The above equations contain the unknowns m_y, m_z, c and k. These can be found through the usual iteration procedure. Thus, if initial values are assigned by m_y, m_z, c and k, then a joint, Gaussian distribution for q_{01}, q_{02} and its derivatives can be calculated, from equations (8.209) and (8.210). This enables new values of c and k to be found, from equations (8.211) and (8.212). Then, new values of m_y and m_z to be calculated, from equations (8.214) and (8.215), and so on.

The calculation involved in computing the expectations in equations (8.211), (8.212) and (8.215) appears, at first sight, to be laborious. However, in many cases, it simplifies drastically. This is because the function $G(\dot{q}_1, q_2)$ is usually such that equation (8.215) is uniquely satisfied by

$$m_z = 0 \tag{8.216}$$

For example, for the curvilinear model, equation (8.215) becomes

$$\gamma E_1 + \nu E_2 = 0 \tag{8.217}$$

where

$$E_1 = E\{|\dot{q}_{01}|(q_{02} + m_z)|q_{02} + m_z|^{n-1}\} \tag{8.218}$$

$$E_2 = E\{\dot{q}_{01}|q_{02} + m_z|^n\} \tag{8.219}$$

As Baber has pointed out (Baber, 1984), through a consideration of odd and even functions, it can be seen by inspection that $m_z = 0$ will satisfy equation (8.217). Although it is difficult to prove analytically that this is a unique solution, numerical evaluation of E_1 and E_2 indicates that the left-hand side of equation (8.217) is monotonic with m_z.

If equation (8.216) is valid then, from equation (8.214), it follows immediately that

$$m_y = \frac{m_f}{\alpha} \tag{8.220}$$

This shows that the effect of the non-zero of the input mean is simply to shift the average centre position of the loop from the origin to a new position. This confirms the assumption made earlier, to this effect, in Section 8.2.4, in the context of the averaging position. Equation (8.220) is a special case of equation (8.70), and the shift effect is as shown graphically in Figure 8.4.

Furthermore, with $m_z = 0$, the evaluation of c and k is identical to the case where $m_f = 0$. Therefore, the expressions given earlier, in Section 8.3.3, are still valid. It follows that the dynamic response due to $f_0(t)$ can be superimposed on the static effect, due to m_f, as given by equation (8.220), to yield the total dynamic response to a non-zero mean $f(t)$.

A similar conclusion applies in the case of the bilinear hysteretic oscillator, confirming the result given earlier, in the context of the averaging method (see equation (8.103)).

8.3.5 Biaxial hysteretic restoring forces

For two-dimensional structures under biaxial excitations the interaction of the restoring forces in two orthogonal directions can be significant. For example, a cantilevered reinforced concrete column, as shown in Figure 8.20, can vibrate in the two directions, x and y, as shown. It has been shown experimentally (Takizawa and Aoyama, 1976) that the hysteretic restoring forces in such a column in the x and y directions, are coupled.

Consider a simple, two degree of freedom model, where q_x and q_y are generalized coordinates relating to vibration in orthogonal directions (e.g., q_x and q_y could represent the first modes of vibration, in the x and y directions, in the case of a cantilevered column). If h_x, h_y are the hysteretic restoring forces in the x and y directions, the equations of motion take the general form

$$\mathbf{m}\ddot{\mathbf{q}} + \mathbf{c}\dot{\mathbf{q}} + \mathbf{h} = \mathbf{Q} \tag{8.221}$$

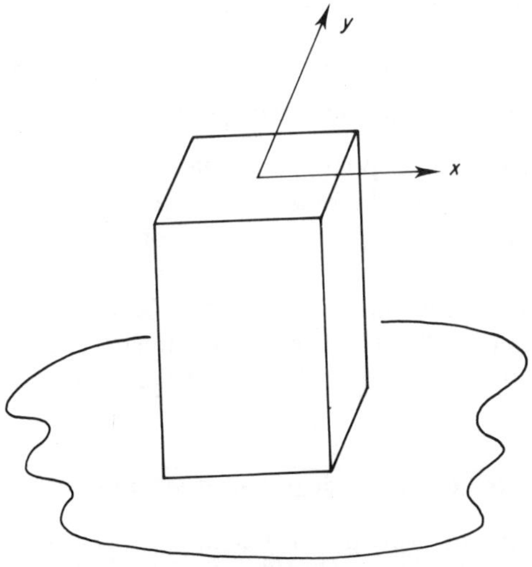

Figure 8.20
Typical structure exhibiting biaxial hysteresis

where **m** and **c** are 2×2 mass and damping matrices and

$$\mathbf{h} = \begin{bmatrix} h_x \\ h_y \end{bmatrix} \tag{8.222}$$

Following Park *et al.* (1986), it is convenient to introduce the transformations

$$q'_y = \frac{k_y}{k_x}\frac{S_x}{S_y} q_y \tag{8.223}$$

$$h'_y = \frac{S_x}{S_y} h_y \tag{8.224}$$

where k_x, k_y are the pre-yielding stiffnesses in the x and y directions, and S_x, S_y are the corresponding yielding strengths, in these two directions, measured under unusual loading conditions. The equation of motion retains the form of equation (8.221), but **q** and **h** are replaced by **q**′ and **h**′ where

$$\mathbf{q}' = \begin{bmatrix} q_x \\ q'_y \end{bmatrix} \quad \mathbf{h}' = \begin{bmatrix} h_x \\ h'_y \end{bmatrix} \tag{8.225}$$

It has been shown by Park *et al* (1986), that **h**′ can be modelled as

$$\mathbf{h}' = \alpha k_x \mathbf{q}' + (1 - \alpha) k_x \mathbf{z} \tag{8.226}$$

where

$$\mathbf{z} = \begin{bmatrix} z_x \\ z_y \end{bmatrix} \tag{8.227}$$

and z_x, z_y satisfy the following coupled differential equations

$$\dot{z}_x = A\dot{q}_x - \gamma|\dot{q}_x z_x| z_x - \nu\dot{q}_x z_x^2 - \gamma|\dot{q}'_y z_y| z_x - \nu\dot{q}'_y z_x z_y \tag{8.228}$$

$$\dot{z}_y = A\dot{q}'_y - \gamma|\dot{q}'_y z_y| z_y - \nu\dot{q}'_y z_y^2 - \nu|\dot{q}_x z_x| z_y - \nu\dot{q}_x z_x z_y \tag{8.229}$$

Through a conversion to polar coordinates one can show that equations (8.228) and (8.229) reduce to the uniaxial model given by equation (8.122), with $n = 2$; thus these equations represent an isotropic, two-dimensional model of hysteresis.

As in the uniaxial case, the equations can be combined and cast into the general form of equation (8.124). The normal linearization procedure, discussed in Chapter 6, can then be applied. Details of the analysis are given by Park *et al.* (1986).

8.3.6 Multi-degree of freedom systems

There is no difficulty, in principle, in linearizing MDOF systems containing a number of hysteretic elements, using the general methodology outlined in this Chapter, and in Chapter 6.

As an illustration the specific MDOF system shown in Figure 8.21 will be

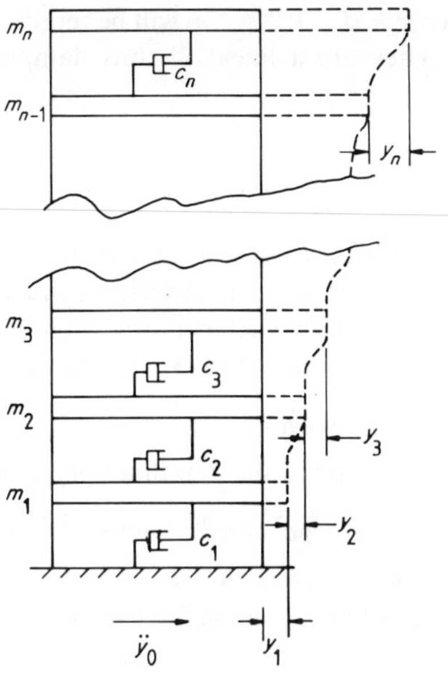

Figure 8.21
Shear building structure. Schematic representation of structure.

discussed here (see also Baber and Wen, 1981; Chang *et al.*, 1986). It consists of a n storey shear beam structure, where all the mass is assumed to be in the form of lumped masses $m_1, m_2, \ldots, m_n$, located at the floors, and the walls provide restoring forces, here taken to be hysteretic in nature. Only motion in the plane shown is considered, and this can be described in terms of the horizontal displacements, $y_1, y_2, \ldots, y_n$, of the floors, as shown. This model is a generalization of the simple, single degree of freedom system (see Figure 4.3) and the two degree of freedom system (see Figure 4.6a) considered earlier. Dynamic response is assumed to occur as a result of random ground acceleration $\ddot{y}_0$.

It is convenient to introduce the relative displacement variables

$$q_i = y_i - y_{i-1} \quad i = 1, 2, \ldots, n \tag{8.230}$$

and to denote the total restoring force, due to the relative deformation q_i, as h_i $(i = 1, 2, \ldots, n)$. Through a consideration of a free-body diagram for the ith floor, it is evident that the equations of motion can be written as

$$m_i \ddot{y}_i + h_i - h_{i+1} = 0 \tag{8.231}$$

or, in terms of relative displacements

$$m_i \left(\sum_{j=1}^{i} \ddot{q}_j + \ddot{y}_0 \right) + h_i - h_{i+1} = 0 \tag{8.232}$$

Here the restoring force h_i $(i = 1, 2, \ldots, n)$ will be represented as a combination of three components. They are a linear viscous damping term, $c_i\dot{q}_i$, a linear stiffness term, $\alpha_i k_i q_i$ and a hysteretic term, $(1 - \alpha_i)k_i z_i$

Thus,

$$h_i = c_i\dot{q}_i + \alpha_i k_i q_i + (1 - \alpha_i)k_i z_i \tag{8.233}$$

where α_i is the rigidity ratio at the ith level, which, as in the single degree of freedom case, determines the ratio of post yield to pre-yield stiffness.

The hysteretic force z_i can modelled as a first-order differential equation, as discussed previously. Thus (see equation (8.120)) one can write

$$\dot{z}_i = G_i(\dot{q}_i, z_i) \quad i = 1, 2, \ldots, n \tag{8.234}$$

where G_i could, for example, be of the curvilinear type, i.e.

$$G_i(\dot{q}_i, z_i) = -\gamma_i|\dot{q}_i|z_i|z_i|^{n_i - 1} - v_i q_i|z_i|^{n_i} + A_i q_i \tag{8.235}$$

The equations of motion, (8.232) to (8.234), can be written in the usual matrix form, given by equation (8.124), if one defines the extended coordinate vector, $\mathbf{q}$, as

$$\mathbf{q} = \begin{bmatrix} \mathbf{q}' \\ \mathbf{z} \end{bmatrix} \tag{8.236}$$

where

$$\mathbf{q}' = \begin{bmatrix} q_1 \\ q_2 \\ \vdots \\ q_n \end{bmatrix} \qquad \mathbf{z} = \begin{bmatrix} z_1 \\ z_2 \\ \vdots \\ z_n \end{bmatrix} \tag{8.237}$$

Then

$$\mathbf{M} = \begin{bmatrix} \mathbf{m} & \mathbf{0} \\ \mathbf{0} & \mathbf{0} \end{bmatrix} \tag{8.238}$$

where

$$\mathbf{m} = \begin{bmatrix} m_1 & 0 & \cdots & 0 \\ m_2 & m_2 & \cdots & 0 \\ & & \cdots & \\ m_n & m_n & \cdots & m_n \end{bmatrix} \tag{8.239}$$

Further,

$$\mathbf{C} = \begin{bmatrix} \mathbf{c} & \mathbf{0} \\ \mathbf{0} & \mathbf{I} \end{bmatrix} \tag{8.240}$$

where

$$\mathbf{c} = \begin{bmatrix} c_1 & 0 & \cdots & 0 \\ 0 & c_2 & \cdots & 0 \\ 0 & & \cdots & \\ 0 & 0 & \cdots & c_n \end{bmatrix} \tag{8.241}$$

Also,

$$\mathbf{K} = \begin{bmatrix} \mathbf{k}_1 & \mathbf{k}_2 \\ \mathbf{0} & \mathbf{0} \end{bmatrix} \tag{8.242}$$

where

$$\mathbf{k}_1 = \begin{bmatrix} a_1 & -a_2 & \cdots & & 0 \\ 0 & a_2 & -a_3 & \cdots & 0 \\ & & \cdots & & \\ \cdots & & a_{n-1} & & -a_n \\ \cdots & & & & a_n \end{bmatrix} \tag{8.243}$$

and

$$\mathbf{k}_2 = \begin{bmatrix} b_1 & -b_2 & \cdots & 0 \\ 0 & b_2 & \cdots & 0 \\ & & \cdots & \\ \cdots & & b_{n-1} & -b_n \\ \cdots & & & b_n \end{bmatrix} \tag{8.244}$$

with

$$\begin{aligned} a_i &= \alpha_i k_i \\ b_i &= (1-\alpha_i) k_i \end{aligned} \qquad i = 1, 2, \ldots, n \tag{8.245}$$

Also

$$\mathbf{\Phi} = \begin{bmatrix} \mathbf{0} \\ \mathbf{G} \end{bmatrix} \tag{8.246}$$

where

$$\mathbf{G} = \begin{bmatrix} -G_1(\dot{q}_1, z_1) \\ -G_2(\dot{q}_2, z_2) \\ \vdots \\ -G_n(\dot{q}_n, z_n) \end{bmatrix} \tag{8.247}$$

and

$$\mathbf{Q} = \begin{bmatrix} \mathbf{F} \\ \mathbf{0} \end{bmatrix} \tag{8.248}$$

where

$$\mathbf{F} = \begin{bmatrix} -m_1\ddot{y}_0 \\ -m_2\ddot{y}_0 \\ \vdots \\ -m_n\ddot{y}_0 \end{bmatrix} \tag{8.249}$$

The standard linearization technique, described in Chapter 6, can now be applied to yield linearized equations in the form of equation (8.130). Here one has a chain-like system. Thus the calculations are considerably simplified, since one linearizes the elements of $G_i(\dot{q}_i, z_i)$ individually (see Section 6.5.2). Equation (8.234) can be replaced by

$$\dot{z}_i + c_{ei}\dot{q}_i + k_{ei}z_i = 0 \quad i = 1, 2, \ldots, n \tag{8.250}$$

where

$$\begin{aligned} c_{ei} &= -E\left\{\frac{\partial G_i}{\partial \dot{q}_i}\right\} \\ k_{ei} &= -E\left\{\frac{\partial G_i}{\partial z_i}\right\} \end{aligned} \qquad i = 1, 2, \ldots, n \tag{8.251}$$

It follows that, in the linearized equations, expressed by equation (8.130),

$$\mathbf{M}_e = \begin{bmatrix} \mathbf{0} & \mathbf{0} \\ \mathbf{0} & \mathbf{0} \end{bmatrix} \tag{8.252}$$

$$\mathbf{C}_e = \begin{bmatrix} \mathbf{0} & \mathbf{0} \\ \mathbf{c}_e & \mathbf{0} \end{bmatrix} \tag{8.253}$$

where

$$\mathbf{c}_e = \begin{bmatrix} c_{e1} & 0 & \cdots & 0 \\ 0 & c_{e2} & \cdots & 0 \\ & & \cdots & \\ & & \cdots & c_{en} \end{bmatrix} \tag{8.254}$$

Also

$$\mathbf{K}_e = \begin{bmatrix} \mathbf{0} & \mathbf{0} \\ \mathbf{0} & \mathbf{k}_e \end{bmatrix} \tag{8.255}$$

where

$$\mathbf{k}_e = \begin{bmatrix} k_{e1} & 0 & \cdots & 0 \\ 0 & k_{e2} & \cdots & 0 \\ & & \cdots & k_{en} \end{bmatrix} \tag{8.256}$$

The elements of $\mathbf{c}_e$ and $\mathbf{k}_e$ can be evaluated according to equation (8.251). Assuming a joint Gaussian distribution for $\dot{q}_i$ and z_i $(i = 1, 2, \ldots, n)$ these elements can be evaluated in terms of $E\{\dot{q}_i^2\}$, $E\{z_i^2\}$ and $E\{\dot{q}_i z_i\}$. The latter quantities can be determined from a solution of the linearized equations. Thus the usual iteration procedure for solving the combined set of non-linear algebraic equations is applicable. Further details can be found in Baber and Wen (1981) and Chang *et al.* (1986).

8.4 NON-STATIONARY PROBLEMS

So far in this chapter it has been assumed that the response process is stationary. Often, in practice, the response is non-stationary, to some extent, due to one or both of two factors. The first factor is associated with system degradation. Degradation of the restoring force usually gradually increases as the structure experiences repeated stress reversals. The parameters in any hysteretic model must become time dependent, if such an effect is to be accounted for. The second factor is associated with non-stationarity of the excitation. Often the response to earthquake ground acceleration is required in applications. As is well known, earthquake excitation is highly non-stationary.

Both factors can be accommodated fairly readily within the extended differential equation approach described in Section 8.3.

8.4.1 Degrading systems

Here, for simplicity the case of a simple oscillator, governed by equation (8.119), will again be discussed, where z is described by a first-order differential equation of the curvilinear type. Extensions to MDOF systems are straightforward (see Baber and Wen, 1981; Baber and Noori, 1984).

Firstly, the standard curvilinear model, given by equation (8.122), will be generalized slightly, to allow a variety of degradation models to be formulated. Thus $G(\dot{y}, z)$ in equation (8.122) is replaced by

$$G(\dot{y}, z) = \frac{-\lambda\{\gamma|\dot{y}|z|z|^{n-1} + \nu\dot{y}|z|^{n}\} + A\dot{y}}{\eta} \tag{8.257}$$

where λ and η are new parameters. Following Baber and Wen (1981) degradation can be modelled by allowing the parameters λ, η and A to vary as a function of the response duration and severity. As convenient measure of the combined effect of duration and severity is the total energy, $e(t)$, dissipated through hysteresis from $t = 0$ to time t. Specifically $\varepsilon(t)$ can be calculated from the history of y and z, as follows

$$\varepsilon(t) = (1 - \alpha) \int z \, \mathrm{d}y = (1 - \alpha) \int_0^t z \dot{y} \, \mathrm{d}t \tag{8.258}$$

The following simple linear functional relationships have been proposed, which lead to physically realistic models of degradation (Baber and Wen, 1981)

$$\begin{aligned} A(t) &= A_0 - \delta_A \varepsilon(t) \\ \eta(t) &= \eta_0 + \delta_\eta \varepsilon(t) \\ \lambda(t) &= \lambda_0 + \delta_\lambda \varepsilon(t) \end{aligned} \tag{8.259}$$

where A_0, η_0 and λ_0 are initial values and δ_A, δ_λ and δ_η are non-negative parameters. The separate effects of positive values of δ_A, δ_η and δ_λ are shown in Figure 8.22 (taken from Figure 3 of Baber and Wen (1981)).

It is noted that A, η and λ depend on $\dot{y}$ and z. This causes complications in the evaluation of the equivalent linear elements, c and k, in equation (8.136). Thus, according to equations (8.134) and (8.135), together with equation (8.257)

$$c = E\left\{\frac{\partial}{\partial \dot{y}}\left[\frac{-\lambda(\gamma|\dot{y}|z|z|^{n-1} + \nu \dot{y}|z|^n) + A\dot{y}}{\eta}\right]\right\} \tag{8.260}$$

$$k = E\left\{\frac{\partial}{\partial z}\left[\frac{-\lambda(\gamma|\dot{y}|z|z|^{n-1} + \nu \dot{y}|z|^n) + A\dot{y}}{\eta}\right]\right\} \tag{8.261}$$

This difficulty can be overcome by recognizing that the time-dependent parameters are normally *slowly varying*. Thus, at least as a first approximation, one can replace A, η and λ by their expected values, m_A, m_η and m_λ, respectively, where, from equation (8.259)

$$\begin{aligned} m_A(t) &= A_0 - \delta_A m_\varepsilon(t) \\ m_\eta(t) &= \eta_0 + \delta_\eta m_\varepsilon(t) \\ m_\lambda(t) &= \lambda_0 + \delta_\lambda m_\varepsilon(t) \end{aligned} \tag{8.262}$$

and

$$m_\varepsilon(t) = E\{\varepsilon(t)\} \tag{8.263}$$

Moreover, from equation (8.258),

$$\dot{m}_E(t) = (1 - \alpha) E\{z \dot{y}\} \tag{8.264}$$

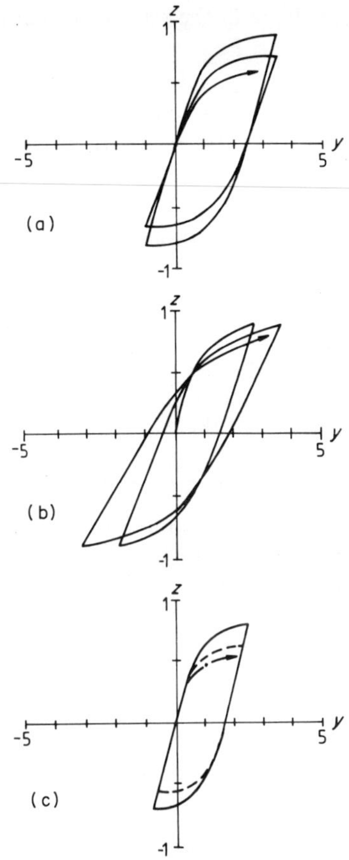

Figure 8.22
Typical degrading hysteresis loops (reproduced from Baber and Wen (1981) by permission of the American Society of Civil Engineers). (a) $\delta_A = 0.05$, (b) $\delta_\eta = 0.20$, (c) $\delta_\lambda = 0.20$

$$c = \frac{m_A(t) - m_\lambda(t)\left[\gamma E\left\{\dfrac{\partial |\dot{y}|}{\partial \dot{y}} z|z|^{n-1}\right\} + \nu E\{|z|^n\}\right]}{m_\eta(t)} \tag{8.265}$$

$$k = \frac{-m_\lambda(t)}{m_\eta(t)}\left[\gamma E\left\{|\dot{y}|\frac{\partial}{\partial z}(|z|^{n-1}z)\right\} + \nu E\left\{\dot{y}\frac{\partial}{\partial z}(|z|^n)\right\}\right] \tag{8.266}$$

If $\dot{y}$ and z are taken to be jointly Gaussian the expectations in equations (8.264) and (8.265) can now be evaluated in closed form, as for the non-degrading case. Thus a complete set of non-linear differential equations is obtained, which can be integrated numerically, with the parameter values updated at each time step.

It is worth noting here that, very recently, further generalizations of the basic Bouc–Wen model have been proposed, which incorporate hysteretic loop pinching (Baber and Noori, 1984, 1985; Noori *et al.* 1986).

8.4.2 Non-stationary excitation

If system degradation is incorporated into the hysteresis model then, since a numerical integration of the governing differential equations is required, it is a relatively small increase in complexity to allow for non-stationarity in the excitation process (e.g. see Kimura *et al.* 1983).

A normal approach is to model the excitation as

$$f(t) = \chi(t)w(t) \tag{8.267}$$

where $\chi(t)$ is a deterministic, modulating function, and $w(t)$ is a stationary random process. The latter is usually taken to be white noise, or filtered white noise. Frequently the modulating function

$$\chi(t) = c[\exp(-\alpha t) - \exp(-\beta t)] \tag{8.268}$$

is adopted in numerical studies.

The general procedure for dealing with such non-stationary problems is given in Chapter 7.

Chapter 9

Relaxation of the Gaussian response assumption

9.1 INTRODUCTION

The statistical linearization techniques considered earlier in this book, in Chapters 5 to 8, are all based on the concept of using an equivalent linear system to compute the necessary expectations which occur in the expressions for the equivalent linear parameters. Clearly, the response of the equivalent linear system is Gaussian, if the excitation is Gaussian, whereas the response of the original non-linear system is non-Gaussian. There is, therefore, an inherent error in the evaluation of the parameters in the equivalent linear model.

In this chapter three different approaches to relaxing the Gaussian response assumption, contained in the normal statistical linearization method, are presented and illustrated through the use of specific examples. The first two of these approaches have the advantage that they can give information on the deviation of the response process distribution from Gaussianity. As shown later in this chapter, reliability statistics are sensitive to the probability distribution of the response, in its extreme 'tails'. Thus, methods of predicting deviations from Gaussianity, due to the effect of non-linearities, are likely to lead to much improved reliability estimates. The third method, whilst not leading to predictions of response probability distributions, does offer an entirely different approach to obtaining equivalent linear models, which does not depend on a priori knowledge of the form of the non-linearities present. Moreover, it is directly applicable in experimental situations, where only measured sample functions of the excitation and response processes are available.

9.2 STATISTICAL LINEARIZATION AND GAUSSIAN CLOSURE

As a first step it is convenient to reformulate the basic results of the normal statistical linearization procedure, described earlier in this book, in terms of state variables.

Starting with the fairly general equation of motion of an n degree of freedom non-linear system, as given by equation (6.59), it is observed that this can be recast as

$$\dot{\mathbf{z}} = \mathbf{F}(\mathbf{z}) + \mathbf{f}(t) \tag{9.1}$$

where

$$\mathbf{z} = \begin{bmatrix} \mathbf{q} \\ \dot{\mathbf{q}} \end{bmatrix} \quad \mathbf{f}(t) = \begin{bmatrix} \mathbf{0} \\ \mathbf{M}^{-1}\mathbf{Q}(t) \end{bmatrix} \tag{9.2}$$

and

$$\mathbf{F}(\mathbf{z}) = \begin{bmatrix} \dot{\mathbf{q}} \\ -\mathbf{M}^{-1}\mathbf{C}\dot{\mathbf{q}} - \mathbf{M}^{-1}\mathbf{K}\mathbf{q} - \mathbf{M}^{-1}\mathbf{g}(\mathbf{q}, \dot{\mathbf{q}}) \end{bmatrix} \tag{9.3}$$

Clearly, equation (9.1) is a first-order equation of order $N = 2n$. For simplicity it will be assumed, henceforth, that the non-linearity, $\mathbf{g}(\mathbf{q}, \dot{\mathbf{q}})$ is such that its elements can be expressed, at least approximately, as a summation of terms consisting of powers of q and $\dot{q}$. Thus each term is of the form

$$\alpha \prod_{i=1}^{n} \prod_{j=1}^{n} (q_i)^{r_i} (\dot{q}_j)^{s_j} \tag{9.4}$$

where α is a constant, and r_i and s_j are integers.

It will also be assumed, for simplicity, in the following that $\mathbf{f}(t)$ is a non-stationary white noise process. As noted earlier, if the input to the system is non-white, then the equations of motion can often be manipulated into the form of equation (9.1), through the introduction of a 'shaping filter', which operates on white noise to produce the desired input. If a zero-mean vector process, $\mathbf{f}_0(t)$, is defined by

$$\mathbf{f}_0(t) = \mathbf{f}(t) - \mathbf{m}_f(t) \tag{9.5}$$

where

$$\mathbf{m}_f(t) = E\{\mathbf{f}(t)\} \tag{9.6}$$

then the covariance matrix of $\mathbf{f}_0(t)$ takes the form

$$E\{\mathbf{f}_0(t)\mathbf{f}_0^{\mathrm{T}}(\tau)\} = \mathbf{w}_{f_0}(t, \tau) = \mathbf{D}(t)\delta(t - \tau) \tag{9.7}$$

where $\mathbf{D}(t)$ is a real, symmetric, non-negative matrix.

Statistical linearization consists of replacing equation (9.1) by the linear differential relationship

$$\dot{\mathbf{z}} = \mathbf{a}(t) + \mathbf{L}(t)\mathbf{z}_0 + \mathbf{f}(t) \tag{9.8}$$

where $\mathbf{z}_0(t)$ is a zero-mean process, defined by

$$\mathbf{z}_0(t) = \mathbf{z}(t) - \mathbf{m}(t) \tag{9.9}$$

and

$$\mathbf{m}(t) = E\{\mathbf{z}(t)\} \tag{9.10}$$

In equation (9.8), $\mathbf{a}(t)$ and $\mathbf{L}(t)$ are to be determined such that the error, or 'equation deficiency' between equations (9.1) and (9.8), given by

$$\boldsymbol{\varepsilon} = \mathbf{F}(\mathbf{z}) - \mathbf{a}(t) - \mathbf{L}(t)\mathbf{z}_0 \tag{9.11}$$

is minimized. Minimizing, as usual, in a least-square sense, the values in $\mathbf{a}(t)$ and $\mathbf{L}(t)$ which reduce

$$E\{\boldsymbol{\varepsilon}^{\mathrm{T}}\boldsymbol{\varepsilon}\}$$

to its lowest possible value are sought. Routine calculation, through differentiation, shows that minimization is achieved if

$$\mathbf{L}(t) = E\{\mathbf{F}(\mathbf{z})\mathbf{z}_0^{\mathrm{T}}\}\mathbf{V}^{-1}(t) \tag{9.12}$$

where

$$\mathbf{V}(t) = E\{\mathbf{z}_0\mathbf{z}_0^{\mathrm{T}}\} \tag{9.13}$$

is the covariance matrix for $\mathbf{z}(t)$, and

$$\mathbf{a}(t) = E\{\mathbf{F}(\mathbf{z})\} \tag{9.14}$$

In the scalar case, the results reduce to the equations given in Section (5.2.1) (see equations (5.7) and (5.8)).

It is interesting, at this stage, to examine the differential equations governing the mean vector, $\mathbf{m}$, and the covariance matrix, $\mathbf{V}$. Taking expectations of equation (9.1), and interchanging the operations of differentiation and expectation, one has

$$\dot{\mathbf{m}} = E\{\mathbf{F}(\mathbf{z})\} + \mathbf{m}_f(t) \tag{9.15}$$

This may be compared with the corresponding result derived from the equivalent linear equation, as given by equation (9.8). Taking expectations of this latter equation gives

$$\dot{\mathbf{m}} = \mathbf{a}(t) + \mathbf{m}_f(t) \tag{9.16}$$

However, $\mathbf{a}(t)$ is evaluated according to equation (9.14). It follows that equations (9.15) and (9.16) are identical. An important consequence of this is that equation (9.16) will give the exact variation of the mean response with time, if the true probability distribution of $\mathbf{z}(t)$ is used in the evaluation of the expectation in equation (9.14).

One can examine the evolution of the covariance matrix, $\mathbf{V}$, in a similar fashion. It is possible to derive an exact expression governing $\mathbf{V}$, from the original non-linear equation, (9.1), using the apparatus of Markov process theory. The result is (e.g. see Jazwinskii, 1970, Theorem 6.2)

$$\dot{\mathbf{V}} = E\{\mathbf{F}(\mathbf{z})\mathbf{z}_0^{\mathrm{T}}\} + E\{\mathbf{z}_0\mathbf{F}^{\mathrm{T}}(\mathbf{z})\} + \mathbf{D} \tag{9.17}$$

This can be rewritten as

$$\dot{\mathbf{V}} = E\{\mathbf{F}(\mathbf{z})\mathbf{z}_0^{\mathrm{T}}\}\mathbf{V}^{-1}\mathbf{V} + \mathbf{V}\mathbf{V}^{-1}E\{\mathbf{z}_0\mathbf{F}^{\mathrm{T}}(\mathbf{z})\} + \mathbf{D} \tag{9.18}$$

Now, using expression (9.12) for $\mathbf{L}(t)$, equation (9.18) can be simplified to

$$\dot{\mathbf{V}} = \mathbf{L}\mathbf{V}^{\mathrm{T}} + \mathbf{V}\mathbf{L}^{\mathrm{T}} + \mathbf{D} \tag{9.19}$$

since $\mathbf{V}$ is a symmetric matrix. This equation is identical to that obtained from the equivalent linear system, governed by equation (9.8) (see, for example, equation (4.155)). It follows that, if the true distribution of $\mathbf{z}(t)$ is used to evaluate $\mathbf{L}$, then the evolution of $\mathbf{V}$ with time, as deduced from the equivalent linear system, will be exact.

One can conclude, therefore, that the equivalent linear system, given by equation (9.8), will give the exact evolution of both $\mathbf{m}$ and $\mathbf{V}$ with time, provided that the exact distribution of $\mathbf{z}(t)$ is used to compute $\mathbf{a}$ and $\mathbf{L}$. Of course, this distribution is not known, in general. However, the multi-dimensional joint probability density function, $f(\mathbf{z})$, for $\mathbf{z}(t)$ can be written as an expansion in terms of $f_{\mathrm{G}}(\mathbf{z})$, the N-dimensional Gaussian density function for $\mathbf{z}(t)$, given by (see equation (3.49))

$$f_{\mathrm{G}}(\mathbf{z}) = \frac{1}{(2\pi)^{N/2}|\mathbf{V}|^{1/2}} \exp(-\tfrac{1}{2}\mathbf{z}_0^{\mathrm{T}}\mathbf{V}^{-1}\mathbf{z}_0) \tag{9.20}$$

Thus, as shown in Section 3.5.2, one has an expansion of the form

$$f(\mathbf{z}) = f_{\mathrm{G}}(\mathbf{z}) + \text{higher order terms} \tag{9.21}$$

where the coefficients in such an expansion can be related to the moments

$$m_{k_1 \ldots k_N} = E\left\{\prod_{i=1}^{N} z_i^{k_i}\right\} \tag{9.22}$$

where

$$K = k_1 + k_2 + \cdots + k_N \tag{9.23}$$

is the order of the moment.

The mean vector, $\mathbf{m}$, and covariance matrix, $\mathbf{V}$, can be expressed in terms of first- and second-order moments only. Suppose that an expansion of the form of equation (9.21) is used to evaluate $\mathbf{m}$ and $\mathbf{V}$, according to the exact relationships given by equations (9.15) and (9.17). Due to the assumed power-law form of the non-linearities present, one finds that $\mathbf{m}$ and $\mathbf{V}$ are dependent on higher order moments, i.e. moments not contained in $\mathbf{m}$ and $\mathbf{V}$, of third or higher order. This involvement of higher order moments in the differential equations for $\mathbf{m}$ and $\mathbf{V}$ is directly attributable to the presence of non-linear terms in the equation of motion. It is possible to derive equations for, say, the third-order moments, but in general these are found to depend on moments of order higher than a third, and so on. Thus one is led to the well known 'closure problem (Nigam, 1983). Specifically, if the hierarchy of moment equations is truncated, at some level, then there are more unknowns than equations, and a solution is impossible.

It is possible to force the system of moment equations to 'close' by simply truncating an expansion for $f(\mathbf{z})$, of the form of equation (9.21), after a suitable number of terms. As will be demonstrated later, this enables higher order moments to be approximated in terms of lower order moments. By this, or alternative similar means, one can supplement the moment equations generated from the equations of motion to the point where there are a sufficient number of equations to solve for the unknown, non-zero, lower order moments.

The lowest possible level at which closure leads to meaningful results is Gaussian closure. This consists of approximating $f(\mathbf{z})$ as $f_G(\mathbf{z})$, the Gaussian density function, i.e. taking only the very first term in equation (9.21). The expectations in equations (9.15) and (9.17) can then be determined entirely in terms of $\mathbf{m}$ and $\mathbf{V}$. It follows that these equations represent a closed set of relationships for $\mathbf{m}$ and $\mathbf{V}$, and can be solved.

It has previously been pointed out that the differential equations for $\mathbf{m}$ and $\mathbf{V}$, deduced from the equivalent linear equation, given by equation (9.8), are of exactly the same form as the corresponding exact equations. Therefore, the evaluation of $\mathbf{a}$ and $\mathbf{L}$, in equation (9.8), according to a Gaussian assumption for $\mathbf{z}$, corresponds exactly with the procedure of using Gaussian closure to evaluate $\mathbf{m}$ and $\mathbf{V}$, according to equations (9.15) and (9.17). In other words, the standard statistical linearization procedure, described so far in this book, is exactly the same as the Gaussian closure procedure for evaluating the first and second moments of the response (for the class of systems under consideration).

It follows that any non-Gaussian closure technique, derived by including higher order terms in an expansion of $f(\mathbf{z})$, is essentially a generalization of the standard statistical linearization method and can be expected to lead to an improvement in accuracy.

9.2.1 An example

To illustrate the Gaussian closure approach, the case of a Duffing oscillator, with stationary white noise excitation, will once again be considered. If non-linear, cubic damping is incorporated into the model, in addition to the usual cubic stiffness term, then the appropriate equation of motion is given by equation (5.138). Here it is convenient to simplify this somewhat by introducing the non-dimensional time, $\tau = \omega_n t$, in place of t, and the non-dimensional displacement

$$Y = \frac{y}{\sigma_{yo}} \tag{9.24}$$

in place of y. In this equation σ_{yo} is the standard deviation of the response, in the linear case (see equation (5.56)). An alternate form of equation (5.138) is then

$$\ddot{Y} + 2\zeta(\dot{Y} + \mu\dot{Y}^3) + Y(1 + \rho Y^2) = (4\zeta)^{1/2} f_n(\tau) \tag{9.25}$$

where

$$f_n(\tau) = \frac{f(t)}{(2\pi S_0 \omega_n)^{1/2}} \tag{9.26}$$

Further, S_0 is the constant spectral level associated with $f(t)$ and differentiation in equation (9.25) is now with respect to τ. $f_n(\tau)$ has 'unit strength', i.e.

$$E\{f_n(\tau) f_n(\tau + \tau')\} = \delta(\tau') \tag{9.27}$$

and the standard deviation of both Y and $\dot{Y}$ is unity, in the linear case ($\mu = \rho = 0$). The parameters ζ, μ and ρ in equation (9.25) are related to the corresponding parameters in equation (5.138), as follows

$$\beta = 2\zeta\omega_n \tag{9.28}$$

$$\mu = \sigma_{yo}^2 \eta \tag{9.29}$$

$$\rho = \sigma_{yo}^2 \lambda \tag{9.30}$$

Equation (9.25) can be recast into the form of equation (9.1) where

$$\mathbf{z} = \begin{bmatrix} z_1 \\ z_2 \end{bmatrix} = \begin{bmatrix} Y \\ \dot{Y} \end{bmatrix} \quad \mathbf{f}(\tau) = \begin{bmatrix} 0 \\ (4\zeta)^{1/2} f_n(\tau) \end{bmatrix} \tag{9.31}$$

and

$$\mathbf{F}(\mathbf{z}) = \begin{bmatrix} F_1(z) \\ F_2(z) \end{bmatrix} = \begin{bmatrix} z_2 \\ -2\zeta z_2 - 2\zeta\mu z_2^3 - z_1 - \rho z_1^3 \end{bmatrix} \tag{9.32}$$

Here, for simplicity, it will be assumed that the mean of $\mathbf{f}(\tau)$ is zero. Thus $\mathbf{m}_f = \mathbf{0}, \mathbf{f}(\tau) = \mathbf{f}_0(\tau)$ and the covariance matrix of $\mathbf{f}_0(\tau)$ is of the form of equation (9.7); here $\mathbf{D}$ is time-invariant and is given by

$$\mathbf{D} = \begin{bmatrix} 0 & 0 \\ 0 & 4\zeta \end{bmatrix} \tag{9.33}$$

The mean vector, $\mathbf{m}$, of the response is governed, exactly, by equation (9.15). In this example, using equation (9.32), one obtains

$$\dot{m}_1 = E\{z_2\} \tag{9.34}$$

$$\dot{m}_2 = -2\zeta E\{z_2\} - 2\zeta\mu E\{z_2^3\} - E\{z_1\} - \rho E\{z_1^3\} \tag{9.35}$$

Similarly, the covariance matrix, $\mathbf{V}$ is governed exactly, by equation (9.17). In this case one obtains

$$\begin{bmatrix} \dot{v}_{11} & \dot{v}_{12} \\ \dot{v}_{21} & \dot{v}_{22} \end{bmatrix} = \begin{bmatrix} E\{F_1 z_1\} & E\{F_1 z_2\} \\ E\{F_2 z_1\} & E\{F_2 z_2\} \end{bmatrix} + \begin{bmatrix} E\{F_1 z_1\} & E\{F_2 z_1\} \\ E\{F_1 z_2\} & E\{F_2 z_2\} \end{bmatrix} + \begin{bmatrix} 0 & 0 \\ 0 & 4\zeta \end{bmatrix} \tag{9.36}$$

This result can be written as a set of differential equations, as follows (using equation (9.32) again)

$$\dot{v}_{11} = 2E\{z_1 z_2\} \tag{9.37}$$

$$\dot{v}_{12} = \dot{v}_{21} = E\{z_2^2\} - 2\zeta E\{z_1 z_2\} - 2\zeta\mu E\{z_1 z_2^3\} - E\{z_1^2\} - \rho E\{z_1^4\} \tag{9.38}$$

$$\dot{v}_{22} = -4\zeta E\{z_2^2\} - 4\zeta\mu E\{z_2^4\} - 2E\{z_1 z_2\} - 2\rho E\{z_1^3 z_2\} + 4\zeta \tag{9.39}$$

These equations for the elements of **m** and **V** typify the general closure problem in non-linear stochastic dynamics. The following 'higher-order' moments arise in the equations

$$m_{13} = E\{z_1 z_2^3\} = E\{Y\dot{Y}^3\} \tag{9.40}$$

$$m_{40} = E\{z_1^4\} = E\{Y^4\} \tag{9.41}$$

$$m_{04} = E\{z_2^4\} = E\{\dot{Y}^4\} \tag{9.42}$$

$$m_{31} = E\{z_1^3 z_2\} = E\{Y^3\dot{Y}\} \tag{9.43}$$

To obtain solutions from the above differential equations it is necessary to 'close' the equations by approximating the higher order moments in terms of the elements of **m** and **V**. This can be achieved by Gaussian closure, whereby the density function of **z**, **f**(**z**) is approximated by the Gaussian form of equation (9.20).

The necessary results for m_{13}, m_{40}, m_{04} and m_{31} can be obtained by using the moment–cumulant relationship given by equation (3.37), in Chapter 3. Noting that, for a Gaussian joint distribution, cumulants of order greater than two are zero, equation (3.37) reduces to

$$\begin{aligned} m_4(\eta_j, \eta_k, \eta_l, \eta_m) &= 3\{\kappa_2(\eta_j, \eta_k)\kappa_2(\eta_l, \eta_m)\}_s \\ &\quad + 6\{\kappa_1(\eta_j)\kappa_1(\eta_k)\kappa_2(\eta_l, \eta_m)\}_s \\ &\quad + \kappa_1(\eta_j)\kappa_1(\eta_k)\kappa_1(\eta_l)\kappa_1(\eta_m) \end{aligned} \tag{9.44}$$

Setting $\eta_j = z_1, \eta_k = \eta_l = \eta_m = z_2$, equation (9.44) gives

$$\begin{aligned} m_{13} &= m_4(z_1, z_2, z_2, z_2) \\ &= 3\kappa_2(z_1, z_2)\kappa_2(z_2, z_2) + 5\kappa_1(z_1)\kappa_1(z_2)\kappa_1(z_2, z_2) \\ &\quad + \kappa_1^2(z_1)\kappa_2(z_1, z_2) + \kappa_1(z_1)\kappa_1^3(z_2). \end{aligned} \tag{9.45}$$

Now, recognizing that (see equation (3.41))

$$\begin{aligned} \kappa_1(z_1) &= m_1 \quad & \kappa_2(z_1, z_1) &= v_{11} \\ \kappa_1(z_2) &= m_2 \quad & \kappa_2(z_1, z_2) &= v_{12} \\ & & \kappa_2(z_2, z_2) &= v_{22} \end{aligned} \tag{9.46}$$

it is found that equation (9.45) can be expressed as

$$m_{13} = 3v_{12}v_{22} + 5m_1 m_2 v_{22} + m_1^2 v_{12} + m_1 m_2^3 \tag{9.47}$$

Similarly, expressions for the other fourth-order moments in equations (9.37) to (9.39) can be found in terms of the elements of **m** and **V**. Hence one can obtain a closed set of differential equations for m_1, m_2, v_{11}, v_{12} and v_{22} which can be solved numerically to obtain the evolution of **m** and **V** with time. The result will be identical to that obtained by applying the standard statistical linearization method, in the non-stationary case (see Chapter 7).

Here the particular case of the stationary response, which is attained after the transient response has decayed away will be investigated in some detail. Then $\dot{\mathbf{m}} = \dot{\mathbf{V}} = \mathbf{0}$, and the response is symmetrical, about a zero mean. It follows that equations (9.34) and (9.35) are automatically satisfied. Moreover (see equation (3.91))

$$m_{11} = E\{z_1 z_2\} = E\{Y\dot{Y}\} = 0 \tag{9.48}$$

so that equation (9.37) is also automatically satisfied. In addition

$$m_{31} = E\{z_1^3 z_2\} = E\{Y^3\dot{Y}\} = \frac{1}{4}\frac{\mathrm{d}}{\mathrm{d}t}E\{Y^4\} = 0 \tag{9.49}$$

Writting $\sigma_Y^2 = E\{z_1^2\}, \sigma_{\dot{Y}}^2 = E\{z_2^2\}$, equations (9.38) and (9.39) can now be simplified to the following

$$0 = \sigma_{\dot{Y}}^2 - 2\zeta\mu E\{Y\dot{Y}^3\} - \sigma_Y^2 - \rho E\{Y^4\} \tag{9.50}$$

$$0 = -4\zeta\sigma_{\dot{Y}}^2 - 4\zeta\mu E\{\dot{Y}^4\} + 4\zeta \tag{9.51}$$

The fourth-order moments in the above can be related to second moments, using Gaussian closure. Thus, with the mean zero, equation (9.45) reduces to

$$m_{13} = E\{Y\dot{Y}^3\} = 0 \tag{9.52}$$

Similarly, from equation (3.37), one obtains

$$m_{04} = E\{\dot{Y}^4\} = 3\sigma_{\dot{Y}}^2 \tag{9.53}$$

and

$$m_{40} = E\{Y^4\} = 3\sigma_Y^2 \tag{9.54}$$

These results can, of course, also be obtained by direct evaluation of the moments, using the appropriate one-dimensional Gaussian density function. Combining equations (9.50) to (9.54) gives

$$\sigma_{\dot{Y}}^2 - \sigma_Y^2 - 3\rho\sigma_Y^4 = 0 \tag{9.55}$$

$$\sigma_{\dot{Y}}^2 + 3\mu\sigma_{\dot{Y}}^4 - 1 \quad = 0 \tag{9.56}$$

The second of these equations can be solved for $\sigma_{\dot{Y}}^2$, independently of the first. Hence

$$\sigma_{\dot{Y}}^2 = \phi(\mu) \tag{9.57}$$

where

$$\phi(x) = \frac{(1 + 12x)^{1/2} - 1}{6x} \tag{9.58}$$

Hence, the first equation, (9.55) can be solved, to give

$$\sigma_{\dot{Y}}^2 = \phi[\rho\phi(\mu)]\phi(\mu) \tag{9.59}$$

where the function $\phi(\ )$ defined by equation (9.58) occurs again.

These results are, of course, in complete agreement with those obtained earlier, in Chapter 5, using the standard statistical linearization procedure (see equations (5.144) and (5.146)).

It is noted that, in the case of linear damping, where $\mu = 0$, equations (9.57) and (9.59) reduce to

$$\sigma_Y^2 = 1 \quad \sigma_{\dot{Y}}^2 = \phi(\rho) \tag{9.60}$$

9.3 NON-GAUSSIAN CLOSURE

The Gaussian closure method can be refined by including higher order terms in an expansion for $f(\mathbf{z})$, of the form given by equation (9.21). To obtain a sufficient number of equations, for the increased number of unknowns, it is then necessary to generate moment equations for moments of order higher than two.

9.3.1 Moment equations

Appropriate moment equations can be generated fairly readily by returning to the original equations of motion, as given by equation (9.1). Here, and henceforth, for simplicity it will be assumed that the mean of $f(t)$ is zero.

If $g(\mathbf{z})$ is a scalar function of $\mathbf{z}$, one can use the following important result from Markov process theory (e.g., see Jazwinskii, 1970)

$$E\{\dot{g}(\mathbf{z})\} = E\{\mathbf{h}^{\mathrm{T}}\mathbf{F}\} + \tfrac{1}{2}\operatorname{tr} E\{\mathbf{DH}\} \tag{9.61}$$

Here $\mathbf{h}$ is the gradient vector of g, i.e.

$$\mathbf{h} = \nabla g = \left[\frac{\partial g}{\partial z_1}, \frac{\partial g}{\partial z_2}, \ldots, \frac{\partial g}{\partial z_N}\right]^{\mathrm{T}} \tag{9.62}$$

Further, $\mathbf{H}$ is the Jacobian matrix of second partial derivatives of $\mathbf{g}$, i.e.

$$\mathbf{H} = \begin{bmatrix} \dfrac{\partial^2 g}{\partial z_1 \partial z_1} & \dfrac{\partial^2 g}{\partial z_1 \partial z_2} & \cdots & \dfrac{\partial^2 g}{\partial z_1 \partial z_N} \\ \dfrac{\partial^2 g}{\partial z_2 \partial z_1} & \dfrac{\partial^2 g}{\partial z_2 \partial z_2} & \cdots & \\ \vdots & & & \\ \dfrac{\partial^2 g}{\partial z_N \partial z_1} & \cdots & & \dfrac{\partial^2 g}{\partial z_N \partial z_N} \end{bmatrix}, \tag{9.63}$$

and the symbol tr denotes the trace of the matrix $\mathbf{A} = E\{\mathbf{DH}\} = [A_{ij}]$, i.e.

$$\operatorname{tr}\mathbf{A} = A_{11} + A_{22} + \cdots + A_{NN} \tag{9.64}$$

With appropriate choices for the function g, in equation (9.61), all the necessary moment equations can be generated. Thus, as a check, if

$$g(\mathbf{z}) = z_i \tag{9.65}$$

then

$$\mathbf{h} = [0, 0, \ldots, 1, \ldots, 0]^{\mathrm{T}} \tag{9.66}$$

where the non-zero element is the ith. Moreover, $\mathbf{H}$ is a zero matrix. Hence, from equation (9.61),

$$\dot{m}_i = E\{\dot{z}_i\} = E\{F_i\} \tag{9.67}$$

This result, of course, is derivable directly by taking expectations of the original matrix equation (see equation (9.15), with here $\mathbf{m}_f = \mathbf{0}$).

More generally, if one lets

$$g(\mathbf{z}) = \sum_{i=1}^{N} (z_i)^{k_i} \tag{9.68}$$

where $k_i (i = 1, 2, \ldots, N)$ are positive integers, then a substitution of this expression into equation (9.61) gives, for the Kth order moment $m_{k_1} \cdots m_{k_N}$, defined by equation (9.22) the following result

$$\dot{m}_{k_1 \ldots k_N} = \sum_{j=1}^{N} k_j \left[E\left\{ \sum_{i=1}^{N} \frac{F_j (z_i)^{k_i}}{z_j} \right\} \right] + \frac{1}{2} \sum_{j=1}^{N} \sum_{i=1}^{N} \gamma_{ji} m_{\rho_1 \ldots \rho_N} D_{ji} \tag{9.69}$$

where

$$\begin{aligned} \gamma_{ji} &= k_j k_i && \text{if } j \neq i \\ & \quad k_j k_{j-1} && \text{if } j = i \end{aligned} \tag{9.70}$$

and

$$\begin{aligned} \rho_1 \cdots \rho_N &= k_1 \cdots k_{j-1} \cdots k_{i-1} \cdots k_N && \text{if } j \neq i \\ &= k_1 \cdots k_{j-2} \cdots k_N && \text{if } j = i \end{aligned} \tag{9.71}$$

Thus, for example, if $N = 2$, the second-order moment equations are obtained as follows

(a) $k_1 = 1 \quad k_2 = 1$

$$\dot{m}_{11} = E\{F_1 z_2\} + E\{F_2 z_1\} \tag{9.72}$$

(b) $k_1 = 2 \quad k_2 = 0$

$$\dot{m}_{20} = 2E\{F_1 z_1\} \tag{9.73}$$

(c) $k_1 = 0 \quad k_2 = 2$

$$m_{02} = 2E\{F_2 z_2\} + 4\zeta \tag{9.74}$$

9.3.2 Closure techniques

To close the moment equations, by expressing unknown higher order moments in terms of lower order moments, one can simply truncate the expansion for $f(\mathbf{z})$, at an appropriate level. As shown in Chapter 3, a useful expansion for $f(\mathbf{z})$, in terms of $f_G(\mathbf{z})$, is obtained using the concept of quasi-moments (see equation (3.69)). The 'quasi-moment neglect' closure technique involves simply setting all quasi-moments above a certain level to zero. Referring again to equation (3.69) and noting that $\mathbf{x}$ is here to be identified with $\mathbf{z}$, and $n = N$, a truncated form of the expansion, suitable for the present application, is

$$f_M(\mathbf{z}) = \left\{ 1 + \sum_{s=3}^{M} \frac{1}{s!} \sum_{j,k,\ldots,m=1}^{N} b_{k_1 \ldots k_m} H_{jk\ldots m}(\mathbf{z} - \mathbf{m}) f_G(\mathbf{z}) \right\} \tag{9.75}$$

In this equation

$$b_{k_1 \ldots k_m} = b_s(\underbrace{z_1, \ldots, z_1}_{k_1 \text{ times}}, \underbrace{z_2, \ldots, z_2}_{k_2 \text{ times}}, \ldots, \underbrace{z_m, \ldots, z_m}_{k_m \text{ times}}) \tag{9.76}$$

similarly equation (3.29) and $f_G(\mathbf{z})$ is given by equation (9.20). In equation (9.75) all quasi-moments of order greater than M have been set to zero. This will be called Mth-order closure. If $M = 2$, the interpretation is that the whole of the summation term in equation (9.75) is zero, i.e. $f_2(\mathbf{z}) = f_G(\mathbf{z})$ and one has the case of Gaussian closure.

Using equation (9.75) one can obtain expressions for the expectation of any function of $\mathbf{z}$, $g(\mathbf{z})$, say. Thus

$$E\{g(\mathbf{z})\} = \int_{-\infty}^{\infty} \cdots \int_{-\infty}^{\infty} g(\mathbf{z}) f_M(\mathbf{z})\, d\mathbf{z} \tag{9.77}$$

can be evaluated in terms of quasi-moments, b_s, from $s = 3$ to $s = M$. Hence using relationships such as equations (3.34) to (3.37) and (3.61) to (3.64), $E\{g(\mathbf{z})\}$ can be expressed in terms of the moments m_s from $s = 1$ to $s = M$. In line with the original assumption that all the non-linearities involved in the equation are of power-law type, it follows that the only expectations which are required are moments. Thus $g(\mathbf{z})$ is of the form

$$g(\mathbf{z}) = z_1^{k_1} z_2^{k_2} \cdots z_N^{k_N} \tag{9.78}$$

Then $E\{g(\mathbf{z})\} = m_{k_1 \ldots k_N}$, according to the definition of $m_{k_1 \ldots k_N}$, given by equation (9.22).

The task of evaluating the moments, using equations (9.75), (9.77) and (9.78) is, at first sight, daunting. Fortunately, however, this task can be circumvented by using the established relationships between the quasi-moments, b_s, and the cumulants, κ_s, (e.g. see equations (3.61) to (3.64)) and between cumulants and moments (e.g. see equations (3.34) to (3.37)). In this way a direct set of relationships between b_s and m_s ($s = 3, 4, \ldots$) can be derived. With b_s set to zero, for $s > M$, moments of order greater than M can all be expressed in terms of moments less than, or equal to, M.

As an example of this procedure, the specific case where $\mathbf{z}$ has a symmetric distribution about $\mathbf{z} = \mathbf{0}$ will be considered. Then the odd order moments, cumulants and quasi-moments will all be zero. For moments of sixth order, equation (3.48) reduces to

$$m_6 = \kappa_6 + 15\{\kappa_2\kappa_4\}_s + 15\{\kappa_2\kappa_2\kappa_2\}_s \tag{9.79}$$

where the arguments of the cumulant functions are omitted, and taken as understood. If fourth-order closure, based on b_s is adopted then $b_s = 0$ for $s > 4$. Hence, from equation (9.79), using the fact that here $b_6 = \kappa_6$ (see equation (3.63))

$$m_6 = 15\{k_2k_4\}_s + 15\{k_2k_2k_2\}_s \tag{9.80}$$

From equation (9.80) expressions for all the sixth-order moments

$$m_{jk} = E\{z_1^j z_2^k\} \tag{9.81}$$

can be derived, in terms of lower order moments. Thus, for example, if $j = 6$ and $k = 0$, in equation (9.81), then, from equation (9.80),

$$m_{60} = 15\kappa_{20}\kappa_{40} + 15\kappa_{20}^3 \tag{9.82}$$

Furthermore, from equations (3.44) and (3.46),

$$\kappa_{20} = m_{20} \tag{9.83}$$

$$\kappa_{40} = m_{40} - 3k_{20}^2 \tag{9.84}$$

$$= m_{40} - 3m_{20}^2$$

Combining equations (9.82) to (9.84) results in the following expression for m_{60}

$$m_{60} = 15m_{20}(m_{40} - 3m_{20}^2) + 15m_{20}^3 = 15m_{20}m_{40} - 30m_{20}^2 \tag{9.85}$$

Similar expressions to equation (9.83) can be derived, from equation (9.80), for all the other sixth-order moments m_{jk} ($j + k = 6$). The moment m_{51} is actually zero, since (similar to equation (9.49)),

$$m_{51} = E\{z_1^5 z_2\} = E\{Y^5 \dot{Y}\} = \frac{1}{6}\frac{\mathrm{d}}{\mathrm{d}t}E\{Y^6\} = 0 \tag{9.86}$$

Expressions for all the non-zero moments, including m_{60}, are summarized in Table 9.1.

As an alternative to the above technique, closure can be based entirely on cumulants. Thus, for Mth order closure, one can simply set all cumulants, κ_s, for which $s > M$, to zero. Although the basis for this method appears to be rather more arbitrary than closure based on quasi-moments, in many cases

Table 9.1
Moment relationships

Sixth order moments	*Expressions in terms of fourth order and second order moments*
m_{60}	$15m_{20}m_{40} - 30m_{20}^3$
m_{06}	$15m_{02}m_{04} - 30m_{02}^3$
m_{15}	$10m_{02}m_{13}$
m_{42}	$6m_{20}m_{22} - 6m_{02}m_{20}^2 + m_{02}m_{40}$
m_{24}	$6m_{02}m_{22} - 6m_{20}m_{02}^2 + m_{20}m_{04}$
m_{33}	$3m_{20}m_{13}$

identical, or virtually identical, results will be obtained, particularly if the closure order is relatively low. This follows from the fact that cumulants, κ_s, and quasi-moments, b_s, are identical, up to order five (see equations (3.61) to (3.64)). Moreover, for a symmetrical distribution, with zero mean, $b_s = \kappa_s$ from $s = 3$ to $s = 7$. Thus, for example, the results given in Table 9.1 apply for both the quasi-moment and cumulant closure of fourth order.

9.3.3 An example

To illustrate non-Gaussian closure the case of an oscillator with linear-plus-cubic damping and stiffness, with stationary, zero-mean white noise excitation, will once again be considered. As before it is convenient to work with the normalized form of the equation of motion, as given by equation (9.25).

Applying the general result given by equation (9.69) to this case, where here $N = 2$ and $\mathbf{z}$, $\mathbf{F}$ and $\mathbf{D}$ are given by equations (9.31), (9.32) and (9.33), respectively, the following set of equations are obtained for the moments given by equation (9.81) (see also Wu and Lin, 1984, for the case where $\mu = 0$).

First order (j + k = 1)

$$\dot{m}_{10} = m_{01} \tag{9.87}$$

$$\dot{m}_{01} = -2\zeta m_{01} - 2\zeta\mu m_{03} - m_{10} - \rho m_{30} \tag{9.88}$$

Second order (j + k = 2)

$$\dot{m}_{20} = 2m_{11} \tag{9.89}$$

$$\dot{m}_{11} = m_{02} - 2\zeta m_{11} - 2\zeta\mu m_{13} - m_{20} - \rho m_{40} \tag{9.90}$$

$$\dot{m}_{02} = -4\zeta m_{02} - 4\zeta\mu m_{04} - 2m_{11} - 2\rho m_{31} + 4\zeta \tag{9.91}$$

Third order (j + k = 3)

$$\dot{m}_{30} = 3m_{21} \tag{9.92}$$

$$\dot{m}_{21} = 2m_{12} - 2\zeta m_{21} - 2\zeta\mu m_{23} - m_{30} - \rho m_{50} \tag{9.93}$$

$$\dot{m}_{12} = m_{03} - 4\zeta m_{12} - 4\zeta\mu m_{14} - 2m_{21} - 2\rho m_{41} + 4\zeta m_{10} \tag{9.94}$$

$$\dot{m}_{03} = -6\zeta m_{03} - 6\zeta\mu m_{05} - 3m_{12} - 3\rho m_{32} + 12\zeta m_{01} \tag{9.95}$$

Fourth order (j + k = 4)

$$\dot{m}_{40} = 4m_{31} \tag{9.96}$$

$$\dot{m}_{31} = 3m_{22} - 2\zeta m_{31} - 2\zeta\mu m_{33} - m_{40} - \rho m_{60} \tag{9.97}$$

$$\dot{m}_{22} = 2m_{13} - 4\zeta m_{22} - 4\zeta\mu m_{24} - 2m_{31} - 2\rho m_{51} + 4\zeta m_{20} \tag{9.98}$$

$$\dot{m}_{13} = m_{04} - 6\zeta m_{13} - 6\zeta\mu m_{15} - 3m_{22} - 3\rho m_{42} + 12\zeta m_{11} \tag{9.99}$$

$$\dot{m}_{04} = -8\zeta m_{04} - 8\zeta\mu m_{06} - 4m_{13} - 4\rho m_{33} + 24\zeta m_{02} \tag{9.100}$$

Equations for higher order moments can be derived in a similar fashion, but the results will not be given here.

An inspection of equations (9.87) to (9.100) reveals that the following moments are involved, which are of order higher than 4

$$m_{23}, m_{50}, m_{14}, m_{41}, m_{05}, m_{32}, m_{33}, m_{60}, m_{24}, m_{51}, m_{15}, m_{42}, m_{06}$$

All these quantities can be expressed in terms of quasi-moments, up to order 6, using the relationships given in Chapter 3. With closure at the fourth order, one simply sets quasi-moments of order 5 and 6 to zero. Then all the moments listed above can be expressed in terms of moments up to order 4. In this way equations (9.87) to (9.100) reduce to 14 non-linear, simultaneous, differential equations in 14 unknowns. These equations can be solved by means of a standard numerical integration algorithm.

The details of the analysis, for the non-stationary, transient response, are lengthy and will not be given here. Instead attention will be concentrated on the case of stationary response, where

$$\dot{m}_{jk} = 0 \quad j + k = 1, 2, \ldots \tag{9.101}$$

Here the analysis becomes much simpler. Firstly, it follows from equation (9.101) that equations (9.87) to (9.100) reduce to algebraic equations. Moreover, since the stationary response has a distribution which is symmetrical about a zero mean, it follows that all odd moments are zero. Thus

$$m_{10} = m_{01} = m_{30} = m_{21} = m_{12} = m_{03} = 0 \tag{9.102}$$

and equations (9.87), (9.88), and (9.92) to (9.95) are automatically satisfied. In addition, as previously shown, $m_{11} = m_{31} = m_{51} = 0$. Hence, the moment equations, up to order 4, reduce to the following

$$m_{02} - 2\zeta\mu m_{13} - m_{20} - \rho m_{40} = 0 \tag{9.103}$$

$$-4\zeta m_{02} - 4\zeta\mu m_{04} + 4\zeta = 0 \tag{9.104}$$

$$3m_{22} - 2\zeta\mu m_{33} - m_{40} - \rho m_{60} = 0 \tag{9.105}$$

$$2m_{13} - 4\zeta m_{22} - 4\zeta\mu m_{24} + 4\zeta m_{20} = 0 \tag{9.106}$$

$$m_{04} - 6\zeta m_{13} - 6\zeta\mu m_{15} - 3m_{22} - 3\rho m_{42} = 0 \tag{9.107}$$

$$-8\zeta m_{04} - 8\zeta\mu m_{06} - 4m_{13} - 4\rho m_{33} + 24\zeta m_{02} = 0 \tag{9.108}$$

To apply Gaussian closure to the above equations, one sets all quasi-moments to zero ($b_3 = b_4 = \cdots = 0$). Alternatively, and equivalently, cumulants of order greater than two may be set to zero. Hence, as before, equations (9.52) to (9.54) apply and equations (9.103) and (9.104) reduce to the closed pair

$$m_{02} - m_{20} - 3\rho m_{20}^2 = 0 \tag{9.109}$$

$$m_{02} + 3\zeta\mu m_{02}^2 - 1 = 0 \tag{9.110}$$

These results are, of course, identical to those given earlier, for Gaussian closure (see equations (9.55) and (9.56)), with a different notation. The resulting expression for $m_{20} = \sigma_Y^2$, given by equation (9.59), can be expanded in powers of μ and ρ, using the following expansion for the function $\phi(x)$, defined by equation (9.58)

$$\phi(x) = 1 - 3x + 18x^2 + \text{higher order terms} \tag{9.111}$$

Hence

$$\sigma_Y^2 = 1 - 3(\mu + \rho) + 18(\mu^2 + \rho^2 + \mu\rho) + \text{higher order terms} \tag{9.112}$$

Thus, for the special case of linear damping ($\mu = 0$) equation (9.112) gives

$$\sigma_Y^2 = 1 - 3\mu + 18\mu^2 + \text{higher order terms} \tag{9.113}$$

whereas, for linear stiffness ($\rho = 0$)

$$\sigma_Y^2 = 1 - 3\rho + 18\rho^2 + \text{higher order terms} \tag{9.114}$$

If terms of second order in μ and ρ are neglected then equation (9.112) reduces to a result given earlier, in Chapter 5 (see equation (5.154)). For large values of μ and ρ, the following asymptotic expressions for $\phi(x)$ can be used

$$\phi(x) \to (3x)^{-1/2} \quad x \to \infty \tag{9.115}$$

Hence

$$\sigma_Y^2 \to \frac{1}{3^{3/4}\mu^{1/4}\rho^{1/2}} \quad \mu, \rho \to \infty \tag{9.116}$$

If $\mu = 0$, corresponding to linear damping, then

$$\sigma_Y^2 \to \frac{1}{(3\rho)^{1/2}} = \frac{0.5774}{\rho^{1/2}} \tag{9.117}$$

whereas, if $\rho = 0$ corresponding to linear stiffness,

$$\sigma_Y^2 \to \frac{1}{(3\mu)^{1/2}} = \frac{0.5774}{\mu^{1/2}} \tag{9.118}$$

To obtain an improved estimate of σ_y^2, and other low-order moments, one can simply apply a fourth-order closure technique to the moment equations. Inspection of equations (9.103) to (9.108) shows that the following moments, of order higher than 4, are involved: $m_{60}, m_{33}, m_{24}, m_{42}, m_{15}$. These are all of sixth order, and can be expressed in terms of lower-order moments using either quasi-moment as cumulant closure. In this case both closure methods lead to the same relationships, given in Table 9.1. These relationships can be added to equations (9.103) to (9.108) to yield a total of eleven equations, in eleven unknowns. Setting

$$\begin{aligned} A &= m_{20} \quad B = m_{02} \quad C = m_{40} \\ D &= m_{22} \quad E = m_{13} \quad F = m_{04} \end{aligned} \tag{9.119}$$

and substituting for the appropriate expressions for the sixth-order moments into equations (9.103) to (9.108), the following set of algebraic equations is obtained

$$B - 2\zeta\mu E - A - \rho C = 0 \tag{9.120}$$

$$-B - \mu F + 1 = 0 \tag{9.121}$$

$$3D - 6\zeta\mu AE - C - \rho(15AC - 30A^3) = 0 \tag{9.122}$$

$$2E - 4\zeta D - 4\zeta\mu(6BD - 6AB^2 + AF) + 4\zeta A = 0 \tag{9.123}$$

$$F - 6\zeta E - 60\zeta\mu BE - 3D - 3\rho(6AD - 6BA^2 - BC) = 0 \tag{9.124}$$

$$-2\zeta F - 2\zeta\mu(15BF - 30B^3) - E - \rho(3AE) + 6\zeta B = 0 \tag{9.125}$$

The above closed set of equations for $A, B, \ldots, F$ must, in general, be solved numerically. However, for the special case of linear damping, where $\mu = 0$, some useful analytical reduction can be achieved. Thus, when $\mu = 0$, $B = 1$, from equation (9.121). Further, the remaining equations simplify to

$$1 - A - \rho C = 0 \tag{9.126}$$

$$3D - C - 15\rho A(C - 2A^2) = 0 \tag{9.127}$$

$$2E - 4\zeta(D - A) = 0 \tag{9.128}$$

$$F - 6\zeta E - 3D - 3\rho(6AD - 6A^2 + C) = 0 \tag{9.129}$$

$$-2\zeta F - E - 3\rho AE + 6\zeta = 0 \tag{9.130}$$

By inspection, it is observed that a solution is obtained if

$$A = D \quad E = 0 \quad F = 3 \tag{9.131}$$

Then equations (9.126) and (9.127) can be solved for A and C. It follows that

A is governed by the equation

$$30\rho^2 A^3 + 15\rho A^2 + (1 - 12\rho)A - 1 = 0 \tag{9.132}$$

and

$$C = \frac{1 - A}{\rho} \tag{9.133}$$

It is interesting to note, from equation (9.132) that $\sigma_Y^2 = A$ depends only on the non-linearity parameter ρ and (as in the case of Gaussian closure) is not explicitly dependent on ζ. For arbitrary values of ρ, equation (9.132) must be solved numerically. For small ρ the following asymptotic expression for σ_y^2 can be derived from equation (9.132) (see Wu and Lin, 1984)

$$\sigma_Y^2 = 1 - 3\rho + 24\rho^2 - 297\rho^3 + 453\rho^4 \tag{9.134}$$

Comparison of this result with the corresponding result for Gaussian closure (see equation (9.114)) shows that there is only agreement up to terms of order ρ. For large ρ, equation (9.132) yields the asymptotic result

$$\sigma_Y^2 \to \frac{0.6325}{\rho^{1/2}} \tag{9.135}$$

This, again, can be compared with the statistical linearization result equation (9.117).

The preceding analysis (for the linear damping case), can be extended to closure at sixth order. Thus, by neglecting *cumulants* above the sixth order it can be shown that (e.g. see Wu and Lin, 1984) $A = \sigma_Y^2$ is governed by the equation

$$630\rho^3 A^4 + 420\rho^2 A^3 + (63 - 336\rho)\rho A^2 + (1 - 90\rho)A - (1 - 30\rho) = 0 \tag{9.136}$$

This yields the following expansion for small ρ

$$\sigma_Y^2 = 1 - 3\rho + 24\rho^2 - 297\rho^3 + 489\rho^4 - 100\,278\rho^5 + \cdots \tag{9.137}$$

Comparing this with equation (9.134) it is seen that the two expressions are in agreement for terms up to order ρ^3. For large ρ, equation (9.136) gives the asymptotic result

$$\sigma_Y^2 \to \frac{0.6480}{\rho^{1/2}} \tag{9.138}$$

This can be compared with equation (9.135), corresponding to fourth-order closure, and equation (9.117) corresponding to Gaussian closure.

Figure 9.1 shows the variation of σ_Y^2 with ρ, according to Gaussian closure (statistical linearization), fourth-order closure and sixth-order (cumulant) closure. It is observed that the difference between the sixth-order and fourth-order results is very small, indicating that little improvement in accuracy is achieved by progressing beyond fourth-order closure. In Chapter 10 those

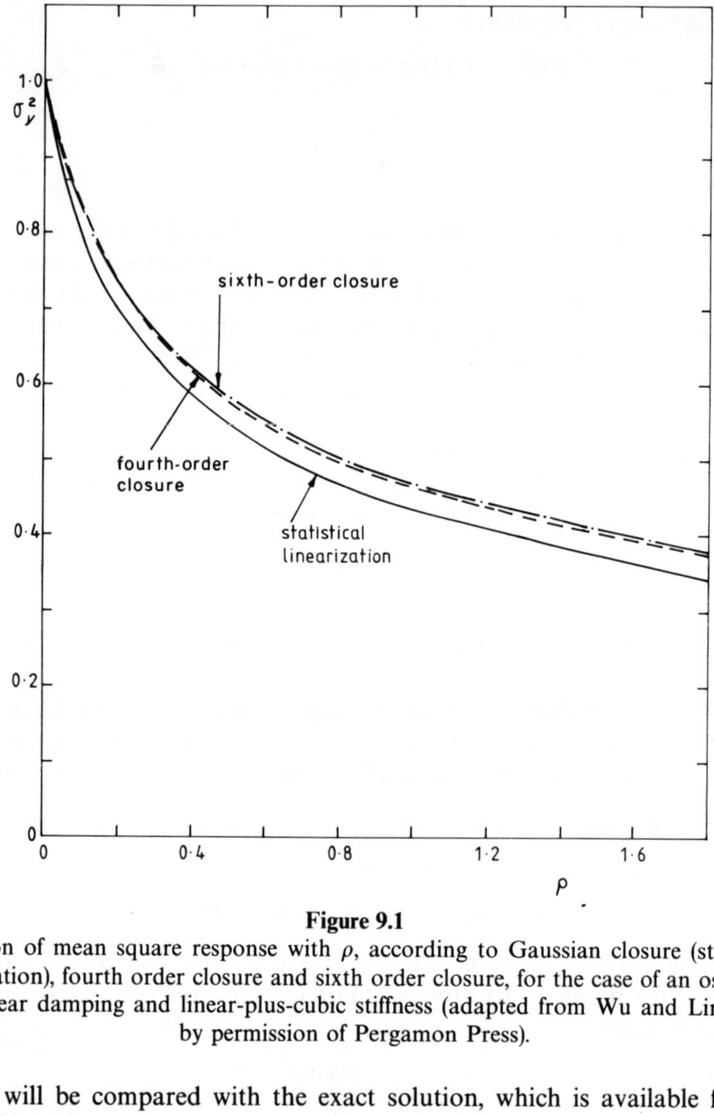

Figure 9.1
Variation of mean square response with ρ, according to Gaussian closure (statistical linearization), fourth order closure and sixth order closure, for the case of an oscillator with linear damping and linear-plus-cubic stiffness (adapted from Wu and Lin (1984) by permission of Pergamon Press).

results will be compared with the exact solution, which is available for this particular case.

Bover (1978) has applied both the quasi-moment and cumulant closure techniques to the present example, in the case of linear damping. As the closure order was increased he found that the results from the two methods began to differ, quasi-moment closure tending to become slightly more accurate than cumulant closure, when results were compared with the exact solution available for this case (see Chapter 10).

Returning to the more general case of non-linear damping and stiffness, then,

Table 9.2
Non-Gaussian closure, (fourth order) $\rho = 0$ (linear stiffness)

$\mu = 0.5$	$\zeta = 0.01$	$\zeta = 0.05$	$\zeta = 0.20$
A	0.5680	0.5687	0.5736
B	0.5679	0.5682	0.5703
C	0.8646	0.8742	0.9419
D	0.2882	0.2916	0.3119
E	-1.860×10^{-3}	-8.653×10^{-3}	-1.654×10^{-2}
F	0.8641	0.8635	0.8593
$\mu = 1.0$			
A	0.4543	0.4556	0.4614
B	0.4542	0.4548	0.4572
C	0.5464	0.5594	0.6178
D	0.1821	0.1861	0.2040
E	-1.804×10^{-3}	-7.858×10^{-3}	-1.037×10^{-2}
F	0.5458	0.5452	0.5428
$\mu = 5.0$			
A	0.2416	0.2445	0.2477
B	0.2415	0.2427	0.2441
C	0.1528	0.1666	0.1825
D	0.0509	0.0551	0.0599
E	-1.232×10^{-3}	-3.515×10^{-3}	-1.806×10^{-3}
F	0.1517	0.1515	0.1512

as pointed out earlier, a numerical solution of the appropriate moment equations (equations (9.120) to (9.125)), for fourth-order closure, is necessary. This appears to be true even in the special case of linear stiffness, but non-linear damping. Fortunately, standard library routines are readily available for solving simultaneous non-linear algebraic equations.

Table 9.2 shows a typical set of results obtained by solving equations (9.120) to (9.125) numerically, for the case of linear stiffness ($\rho = 0$). Unlike the linear damping case, σ_Y^2 is found to depend on ζ, as well as the non-linearity parameter (here μ). However, as the results in Table 9.2 show, the values of the moments, $A, B, \ldots, F$, asymptotically approach, limiting values as $\zeta \to 0$, and for small ζ, say $\zeta < 0.05$, the moment values are insensitive to the value of ζ. Moreover, the numerical results reveal that, as $\zeta \to 0$.

$$\begin{aligned} A \to B \quad C \to F \\ E \to O \quad 3D \to C \end{aligned} \tag{9.139}$$

Figure 9.2 shows the variation of σ_Y^2 with μ, according to Gaussian closure,

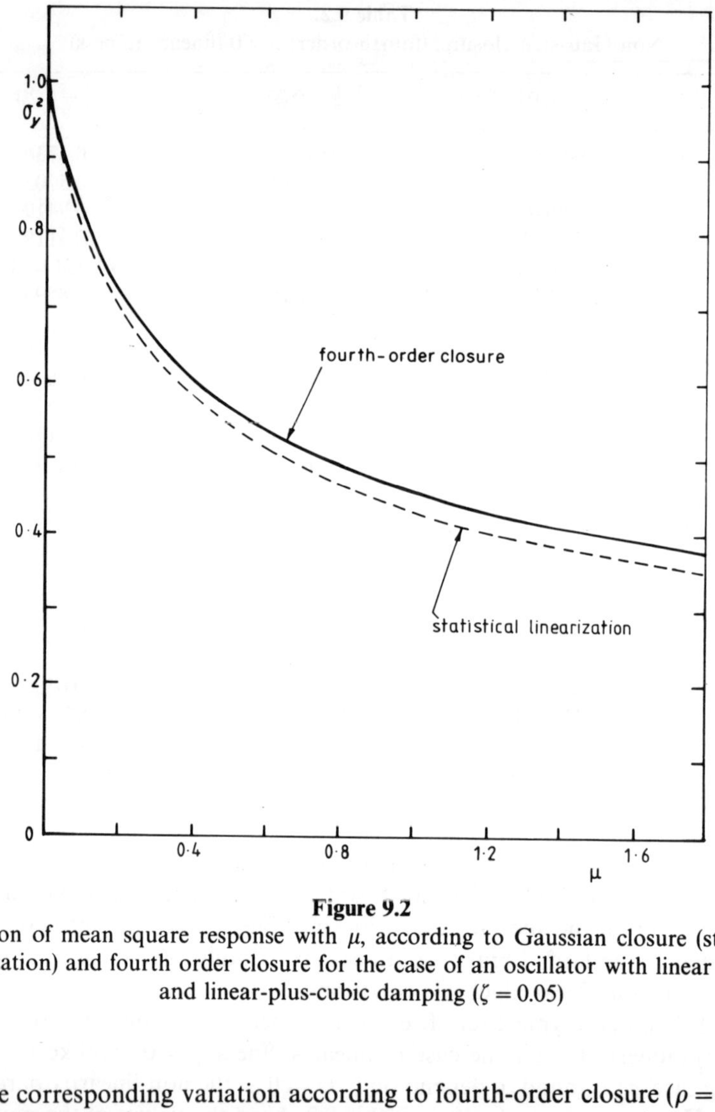

Figure 9.2
Variation of mean square response with μ, according to Gaussian closure (statistical linearization) and fourth order closure for the case of an oscillator with linear stiffness and linear-plus-cubic damping ($\zeta = 0.05$)

and the corresponding variation according to fourth-order closure ($\rho = 0$). It is observed that, similar to the result for linear damping, shown in Figure 9.1, there is a small but significant difference between the second-order and fourth-order closure results. Specifically the effect of including higher-order moments is to raise the value of σ_Y^2. One can expect, in the light of the results shown in Figure 9.1, that, for the linear stiffness case, the difference between fourth-order and sixth-order closure results will be very small.

Later in this chapter the results for the linear stiffness case, shown graphically in Figure 9.2, will be compared with corresponding results for this case, obtained by other methods of releasing the Gaussian response assumption, inherent in the normal statistical linearization procedure.

Finally, to illustrate the method of estimating probability density functions for the response, from a knowledge of the computed moments, corresponding to closure at some level M, the probability density function of the amplitude response, $Y = z_1$, will be considered here.

The appropriate expansion of this one-dimensional density function, in terms of quasi-moments, is given earlier, in Chapter 3 (see equation (3.70)). For Mth-order closure, $b_s = 0$ for $s > M$, and the truncated approximation for $f(Y)$ is as follows

$$f(Y) = \left[1 + \sum_{s=3}^{M} \frac{1}{s!}\frac{b_s}{\sigma^s} H_s\left(\frac{Y-m}{\sigma}\right)\right] f_G(Y) \tag{9.140}$$

where $f_G(Y)$ is the Gaussian density function (see equation (3.74)) and $\sigma \equiv \sigma_Y$.

In the present example, focusing attention on the stationary case again, the odd moments and quasi-moments are zero, as previously noted. Hence equation (9.140) reduces to

$$f(Y) = \left[1 + \sum_{s=4,6\ldots}^{M} \frac{1}{s!}\frac{b_s}{\sigma^s} H_s\left(\frac{Y}{\sigma}\right)\right] f_G(Y) \tag{9.141}$$

where

$$f_G(Y) = \frac{1}{(2\pi)^{1/2}\sigma} \exp\left(-\frac{Y^2}{2\sigma^2}\right) \tag{9.142}$$

Expressions for the first six Hermite polynomials, H_s, are given by equation (3.75). In the particular case of fourth-order closure ($M = 4$), using equation (3.75), it is found that equations (9.141) and (9.142) reduce to the following

$$f(\lambda) = \frac{1}{(2\pi)^{1/2}}\left(1 + \frac{b_4}{24\sigma^4}(\lambda^4 - 6\lambda^2 + 3)\right)\exp\left(-\frac{\lambda^2}{2}\right) \tag{9.143}$$

where

$$\lambda = Y/\sigma \tag{9.144}$$

and $f(\lambda) = f(Y)\mathrm{d}Y/\mathrm{d}\lambda = \sigma f(Y)$ is the density function of λ, rather than Y.

The quasi-moment, b_4, in equation (9.143) is, from equations (3.73) and (3.77)

$$b_4 = b_{40} = b_4(Y, Y, Y, Y) \tag{9.145}$$

and since $b_4 = \kappa_4$, or $b_{40} = \kappa_{40}$, equation (9.84) shows that

$$b_4 = m_{40} - 3m_{20}^2 = C - 3A^2 \tag{9.146}$$

where the notation of equation (9.119) is used again. Thus, noting that $\sigma^2 = A^2$,

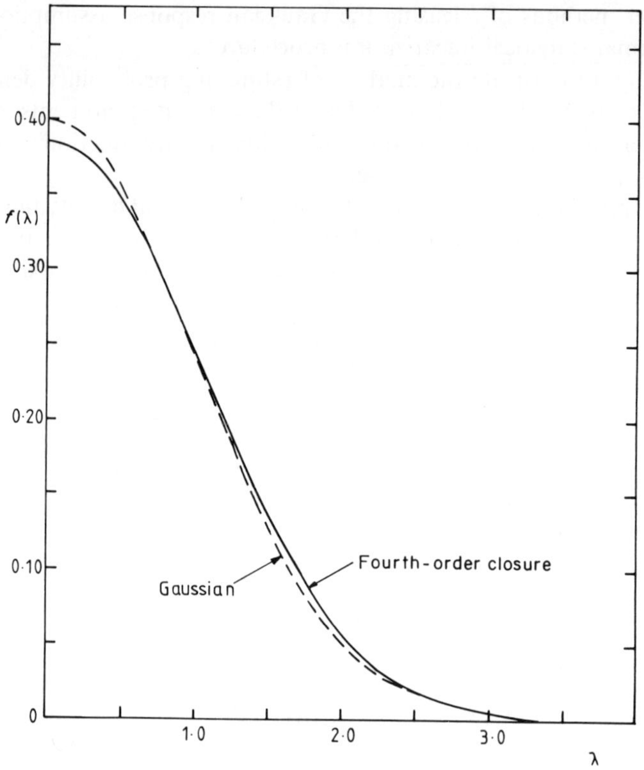

Figure 9.3
Probability density function for the displacement response of an oscillator with linear damping and linear-plus-cubic stiffness ($\rho = 1$). Comparison between the Gaussian distribution and the fourth-order closure result

equation (9.143) can be expressed as

$$f(\lambda) = \frac{1}{(2\pi)^{1/2}}\left[1 + \frac{1}{24}\left(\frac{C}{A^2} - 3\right)(\lambda^4 - 6\lambda^2 + 3)\right]\exp\left(-\frac{\lambda^2}{2}\right) \quad (9.147)$$

Clearly, in the case of Gaussian closure, $C = 3A^2$ and equation (9.147) reduces to

$$f(\lambda) = \frac{1}{(2\pi)^{1/2}}\exp\left(-\frac{\lambda^2}{2}\right) \quad (9.148)$$

Figure 9.3 shows the variation of $f(\lambda)$ with λ, according to equations (9.147) and (9.148), for the case of linear damping, with $\rho = 1$. A similar comparison is shown in Figure 9.4, for the case of linear stiffness, with $\mu = 1$. The results in Figure 9.4 were obtained for $\zeta = 0.05$. However, from equation (9.133) it is seen that the result shown in Figure 9.3 is actually independent of ζ.

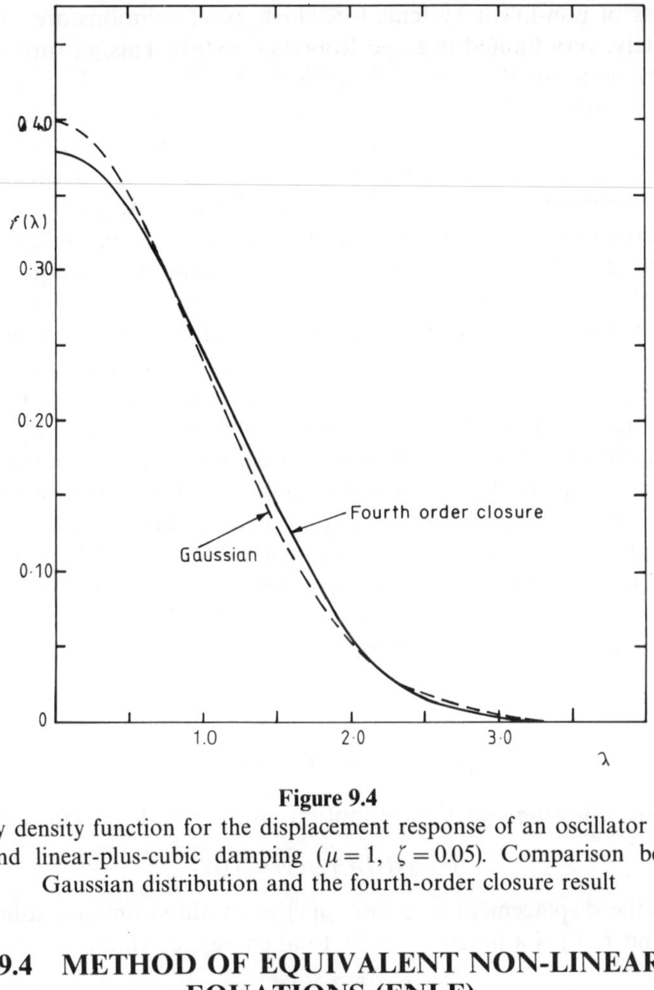

Figure 9.4
Probability density function for the displacement response of an oscillator with linear stiffness and linear-plus-cubic damping ($\mu = 1$, $\zeta = 0.05$). Comparison between the Gaussian distribution and the fourth-order closure result

9.4 METHOD OF EQUIVALENT NON-LINEAR EQUATIONS (ENLE)

Standard statistical linearization consists of replacing the original set of non-linear differential equations with an equivalent *linear* set, where the difference between the two sets is minimized, in a mean square sense. However, as previously noted, in the introduction to Chapter 6, the replacement, or auxiliary, set of equations need not necessarily be linear. Instead one can use a replacement set of *non-linear* equations, which belong to a class of problems which can be solved exactly. This approach will be referred to here as the equivalent non-linear equation (ENLE) method. The basic idea was originally suggested by Caughey, about two decades ago, and has been applied to particular problems by Kirk (1974), Lutes (1970b) and Caughey (1986a).

The class of non-linear systems for which exact solutions are available is, unfortunately, very limited (e.g. see Roberts (1981a)). This, in turn does place some restrictions on the range of applicability of the ENLE method. This restriction is particularly severe in the case of MDOF systems. Even for SDOF systems the only exact solution which is available relates to the stationary response an oscillator with non-linear stiffness and a very particular type of non-linear damping, excited by white noise. However, despite this limitation, very useful results can be obtained by the ENLE method. A particular advantage of the method is that, unlike the statistical linearization technique, it enables one to predict deviations from a Gaussian response distribution, due to the presence of non-linearities. The ability to predict non-Gaussian response distributions suggests that the method is more accurate than statistical linearization, and specific examples certainly support this conjecture. In particular, the ENLE method, where it is applicable, is very likely to lead to improved estimates of system reliability, since such estimates are very sensitive to the precise shape of the response distribution, at high amplitude levels.

To date the method has only been applied to stationary response of SDOF systems and, accordingly, only this class of problems will be considered in detail here. It is noted, however, that there appears to be no reason, in principle, why some (may be rather limited) results can not be obtained for the stationary response of MDOF systems, using the exact specific solutions which are available.

9.4.1 Exact solution

Consider the following equation of motion for an SDOF system

$$\ddot{y} + \dot{y}H(E) + g(y) = f(t) \tag{9.149}$$

where y is the displacement response, $g(y)$ is an antisymmetric stiffness non-linearity, and $H(E)$ is a function of the total energy, or Hamiltonian function

$$E = \tfrac{1}{2}\dot{y}^2 + V(y) \tag{9.150}$$

Here the first term on the right-hand side of the above expression represents the kinetic energy of the oscillator and

$$V(y) = \int_0^y g(\xi)\,\mathrm{d}\xi \tag{9.151}$$

is the potential energy function. It will be assumed, henceforth, that the excitation process, $f(t)$, has zero mean.

Suppose that the excitation process, $f(t)$, in equation (9.149) is approximated as an ideal, stationary white noise process. Thus

$$E\{f(y)f(t+\tau)\} = 2\pi S_0\delta(\tau) \tag{9.152}$$

where S_0 is the constant spectral level of $f(t)$. Then, as Caughey (1964) has shown, an exact solution for the probability density function, $f(y, \dot{y})$, of the joint stationary process, $[y(t)\dot{y}(t)]$, can be derived. The result is

$$f(y, \dot{y}) = C \exp\left(-\frac{1}{\pi S_0} \int_0^E H(\xi)\, d\xi \right) \tag{9.153}$$

where C is a normalization constant, determined by the condition

$$\int_{-\infty}^{\infty} \int_{-\infty}^{\infty} f(y, \dot{y})\, dy\, d\dot{y} = 1 \tag{9.154}$$

In the special case of linear damping, but non-linear stiffness, one can set $H(E) = \beta$, a constant, in equation (9.149). Equation (9.153) then reduces to the following well known exact result, originally due to Kramers (1940)

$$f(y, \dot{y}) = C \exp\left(-\frac{\beta E}{\pi S_0} \right) \tag{9.155}$$

If both the damping and stiffness are linear one can set

$$g(y) = \omega_n^2 y \tag{9.156}$$

where ω_n is the undamped natural frequency. Equation (9.153) then simplifies to

$$f(y, \dot{y}) = C \exp\left[-\frac{1}{2}\left(\frac{y^2}{\sigma_y^2} + \frac{\dot{y}^2}{\sigma_{\dot{y}}^2} \right) \right] \tag{9.157}$$

where

$$\sigma_y^2 = \frac{\pi S_0}{\beta \omega_n^2} \qquad \sigma_{\dot{y}}^2 = \frac{\pi S_0}{\beta} \tag{9.158}$$

and C is given by

$$C = (2\pi \sigma_y \sigma_{\dot{y}})^{-1} \tag{9.159}$$

Equation (9.157) is recognized as a bivariate Gaussian probability density function and the expressions for σ_y and $\sigma_{\dot{y}}$, the standard deviations of y and $\dot{y}$, respectively, agree with those obtained earlier, in Chapter 4, by different methods. Equation (9.157) also shows that, for the stationary response case, y and $\dot{y}$ are statistically independent processes. Again this agrees with earlier findings. It is interesting to note, from equation (9.155), that y and $\dot{y}$ are still statistically independent if non-linear stiffness is introduced, but, from equation (9.153), not independent in the case of non-linear damping.

Returning to the more general case, where $f(y, \dot{y})$ is given by equation (9.153), it is observed that many statistical parameters, and related response distributions, can be deduced from this result. In particular the probability density function, $f(A)$, of the amplitude response process $A(t)$ can be found, where $A(t)$ is defined by

$$V(A) = E \tag{9.160}$$

From equation (9.153) it can be shown that (Caughey and Ma, 1982)

$$f(A) = C_1 g(A) \exp\left(-\frac{1}{\pi S_0} \int_0^{V(A)} H(\xi)\,\mathrm{d}\xi \right) T(A) \tag{9.161}$$

where C_1 is another normalization constant and $T(A)$ is the period of free oscillations of the undamped oscillator

$$\ddot{y} + g(y) = 0 \tag{9.162}$$

with initial conditions

$$y(0) = A \quad \dot{y}(0) = 0 \tag{9.163}$$

In the case of linear stiffness, but non-linear damping, where $g(y)$ is given by equation (9.156), it is clear that $T(A)$ is a constant. That is

$$T(A) = \frac{2\pi}{\omega_{\mathrm{n}}} \tag{9.164}$$

Hence, from equation (9.161)

$$f(A) = C_2 A \exp\left(-\frac{1}{\pi S_0} \int_0^{V(A)} H(\xi)\,\mathrm{d}\xi \right) \tag{9.165}$$

where C_2 is another normalization constant. Also, in this case, from equation (9.160)

$$\frac{\omega_{\mathrm{n}}^2 A^2}{2} = \frac{\dot{y}^2}{2} + \frac{\omega_{\mathrm{n}}^2 y^2}{2} \tag{9.166}$$

or

$$A = \left(\frac{\dot{y}^2}{\omega_{\mathrm{n}}^2} + y^2 \right)^{1/2} \tag{9.167}$$

Hence one can write

$$y = A \cos(\omega_{\mathrm{n}} t + \theta) \tag{9.168}$$

$$\dot{y} = -\omega_{\mathrm{n}} A \sin(\omega_{\mathrm{n}} t + \theta) \tag{9.169}$$

It is noted that the process $A(t)$ is, in this case, identical to that introduced in Chapter 8, in the context of hysteretic systems (see equations (8.11) and (8.12)). Moreover, for the linear stiffness case, the density function, $f(A, \theta)$ of the joint process $[A(t), \theta(t)]$ is related to $f(y, \dot{y})$ as follows

$$f(y, \dot{y})\,\mathrm{d}y\,\mathrm{d}\dot{y} = f(A, \theta)\,\mathrm{d}A\,\mathrm{d}\theta \tag{9.170}$$

From equations (9.168) or (9.169), one has the usual polar transformation relationship

$$\mathrm{d}y\,\mathrm{d}\dot{y} = A\,\mathrm{d}A\,\mathrm{d}\theta \tag{9.171}$$

Combining equations (9.153), (9.170) and (9.171) yields

$$f(A,\theta) = CA\exp\left(-\frac{1}{\pi S_0}\int_0^{V(A)} H(\xi)\,\mathrm{d}\xi\right) \tag{9.172}$$

and, from equation (9.165)

$$f(A,\theta) = C_3 f(A) \tag{9.173}$$

Since

$$f(A) = \int_0^{2\pi} f(A,\theta)\,\mathrm{d}\theta \tag{9.174}$$

it is evident that the normalization constant, C_3, is equal to $1/2\pi$. Thus

$$f(A,\theta) = \frac{f(A)}{2\pi} \tag{9.175}$$

with the implication that the phase angle, θ, is uniformly distributed between 0 and 2π.

If the damping is also linear, $H(E) = \beta$ and equation (9.165) reduces to

$$f(A) = C_2 A\exp\left(-\frac{\beta\omega_{\mathrm{n}}^2 A^2}{2\pi S_0}\right) \tag{9.176}$$

or

$$f(A) = \frac{A}{\sigma_y^2}\exp\left(-\frac{A^2}{2\sigma_y^2}\right) \tag{9.177}$$

This is the well known Rayleigh density function, also introduced in Chapter 8 (see equation (8.44)).

9.4.2 Equivalent non-linear equations

A wide class of non-linear oscillators with white noise excitation are governed by an equation of motion of the form

$$\ddot{y} + b(y,\dot{y}) + g(y) = f(t) \tag{9.178}$$

where $b(y,\dot{y})$ is a general type of non-linear damping. Unfortunately, in nearly every case of practical interest, $b(y,\dot{y})$ can not be expressed as $\dot{y}H(E)$, as indicated in equation (9.149). For example, consider the case of simple, quadratic damping, where

$$b(y,\dot{y}) \propto |\dot{y}|\dot{y} \tag{9.179}$$

It is clear that this can not be expressed as $\dot{y}H(E)$, even in the case of linear stiffness.

To overcome this difficulty, one can proceed by replacing equation (9.178) with an equivalent non-linear equation, of the form of equation (9.149), where the error term

$$\varepsilon = b(y, \dot{y}) - \dot{y}H(E) \tag{9.180}$$

is minimized, in a suitable fashion. As with statistical linearization, it seems best to minimize the mean square of ε, $E\{\varepsilon^2\}$.

If $b(y, \dot{y})$ has a simple form then it is possible to proceed by setting

$$H(E) = cH_0(E) \tag{9.181}$$

where $H_0(E)$ is a suitably chosen function of E and c is a constant. The value of c can be determined from the minimization condition

$$\frac{\mathrm{d}\,E\{\varepsilon^2\}}{\mathrm{d}c} = 0 \tag{9.182}$$

Hence, from equations (9.180) and (9.182), one obtains

$$c = \frac{E\{\dot{y}H_0(E)b(y, \dot{y})\}}{E\{\dot{y}^2H_0^2(E)\}} \tag{9.183}$$

With $H_0(E) = 1$, equation (9.183) reduces to

$$c = \frac{E\{\dot{y}b(y, \dot{y})\}}{E\{\dot{y}^2\}} \tag{9.184}$$

which is the usual standard statistical linearization expression for the damping coefficient (see equation (5.120)), noting that here the non-linear stiffness term may be retained. However, one can expect accuracy to improve if $H_0(E)$ is chosen such that its form matches that of $b(y, \dot{y})$, as closely as possible.

If the damping is of power-law type then, generally, $b(y, \dot{y})$ may be expressed as

$$b(y, \dot{y}) = b_{rs}|y^r\dot{y}^s|\,\mathrm{sgn}\,(\dot{y}) \tag{9.185}$$

where b_{rs} is a constant, r and s are integers and $\mathrm{sgn}\,(\dot{y})$ is the signum function, such that $\mathrm{sgn}\,(\dot{y}) = -1$ if $\dot{y} < 0$ and $+1$ if $\dot{y} > 0$. Taking $\dot{y}$ to be the same order as y, the above damping term is of order y^{r+s}. This may be compared with E, which is of order y^2. To ensure that $\dot{y}H(E)$ is of the same order in y as $b(y, \dot{y})$, in equation (9.185), one can select

$$H_0(E) \equiv H_{rs}(E) = (2E)^{(r+s-1)/2} \tag{9.186}$$

Other choices of $H_0(E)$ are, of course, possible but experience with specific examples indicates that equation (9.186) leads to very good approximations.

It is now possible to evaluate the coefficient c, according to equation (9.183). One has

$$c \equiv c_{rs} = \frac{b_{rs} E\{\dot{y}(2E)^{(r+s-1)/2} |y^r \dot{y}^s| \operatorname{sgn}(\dot{y})\}}{E\{\dot{y}^2 (2E)^{r+s-1}\}} \tag{9.187}$$

The expectations can be calculated using the expression for the joint density function, $f(y, \dot{y})$, given by equation (9.153). Thus

$$E\{\dot{y}^2(2E)^{r+s-1}\} = \int_{-\infty}^{\infty} \int_{-\infty}^{\infty} \dot{y}^2 (2E)^{r+s-1} f(y, \dot{y})\, \mathrm{d}y\, \mathrm{d}\dot{y} \tag{9.188}$$

and similarly for the expectation in the numerator of equation (9.187). Once c is determined then, from equations (9.153) and (9.187)

$$f(y, \dot{y}) = C \exp\left(-\frac{c}{\pi S_0} \int_0^E H_0(\xi)\, \mathrm{d}\xi \right) \tag{9.189}$$

can be used as an approximation to the true probability density function for y and $\dot{y}$, in the stationary case.

The foregoing method can be extended fairly easily to cope with more complex forms of $b(y, \dot{y})$. Suppose $b(y, \dot{y})$ consists of a sum of terms, each of the form given by equation (9.185). Then

$$b(y, \dot{y}) = \sum_{r,s}\sum b_{rs} |y^r \dot{y}^s| \operatorname{sgn}(\dot{y}) \tag{9.190}$$

where r, s belong to a set, Ω. Further in line with the preceding discussion for the single term case, it is sensible to choose $H(E)$ as follows

$$H(E) = \sum_{r,s}\sum c_{rs} H_{rs}(E) \tag{9.191}$$

where H_{rs} is defined by equation (9.186) and the summation range, Ω, is identical to that in equation (9.190).

The error ε, defined by equation (9.180) is now given by

$$\varepsilon = \sum_{r,s}\sum [b_{rs} |y^r \dot{y}^s| \operatorname{sgn}(\dot{y}) - c_{rs} \dot{y} H_{rs}(E)] \tag{9.192}$$

The mean square of this error can be minimized with respect to all the coefficients, c_{rs}. Thus one sets

$$\frac{\partial E\{\varepsilon^2\}}{\partial c_{rs}} = 0 \tag{9.193}$$

over the whole range, Ω, of r and s. Using the fact that the linear operations of differentiation and expectation may be commuted, equation (9.193) can be rewritten as

$$E\left\{\frac{\partial \varepsilon^2}{\partial c_{rs}}\right\} = 0 \tag{9.194}$$

Hence, from equation (9.192), one obtains a set of equations of the form

$$E\{\dot{y}H_{rs}\sum_{i,j}\sum [b_{ij}|y^i\dot{y}^j|\operatorname{sgn}(\dot{y}) - c_{ij}\dot{y}H_{ij}]\} = 0 \quad r, s \subset \Omega \tag{9.195}$$

The expectation in this last equation can be evaluated using the expression for $f(y, \dot{y})$ given by equation (9.153), where $H(E)$ is given by equation (9.191). Hence a set of algebraic equations for the coefficients c_{rs} can be generated, which can be solved numerically.

9.4.3 Oscillators with linear stiffness and non-linear damping

The calculations necessary to obtain the equivalent non-linear equation are much simplified in the special case of linear stiffness, but non-linear damping.

Consider the case where $b(y, \dot{y})$ is given by the power-sum form of equation (9.190). Then, it has been shown that the coefficients c_{rs} in equation (9.191) are determined by the set of equations given by equation (9.195). For linear stiffness, $g(y) = \omega_n^2 y$, y and $\dot{y}$ can be expressed in polar form, as given by equations (9.168) and (9.169). Moreover

$$E = \frac{\omega_n^2 A^2}{2} \tag{9.196}$$

and hence, from equation (9.186)

$$H_{rs}(E) = (\omega_n A)^{r+s-1} \tag{9.197}$$

Substituting these expression into equation (9.195) yields

$$E\left\{A^{r+s}\sum_{i,j}\sum(\omega_n A)^{i+j}[b_{ij}|\cos^i \aleph \sin^{j+1}\aleph| - c_{ij}\sin^2\aleph]\right\} = 0 \quad r, s \subset \Omega \tag{9.198}$$

where

$$\aleph = \omega_n t + \theta \tag{9.199}$$

The evaluation of the expectation here is simplified considerably if use is made of the fact that A and θ (and hence $\aleph$) are statistically independent. Hence equation (9.198) can be expressed as

$$\sum_{i,j}\sum E\{A^{r+s}\omega_n^j A^{i+j}\}E\{b_{ij}|\cos^i\aleph \sin^{j+1}\aleph| - c_{ij}\omega_n^i \sin^2\aleph\} = 0 \quad r, s \subset \Omega \tag{9.200}$$

This set of equations is satisfied by choosing

$$c_{rs} = b_{rs}\frac{E\{|\cos^r \aleph \sin^{s+1}\aleph|\}}{\omega_{\mathrm{n}}^r E\{\sin^2\aleph\}} \qquad r, s \subset \Omega \tag{9.201}$$

Moreover, since θ, and hence χ, is uniformly distributed, equation (9.201) can be written, alternatively, as

$$c_{rs} = \frac{b_{rs}}{\omega_{\mathrm{n}}^r}\frac{\displaystyle\int_0^{\pi/2}\cos^r\chi\cos^{s+1}\chi\,\mathrm{d}\chi}{\displaystyle\int_0^{\pi/2}\sin^2\chi\,\mathrm{d}\chi} \tag{9.202}$$

The integral in the numerator can be expressed in terms of the Gamma function, $\Gamma(\ )$. Hence

$$c_{rs} = \frac{2b_{rs}}{\pi\omega_{\mathrm{n}}^r}\frac{\Gamma\left(\dfrac{s+2}{2}\right)\Gamma\left(\dfrac{r+1}{2}\right)}{\Gamma\left(\dfrac{r+s+3}{2}\right)} \tag{9.203}$$

Once c_{rs} is determined then the probability density function for $A(t)$ may be found from equation (9.165), where here $H(E)$ is given by equation (9.191). Therefore

$$f(A) = C_2 A \exp\left(-\frac{1}{\pi S_0}\sum_{r,s}\sum c_{rs}\int_0^{V(A)} H_{rs}(\xi)\,\mathrm{d}\xi\right) \tag{9.204}$$

From this, the joint density function $f(A, \theta)$ may be determined from equation (9.175), and the joint density of $f(y, \dot{y})$ from (see equations (9.170) or (9.171))

$$f(y, \dot{y}) = \frac{f(A, \theta)}{A} = \frac{f(A)}{2\pi A} \tag{9.205}$$

These relationships facilitate the evaluation of the moments of $y(t)$ and $\dot{y}(t)$. For example, the mean square of y, σ_y^2, is given by

$$\sigma_y^2 = \int_{-\infty}^{\infty}\int_{-\infty}^{\infty} y^2 f(y, \dot{y})\,\mathrm{d}y\,\mathrm{d}\dot{y} \tag{9.206}$$

Writing $y = A\cos\theta$, and replacing $f(y, \dot{y})\,\mathrm{d}y\,\mathrm{d}\dot{y}$ by $f(A, \theta)\,\mathrm{d}A\,\mathrm{d}\theta$, one has

$$\sigma_y^2 = \int_{-0}^{2\pi}\int_0^{\infty} A^2\cos^2\theta\, f(A, \theta)\,\mathrm{d}A\,\mathrm{d}\theta = \frac{1}{2}\int_0^{\infty} A^2 f(A)\,\mathrm{d}A = \frac{1}{2}E\{A^2\} \tag{9.207}$$

Here use has been made of equation (9.205). For the special case where $A(t)$ is Rayleigh distributed, this result was obtained earlier, in Chapter 8, (see equation (8.46)).

9.4.4 Oscillators with quadratic damping

As a simple specific example of the analysis presented in the preceding section, consider an oscillator with quadratic damping and linear stiffness. The appropriate equation of motion is

$$\ddot{y} + \lambda|\dot{y}|\dot{y} + \omega_n^2 y = f(t) \tag{9.208}$$

where λ is the quadratic damping coefficient. Thus $b(y, \dot{y}) = \lambda|\dot{y}|\dot{y}$ and comparing this with the general form of equation (9.190) it is seen that there is only one relevant term in the summation, namely

$$b(y, \dot{y}) = b_{02}|\dot{y}^2|\operatorname{sgn}(\dot{y}) \tag{9.209}$$

where $b_{02} = \lambda$. Thus the appropriate form for $H(E)$ is, from equations (9.186) and (9.19),

$$H(E) = c_{02}H_{02}(E) = c_{02}(2E)^{1/2} \tag{9.210}$$

Further, the equivalent non-linear equation is as follows

$$\ddot{y} + c_{02}\dot{y}(2E)^{1/2} + \omega_n^2 y = f(t) \tag{9.211}$$

The coefficient c_{02} is readily obtained by using equation (9.203), with $r = 0, s = 2$. Hence

$$c_{02} = \frac{2\lambda}{\pi}\frac{\Gamma(2)\Gamma(\frac{1}{2})}{\Gamma(\frac{5}{2})} = \frac{8\lambda}{3\pi} \tag{9.212}$$

Moreover, using equations (9.204) and (9.212), and evaluating the normalization constant C, the following explicit expression for the density function of $A(t)$ is obtained

$$f(A) - \frac{3A}{\Gamma(\frac{2}{3})}\left(\frac{8\lambda\omega_n^3}{9\pi^2 S_0}\right)^{2/3} \exp\left(\frac{8\lambda\omega_n^3}{9\pi^2 S_0}A^3\right) \tag{9.213}$$

It follows, from equations (9.167) and (9.205) that

$$f(y, \dot{y}) = \frac{3}{2\pi\Gamma(\frac{2}{3})}\left(\frac{8\lambda\omega_n^3}{9\pi^2 S_0}\right)\exp\left(-\frac{8\lambda\omega_n}{9\pi^2 S_0}(\dot{y}^2 + y^2\omega_n^2)^{3/2}\right) \tag{9.214}$$

Using equation (9.207) and performing the necessary integration one finds that

$$\sigma_y^2 = \frac{\Gamma(\frac{4}{3})}{2\omega_n^2\Gamma(\frac{2}{3})}\left(\frac{3\pi}{2\sqrt{2}}\right)^{4/3}\left(\frac{S_0}{\lambda}\right)^{2/3} = \frac{1.6417}{\omega_n^2}\left(\frac{S_0}{\lambda}\right)^{2/3} \tag{9.215}$$

The results may be compared with the corresponding results obtained by the standard statistical linearization method. In this technique one replaces equation (9.208) by

$$\ddot{y} + \beta_{eq}\dot{y} + \omega_n^2 y = f(t) \tag{9.216}$$

where, from equations (5.204) and (5.207), here

$$\beta_{\text{eq}} = \left(\frac{8}{\pi}\right)^{1/2} \lambda \sigma_{\dot{y}} \tag{9.217}$$

From equation (9.216)

$$\sigma_{\dot{y}}^2 = \frac{\pi S_0}{\beta_{\text{eq}}} \quad \sigma_y^2 = \frac{\pi S_0}{\beta_{\text{eq}} \omega_{\text{n}}^2} \tag{9.218}$$

Hence one finds that

$$\sigma_y^2 = \frac{\pi}{2\omega_{\text{n}}^2} \left(\frac{S_0}{\lambda}\right)^{2/3} = \frac{1.5708}{\omega_{\text{n}}^2} \left(\frac{S_0}{\lambda}\right)^{2/3} \tag{9.219}$$

Moreover, the amplitude response determined by relying on equation (9.216) is Rayleigh distributed, and the density function is thus given by equation (8.44).

The mean square estimates given by equations (9.215) and (9.219) are seen to be in very close agreement, the difference being approximately 4%. However, the probability density function for the displacement amplitude response, given by equation (9.213) differs significantly from the Rayleigh distribution, in the 'tail' of the distribution. For the purpose of a graphical comparison it is convenient to examine the density function of the normalized amplitude response displacement,

$$\alpha = \frac{A}{\sigma_y} \tag{9.220}$$

From equations (9.213) and (9.215) one finds that

$$f(\alpha) = \frac{3k^{2/3}\alpha}{\Gamma(\frac{2}{3})} \exp(-k\alpha^3) \tag{9.221}$$

where

$$k = \left(\frac{\Gamma(\frac{4}{3})}{2\Gamma(\frac{2}{3})}\right)^{3/2} \tag{9.222}$$

For the corresponding Rayleigh distribution one has

$$f(\alpha) = \alpha \exp\left(-\frac{\alpha^2}{2}\right) \tag{9.223}$$

Figure 9.5 shows a comparison between these density functions, plotted on log–linear scale. There is a marked deviation in the probability of the response reaching high amplitudes, according to the two results. One would expect, on physical grounds, that the non-linearity in damping would have the effect of inhibiting large amplitude displacement, and this effect is certainly exhibited by the non-Gaussian result.

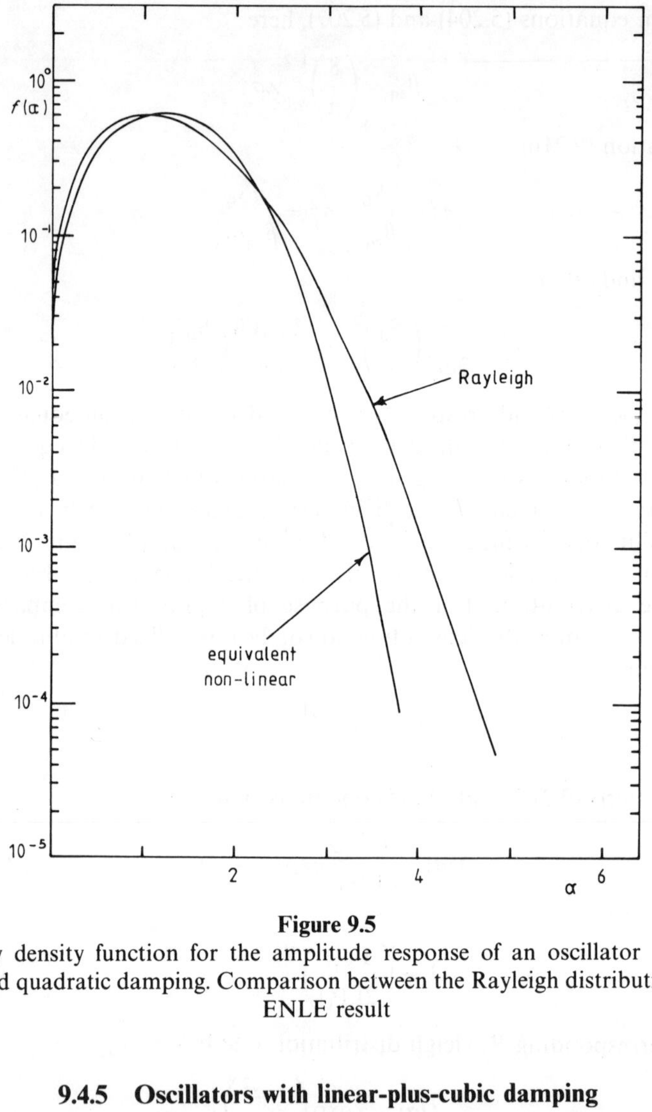

Figure 9.5
Probability density function for the amplitude response of an oscillator with linear stiffness and quadratic damping. Comparison between the Rayleigh distribution and the ENLE result

9.4.5 Oscillators with linear-plus-cubic damping

As another example the case of an oscillator with linear-plus-cubic damping, but linear stiffness, will be considered. A suitable, normalized form of the equation of motion has already been given (see equation (9.25)). Here $\rho = 0$, so

$$\ddot{Y} + 2\zeta(\dot{Y} + \mu \dot{Y}^3) + Y = (4\zeta)^{1/2} f_n(\tau) \qquad (9.224)$$

where $f_n(\tau)$, defined by equation (9.26), has unit strength (see equation (9.27)).
Comparing the damping term $2\zeta(\dot{Y} + \mu \dot{Y}^2)$ with the general expression of

equation (9.190) it is seen that here there are two relevant terms in the summation, i.e.

$$b(y, \dot{y}) = b_{01}\dot{y} + b_{03}\dot{y}^3 \tag{9.225}$$

where

$$b_{01} = 2\zeta \quad b_{03} = 2\zeta\mu \tag{9.226}$$

The appropriate form of $H(E)$ is thus, from equations (9.186) and (9.191)

$$H(E) = c_{01} + 2c_{03}E \tag{9.227}$$

Hence the equivalent non-linear equation is as follows

$$\ddot{Y} + (c_{01} + 2c_{03}E)\dot{Y} + Y = (4\zeta)^{1/2} f_n(\tau) \tag{9.228}$$

The coefficients c_{01} and c_{03} are readily determined from equation (9.203). Hence

$$c_{01} = 2\zeta \quad c_{03} = \frac{2}{\pi} b_{03} \frac{\Gamma(\frac{5}{2})\Gamma(\frac{1}{2})}{\Gamma(2)} = \frac{3\zeta\mu}{2} \tag{9.229}$$

Employing equation (9.204), the following expression for $f(A)$ can now be written (using the fact that here $\pi S_0 = 2\zeta$)

$$f(A) = C_2 A \exp\left[-\left(\frac{A^2}{2} + \frac{3\mu A^4}{16} \right) \right] \tag{9.230}$$

The normalization constant C_2 here is expressible in terms of the error function; thus

$$C_2 = [\pi^{1/2} \theta \exp(\theta^2)(1 - \exp\theta)]^{-1} \tag{9.231}$$

where

$$\theta = \left(\frac{1}{3\mu} \right)^{1/2} \tag{9.232}$$

As μ becomes large, C_2 approaches the asymptotic limit

$$C_2 \to 1/(\pi^{1/2}\theta) \quad \mu \to \infty \tag{9.233}$$

Hence, by integration, using equation (9.207) again, one obtains

$$\sigma_y^2 = 2\theta\{[\pi^{1/2} \exp(\theta^2)(1 - \exp\theta)]^{-1} - \theta\} \tag{9.234}$$

and as $\mu \to \infty$ one finds that

$$\sigma_y^2 \to \frac{2}{\pi^{1/2}} \theta \tag{9.235}$$

Once again the joint density of Y and $\dot{Y}$, $f(Y, \dot{Y})$, can be determined from equation (9.205). Hence

$$f(Y, \dot{Y}) = \frac{C_2}{2\pi} \exp\left[-\frac{1}{2}\left((\dot{Y}^2 + Y^2) + \frac{3\mu}{8}(\dot{Y}^2 + Y^2)^2 \right) \right] \tag{9.236}$$

Table 9.3
Linear-plus cubic damping. Mean-square estimates

μ	*Statistical linearization*	*Fourth order closure* ($\zeta = 0.05$)	*ENLE*
1	0.4342	0.4556	0.4603
2	0.3333	0.3537	0.3584
3	0.2824	0.3017	0.3058
4	0.2500	0.2683	0.2720
5	0.2270	0.2445	0.2479
6	0.2096	0.2263	0.2294
7	0.1957	0.2219	0.2147
8	0.1844	0.2000	0.2025
9	0.1748	0.1899	0.1923
10	0.1667	0.1813	0.1835

From this, $f(Y)$ can be determined by integration, i.e.

$$f(Y) = \int_{-\infty}^{\infty} f(Y, \dot{Y})\,\mathrm{d}\dot{Y} \tag{9.237}$$

Hence

$$f(Y) = \frac{C_2 h(Y)}{2\pi} \exp\left[-\frac{1}{2}\left(Y^2 + \frac{3\mu}{8} Y^4 \right)\right] \tag{9.238}$$

where

$$h(Y) = 2\int_0^{\infty} \exp\left[-\frac{1}{2}\left(x^2 + \frac{3\mu}{8}(x^4 + 2x^2 Y^2) \right)\right] \mathrm{d}x \tag{9.239}$$

This last integral can be evaluated most easily by numerical means.

The above results for σ_Y^2 and $f(Y)$ can be compared with the corresponding results for this system obtained earlier in this chapter, using closure methods, and with results from standard statistical linearization. For σ_Y^2 a comparison of values, for various values of μ, is given in Table 9.3. It is seen that there is very close agreement between fourth-order closure results and ENLE results, the difference being approximately 1%.

Figure 9.6 shows a typical comparison between the density function $f(\lambda) = \sigma f(Y)$ (where λ is the scaled response variable, defined by equation (9.144)) computed from equation (9.238) and the corresponding closure results, i.e. second-order (Gaussian) closure and the fourth-order closure result given by equation (9.143). Here $\mu = 1, \zeta = 0.05$ and the graph is a log–linear. It is noted that the ENLE result is independent of the ζ value, whereas the fourth-order closure result is insensitive to, but not entirely independent of, this value (e.g. see Table 9.2). Once again it is seen that the ENLE result agrees fairly well with the fourth-order closure result. In fact, the fourth-order result does

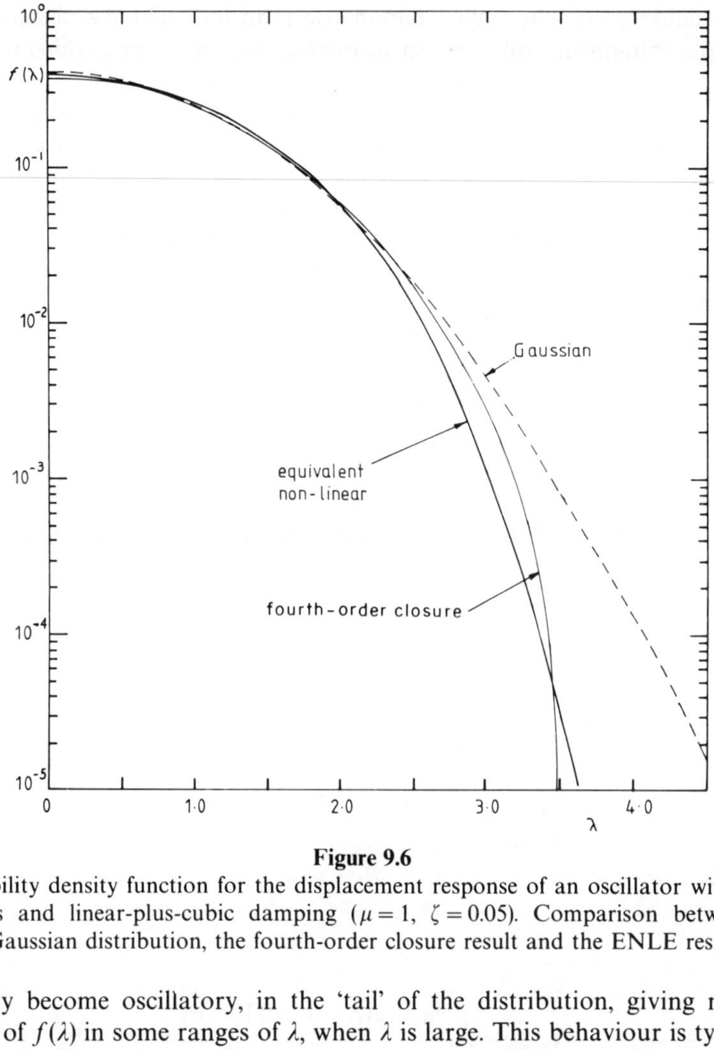

Figure 9.6
Probability density function for the displacement response of an oscillator with linear stiffness and linear-plus-cubic damping ($\mu = 1$, $\zeta = 0.05$). Comparison between the Gaussian distribution, the fourth-order closure result and the ENLE result

actually become oscillatory, in the 'tail' of the distribution, giving negative values of $f(\lambda)$ in some ranges of λ, when λ is large. This behaviour is typical of density functions generated by cumulant, or quasi-moment closure (e.g. see Roberts (1966)).

In the following chapter a comparison of these results with digital simulation estimates will also be made.

9.4.6 An alternative approach

In the preceding sections the error difference, ε, between equations (9.178) and (9.149), as given by equation (9.180), was minimized, in a mean square sense.

This was achieved by choosing a parametric form for $H(E)$ (e.g. see equation (9.191)) and minimizing with respect to these parameters, c_{rs}, in this form.

An alternative, non-parametric, approach is possible which is appropriate when the damping is light, i.e. when the energy dissipated per 'cycle', due to damping is, on average, a small fraction of the average energy in a cycle of oscillation. In this case one can treat the total energy, E, as approximately constant, over one cycle of oscillation. More precisely, over an interval of time, $T(E)$, corresponding to the period of free, undamped oscillations, one can assume that $H(E)$ is approximately constant, if the damping is light. Setting $b(y, \dot{y})$ and $f(t)$ to zero, in equation (9.178), the period of free oscillation is found to be given by

$$T(E) = 4 \int_0^b \frac{\mathrm{d}y}{[2(E - V(y))]^{1/2}} \tag{9.240}$$

where b is such that

$$V(b) = E \tag{9.241}$$

Returning to the expression for the equation error ε, given by equation (9.180), the error integral,

$$I = \int_0^{T(E)} \varepsilon^2 \, \mathrm{d}t = \int_0^{T(E)} [b(y, \dot{y}) - \dot{y}H(E)]^2 \, \mathrm{d}t \tag{9.242}$$

can be minimized with respect to $H(E)$, where the latter is treated as a constant. This yields the following expression for $H(E)$

$$H(E) = \frac{\int_0^{T(E)} b(y, \dot{y})\dot{y} \, \mathrm{d}t}{\int_0^{T(E)} \dot{y}^2 \, \mathrm{d}t} \tag{9.243}$$

or

$$H(E) = \frac{\int_0^b b(y, [2(E - V(y))]^{1/2}) \, \mathrm{d}y}{\int_0^b [2(E - V(y))]^{1/2} \, \mathrm{d}y} \tag{9.244}$$

A combination of equation (9.244) with equation (9.153) now gives an approximate expression for $f(y, \dot{y})$. This expression will become increasingly accurate as the magnitude of the damping is progressively reduced. The integrals in equation (9.244) can be evaluated, either analytically or numerically, for specific functions, $b(y, \dot{y})$ and $g(y)$ in the equation of motion, given by equation (9.178).

It is interesting that equations (9.243) or (9.244) can also be derived from energy considerations. Suppose that the energy dissipations per cycle, due to the term $b(y, \dot{y})$, in equation (9.178), and to the term $H(E)\dot{y}$ in equation (9.149), are equated. Thus

$$\int_0^{T(E)} b(y, \dot{y})\dot{y}\,\mathrm{d}t = \int_0^{T(E)} [\dot{y}H(E)]\dot{y}\,\mathrm{d}t \tag{9.245}$$

Then, if $H(E)$ is treated as constant, it is clear that equation (9.245) leads immediately to an equation for $H(E)$ which is identical to equation (9.243).

It is worth noting here that this approach to evaluating $H(E)$ has similarities to the method discussed earlier, in Section 8.2.1, for evaluating the amplitude dependent damping term, $\beta(A)$, for a hysteretic oscillator (see equation (8.28)). Indeed, $H(E)$, in the present analysis can be regarded as a damping coefficient, which is dependent on the energy level E, and hence on the amplitude level, A, where A is related to E by equation (9.160). In the case of linear stiffness there is a complete correspondence between $\beta(A)$ and $H(E)$.

Returning to the approximate solution for $f(y, \dot{y})$ obtained by combining equation (9.244) with equation (9.153), it is interesting to note that, as pointed out by Hennig and Roberts (1986), this result is identical to that obtained by the generalized method of stochastic averaging, originally proposed by Stratonovich (1964) (see also Roberts and Spanos, 1986). The latter method can be shown, by various physical arguments (Roberts, 1976b, 1978d) to become increasingly accurate as the damping becomes lighter. Further, it has been shown, by a rigorous mathematical argument, to be (under certain conditions) asymptotically exact as the magnitude of the damping approaches zero (Zhu, 1983; Khasminskii, 1968).

In many cases the non-parameric method described here will give a result which is identical to that obtained by the parametric method described in Section 9.4.2. In particular, in the special case of linear stiffness, the two methods yield identical results when $b(y, \dot{y})$ is of the general form given by equation (9.190). This is, perhaps, surprising in view of the fact that the non-parametric method depends on the assumption of light damping, whereas the parametric technique does not.

To prove that the two methods lead to identical results, for linear stiffness, the expression for $b(y, \dot{y})$, given by equation (9.190), can be substituted into equation (9.243). Thus, one obtains

$$H(E) = \sum_{r,s}\sum \frac{b_{rs} \displaystyle\int_0^{2\pi/\omega_\mathrm{n}} |y^r \dot{y}^s| \operatorname{sgn}(\dot{y})\dot{y}\,\mathrm{d}t}{\displaystyle\int_0^{2\pi/\omega_\mathrm{n}} \dot{y}^2\,\mathrm{d}t} \tag{9.246}$$

Here the fact that, with a linear stiffness $g(y) = \omega_\mathrm{n}^2 y$, $T(E) = 2\pi/\omega_\mathrm{n}$, a constant, has been used. Now using the expressions (9.168) and (9.169) for y and $\dot{y}$, in the above, leads to

$$H(E)=\sum_{r,s}\sum b_{rs}(A)^{r+s-1}\omega_{n}^{r}\frac{\int_{0}^{\pi/2}\cos^{r}\chi\cos^{s+1}\chi\,d\chi}{\int_{0}^{\pi/2}\sin^{2}\chi\,d\chi} \tag{9.247}$$

This is identical to the result obtained earlier, (see equations (9.191), (9.197) and (9.202)).

9.5 RELIABILITY ESTIMATION

Here two types of failure which can occur in mechanical systems and structures will be briefly discussed, together with the problem of estimating statistics relating to these failure modes. As pointed out earlier, such statistics are usually very sensitive to the character of the 'tails' of the response distribution. Thus the Gaussian approximation, inherent in the usual statistical linearization approximation, can lead to significant errors in estimating system reliability. The method of non-Gaussian closure, and the ENLE method, described earlier in the chapter, both predict the effect of non-linearities on the shape of the response distribution and should lead to more accurate estimation of failure statistics.

9.5.1 First passage probability

In many practical applications it is often required to estimate the probability that the system response stays within a safe region, within a specified interval of time. The determination of such a probability is usually called the 'first-passage problem' and has been extensively studied, during the last two decades (e.g. see Roberts, 1986).

The outer limits of the safe region can be defined by a suitable 'barrier'. The first-passage problem may then be stated as follows: find the probability, $P(t)$, that a response process, $y(t)$, crosses some initial barrier (i.e., exits from the safe region) at least once in a given interval of time, 0–t. Clearly, $P(t)$ will depend on the initial condition of the response. If $W(t)$ is the 'survival', probability that $y(t)$ stays *within* the safe region in the interval 0–t, then

$$W(t)=1-P(t) \tag{9.248}$$

In practice there are two types of barriers which are of primary importance. The first of these, the 'single-sided' barrier, is such that any value of $y(t)$ less than a fixed amplitude level, b say, is safe. Thus, first-passage failure occurs when $y(t)$ first exceeds the value b. The second type is the 'double-sided' barrier, such that, for safe operation, $|y(t)|$ is less than b.

Associated with $P(t)$ is the first-passage density function, $p(t)$, where $p(t)\,dt$ is the probability density function for the time, T, to first-passage failure. The moments of this time are given by

$$M_n = E\{T^n\} = \int_0^\infty t^n p(t)\,\mathrm{d}t \tag{9.249}$$

Of these moments the first, which is simply the mean time to first-passage failure, is by far the most important; it will be denoted, alternatively, as $\bar{T}$. Thus

$$\bar{T} = M_1 = \int_0^\infty tp(t)\,\mathrm{d}t \tag{9.250}$$

Although complete, exact analytical solutions for $p(t)$ are generally unavailable, its asymptotic behaviour, in the case where the excitation is stationary, is well understood. As shown earlier, for stationary excitation the response process, $y(t)$, will approach stationarity as the elapsed time becomes large, irrespective of the initial conditions. Under these conditions it can be shown that, subject to fairly mild restrictions (Roberts, 1974; Kuznetsov *et al.*, 1965)

$$P(t) \to 1 - \exp(-\alpha t) \quad \text{as } t \to \infty \tag{9.251}$$

where α is called the 'limiting decay rate'. When $\bar{T}$ is very large, equation (9.251) is a good approximation for the entire distribution and it follows that

$$\alpha \to \frac{1}{\bar{T}} \quad \text{as } \bar{T} \to \infty \tag{9.252}$$

Thus, for very long mean times to failure, the single statistic $\bar{T}$ can be used to form an approximation to $P(t)$.

It has been rigorously proved by Cramer (1966) and others that, for a normally distributed process, the distribution of barrier crossings on the time axis, from within the safe domain to the unsafe domain, is asymptotically Poisson distributed as the critical level, b, becomes very large, i.e.

$$P(t) \to 1 - \exp(-\nu t) \quad \text{as } b \to \infty \tag{9.253}$$

where ν is the average number of barrier crossings per unit time. This implies that

$$\alpha \to \nu \quad \text{as } b \to \infty \tag{9.254}$$

There is some evidence to suggest that equation (9.254) may also be valid for non-Gaussian response processes, which occur when non-linear systems are randomly excited (e.g. see Dunne and Wright, 1985; Roberts, 1978b, 1978c). Also one can argue, on physical grounds, that for high barrier levels crossings are rare events and in the limit one can expect these to have a Poisson distribution. However, rigorous proofs for this more general situation do not appear to exist as yet.

The average rate of barrier crossings is easily found by using Rice's formula (Rice, 1944). Thus, for a single-sided barrier

$$\nu = \int_0^{\infty} \dot{y} f(b, \dot{y})\, \mathrm{d}\dot{y} \tag{9.255}$$

where $f(y, \dot{y})$ is the joint density function for y, and $\dot{y}$. For a double-sided barrier ν is double that for a single-sided barrier, if $y(t)$ is a symmetrically distributed about zero mean.

For the case of linear and non-linear oscillators it has been demonstrated that the rate at which $\alpha \to \nu$ depends critically on the band-width of the response (Roberts, 1976a). For wide-band response corresponding to large damping, $\alpha = \nu$ is a good approximation, with $\alpha < \nu$. For narrow-band response corresponding to light damping, $\alpha < \nu$ again, and the rate at which the asymptote, indicated by equation (9.254), is reached is slower. Thus, over the whole range of band-width, and barrier level, the estimation of $P(t)$ using equation (9.253) rather than equation (9.251) (which is equivalent to approximating α by ν) is *conservative*, i.e. the probability of failure within a specified time interval is overestimated. Hence, from a practical viewpoint, the estimation of $P(t)$ based on equation (9.253), with ν calculated from equation (9.255), yields an approximate, upper-bound estimate of $P(t)$ which can be used for design purposes.

It has been demonstrated earlier in this chapter that the joint density $f(y, \dot{y})$ can be estimated, approximately, for a non-linear oscillator by using either the non-Gaussian closure approach, or the equivalent non-linearization method. Therefore, either of these techniques, in combination with equations (9.253) and (9.255) provides a practical means of assessing the influence of non-linearities on first-passage failure probabilities.

9.5.2 Fatigue life

Although fatigue failure under random loading conditions is still far from understood, at a detailed microscopic level, it is possible to form rough estimates of statistics relating to fatigue life by using the concept of cumulative damage. This failure model assumes that every 'cycle' in the response inflicts an increment of damage, the magnitude of which is dependent on the amplitude of the cycle. It further assumes that the incremental damage due to each cycle linearly accumulates until the total damage reaches a critical level, at which failure occurs.

Consider an oscillator with light, non-linear damping, such that the displacement response is narrow-band, i.e. sample functions of the response consist of well defined cycles. If, for simplicity, it is assumed that the stiffness is linear then, to a good approximation, the expected number of cycles in a period of time, of duration T, is simply $\omega_n T/2\pi$, where ω_n is the undamped natural frequency of oscillation. Of these cycles a fraction $f(A)\,\mathrm{d}A$ will have peak amplitudes between A and $A+\mathrm{d}A$, where $f(A)$ is the probability density function of peak amplitudes. Each cycle in the range A to $A+\mathrm{d}A$ will cause an increment of damage, according to the cumulative damage hypothesis. If this incremental damage is denoted $\Delta(A)$, then the total, expected, linearly accumulated damage due to cycles in the range A, $A+\mathrm{d}A$ is $(\omega_n T/2\pi)p(A)\,\mathrm{d}A\,\Delta(A)$.

Integrating this over the whole range of A gives the following expression for the expectation of the total damage, $D(T)$, in the interval T

$$E\{D(T)\} = \frac{\omega_n T}{2\pi}\int_0^\infty p(A)\Delta(A)\,\mathrm{d}A \tag{9.256}$$

It should be noted that the derivation of equation (9.256) contains a number of implicit assumptions and approximations (e.g. see Lin, 1967) and, at best, is heuristic in nature. However, bearing in mind the damage hypothesis upon which it is based is also far from 'exact', the level of rigour involved in deriving equation (9.256) is appropriate in these circumstances.

The evaluation of the integral in equation (9.256) requires a knowledge of the functions $p(A)$ and $\Delta(A)$. The former can be obtained by using the methods described earlier in this chapter, for oscillators with non-linear damping. The latter can be estimated by relying on fatigue data relating to regular, sinusoidal response. Let S be a fixed stress amplitude, and N the experimentally determined number of cycles to failure. Then, as is very well known, if S is plotted against N, on log–log scales, for many materials the results conform reasonably well to a straight line. Thus one can write

$$NS^b = c \tag{9.257}$$

where c is a constant which depends on the material and the exponent b ranges from 5 to 20, for various materials.

In view of the lack of understanding regarding fatigue failure under random loading, it is expedient to extrapolate from the regular to the irregular loading case. For stress cycles of amplitude S, the number of cycles to failure is, from equation (9.257), equal to cS^{-b}. Thus, according to the linear cumulative damage hypothesis, one cycle, of stress S, will cause an incremental damage

$$\Delta(S) = 1/(cS^{-b}) = \frac{S^b}{c} \tag{9.258}$$

or

$$\Delta(A) = \frac{(\lambda A)^b}{c} \tag{9.259}$$

where it has been assumed that stress is linearly related to amplitude, i.e.

$$S = \lambda A \tag{9.260}$$

where λ is a constant. On substituting this expression into equation (9.256) one obtains

$$E\{D(T)\} = \frac{\omega_n T}{2\pi}\frac{\lambda^b}{c}\int_0^{\infty} A^b f(A)\,dA \tag{9.261}$$

It is noted that, with $\Delta(S)$ scaled as in equation (9.258), the critical value of D, for failure, is unity.

Although it is possible to examine other statistics relating to fatigue, such as the variance of $D(T)$, (e.g. see Crandall and Mark, 1963; Lin, 1967; Zhu and Lei, 1989) here the discussion will not be widened further. The above is deemed sufficient to show how the theory developed earlier in this chapter can be employed to study fatigue life under random loading.

9.5.3 An example

To illustrate the use of the foregoing results for addressing first-passage and fatigue failure, the specific case of an oscillator with linear-plus-cubic damping, linear stiffness and white noise excitation will again be revisited. The appropriate stationary distribution results, as derived by the method of equivalent non-linear equations, will be used for this purpose as these are more accurate than the corresponding non-Gaussian closure results (see Chapter 10).

Considering first-passage failure, initially, one can, in the first instance combine equations (9.236) and (9.255) to calculate the average rate of single-sided barrier crossings. If

$$\eta = \frac{b}{\sigma_{yo}} \tag{9.262}$$

is a non-dimensional barrier height then one obtains the following result for ν (see also Roberts, 1977)

$$\frac{\nu}{\nu_0} = \exp\left(\frac{\eta^2}{2}\right)\frac{[1 - \exp(\theta + (\eta^2/4\theta))]}{(1 - \exp\theta)} \tag{9.263}$$

where ν_0 is the linear result ($\mu = 0$), and θ is given by equation (9.232). When $\eta = 0$ this result reduces to

$$\nu = \nu_0 = \frac{\omega_n}{2\pi} \tag{9.264}$$

Equation (9.263) shows that, according to the equivalent non-linearization method, the crossing rate depends only on two parameters, i.e. η and θ; it does not depend on the damping parameter ζ.

The result given by equation (9.263) can be compared with the corresponding result from the method of statistical linearization. This latter implies a Gaussian distribution for the response process, i.e.

$$f(Y, \dot{Y}) = \frac{1}{2\pi\sigma_Y\sigma_{\dot{Y}}} \exp\left[-\frac{1}{2}\left(\frac{Y^2}{\sigma_Y^2} + \frac{\dot{Y}^2}{\sigma_{\dot{Y}}^2}\right)\right] \tag{9.265}$$

since Y and $\dot{Y}$ are uncorrelated, for a stationary process. The symbols σ_Y and

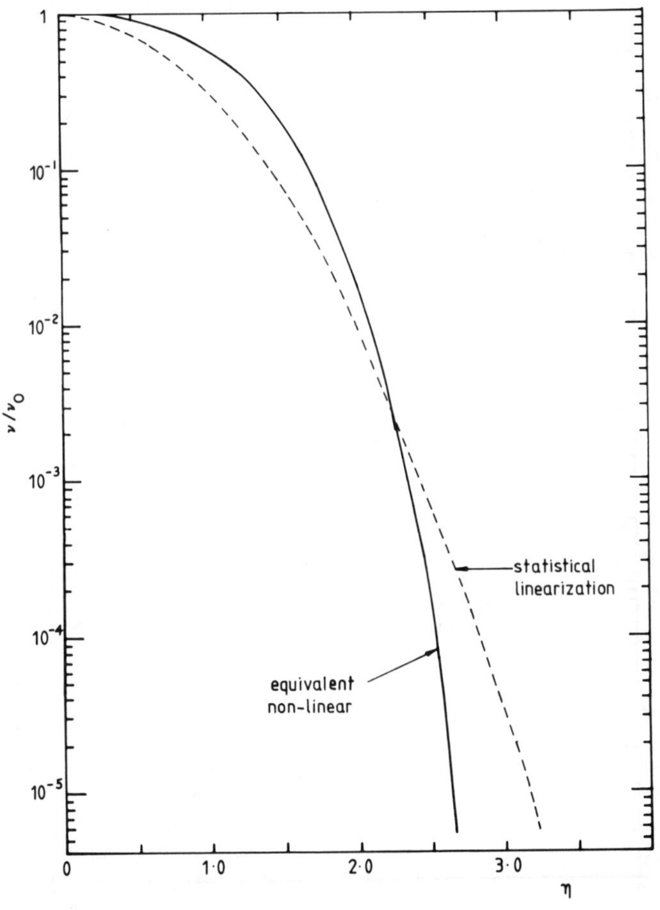

Figure 9.7
Variation of the average barrier crossing rate with barrier height for an oscillator with linear stiffness and linear-plus-cubic damping ($\mu = 1$). Comparison between statistical linearization and ENLE results

$\sigma_{\dot{Y}}$ denote the standard deviations of $\dot{Y}(t)$ and $Y(t)$, respectively, and may be found from equations (9.57) and (9.59) (with $\rho = 0$). Thus, according to statistical linearization (or Gaussian closure)

$$\sigma_Y^2 = \sigma_{\dot{Y}}^2 = \phi(\mu) \tag{9.266}$$

where $\phi(\ )$ is the function defined by equation (9.58). Substituting equation (9.265) into equation (9.255), and integrating, the following result is obtained

$$\frac{\nu}{\nu_0} = \exp\left(-\frac{\eta^2}{2\sigma_Y^2}\right) \tag{9.267}$$

Figure 9.7 shows some typical comparisons between the variation of ν/ν_0 with

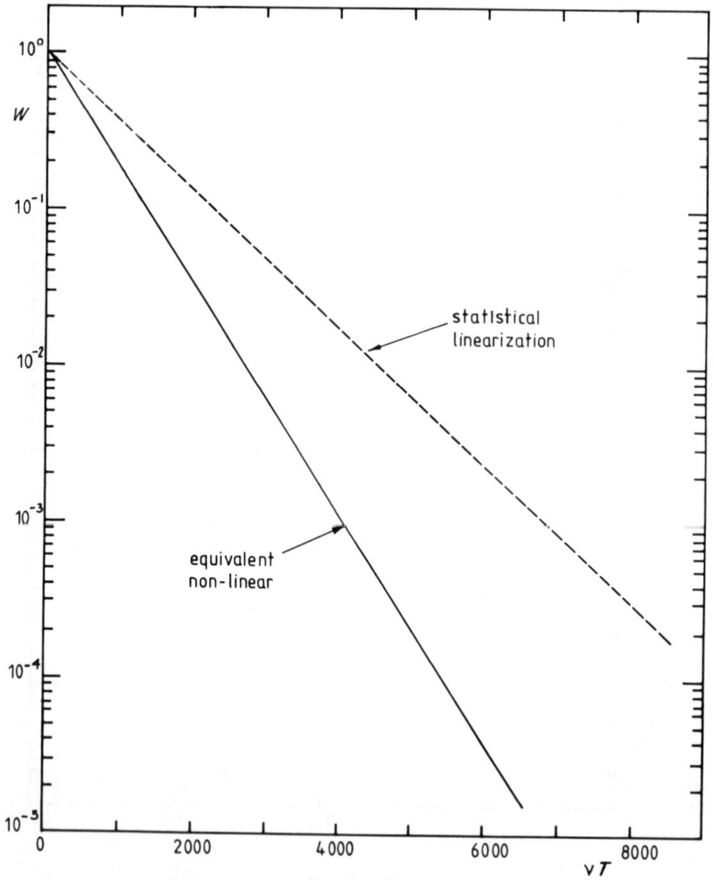

Figure 9.8
Variation of the probability of survival with elapsed time for an oscillator with linear stiffness and linear-plus-cubic damping ($\mu = 0.5, \eta = 2.75$). Comparison between statistical linearization and ENLE results

η according to the ENLE method, and the corresponding variations according to statistical linearization, here for the value $\mu = 1$. It is seen that, for large barrier levels, the ENLE result shows the tendency of the non-linear damping to significantly reduce the rate at which the barrier is crossed. This tendency is also, of course, reflected in the distribution of the amplitude response (e.g. see Figure 9.6).

From a knowledge of v, the probability of first-passage failure, $P(t)$ or, alternatively the probability of survival, $W(t)$, can be estimated, according to equation (9.253). Figure 9.8 shows a typical variation of $W(t)$ with non-dimensional time, for $\eta = 2.75$ and $\mu = 1$, according to statistical linearization and to the ENLE method. Clearly, the effect of non-linearity in damping, which is accounted for in the ENLE result, has a dramatic affect on the magnitude of the survival probability. Thus, the statistical linearization can lead to grossly over-conservative estimates of failure probabilities.

Finally, the evaluation of the expected cumulative fatigue damage, $E\{D(T)\}$, can be carried out, for the present example, by combining equation (9.230) with equation (9.261). Hence,

$$E\{D(T)\} = C_2 \frac{\omega_n T}{2\pi} \frac{\lambda^b}{c} \int_0^\infty A^{b+1} \exp\left[-\left(\frac{A^2}{2} + \frac{3\mu A^4}{16}\right)\right] \mathrm{d}A \qquad (9.268)$$

where C_2 is given by equation (9.231). The integral in this expression is most readily evaluated by straightforward numerical means.

Equation (9.268) may be compared with the corresponding result obtained by statistical linearization. This leads to the adoption of a Rayleigh distribution for $A(t)$, given by (see equation (9.177))

$$f(A) = \frac{A}{\sigma_y^2} \exp\left(-\frac{A^2}{2\sigma_y^2}\right) \qquad (9.269)$$

where σ_y is given by equation (9.266). Hence, combining equation (9.265) with equation (9.261) one has

$$\begin{aligned} E\{D(T)\} &= \frac{\omega_n T}{2\pi} \frac{\lambda^b}{c} \frac{1}{\sigma_y^2} \int_0^\infty A^{b+1} \exp\left(-\frac{A^2}{2\sigma_y^2}\right) \mathrm{d}A \\ &= \frac{\omega_n T}{2\pi} \frac{\lambda^b}{c} (2\sigma_y)^{b/2} \Gamma\left(1 + \frac{b}{2}\right) \end{aligned} \qquad (9.270)$$

The ratio R of expected damage, according to the ENLE method, as given by equation (9.268), to expected damage according to statistical linearization, as given by equation (9.270), is

$$R = \frac{C_2 \int_0^\infty A^{b+1} \exp\left[-\left(\frac{A^2}{2} + \frac{3\mu A^4}{16}\right)\right] \mathrm{d}A}{(2\sigma_y)^{b/2} \Gamma(1 + b/2)} \qquad (9.271)$$

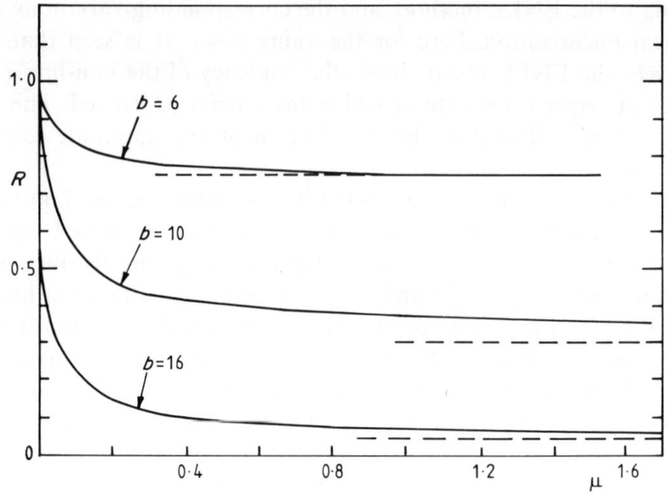

Figure 9.9
Variation of the ratio R with μ for an oscillator with linear stiffness and linear-plus-cubic damping, for various values of b. (---asymptotes)

It can easily be shown that R has the following asymptotes

$$R \to 1 \quad \text{as } \mu \to 0 \tag{9.272}$$

and

$$R \to \frac{(2)^{b/2}}{\pi^{1/2}} \frac{\Gamma\left(\dfrac{b+2}{4}\right)}{\Gamma\left(\dfrac{b+2}{2}\right)} \quad \text{as } \mu \to \infty \tag{9.273}$$

Thus, for large μ, R approaches an asymptotic value which depends only on b.

Figure 9.9 shows the variation of R with μ, for various levels of b. Again, due to the amplitude limiting effect of non-linear damping, as accounted for in the ENLE method (but not by the statistical linearization method) it is to be expected that R will be less than unity. As b increases the damage rate becomes more sensitive to the distribution of the response at extreme amplitudes. This accounts for the decrease of R with increasing b, for a fixed value of μ. Clearly, for large values of b the statistical linearization estimate of the damage rate is overly conservative.

9.6 PARAMETRIC IDENTIFICATION

So far, throughout this book the parameters in equivalent linear models have been determined by minimizing the mean square of the equation deficiency, usually denoted $\boldsymbol{\varepsilon}$. Thus, in Section 9.2, the matrices $\mathbf{a}(t)$ and $\mathbf{L}(t)$ are determined by minimizing $E\{\boldsymbol{\varepsilon}^{\mathrm{T}}\boldsymbol{\varepsilon}\}$. As shown earlier this procedure needs a knowledge of the probability distribution of the response, to evaluate the required expectations (see equations (9.12) and (9.14)).

There is, however, an alternative approach which does not require a knowledge of the response probability distribution (see Kozin, 1988). Suppose that, for the system governed by equation (9.1), sample functions, or realizations, of both the response process $\mathbf{z}(t)$ and excitation process $\mathbf{f}(t)$ are available, over a fixed interval of time, 0–T. These sample functions will be denoted by $\mathbf{z}_i(t)$ and $\mathbf{f}_i(t)$, respectively. One can then attempt to fit, in some optimum way, an equivalent linear equation of motion to the data represented by the sample functions. More precisely, one can attempt to find the parameters in the linear model which will give a best fit to the data. This is a problem in the field of parametric identification, a subject which has a very extensive literature (e.g. see Aström and Eykhoff, 1971; Goodwin and Payne, 1977).

Here, for simplicity, attention will be restricted to the case where both the excitation and response processes are stationary, with zero mean. Then, from equation (9.8), a suitable equivalent linear differential relationship between $z_i(t)$ and $f_i(t)$ is as follows

$$\dot{\mathbf{z}}_i = \mathbf{L}\mathbf{z}_i(t) + \mathbf{f}_i(t) \tag{9.274}$$

The symbol $\mathbf{L}$ is an $N \times N$ matrix of time-independent parameters, which are to be determined such that equation (9.274) fits the data, in the range 0–T, in some optimum sense. If an error term

$$\boldsymbol{\varepsilon}(t) = \dot{\mathbf{z}}_i - \mathbf{L}\mathbf{z}_i(t) - \mathbf{f}_i(t) \tag{9.275}$$

is defined then it is natural to seek an $\mathbf{L}$ matrix which minimizes the average value of the scalar $\boldsymbol{\varepsilon}^{\mathrm{T}}(t)\boldsymbol{\varepsilon}(t)$. Here the average is interpreted as a time average, over the interval $(0 < t < T)$ (rather than an expectation, or ensemble average, as in earlier parts of this book). Hence, one seeks $\mathbf{L}$ which minimizes

$$\frac{1}{T}\int_0^T \boldsymbol{\varepsilon}^{\mathrm{T}}(t)\boldsymbol{\varepsilon}(t)\,\mathrm{d}t \tag{9.276}$$

On combining equations (9.275) and (9.276) it can be shown that the solution of the minimization problem is as follows (see also Kozin, 1986):

$$\begin{aligned}\mathbf{L}_i(T) = &-\left(\frac{1}{T}\int_0^T \mathbf{f}_i(t)\mathbf{z}_i^{\mathrm{T}}(t)\,\mathrm{d}t\right)\left(\frac{1}{T}\int_0^T \mathbf{z}_i(t)\mathbf{z}_i^{\mathrm{T}}(t)\,\mathrm{d}t\right)^{-1}\\ &+\left(\frac{1}{T}\int \dot{\mathbf{z}}_i(t)\mathbf{z}_i^{\mathrm{T}}(t)\,\mathrm{d}t\right)\left(\frac{1}{T}\int_0^T \mathbf{z}_i(t)\mathbf{z}_i^{\mathrm{T}}(t)\,\mathrm{d}t\right)^{-1}\end{aligned} \tag{9.277}$$

As indicated, the elements of $\mathbf{L}$, computed in this manner, will depend on the particular sample functions, as referenced by i, and on the duration, T, of the sample functions.

If the additional assumption is now made that $\mathbf{z}(t)$ and $\mathbf{f}(t)$ are both stationary and jointly ergodic in correlation (see Chapter 3), then it is possible to obtain an analytical result for the limit of $\mathbf{L}_i(T)$, as $T \rightarrow \infty$. Thus

$$\lim_{T \to \infty} \frac{1}{T} \int_0^T \mathbf{f}_i(t)\mathbf{z}_i^{\mathrm{T}}(t)\,\mathrm{d}t = E\{\mathbf{f}\mathbf{z}^{\mathrm{T}}\} \tag{9.278}$$

$$\lim_{T \to \infty} \frac{1}{T} \int \mathbf{z}_i(t)\mathbf{z}_i^{\mathrm{T}}(t)\,\mathrm{d}t = E\{\mathbf{z}\mathbf{z}^{\mathrm{T}}\} \tag{9.279}$$

$$\lim_{T \to \infty} \frac{1}{T} \int \dot{\mathbf{z}}_i(t)\mathbf{z}^{\mathrm{T}}(t)\,\mathrm{d}t = E\{\dot{\mathbf{z}}\mathbf{z}^{\mathrm{T}}\} \tag{9.280}$$

Since $\mathbf{z}(t)$ is assumed stationary

$$E\{\dot{\mathbf{z}}\mathbf{z}^{\mathrm{T}}\} = \mathbf{0} \tag{9.281}$$

Hence, from equations (9.277) to (9.281)

$$\lim_{T \to \infty} \mathbf{L}_i(T) = -E\{\mathbf{f}\mathbf{z}^{\mathrm{T}}\}E\{\mathbf{z}\mathbf{z}^{\mathrm{T}}\}^{-1} \tag{9.282}$$

or

$$\lim_{T \to \infty} \mathbf{L}_i(T) = -E\{\mathbf{f}\mathbf{z}\}\mathbf{V}^{-1} \tag{9.283}$$

where $\mathbf{V}$ is the covariance matrix for $\mathbf{z}$, i.e.

$$\mathbf{V} = E\{\mathbf{z}(t)\mathbf{z}^{\mathrm{T}}(t)\} \tag{9.284}$$

Moreover, from equations (9.1) and (9.281)

$$E\{\mathbf{F}(\mathbf{z})\mathbf{z}^{\mathrm{T}}\} = -E\{\mathbf{f}\mathbf{z}^{\mathrm{T}}\} \tag{9.285}$$

Hence equation (9.285) can be rewritten as follows

$$\lim_{T \to \infty} \mathbf{L}_i(T) = E\{\mathbf{F}(\mathbf{z})\mathbf{z}^{\mathrm{T}}\}\mathbf{V}^{-1} \tag{9.286}$$

On comparing this with the result given by equation (9.12) it is seen that the limit of $\mathbf{L}_i(T)$, as $T \rightarrow \infty$, is identical to the matrix $\mathbf{L}$ determined by the normal statistical linearization method. It is noted that, if $\mathbf{z}(t)$ and $\mathbf{f}(t)$ are strictly ergodic (e.g. see Parzen, 1962) then $\mathbf{L}_i(T)$ converges to the normal statistical linearization result with probability one (see Kozin, 1988).

Returning to equation (9.277) it is evident that the evaluation of $\mathbf{L}_i(T)$ does not require a knowledge of either the non-linearity function $\mathbf{F}(\mathbf{z})$, or the probability distribution of the response. Moreover, as $T \rightarrow \infty$, $\mathbf{L}_i(T)$ converges to the *true* value of $\mathbf{L}(t)$, i.e. to the value of $\mathbf{L}(t)$ in equation (9.12), if the exact probability distribution of the response where used to evaluate the required expectations. In practice, of course, the method is only useful if the convergence, with respect to the period T, is reasonably fast.

The above analysis gives an alternative approach to statistical linearization. Given an analytically defined non-linear system, with a random input of known power spectrum, one can proceed as follows:

(a) simulate a sample function of the excitation process, $\mathbf{f}(t)$, over an interval $0 < t < T$.
(b) compute a corresponding sample function of the response, over the same interval of time, by numerically solving equation (9.1).
(c) compute the elements of $\mathbf{L}_i(T)$, according to equation (9.277).
(d) compute statistics of the response, from the equivalent linear equation of motion (see equation (9.274)), using the standard linear theory given earlier in this book.

In the above procedure it is important to ensure that both the excitation and response sample functions relate to stationary conditions, i.e. there are no transients in the sample functions of $\mathbf{f}_i(t)$ and $\mathbf{z}_i(t)$.

Alternatively, in an experimental situation, where $\mathbf{f}_i(t)$ and $\mathbf{z}_i(t)$ are measurable, but the form of non-linear equation of motion is unknown, one can apply equation (9.277) to obtain an optimum linear model.

It is noted that since $\mathbf{L}(t)$ converges to the true value of $\mathbf{L}$, as $T \to \infty$, then the covariance matrix $\mathbf{V}$, computed from the equivalent linear system, will converge to its true value, since, as shown in Section 9.2, $\mathbf{V}$ will be exact if $\mathbf{L}$ is exact.

A disadvantage of equation (9.277), as a means of identifying the parameters in $\mathbf{L}$, is that it is necessary to determine $\dot{\mathbf{z}}_i(t)$ explicitly, in addition to $\mathbf{z}_i(t)$. This will usually require numerical differentiation. In the situation where the method is applied to experimental data the recorded data usually have additive noise, due to measurement errors, and differentiation can significantly magnify this noise. In the following two sections, alternative identification algorithms are briefly described, which do not involve differentiating the response and excitation sample functions. For simplicity these methods are described in the context of a single degree of freedom system. However, extensions to the multi-degree of freedom case are relatively straightforward.

9.6.1 Direct optimization

Consider the case of an oscillator with non-linearity in inertia, damping and stiffness. A very general form of the equation of motion is given by

$$\Phi(y, \dot{y}, \ddot{y}) = f(t) \tag{9.287}$$

where $\Phi(\ )$ is a non-linear function of the displacement response, y, the velocity,

$\dot{y}$, and the acceleration, $\ddot{y}$. As before, it will be assumed that $f(t)$ is a stationary random process with zero mean and that the function is such that $y(t)$, and its derivatives, also have zero mean.

An equivalent linear oscillator has an equation of motion of the form

$$m\ddot{y}^{\mathrm{L}} + c\dot{y}^{\mathrm{L}} + ky^{\mathrm{L}} = f(t) \tag{9.288}$$

where m, c and k are the equivalent mass, damping coefficient and stiffness coefficient, respectively. The subscript L is used to distinguish linear, from non-linear response. Dividing equation (9.283) throughout by m gives

$$\ddot{y}^{\mathrm{L}} + a_1\dot{y}^{\mathrm{L}} + a_2 y^{\mathrm{L}} = bf(t) \tag{9.289}$$

where

$$a = c/m \quad a_2 = k/m \quad b = 1/m \tag{9.290}$$

Clearly

$$a_1 = 2\zeta_{\mathrm{eq}}\omega_{\mathrm{eq}} \quad a_2 = \omega_{\mathrm{eq}}^2 \tag{9.291}$$

where ω_{eq} is the equivalent natural frequency and ζ_{eq} is the equivalent non-dimensional damping factor (see Chapter 5).

Suppose that, over an interval 0–T, sample functions of the excitation process, $f(t)$, and the non-linear system response, $y(t)$, are available; these are denoted $f_i(t)$ and $y_i(t)$, respectively. As noted earlier, these could be obtained either experimentally, or through a numerical solution of the non-linear system equation. Using $f_i(t)$ one can also, by integration of equation (9.284), determine a corresponding variation of $y_i^{\mathrm{L}}(t)$, if a_1, a_2 and b are specified. The optimum values of a_1, a_2 and b can now be determined by minimizing the cost function

$$J = \frac{1}{T}\int_0^T [y_i(t) - y_i^{\mathrm{L}}(t)]^2 \, \mathrm{d}t \tag{9.292}$$

Since $y_i^{\mathrm{L}}(t)$ depends non-linearly on a_1, a_2 and b a numerical search routine will be required to minimize J. For each trial value of a_1, a_2 and b equation (9.285) must be integrated numerically, over the chosen time interval.

This direct, non-linear optimization technique is obviously rather crude and, if generalized to multi-degree of freedom situations, where many parameters are to be the estimated, will certainly be computationally inefficient. However, for simple problems, such as the present one, where only a few parameters are required, the above method is a useful elementary approach to finding the required parameters.

9.6.2 State variable filters

Again, the problem of estimating a_1, a_2 and b, in the equivalent linear oscillator given by equation (9.289) will be addressed. It will now be shown that, through the use of state variable filters (e.g. see Gawthrop, 1986; Gawthrop *et al.*, 1988) the estimation problem can be cast into a standard linear regression problem. Extension to the multi-degree of freedom case (not considered here) is not difficult (see Roberts *et al.*, 1990).

Again, it will be supposed that two sample functions, $f_i(t)$ and $y_i(t)$ are available, over the interval 0–T. Similarly the analysis in Section 9.6, values of a_1, a_2 and b will be sought such that the linear model

$$\ddot{y}_i + a_1 \dot{y}_i + a_2 y_i = b f_i(t) \tag{9.293}$$

yields the best fit to the data.

As an initial step, the terms $c_1 \dot{y}_i$ and $c_2 y_i$ will be added to each side of equation (9.293). On rearranging one then obtains

$$\ddot{y}_i + c_1 \dot{y}_i + c_2 y_i = b f_i(t) - h_1 \dot{y}_i - h_2 y_i \tag{9.294}$$

where

$$h_1 = a_1 - c_1 \quad h_2 = a_2 - c_2 \tag{9.295}$$

Here the parameters c_1 and c_2 will be regarded as *known* constants, which can be set to any desired values. It should be noted, at this stage, that both c_1 and c_2 should be positive; this guarantees that the linear filter, represented by the left-hand side of equation (9.294), is stable. Moreover, the filter should have a pass band which encompasses the system's natural frequency. Within these restrictions the choice of values of c_1 and c_2 is relatively unimportant.

With this rearrangement the unknown parameters are now h_1, h_2 and b. There are, in addition, two unknowns associated with the initial conditions

$$y_i(0) = e_1 \quad \dot{y}_i(0) = e_2 \tag{9.296}$$

In the following, for generality, e_1 and e_2 will be treated as two further parameters, which are to be estimated. It should be noted, however, that in the case of long duration records, initial conditions can usually be ignored, since the effect decays fairly rapidly, in most cases.

An advantage of the rearrangement given by equation (9.294) is that the response, $y_i(t)$, now depends *linearly* on the parameters appearing on the right-hand side of the equation. A linear estimation scheme can thus be formulated fairly readily. For this purpose it is convenient to introduce the following set of three auxiliary equations

$$\ddot{x}_k + c_1 \dot{x}_k + c_2 x_k = u_k \quad k = 1, 2, 3 \tag{9.297}$$

where

$$u_1 = -y_i \quad u_2 = f(t) \quad u_3 = 0 \tag{9.298}$$

On comparing equation (9.294) with equations (9.297) and (9.298) one can deduce that y_i may be related to x_k ($k = 1, 2, 3$) as follows

$$y_i = h_1 \dot{x}_1 + h_2 x_2 + b x_2 + d_1 \dot{x}_3 + d_2 x_3 \tag{9.299}$$

Here the constants d_1 and d_2 are introduced to satisfy the, in general, non-zero initial conditions. If the initial conditions for the auxiliary equations are chosen as

$$\begin{aligned} x_k(0) = \dot{x}_k(0) = 0 \qquad k = 1, 2 \\ x_3(0) = 0 \quad \dot{x}_3(0) = 1 \end{aligned} \tag{9.300}$$

then one finds that

$$\begin{aligned} e_1 &= d_1 \\ e_2 &= d_2 - a_1 d_1 \end{aligned} \tag{9.301}$$

Hence, if required, one can calculate the initial conditions from a knowledge of d_1 and d_2. Clearly, d_1 and d_2 may be substituted for e_1 and e_2; the parameter set to be estimated now becomes h_1, h_2, b, d_1 and d_2.

On introducing the matrix notation

$$\mathbf{x}(t) = \begin{bmatrix} \dot{x}_1 \\ x_1 \\ x_2 \\ \dot{x}_3 \\ x_3 \end{bmatrix} \qquad \boldsymbol{\theta} = \begin{bmatrix} h_1 \\ h_2 \\ b \\ d_1 \\ d_2 \end{bmatrix} \tag{9.302}$$

One can write equation (9.299) compactly as

$$y_i(t) = \mathbf{X}^{\mathrm{T}}(t)\boldsymbol{\theta} \tag{9.303}$$

where $\mathbf{X}^{\mathrm{T}}(t)$ is known as the 'estimation filter state', and $\boldsymbol{\theta}$ is the 'parameter vector'. It is observed that $\mathbf{X}(t)$ does *not* depend on unknown quantities. Given continuous functions of times, $y_i(t)$ and $f_i(t)$, the components of $\mathbf{X}(t)$ can be generated exactly, through a solution of equation (9.297), with initial conditions given by equation (9.300). Of course if the sample functions are measured only at discrete times, Δt, then the generation of $\mathbf{X}(t)$ is necessarily approximate, since some assumption must be made regarding their behaviour between the sample times.

Equation (9.303) is the basis for a straightforward least-square estimation scheme for the parameters in $\boldsymbol{\theta}$. Suppose one considers $y_i(t)$ at the equi-spaced times $t_i = i\Delta t$. Then a least-squares cost function, J, can be defined by

$$J = \sum_{j=0}^{\infty} [\, y_i(t_j) - \mathbf{X}^{\mathrm{T}}(t_j)\boldsymbol{\theta}]^2 \tag{9.304}$$

The unknown parameters can be estimated by finding the vector $\boldsymbol{\theta}$ which minimizes J. It is noted that, in equations (9.298) and (9.304) y_i is the measured, non-linear response. Standard non-recursive algorithms exist for this purpose (e.g. see Aström and Eykhoff, 1971).

Alternatively $\boldsymbol{\theta}$ can be estimated recursively, by successively refining the parameters at each time step. This approach is useful since one can plot the evolution of the parameter estimates with time. One would normally expect the parameter estimates to converge to stable values, after a finite number of time steps. If such a convergence is achieved, for all the parameters, then it is a strong indication that a sufficient amount of data has been processed to achieve physically meaningful results. Thus, one can guard against the possibility of analysing too little data, and avoid analysing more data than necessary.

The recursive algorithm may be summarized as follows: let $\hat{\boldsymbol{\theta}}_N$ be the least square estimate of $\boldsymbol{\theta}$ at time $t_N = N\Delta t$, and denote $\mathbf{X}(t_N) = \mathbf{X}_N$, $y_i(t_N) = y_N$. Further, let

$$\mathbf{S}_N = \sum_{j=0}^{N} \mathbf{X}_j \mathbf{X}_j^{\mathrm{T}} \tag{9.305}$$

Then, at time t_N one computes recursively

$$\sigma_{N+1} = \mathbf{X}_{N+1}^{\mathrm{T}} \mathbf{S}_N^{-1} \mathbf{X}_{N+1} \tag{9.306}$$

$$\mathbf{K}_{N+1} = \frac{1}{(1+\sigma_{N+1})} \mathbf{S}_N^{-1} \mathbf{X}_{N+1} \tag{9.307}$$

$$\hat{e}_{N+1} = y_{N+1} - \mathbf{X}_{N+1}^{\mathrm{T}} \hat{\boldsymbol{\theta}}_N \tag{9.308}$$

$$\hat{\boldsymbol{\theta}}_{N+1} = \hat{\boldsymbol{\theta}}_N + \mathbf{K}_{N+1} \hat{\mathbf{e}}_{N+1} \tag{9.309}$$

$$\mathbf{S}_{N+1}^{-1} = \mathbf{S}_N^{-1} - (1+\sigma_{N+1}) \mathbf{K}_{N+1} \mathbf{K}_{N+1}^{\mathrm{T}} \tag{9.310}$$

These steps can be repeated at t_{N+1}, and so on, enabling $\hat{\boldsymbol{\theta}}_N$ to be generated recursively, for $N = 0, 1, \ldots$

It is, of course, necessary to start the recursive procedure, at time zero, with initial values for $\hat{\boldsymbol{\theta}}_{-1}$ and $\mathbf{S}_{-1}^{-1}$. In principle $\mathbf{S}_{-1} = \mathbf{0}$, so $\mathbf{S}_{-1}^{-1} = \infty$; this corresponds to total uncertainty regarding the parameter values. Also, in principle, $\hat{\boldsymbol{\theta}}_{-1}$ should be undefined. In practice one chooses

$$\hat{\boldsymbol{\theta}}_{-1} = \mathbf{0} \quad \mathbf{S}_{-1}^{-1} = \lambda \mathbf{I} \tag{9.311}$$

where λ is a very large number (typically 10^5) and $\mathbf{I}$ is a unit matrix.

It is noted that specified elements in $\boldsymbol{\theta}$ can be 'locked' to fixed values, in the above recursive scheme. For example, if b is known, *a priori*, one can set the appropriate initializing element of $\boldsymbol{\theta}$ to this value. Then by equating to zero all elements in $\mathbf{S}^{-1}$, associated with the locked element, the value of this element will remain unchanged throughout the recursive estimation procedure.

Gawthrop *et al.* (1988) have described a simple and efficient method of numerically integrating the ancillary equations, using data sampled at finite times, with linear interpolation between samples.

9.6.3 An example

To illustrate the use of the foregoing two identification methods, the case of an oscillator with zero-mean white noise excitation, linear stiffness, and linear-plus-cubic damping, will once again be discussed. An appropriate, non-dimensional equation of motion is given by (see equation (9.224))

$$\ddot{Y} + 2\zeta(\dot{Y} + \mu\dot{Y}^3) + Y = f^*(\tau) \tag{9.312}$$

where $f^*(\tau) = (4\zeta)^{1/2} f_n(\tau)$. This equivalent linear relationship is, from equation (9.289)

$$\ddot{Y}^{\mathrm{L}} + a_1\dot{Y}^{\mathrm{L}} + Y^{\mathrm{L}} = f^*(\tau) \tag{9.313}$$

where a_2 and b are set to unity. The problem is to estimate the parameter a_1 which minimizes the cost function given by equation (9.292).

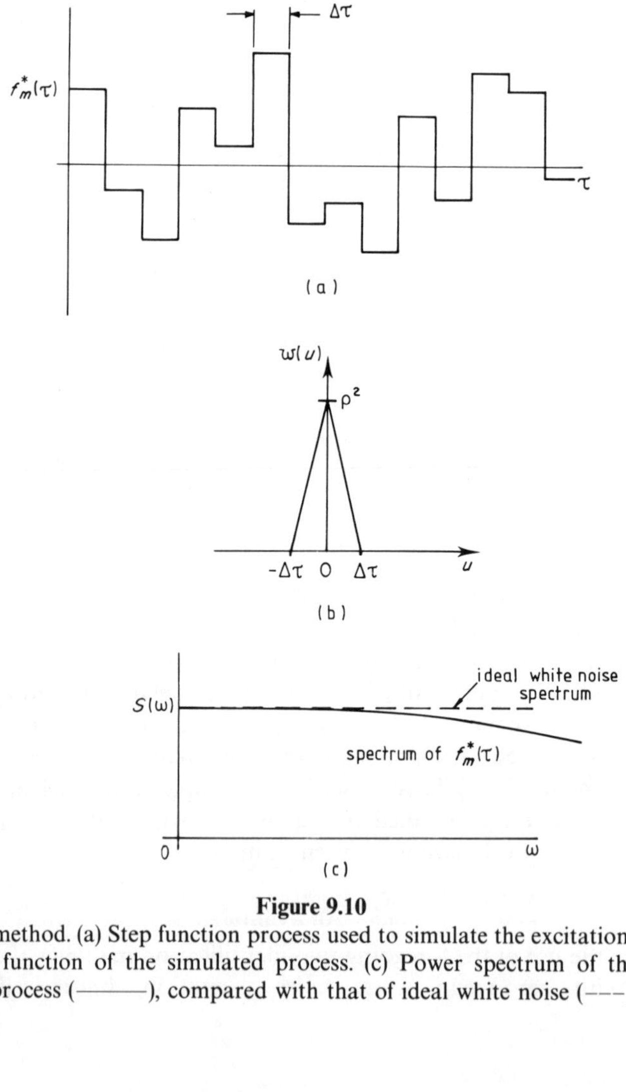

Figure 9.10

Simulation method. (a) Step function process used to simulate the excitation process. (b) Correlation function of the simulated process. (c) Power spectrum of the simulated process (———), compared with that of ideal white noise (–––)

As a first step it is necessary to generate a sample function of $f^*(\tau)$. For an ideal white noise process this is impossible, since $f^*(\tau)$ would have infinite variance. However, $f^*(\tau)$ can be replaced by a non-white process, $f_m^*(\tau)$, which has sample functions in the form of a contiguous function sequence, as shown in Figure 9.10a (see Clough and Penzien, 1975; Spanos, 1980b). Between the equi-spaced time constant $\tau_i = i\Delta\tau$, the sample functions hold constant levels. These levels are governed by a sequence of independent Gaussian random variables with standard deviation, ρ, and mean zero. It is easy to show, that, for such a process the correlation function is of the triangular form shown in Figure 9.10b. This correlation function may be compared with the corresponding delta function form for an ideal white noise, given by

$$E\{f^*(\tau)f^*(\tau+u)\} = I\delta(u) \tag{9.314}$$

where I is the 'strength' of white noise. For sufficiently small $\Delta\tau$ the correlation function for $f_m^*(\tau)$ approximates a delta function. Further, equating the area under the correlation function shown in Figure 9.10b with the area under the correlation function for ideal white noise, as given by equation (9.314), one has

$$\rho^2\Delta\tau = I \tag{9.315}$$

or

$$I = \frac{\rho^2\Delta\tau}{2\pi} \tag{9.316}$$

In the present application, the scaling results in $I = 4\zeta$. Figure 9.10c sketches the shape of the power spectrum of $f_m^*(\tau)$, in comparison with that of ideal white noise.

Once a sample function of $f_m^*(\tau)$ is generated, over an interval, $0 < \tau < T$, corresponding sample functions of $y(\tau)$ and $y^{\mathrm{L}}(\tau)$ can be computed, by integrating equations (9.312) and (9.313), respectively. Hence, the cost function J, in equation (9.292) can be evaluated, for particular choices of ζ, μ and a_1. For the case where $\zeta = 0.05$, $\Delta\tau = 0.05$ and $T = 250$, Figure 9.11 shows the variation of J with a_1, for various μ values. It is seen, as one would expect, that the cost function exhibits a sharp minimum when $\mu = 0$, reducing to zero at $a_1 = 0.1$ (corresponding to the value of 2ζ). As shown in Figure 9.11, as μ is progressively increased, the minimum in the cost function tends to be less sharp, but is still well defined.

From equation (9.313) the mean square of the response is

$$\sigma_Y^2 = \frac{I}{2a_1} = \frac{2\zeta}{a_1} \tag{9.317}$$

Thus estimates of σ_y^2 can be derived directly from the estimate of a_1. These estimates can be compared directly with the values given earlier, as obtained from statistical linearization, fourth-order closure and the method of equivalent non-linear equations (ENLE).

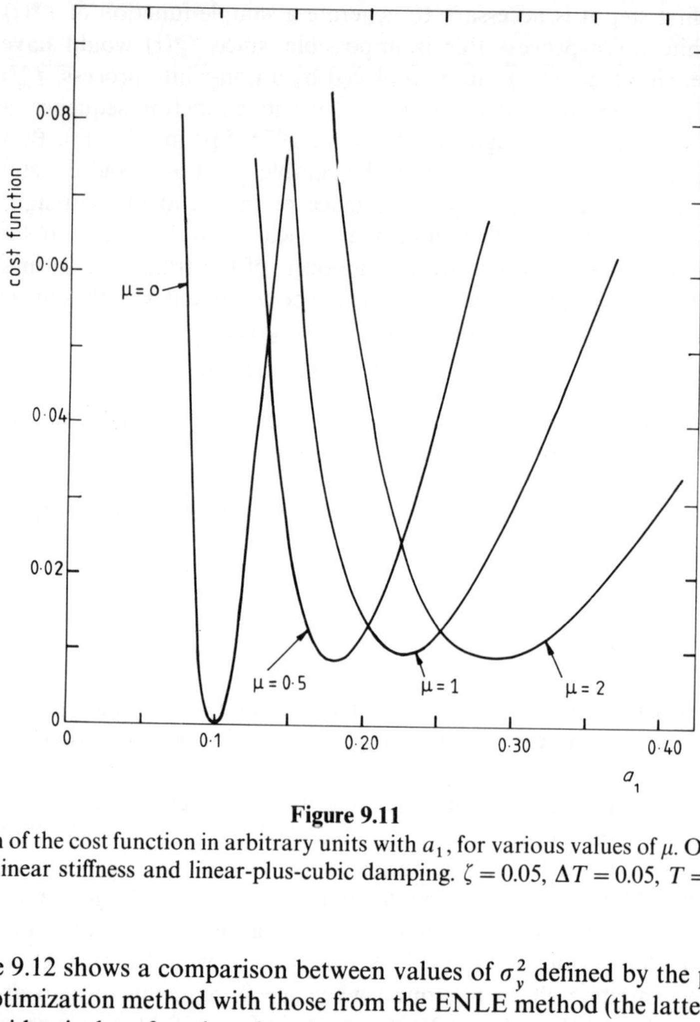

Figure 9.11
Variation of the cost function in arbitrary units with a_1, for various values of μ. Oscillator with linear stiffness and linear-plus-cubic damping. $\zeta = 0.05$, $\Delta T = 0.05$, $T = 250$

Figure 9.12 shows a comparison between values of σ_y^2 defined by the present direct optimization method with those from the ENLE method (the latter being virtually identical to fourth-order closure results, as shown earlier), over the range $0 < \mu < 2$. It is seen that there is good agreement between the two methods of estimating the mean square response.

Finally, some typical results obtained by applying the recursive least square method, based on a state variable filter, are presented. In this case the equivalent linear equation of motion was assumed to be

$$\ddot{Y} + a_1 \dot{Y} + a_2 Y = f^*(\tau) \tag{9.318}$$

Thus, in equation (9.293), b was set to unity (corresponding to the assumption of a known mass) and the parameters a_1 and a_2 in equation (9.318) were estimated. Comparison with equation (9.312) indicates that one expects the a_2

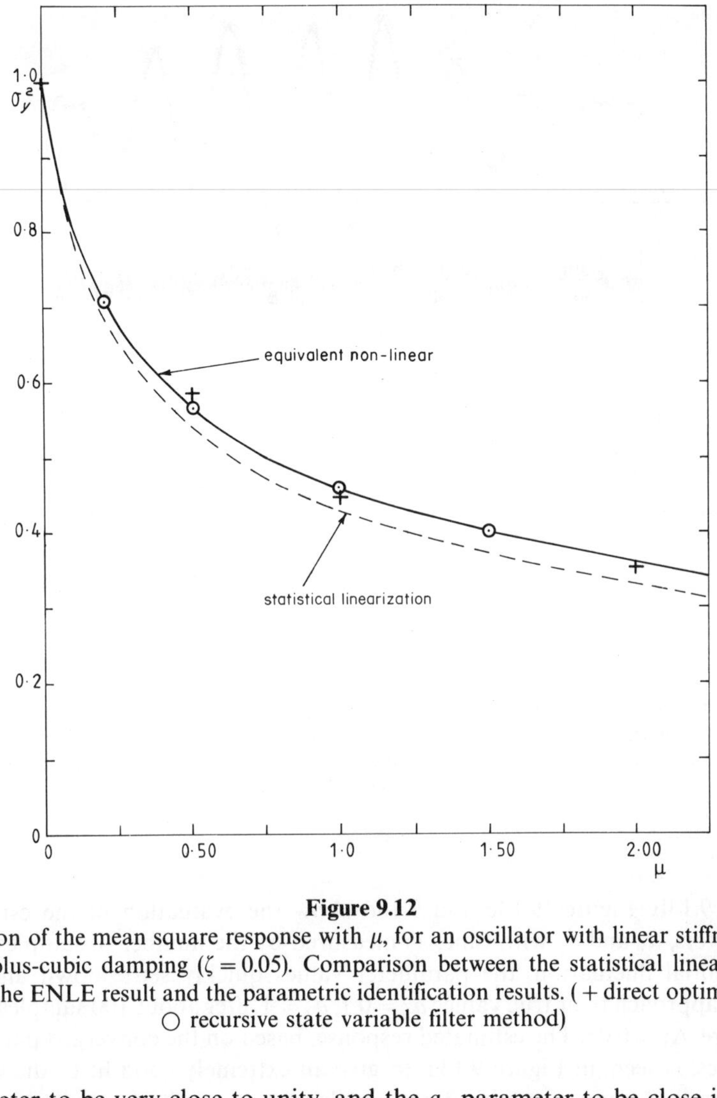

Figure 9.12
Variation of the mean square response with μ, for an oscillator with linear stiffness and linear-plus-cubic damping ($\zeta = 0.05$). Comparison between the statistical linearization result, the ENLE result and the parametric identification results. (+ direct optimization, ○ recursive state variable filter method)

parameter to be very close to unity, and the a_1 parameter to be close in value to that obtained by direct optimization.

To test the method, it was initially applied in the case where $\mu = 0$, in equation (9.312). In this case, of course, there is complete correspondence between equations (9.312) and (9.318); therefore, a_1 should converge to 2ζ and a_2 should converge to unity, with the cost function reducing to an extremely low level. Figure 9.13 shows, graphically, a typical result of applying the method in the linear case, with $\zeta = 0.05$. Figure 9.13a shows the computed response to a simulated sample function of the excitation process, $f_m^*(\tau)$, as shown in

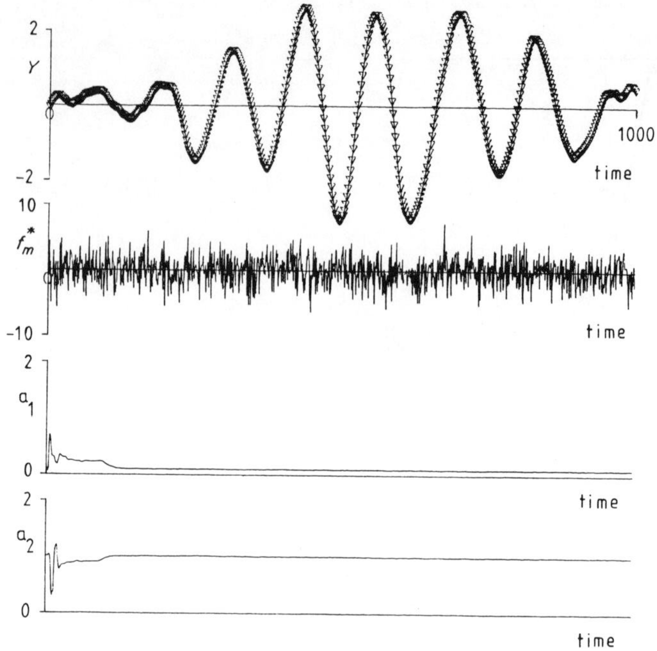

Figure 9.13
Parametric identification results. $\zeta = 0.05$, $\mu = 0$, $\rho = 0$ (a) Response sample function and comparison with estimated response. (∇ simulated data, ——— estimated) (b) Excitation sample function. (c) Variation of estimate of a_1 with time. (d) Variation of estimate of a_2 with time.

Figure 9.13b. Figures 9.13c and 9.13d show the evaluation of the estimated parameters, a_1 and a_2 with time, as measured by the number of samples. After some initial fluctuation, the parameter values quickly settle down and very closely approach their true values $a_1 = 0.1$, $a_2 = 1$ after about 100 samples. Here, as before, $\Delta\tau = 0.05$. The estimated response, based on the converged parameter estimates, is seen, in Figure 9.13a, to give an extremely good fit to the 'data'.

Figure 9.14 shows, graphically, a similar set of results for the case where $\mu = 0.5$, and $\zeta = 0.05$, in equation (9.312). In this case the linear equation of motion, given by equation (9.318) is fitted to a non-linear relationship between $f^*(\tau)$ and $Y(\tau)$. As one might expect, the parameters a_1 and a_2 now take longer to converge to stable values. The fit between the linear response, and the non-linear response based on the converged values of a_1 and a_2, as shown in Figure 9.14a, is seen to be remarkably good, but not, of course, perfect.

Table 9.4 shows a set of results obtained by applying the state variable method to the system described by equation (9.312), for various values of μ, and with

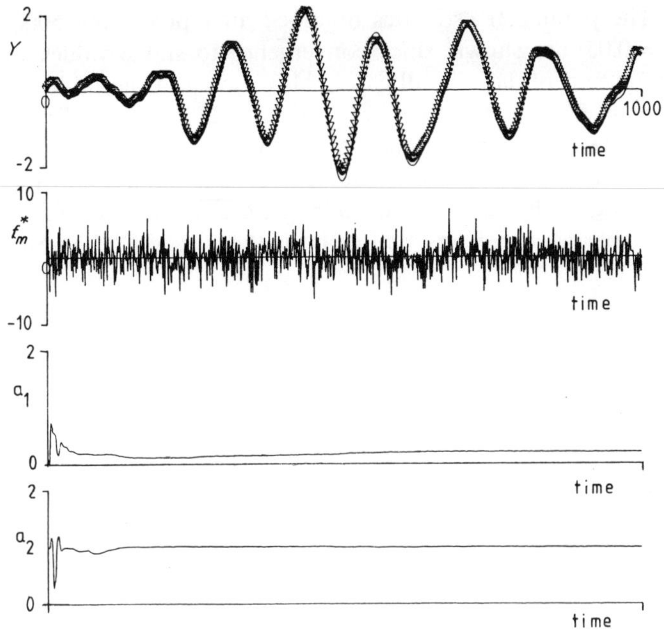

Figure 9.14
Parametric identification results. $\zeta = 0.05$, $\mu = 0.5$, $\rho = 0$. (a) Response sample function and comparison with estimated response. (∇ simulated data, ——— estimated) (b) Excitation sample function. (c) Variation of estimate of a_1 with time. (d) Variation of estimate of a_2 with time.

Table 9.4
Linearization by parametric identification

μ	*Estimated parameters* a_1	a_2	*Estimated mean-square response,* σ_Y^2
0.2	0.141	1.000	0.709
0.5	0.177	1.000	0.565
1.0	0.218	1.000	0.459
1.5	0.250	1.001	0.400
5	0.395	1.000	0.253

$\zeta = 0.05$. The parameter estimates obtained after processing 5000 samples of data ($\Delta\tau = 0.05$) are shown, since convergence to stable values was achieved with this amount of data, in all cases. Also shown in the table are the values of the mean square response, σ_y^2, calculated from the linear model of equation (9.318), using the estimate of a_1 and a_2. As one would expect, in all cases, the estimated values of a_2 are very close to unity, and the equivalent linear damping coefficient, a_1, tends to increase, as μ increases. The mean square results in Table 9.4, agree very well with the theoretical ENLE result, as shown in Figure 9.12.

Chapter 10

Accuracy of statistical linearization

10.1 INTRODUCTION

The accuracy of approximate solutions to non-linear random vibration problems can ordinarily be assessed by either of two approaches. The first of these involves a comparison of the approximate solution, for a particular system, with the corresponding exact theoretical result, obtained by solving the pertinent FPK equation (see Chapter 1). Clearly this approach is dependent on the assumption that the system's response is a Markov process. This is a serious limitation since it effectively restricts consideration to those cases where the excitation processes can be modelled adequately in terms of white noise processes. Moreover, exact solutions to FPK equations are scarce and, especially for multi-degree of freedom systems, of very limited scope. It follows, therefore, that this approach is often inapplicable. The second approach, which is almost indispensible when dealing with complex problems involving, for example, the non-stationary response of multi-degree of freedom systems, is based on a comparison with Monte Carlo simulation results, for a range of representative cases (Shinozuka, 1972; Spanos and Mignolet (1989)).

In this chapter these two approaches will be exemplified by considering some typical, specific non-linear problems. Emphasis will be placed on assessing the accuracy of statistical linearization results but the opportunity is also taken to examine the accuracy of results generated by the 'improved' methods of non-Gaussian closure and equivalent non-linear equations, discussed in Chapter 9.

At this stage it is worth mentioning, in passing, that there is a third approach. This entails developing theoretical bounds on the errors of response statistics generated by approximate methods. Efforts in this direction are, so far, of limited applicability but, for the method of statistical linearization, some results are available (Kolovski, 1966; Budgor, 1976). This is an area which merits further research.

10.2 EXACT SOLUTIONS

Consider an n degree of freedom system, with displacement coordinates $y_1, y_2, \ldots, y_n$, governed by the following set of differential equations

$$\ddot{y}_i + \sum_{j=1}^{n} \beta_{ij}\dot{y}_i H(E) + \frac{\partial V}{\partial y_i} = f_i(t) \quad i = 1, 2, \ldots, n \tag{10.1}$$

where β_{ij} are damping constants and E is the total energy of the system. It is defined by

$$E = \tfrac{1}{2}\dot{\mathbf{y}}^{\mathrm{T}}\dot{\mathbf{y}} + V \tag{10.2}$$

where V is the total potential energy function of the system (which depends on $y_1, y_2, \ldots, y_n$) and $\mathbf{y}$ is the n-vector of displacements; thus

$$\mathbf{y} = [y_1, y_2, \ldots, y_n]^{\mathrm{T}} \tag{10.3}$$

The excitation processes, $f_i(t)$, will be assumed here to be modelled as a set of correlated, stationary white noises, each with zero mean. Thus, the covariance matrix, $\mathbf{R}(\tau)$, of the n-vector

$$\mathbf{f} = [f_1, f_2, \ldots, f_n]^{\mathrm{T}} \tag{10.4}$$

defined by

$$\mathbf{R}(\tau) = E[\mathbf{f}(t)\mathbf{f}^{\mathrm{T}}(t+\tau)] \tag{10.5}$$

is of the form

$$\mathbf{R}(\tau) = \mathbf{D}\delta(\tau) \tag{10.6}$$

where $\mathbf{D} = [D_{ij}]$ is an n × n symmetric matrix of constants.

It can be shown (e.g. see Dimentberg, 1988) that, with the severe restriction that

$$\frac{2\beta_{ij}}{D_{ij}} = \gamma \quad i, j = 1, 2, \ldots, n \tag{10.7}$$

where γ is a constant, the exact stationary solution of the governing FPK equation can be obtained. This solution is as follows

$$f(\mathbf{z}) = C\exp\left(-\gamma\int_0^E H(\xi)\,\mathrm{d}\xi\right) \tag{10.8}$$

Here $\mathbf{z}$ is the $2n$ state variable vector

$$\mathbf{z} = [\mathbf{y}, \dot{\mathbf{y}}]^{\mathrm{T}} \tag{10.9}$$

and $f(\mathbf{z})$ is the joint, stationary probability density function for $\mathbf{z}$. The symbol C denotes a normalization constant which satisfies the condition

$$\int_{-\infty}^{\infty} f(\mathbf{z})\,\mathrm{d}\mathbf{z} = 1 \tag{10.10}$$

In the special case where $n = 1$ corresponding to an SDOF system, equation (10.8) reduces to the result used earlier, in Chapter 9 (see equation (9.153)).

10.2.1 Linear damping

Another important special case is that of linear damping. Then, for an n degree of freedom system, one can set $H(E) = 1$ in equation (10.1), with the consequence that equation (10.8) reduces to

$$\begin{aligned} f(\mathbf{z}) &= C\exp(-\gamma E) \\ &= C\exp[-\gamma(\tfrac{1}{2}\dot{\mathbf{y}}^{\mathrm{T}}\dot{\mathbf{y}} + V)] \end{aligned} \tag{10.11}$$

An integration of $f(\mathbf{z})$ with respect to the velocity vector, $\dot{\mathbf{y}}$, shows that the probability density function of $\mathbf{y}$ is given by

$$f(\mathbf{y}) = C_1\exp(-\gamma V) \tag{10.12}$$

where C_1 is another normalization constant. Similarly, an integration of equation (10.11) with respect to the displacement vector, $\mathbf{y}$, shows that the probability density function of $\dot{\mathbf{y}}$ is Gaussian, and is given by

$$f(\dot{\mathbf{y}}) = c_2\exp(-\tfrac{1}{2}\gamma\dot{\mathbf{y}}^{\mathrm{T}}\dot{\mathbf{y}}) \tag{10.13}$$

where C_2 is another normalization constant. Clearly C_1 and C_2 must satisfy

$$\int_{-\infty}^{\infty} f(\mathbf{y})\,\mathrm{d}\mathbf{y} = \int_{-\infty}^{\infty} f(\dot{\mathbf{y}})\,\mathrm{d}\dot{\mathbf{y}} = 1 \tag{10.14}$$

It is easy to show from these results that

$$E\{\mathbf{y}\dot{\mathbf{y}}^{\mathrm{T}}\} = \mathbf{0} \tag{10.15}$$

and

$$E\{\dot{\mathbf{y}}\dot{\mathbf{y}}^{\mathrm{T}}\} = \frac{1}{\gamma}\mathbf{I} \tag{10.16}$$

where $\mathbf{0}$ and $\mathbf{I}$ are the null and identity matrices, respectively. The covariance matrix for $\mathbf{y}, E\{\mathbf{y}\mathbf{y}^{\mathrm{T}}\}$, depends on the particular form of the potential energy function, V.

10.2.2 Chain-like systems

If the MDOF system governed by equation (10.1) has a simple 'chain-like' structure, and the damping is linear, then the expression given above for $f(\mathbf{y})$ may be considerably simplified.

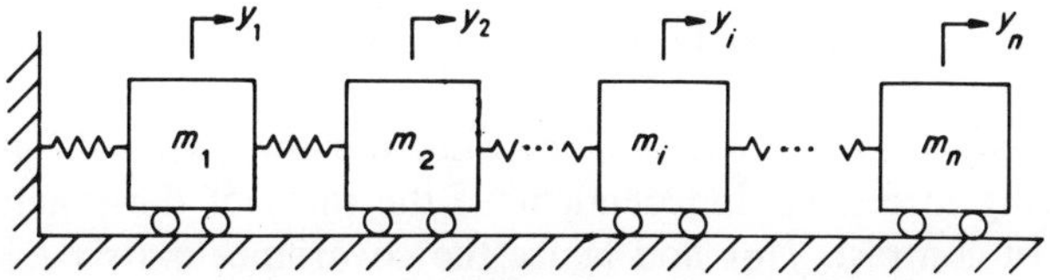

Figure 10.1
n degree of freedom mass–spring system

Consider, for example, the system shown in Figure 10.1. Here there are n masses, and n non-linear springs, interconnected as shown, together with a set of linear damping elements (not shown). The n-vector of the relative displacements, $q_1, q_2, \ldots, q_n$, across the n springs,

$$\mathbf{q} = [q_1, q_2, \ldots, q_n]^{\mathrm{T}} \tag{10.17}$$

may be easily related to the absolute displacement vector, $\mathbf{y}$, through the linear transformation

$$\mathbf{y} = \mathbf{A}\mathbf{q} \tag{10.18}$$

where

$$\mathbf{A} = \begin{bmatrix} 1 & 0 & \cdot & \cdot & \cdot & \cdot & 0 \\ 1 & 1 & \cdot & \cdot & \cdot & \cdot & 0 \\ \cdot & \cdot & \cdot & \cdot & \cdot & \cdot & \cdot \\ 1 & 1 & \cdot & \cdot & \cdot & \cdot & 1 \end{bmatrix} \tag{10.19}$$

Since the total potential energy, V, is simply the sum of the potential energies in the n springs one can write

$$V = \sum_{i=1}^{n} V_i(q_i) \tag{10.20}$$

where V_i is the potential energy in the ith spring. Transforming $f(\mathbf{y})$ and $f(\dot{\mathbf{y}})$ to $f(\mathbf{q})$ and $f(\dot{\mathbf{q}})$, respectively, according to equation (10.18), and using equation (10.20), it is found that

$$f(\mathbf{q}) = C_1 \exp\left(-\gamma \sum_{i=1}^{n} V_i(q_i) \right) \tag{10.21}$$

and

$$f(\dot{\mathbf{q}}) = C_2 \exp(-2\gamma \dot{\mathbf{q}} \mathbf{A}^{\mathrm{T}} \mathbf{A} \dot{\mathbf{q}}) \tag{10.22}$$

The normalization constants are unchanged by the transformation, since the modulus of the appropriate Jacobian is unity (see equation (3.15)).

From equations (10.21) and (10.22) it is easily found that

$$E\{\mathbf{q}\dot{\mathbf{q}}^{\mathrm{T}}\} = \mathbf{0} \tag{10.23}$$

and

$$E\{\dot{\mathbf{q}}\dot{\mathbf{q}}^{\mathrm{T}}\} = \frac{1}{\gamma} A^{-1}(A^{-1})^{\mathrm{T}} \tag{10.24}$$

As for the n-vector $\mathbf{y}$, these results are independent of the form of the non-linearity in the restoring forces, i.e. independent of the form of the potential energy function, V. In contrast, $f(\mathbf{q})$, and hence the covariance matrix $E\{\mathbf{q}\mathbf{q}^{\mathrm{T}}\}$, does depend on V. However, it follows from equation (10.21) that q_i $(i = 1, 2, \ldots, n)$

are statistically independent processes, since

$$f(\mathbf{q}) = \prod_{i=1}^{n} f(q_i) \tag{10.25}$$

where

$$f_i(q_i) = C_i' \exp(-\gamma V_i(q_i)) \quad i = 1, 2, \ldots, n \tag{10.26}$$

and C_1' are normalization constants (see equation (3.11)). Hence the covariance matrix for $\mathbf{q}$ is diagonal; thus

$$E\{q_i q_j\} = 0 \quad i \neq j \tag{10.27}$$

and

$$\sigma_{q_i}^2 = E\{q_i^2\} = \frac{\int_{-\infty}^{\infty} q_i^2 \exp(-\gamma V_i(q_i))\, \mathrm{d}q_i}{\int_{-\infty}^{\infty} \exp(-\gamma V_i(q_i))\, \mathrm{d}q_i} \tag{10.28}$$

The foregoing analysis demonstrates that, for the system under consideration, the exact solution 'uncouples'. Thus, dropping the subscript, $f(q)$, as given in equation (10.26) is the exact probability density function for a SDOF system with the equation of motion

$$\ddot{q} + \beta q + g(q) = f(t) \tag{10.29}$$

where

$$g(q) = \frac{\mathrm{d}V}{\mathrm{d}q} \tag{10.30}$$

$$E\{f(t)f(t+\tau)\} = D\delta(\tau) \tag{10.31}$$

and

$$\gamma = 2\beta/D \tag{10.32}$$

If the system is linearized, according to the statistical linearization method given in Chapter 6, then, because of the chain-like structure, each non-linear spring is essentially linearized separately (see Section 6.5.2). Thus, if

$$g_i(q_i) = \frac{\mathrm{d}V_i(q_i)}{\mathrm{d}q_i} \tag{10.33}$$

is the restoring function for the ith spring, then the equivalent linear spring has a stiffness (see equation (6.51))

$$k_i = E\left\{\frac{\mathrm{d}g_i}{\mathrm{d}q_i}\right\} = E\left\{\frac{\mathrm{d}^2 V_i}{\mathrm{d}q_i^2}\right\} \tag{10.34}$$

The results given by equations (10.23) to (10.27) are still applicable, for the linearized system, and equation (10.28) reduces to

$$\sigma_{q_i}^2 = E\{q_i^2\} = \frac{1}{\gamma k_i} \tag{10.35}$$

10.2.3 First-order systems

In some special problems the restoring term is absent and the analysis can be considerably simplified. In particular, it is sometimes possible to obtain exact solutions for systems with arbitrary non-linear damping.

An example of such a problem has been considered earlier, in Section 5.6.1. For the friction controlled slip of a structure on a foundation, the governing equation (see equation (5.175)) contains only inertial and damping terms, on the left-hand side, and an excitation term, $f(t) = -x_g(t)$, on the right. It is of the form

$$\ddot{q} + d(\dot{q}) = f(t) \tag{10.36}$$

where $d(\dot{q})$ is a non-linear damping function.

If $f(t)$ is a zero-mean white noise process, with a covariance function given by equation (10.31), then an exact stationary solution for probability density function of $\dot{q}$ can easily be found (e.g. see Roberts, 1981a); this is due to the fact that the governing equation for $\dot{q}$ is of first order. The result is

$$f(\dot{q}) = C\exp\left(-\frac{2}{D}\int_0^{\dot{q}} d(\xi)\,d\xi\right) \tag{10.37}$$

10.3 COMPARISON WITH EXACT SOLUTIONS

In this section the exact solutions given by equation (10.8) (for $n = 1$), equations (10.28) and (10.37) will be compared with the corresponding results derived from the statistical linearization method, for specific forms of non-linearity. The basis for comparison will be the standard deviation of the response (and, in some cases, the mean of the response).

In the subsequent discussion, only SDOF systems will be considered. However, as shown in the preceding section, the result given by equation (10.28) is an exact solution for a special class of MDOF systems, in addition to being the exact solution to the SDOF system governed by equation (10.29).

10.3.1 First-order systems

For the problem considered in Section 5.6.1 (and 7.3.1) the equation of motion is governed by equation (5.175). If $f(t) = -\ddot{x}_g(t)$ is a zero-mean white noise process, with a constant spectral level, S_0, then, using equations (10.36) and (10.37), with $D = 2\pi S_0$, the stationary probability density function for the velocity, $\dot{q}$, is given exactly by

$$f(\dot{q}) = C \exp\left(-\frac{\mu g(\dot{q})}{\pi S_0} \right) \tag{10.38}$$

(noting that $D = 2\pi S_0$). By integration, the normalization constant is found to be given by

$$C = \frac{\mu g}{2^{3/2}\pi S_0} \tag{10.39}$$

From equations (10.38) and (10.39) one can obtain an exact expression for $\sigma_{\dot{q}}^2$. Thus

$$\sigma_{\dot{q}}^2 = \int_{-\infty}^{\infty} \dot{q}^2 f(\dot{q})\, \mathrm{d}\dot{q} \tag{10.40}$$

and hence, by integration, one obtains

$$\sigma_{\dot{q}}^2 = \frac{\sqrt{2\pi^2 S_0^2}}{(\mu g)^2} \tag{10.41}$$

This result may be compared with the corresponding result obtained by the method of statistical linearization, see equation (5.185). It is found that

$$R \equiv \frac{\sigma_{\dot{q}}(\text{stat. lin.})}{\sigma_{\dot{q}}(\text{exact})} = \frac{\pi^{1/2}}{2^{3/4}} = 1.053 \tag{10.42}$$

Thus, the percentage error involved in the statistical linearization result is about 5.3%, in the present case.

It is noted that, for the case under discussion, an exact, non-stationary solution is also available (see Ahmadi, 1983). This could be compared with the statistical linearization results for the non-stationary response, deduced by the method given in Chapter 7. Such a comparison will not be undertaken here. One can anticipate, however, that the percentage error involved in the non-stationary standard deviation of the response will be of similar magnitude to that pertaining to the stationary response.

10.3.2 Oscillators with power-law springs

For an oscillator with linear damping and non-linear stiffness, driven by a zero mean white noise process, the appropriate equation of motion is given by equation (10.29). If the restoring term is of the simple power law form

$$g(q) = \alpha |q|^{\nu} \operatorname{sgn}(q) \tag{10.43}$$

where α and ν are positive constants, then, from equations (10.26) and (10.32), the exact stationary probability density function for q is given by

$$f(q) = C\exp\left(-\frac{\alpha\beta\,|q|^{1+\nu}}{\pi S_0(1+\nu)}\right) \tag{10.44}$$

Hence, by integration (see equation (10.28)) it is found that the standard deviation of the response may be expressed as

$$\sigma_q^2 = \left(\frac{(1+\nu)\pi S_0}{\alpha\beta}\right)^{\frac{2}{(1+\nu)}} \frac{\Gamma\left(\frac{3}{1+\nu}\right)}{\Gamma\left(\frac{1}{1+\nu}\right)} \tag{10.45}$$

This result may be compared with the corresponding statistical linearization estimate of σ_q. The equivalent linear equation is

$$\ddot{q} + \beta\dot{q} + \omega_{\text{eq}}^2 q = f(t) \tag{10.46}$$

where

$$\omega_{\text{eq}}^2 = \alpha E\left\{\frac{\mathrm{d}}{\mathrm{d}q}|q|^{\nu}\right\} \tag{10.47}$$

If, as usual, the response is taken to be Gaussian (here with zero mean), then the evaluation of the above expectation leads to (see Appendix A, Table A1)

$$\omega_{\text{eq}}^2 = \frac{\alpha}{\pi^{1/2}} 2^{(1+\nu)/2} \sigma^{\nu-1} \Gamma\left(1 + \frac{\nu}{2}\right). \tag{10.48}$$

Also (see equation (10.35))

$$\sigma_{\text{q}}^2 = \frac{1}{\gamma\omega_{\text{eq}}^2} = \frac{\pi S_0}{\beta\omega_{\text{eq}}^2} \tag{10.49}$$

Hence, a combination of equation (10.48) with equation (10.49) yields the

Table 10.1

ν	R	% *error*
0	0.886	11.4
1	1	0
2	0.971	2.9
3	0.924	7.6
4	0.878	12.2
5	0.837	16.2

following expression for R, as defined by equation (10.42)

$$R = \frac{1}{\sqrt{2}}\left(\frac{\pi^{1/2}}{(1+v)\Gamma\left(1+\frac{v}{2}\right)}\right)^{1/(1+v)}\left(\frac{\Gamma\left(\frac{1}{1+v}\right)}{\Gamma\left(\frac{3}{1+v}\right)}\right)^{1/2} \tag{10.50}$$

Table 10.1 gives values of R, and the percentage error, $100|1-R|$, for integer values of v up to 5. As expected, the percentage error tends to increase as v increases beyond unity, but is well below 20%, for this range of v values.

10.3.3 Duffing oscillators

If $g(q)$ in equation (10.29) is of a linear-plus-cubic form then the system is a Duffing oscillator. The stationary response of Duffing oscillators to zero-mean white noise random excitation has been studied earlier, in Chapters 5 and 9. For the case of linear damping a suitable normalised form of the equation of motion is (see equation (9.25))

$$\ddot{Y} + 2\zeta\dot{Y} + Y(1+\rho Y^2) = (4\zeta)^{1/2} f_n(\tau) \tag{10.51}$$

where Y, ρ, f_n and τ are defined in Section 9.2.1. With this normalization γ, as defined by equation (10.32), is unity.

Applying equation (10.26) again, the exact stationary probability density function for Y is obtained as

$$f(Y) = C\exp\left[-\left(\frac{Y^2}{2}+\frac{\rho Y^4}{4}\right)\right] \tag{10.52}$$

Hence, by integration, the normalization constant is found to be given by

$$C^{-1} = \left(\frac{\pi}{\rho}\right)^{1/2}\exp\left(\frac{1}{8\rho}\right)K_{1/4}\left(\frac{1}{8\rho}\right) \tag{10.53}$$

where $K_{1/4}$ is a modified Bessel function (see Abramowitz and Stegun, 1972). Hence, from equation (10.28), the standard deviation of the response is obtained as

$$\sigma_Y^2 = \left(\frac{\pi\rho}{2}\right)^{1/2}\left(\frac{\rho}{2}\right)^{-3/4} D_{-3/2}\left(\frac{1}{(2\rho)^{1/2}}\right)K_{1/4}^{-1}\left(\frac{1}{8\rho}\right) \tag{10.54}$$

where $D_{-3/2}$ is a parabolic cylinder function (see Abramowitz and Stegun, 1972).

From this exact solution for σ_Y, asymptotic expansions, in terms of ρ, can be derived (e.g. see Wu and Lin, 1984). For small ρ one has

$$\sigma_Y^2 = 1 - 3\rho + 24\rho^2 - 297\rho^3 + 4896\rho^4 - 100\,278\rho^5 \tag{10.55}$$

Table 10.2

Method	*k*	*% error*
Statistical linearization	0.7599	7.6
Fourth-order closure	0.7953	3.3
Sixth-order closure	0.8050	2.1
Exact	0.8222	0

correct to order ρ^5. This result, when compared with that obtained by statistical linearization (see equation (9.114)), shows that the latter is only correct to order ρ. In contrast, the fourth order, non-Gaussian closure result obtained in Chapter 9 (see equation (9.134)), agrees with the exact solution up to order ρ^3. The sixth-order non-Gaussian closure result obtained by Wu and Lin (1984) (see equation (9.137)), is actually in complete agreement with equation (10.55); i.e. correct up to order ρ^5.

For large ρ, σ_Y obeys the asymptotic form

$$\sigma_Y \to \frac{k}{\rho^{1/4}} \quad \rho \to \infty \tag{10.56}$$

Table 10.2 gives the values of k obtained by statistical linearization, compared with corresponding values found by non-Gaussian closure (see Section 9.3.3), and from the exact solution.

Of course, as $\rho \to \infty$, the cubic term is the equation of motion dominates the linear term and the percentage error agrees with that found in the preceding section, for an oscillator with a cubic spring corresponding to $\nu = 3$.

To assess the reliability of the method for intermediate non-linearities the value of σ_Y^2 is plotted against ρ in Figure 10.2. It is observed that both the exact and the statistical linearization solutions decrease as ρ increases, the approximate solution remaining less than the exact solution by an error not exceeding 7.6%, based on σ_Y. Related results for a hardening system of the Duffing type with an asymmetric non-linearity can be found in Spanos (1980).

A similar comparison between the exact solution and the results from fourth-order closure and sixth-order closure (compare Figures 9.1 and 10.2) reveals that closure, at both levels, is in close agreement with the exact solution. Indeed the error over the range of ρ shown is comparable with the error relating to the asymptotic case, $\rho \to \infty$ (see Table 10.2). Therefore, progressing from second order closure (statistical linearization) to fourth-order closure reduces the error by roughly 50% (based on σ_Y). Further, but less significant reduction in error is achieved with sixth-order closure.

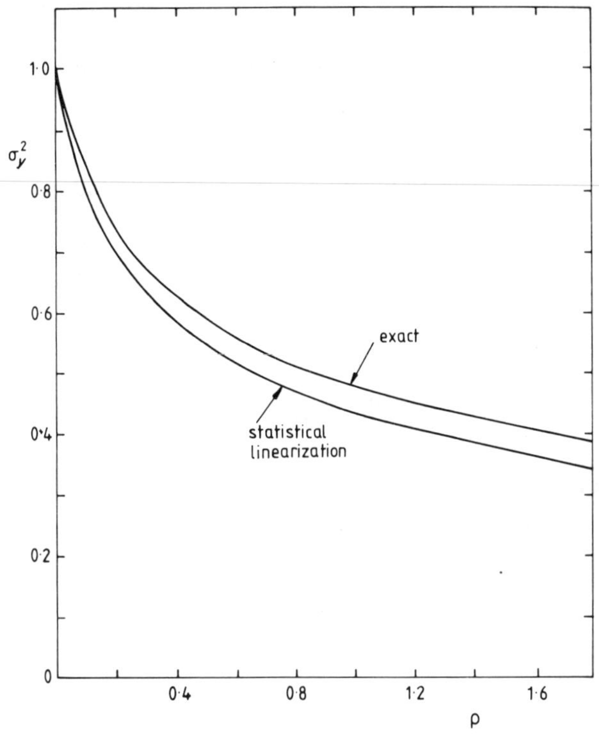

Figure 10.2
Variation of the mean square response with ρ, according to the exact solution and to the statistical linearization solution, for the case of an oscillator with linear damping and linear-plus-cubic stiffness

10.3.4 Oscillators with tangent-law springs

As a third example of an oscillator with non-linear stiffness, the case where $g(q)$ in equation (10.29) is of the form

$$g(q) = \frac{2}{\pi} K \tan^{-1}\left(\frac{\pi k_q}{2K}\right) \tag{10.57}$$

will be considered (see also Iwan and Yang, 1972), where K is the maximum force exerted by the springs. Clearly for small values of q the function $g(q)$ may be approximated by a linear function, of slope k. For greater values of q the non-linearity of equation (10.57) 'softens' and in fact the restoring force becomes practically independent of q.

The potential energy function associated with $g(q)$, as given by equation (10.57), is

$$V(q)=\frac{2K}{\pi}\left\{q\tan^{-1}\left(\frac{\pi k_q}{K}\right)-\frac{K}{\pi k}\ln\left[1+\left(\frac{\pi k}{2K}\right)^2\right]\right\}. \tag{10.58}$$

Hence, using equations (10.26) and (10.28) again, exact expressions for the stationary probability density function for q, and the standard deviation of q, σ_q, may be obtained. In general a numerical quadrature scheme is needed to compute σ_q.

The corresponding statistical linearization result for this system may be obtained by employing the equivalent linear system again, where here

$$\omega_{\text{eq}}^2=\frac{2K}{\pi}E\left\{\frac{\mathrm{d}}{\mathrm{d}q}\tan^{-1}\left(\frac{\pi k_q}{2K}\right)\right\}=kE\left\{\frac{1}{1+\left(\frac{\pi k}{2K}\right)^2 q^2}\right\} \tag{10.59}$$

Assuming a Gaussian distribution for q, the expectation in equation (10.59) is readily evaluated. Hence one finds that

$$\omega_{\text{eq}}^2=\frac{2\,K}{\pi\,\sigma_q}\exp(\mu^2)[1-\operatorname{erf}(\mu)] \tag{10.60}$$

where

$$\mu=\frac{\sqrt{2}K}{\pi k\sigma_q} \tag{10.61}$$

Hence, using equation (10.49) again, the following result is obtained

$$\sigma_q=\frac{\pi}{2}\left(\frac{\pi S_0}{\beta K}\right)\exp(-\mu^2)[1-\operatorname{erf}(\mu)]^{-1} \tag{10.62}$$

In the limiting case where $k/K\to\infty$, then the exact solution for σ_q^2 can be found analytically. The result is

$$\sigma_q=\sqrt{2}\left(\frac{\pi S_0}{\beta K}\right) \tag{10.63}$$

The corresponding asymptotic statistical linearization result is, from equation (10.61),

$$\sigma_q=\left(\frac{\pi}{2}\right)^{1/2}\left(\frac{\pi S_0}{\beta K}\right) \tag{10.64}$$

Note that, in the limit, the tangent function of equation (10.57) becomes proportional to the signum function, i.e.

$$g(q)\to K\,\operatorname{sgn}(y) \tag{10.65}$$

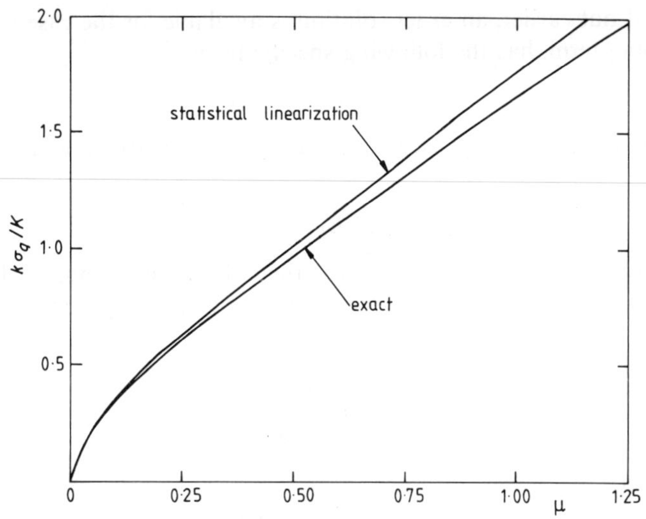

Figure 10.3
Variation of the standard deviation of the response with μ, according to the exact solution and to the statistical linearization solution, for the case of an oscillator with a tangent-law stiffness

Then equations (10.62) and (10.65) correspond to the results given earlier, in Section 10.3.2, for a power-law spring with $\gamma = 0$ (noting that $\alpha = K$). Thus the error of the statistical linearization result for σ_q, in this limiting case, from Table 10.1, about 11.4%.

Figure 10.3 shows the variation of $k\sigma_q/K$ with the parameter

$$\mu = \frac{\pi S_0 k}{\beta K^2} \tag{10.66}$$

according to the exact solution, and to the statistical linearization result. For the considered values, the error in the σ_q calculation, through statistical linearization, increases monotonically with μ. However, this error remains appreciably smaller than its limiting value (11.4%), corresponding to k/K (or μ) approaching infinity, over the μ range shown.

10.3.5 Oscillators with non-linear damping

Oscillators with linear stiffness, and non-linear damping, driven by zero-mean, stationary white excitation will now be considered. An appropriate equation of motion is

$$\ddot{q} + b(q, \dot{q}) + \omega_n^2 q = f(t) \tag{10.67}$$

As pointed out earlier, an exact solution is available for the case where $b(q, \dot{q})$, the damping term, has the following specific form

$$b(q, \dot{q}) = \dot{q}H(E) \tag{10.68}$$

where E is the total energy (see equations (10.1) and (10.2)), with $n = 1$.

If

$$H(E) = cE^{\nu} \tag{10.69}$$

where c and γ are positive constants, then the exact standard deviation of the response, σ_q, can be evaluated fairly easily (see equation (9.207)). The result is

$$\sigma_q^2 = \frac{1}{\omega_n^2}\left(\frac{\pi S_0}{c}\right)^{1/(1+\nu)} h(\nu) \tag{10.70}$$

where

$$h(\nu) = (1+\nu)^{1/(1+\nu)} \frac{\Gamma\left(\dfrac{2}{1+\nu}\right)}{\Gamma\left(\dfrac{1}{1+\nu}\right)} \tag{10.71}$$

Applying the standard statistical linearization method a result of similar form to equation (10.70) is obtained, but now

$$h(\nu) = \left(\frac{1}{(1+\nu)\Gamma(1+\nu)}\right)^{1/(1+\nu)} \tag{10.72}$$

Table 10.3 gives the values of $h(\nu)$, for various ν, according to the exact result, and according to the statistical linearization approximation, together with the percentage error based on σ_q. As expected, the error reduces to zero as ν is reduced towards zero.

Table 10.3

	$h(\nu)$		
ν	*exact*	*Statistical linearization*	% error
0.5	0.8641	0.8271	2.2
1	0.7979	0.7071	5.9
1.5	0.7568	0.6185	9.6
2	0.7288	0.5503	13.1
3	0.6667	0.4518	17.7

10.4 COMPARISON WITH MONTE CARLO SIMULATION RESULTS

As pointed out in the introduction to this chapter, Monte Carlo simulation is often the sole tool available for assessing the accuracy of random vibration solutions generated by approximate methods of analysis, such as statistical linearization. (Shinozuka, 1972; Rubinstein, 1981; Spanos and Mignolet, 1989).

Here results from digital simulation studies will be compared with corresponding statistical linearization results (and, in some cases, with results obtained by the method of equivalent non-linear equations (ENLE)) for a selection of particular problems.

10.4.1 Simulation technique

The basic principle of the digital simulation technique has been discussed earlier, in Chapter 1, and also in Chapter 9 (see Section 9.6.3). Basically, sample functions of the excitation process are generated, over a fixed interval of time and converted into corresponding sample functions of the response process, through a numerical integration of the governing equation of motion. Actually, since the procedure is digital, the sample functions are evaluated only at equi-spaced instants of time.

For cases where the excitation process is stationary, and ergodic, it is only necessary to generate a single sample function of the response process, $y(t)$, from a corresponding single sample function of the excitation process. If $y_i(t_j)$ denotes the computed response sample function, where $t_j = j\Delta t$, Δt is the basic time interval and $j = 1, 2, \ldots N$, then estimates of statistical parameters relevant to the response process can be readily generated by the usual methods. For example, an estimate, $\hat{m}_y$, of the mean response, m_y, is obtained from the equation

$$\hat{m}_y = \frac{1}{N} \sum_{j=1}^{N} y_i(t_j) \tag{10.73}$$

and an estimate, $\hat{\sigma}_y$, of the standard deviation of the response, σ_y, may be obtained as follows

$$\hat{\sigma}_y = \left(\frac{1}{N-1} \sum_{i=1}^{N} [y_i(t_j) - \hat{m}_y]^2 \right)^{1/2} \tag{10.74}$$

A characteristic feature of these estimates is that their accuracy is inversely proportional to the square root of N, the number of samples computed. Thus, to improve the accuracy by a factor of 10 it is necessary to increase the amount of computation by a factor of 100. Precise confidence limits for the estimates $\hat{m}_y$ and $\hat{\sigma}_y$ can be established in the case where the samples are statistically independent. However, in the majority of cases of interest, the system of concern is lightly damped. The consequence of this is that there is a high degree of correlation between successive samples. Thus the equivalent number of independent samples, N_{eq} say, is often substantially less than the actual number of samples. Methods exist for estimating the ratio N/N_{eq} in specific cases (e.g. see Roberts and Dacunha, 1985).

From the sample values $y_i(t_j)$ one can also obtain estimates of the probability density function of the response, by forming histograms, in the normal manner. However, to obtain reliable estimates in the 'tails' of the distribution, where samples occur with low probability, it is necessary to choose very large values for N (typically $N \sim 10^6$).

In the case of systems with non-stationary excitation it is necessary to replace the averaging of single sample functions with ensemble averaging over a large number of sample functions. Thus, if M sample functions, $y_1(t_j), y_2(t_j), \ldots, y_M(t_j)$ are generated, each for $j = 1, 2, \ldots, N$, then an estimate of the mean at $t_j, m(t_j)$, may be obtained from

$$\hat{m}_y(t_j) = \frac{1}{M}\sum_{i=1}^{M} y_i(t_j) \tag{10.75}$$

and the standard deviation of the response at time $t_j, \sigma_y(t_j)$, may be estimated from

$$\hat{\sigma}_y(t_j) = \left[\frac{1}{M-1}\left(\sum_{i=1}^{M} [y_i(t_j) - \hat{m}(t_j)]^2\right)\right]^{1/2} \tag{10.76}$$

Clearly, considerably more computation is required than in the case of stationary, ergodic excitation.

The accuracy of the estimates $\hat{m}_y(t_j)$ and $\hat{\sigma}_y(t_j)$ is easier to assess, in the case of ensemble averaging, since the sample values at a particular t_j will be statistically independent. In fact, for large values of M one has the usual well known results that $\hat{m}(t_j)$ has a standard deviation given by

$$\hat{\sigma}_1(t_j) = \frac{\sigma_y(t_j)}{M^{1/2}} \tag{10.77}$$

whereas $\hat{\sigma}_y(t_j)$ has a standard deviation given by

$$\hat{\sigma}_2(t_j) = \frac{\sigma_y(t_j)}{(2M)^{1/2}} \tag{10.78}$$

It is seen from these equations that the mean estimate has a wider dispersion than the standard deviation estimate, for a fixed value of M (Spanos, 1981b)

For the case of white noise excitation, a method of generating sample functions of the excitation, in an approximate sense, has been described earlier, in some detail (see Section 9.6.3). For non-white excitation processes, one can employ pre-filters and numerically integrate the combined system, consisting of the pre-filter and original system in series (driven by white noise) to obtain sample functions of the response from sample functions of white noise excitation. (Spanos, 1983, 1986). Alternatively, digital filters associated with auto-regressive-moving-average (ARMA) algorithms can be used to simulate realizations of processes with specified spectra (Spanos and Mignolet, 1989).

The foregoing method of digital simulation can be generalized to cope with systems where there are multiple inputs and outputs, without great difficulty. In the following, only cases where there is a single excitation process will be considered.

Confidence limits for the presented simulation results will not be given here. In most cases where the excitation is stationary, so that time averaging is used, very large values of N are used to obtain the results, and the dispersion of the estimates is not significant. For cases where low values of N are used, or where ensemble averaging, rather than time averaging, is used, due to non-stationarity of the excitation, a visual inspection of the graphically presented estimates gives a good indication of their variability.

10.4.2 Oscillators with non-linear damping

As pointed out earlier, exact solutions for oscillators with arbitrary non-linear damping, and linear stiffness do not exist, even in the case of white noise excitation.

Earlier, in Chapter 9 (see Section 9.4.5), the specific case of an oscillator with linear-plus-cubic damping, linear stiffness and stationary white noise excitation was considered (see equation (9.224) for the governing equation of motion). Approximate results for the mean square response, σ_Y^2, obtained by the methods of non-Gaussian closure, equivalent non-linear equations (ENLE) and statistical linearization, were compared for various values of the non-linearity parameter, μ (see Table 9.3 and Figure 9.2). This comparison shows that ENLE results are in close agreement with those obtained by a fourth-order non-Gaussian closure technique.

Digital simulation estimates of σ_Y^2, have also been obtained for this particular case (see Roberts, 1977). Each estimate was derived from a single, computed sample function, $y_i(\tau_j)$, of the response, where N, the number of samples, was equal to 10^6, and the non-dimensional time step, $\Delta\tau$, was set to 0.02. The estimates of σ_Y^2, so obtained, for $\zeta = 0.05$ and $\zeta = 0.5$, are shown in Figure 10.4; here they are compared with the approximate statistical linearization and ENLE solutions.

It is observed that the ENLE result is in very close agreement with the simulation estimates relating to the lower value of ζ. As pointed out earlier, the ENLE result is identical to that obtained by the method of stochastic averaging. Since the latter is known to be asymptotically correct as the damping approaches zero, it is not surprising that the ENLE method is near-exact at $\zeta = 0.05$.

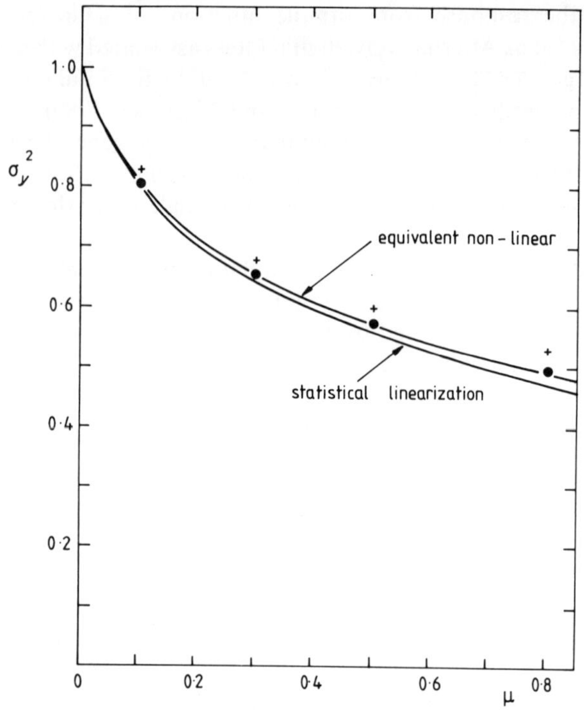

Figure 10.4
Variation of the mean square response with μ, according to statistical linearization, for an oscillator with linear stiffness and linear-plus-cubic damping. Comparison with simulation estimates (● $\zeta = 0.05$, $+\zeta = 0.5$) and with the ENLE result (adapted from Roberts (1977) by permission from Academic press)

The simulation results for $\zeta = 0.5$ lie a little above the ENLE result, indicating that the accuracy begins to deteriorate at this level of damping.

Taking, on the basis of this evidence, the ENLE result to be virtually exact, for the case of light damping corresponding to $\zeta < 0.1$, say it is possible to estimate the error inherent in the statistical linearization approximation. In fact the results in Table 9.3 can be converted into R values (see equation (10.42)) and hence into percentage error estimates, for the statistical linearization results. The results of these calculations, for several values of μ, are given in Table 10.4.

It is clear, from these results, that the percentage error is very small, over a wide range of μ values, and asymptotes towards a value close to 5%, as $\mu \to \infty$. The existence of such an asymptotic limit is not unexpected, in view of its existence in the case of oscillators with non-linear stiffness, and linear damping (see Sections 10.3.2 to 10.3.4). It corresponds, of course, to the limiting situation where the linear damping term becomes negligible, compared with cubic,

Table 10.4

μ	R	% error
1	0.971	2.9
3	0.964	3.6
5	0.957	4.3
7	0.954	4.6
9	0.953	4.7

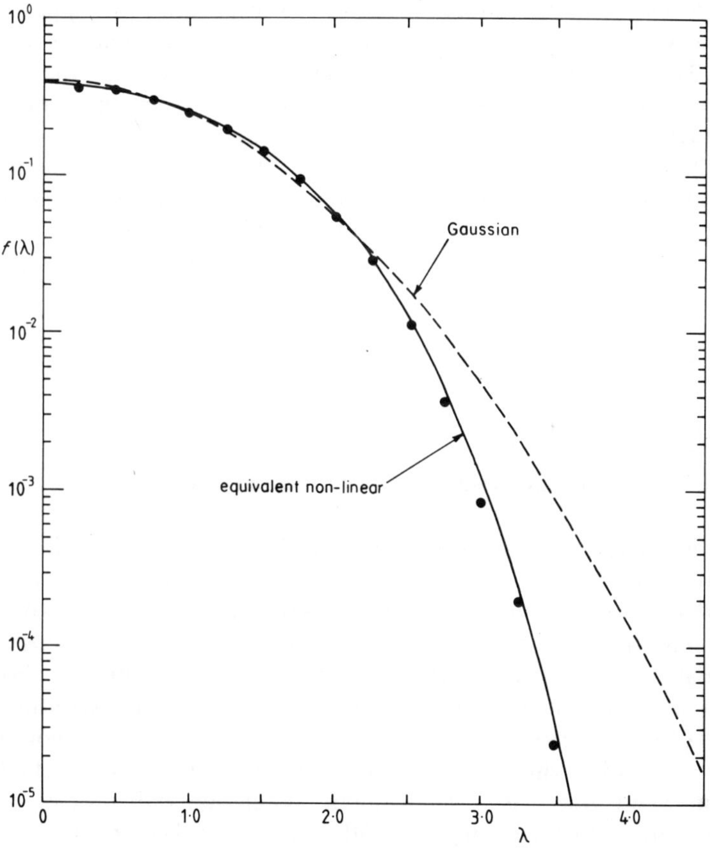

Figure 10.5
Probability density function for the displacement response for an oscillator with linear stiffness and linear-plus-cubic damping ($\mu = 1$, $\zeta = 0.05$). Comparison between the Gaussian distribution, the ENLE result and simulation estimates

non-linear term. One can expect that the percentage error for this limiting case will be similar to that for the oscillator governed by equations (10.67) to (10.69), where $v = 1$, since with this choice of $v, b(q, \dot{q})$ is of order $\dot{q}^3$. Table 10.2 shows that, for $v = 1$, the percentage error (5.9%) is indeed of very similar magnitude to that approached in Table 10.3.

In Roberts (1977), a comparison between simulation estimates of the probability density function of the response, and the corresponding prediction by the ENLE method was presented. Figure 10.5 shows a typical comparison, for the case where $\mu = 1$ and $\zeta = 0.05$ (for simulation). See also Section 9.9.5, where $f(\lambda)$ and λ are defined, and Figure 9.6. Again, in the simulation, $N = 10^6$ and $\Delta\tau = 0.02$. It is observed that the ENLE result is in excellent agreement with the simulation estimates, whereas the corresponding Gaussian distribution (relating to the statistical linearization approximation) is seriously in error in the tail of the distribution. A comparison between Figures 9.6 and 10.5 also reveals that progressing from second-order, Gaussian closure to fourth-order, non-Gaussian closure leads to a significant improvement in accuracy, with respect to the distribution of the response.

Finally, it is noted that, for oscillators with pure quadratic damping and linear stiffness, the results given in Section 9.4.4 also lead to an estimation of the accuracy of the statistical linearization result. Using, again, the fact that the ENLE result is asymptotically exact as the damping approaches zero, it follows from equations (9.215) and (9.219) that the accuracy of the σ_y value obtained by statistical linearization is about 2.2%. This figure agrees closely with that pertaining to the oscillator governed by equations (10.67) to (10.69), where $v = \frac{1}{2}$.

10.4.3 Oscillators with non-linear springs

Earlier, in Chapter 5, statistical linearization results were obtained for the stationary response of oscillators with linear damping and various types of non-linear stiffness. In some cases (see Sections 5.3.6 to 5.3.8) it was found that solutions existed, at least in some regions of the appropriate parameter space, where, strictly, a stationary solution should not exist. In such cases it was suggested that the statistical linearization results relate to short-term statistics of the response, rather than to the more normal, long-term statistics which are obtained from the exact solution of the FPK equation (where this exists). Here digital simulation estimates of oscillator response statistics will be compared with some of the theoretical results found earlier, in Chapter 5.

To begin with, the oscillator governed by equation (5.89) will once again be considered. The calculations described in Section 5.3.6 showed that there is only a finite region in the ρ–Θ parameter plane (see equations (5.88) and (5.94) for

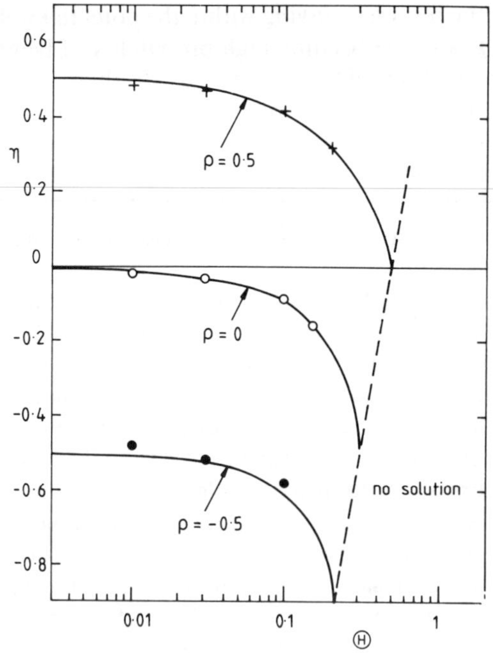

Figure 10.6
Variation of η with Θ, for various values of ρ, (———) and a comparison with simulation estimates (+, ○, ●)

definitions of ρ and Θ) for which solutions exist. In particular, it was found that, for a particular value of ρ, the scaled mean response, η, (see equation (5.86)) varied with Θ according to one of two solutions, up to a limiting value of Θ, at which the solutions coalesced, as shown in Figure 5.4a.

In Figure 10.6 the theoretical results for η are compared with corresponding digital simulation estimates, for three values of ρ. Each simulation estimate is derived from 100 000 successive samples of a single realization of the response, spaced at $\Delta\tau = 0.02$, where $\tau = \omega_n t$. It is seen that, at low values of the input strength parameter Θ the simulation estimates closely follow the theoretical curves corresponding to the first solution. Further, the simulation studies reveal that, for a particular value of ρ, there is a fairly well defined value of Θ, beyond which there is a very low probability of obtaining a bounded sample function, of the duration indicated above. This is certainly in qualitative agreement with the statistical linearization results, which provides a precise boundary, beyond which solutions do not exist (see Figure 10.6).

These results confirm that in this situation, where theoretical, long-term stationary statistics do not exist, the statistical linearization results for stationary response statistics do have a physical meaning. Thus, for low excitation levels, the probability of the response staying within the potential well (see Figure 5.5b) is high and there is a corresponding high probability of being able to generate fairly long sample functions of the response, by the digital simulation techniuqe. Estimates of the mean of the response, from these finite duration sample functions, show good agreement with the predictions from the statistical linearization method, where the latter does give solutions. A similarly good agreement (not shown here) is obtained with the theoretical results for the standard deviation of the response, as measured by the r parameter (see Figure 5.4b).

As one would expect, as the duration of the sample functions generated by digital simulation is decreased, the range of Θ values for which bounded sample functions can be obtained is increased and it is possible to get nearer to the theoretical boundary indicated by the broken line in Figure 10.6. This effect is shown in Figurc 10.7; hcrc the limit of the solutions regime, in the ρ–Θ plane is compared with estimates of this limit, inferred from the behaviour of simulated sample functions. For short duration sample functions ($N = 1000$ and $\Delta\tau = 0.02$) it is found that there is a high probability of being able to obtain bounded sample functions, within the region above the points shown. As one would expect, for longer duration sample functions ($N = 100\,000$, $\Delta\tau = 0.02$) the region in the ρ–Θ plane for which bounded sample functions may be obtained, with high probability, reduces (see Figure 10.7). Of course, the definition of these regimes, based on the behaviour of sample functions, is not, in principle, well

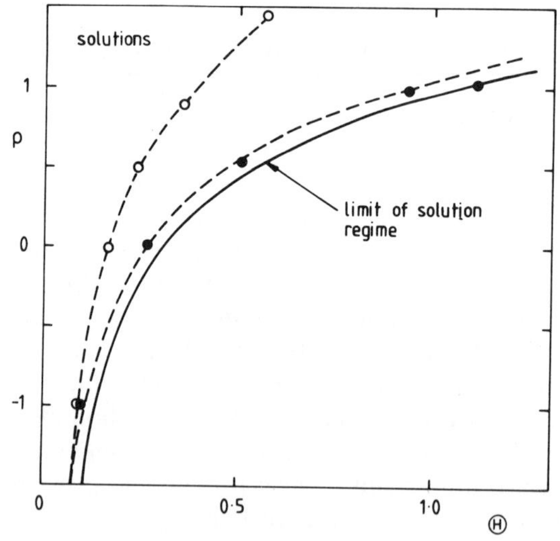

Figure 10.7
Boundary of the solution regime in the $\rho - \Theta$ plane and a comparison with simulation estimates (● $N = 1000$, ○ $N = 100\,000$)

defined, since it depends on one's interpretation of 'high probability'. However, the simulation results indicate that the regions indicated by the results in Figure 10.7 are not sensitive to this interpretation. In other words, there is a rapid transition from a high probability of being able to obtain bounded sample functions, to a low probability, within a fairly small band centred on the limits indicated by the simulated results shown in Figure 10.7.

These conclusions are supported by a comparison of digital simulation estimates with statistical linearization results for an oscillator with two, symmetrically disposed, static equilibrium positions. It was shown earlier, in Section 5.3.8, that for such an oscillator, with a restoring term given by equation (5.96), either three solutions exist, or one solution exists, depending on the value of the input strength parameter β.

In Figure 10.8 digital simulation estimates for the scaled mean square of the response, $\sigma^2 = \lambda\sigma_y^2$, are plotted for various values of β. These simulation estimates relate to a value of ζ of 0.2 and a time step, $\Delta\tau$, of 0.05. Two sets of estimates are shown; for one set $N = 1000$ (very short sample functions) and for the other set $N = 500\,000$ (very long sample functions). Since the estimates from the first set have high variability, due to the low value of N adopted, separate estimates were obtained from several sample functions, at certain low values of β.

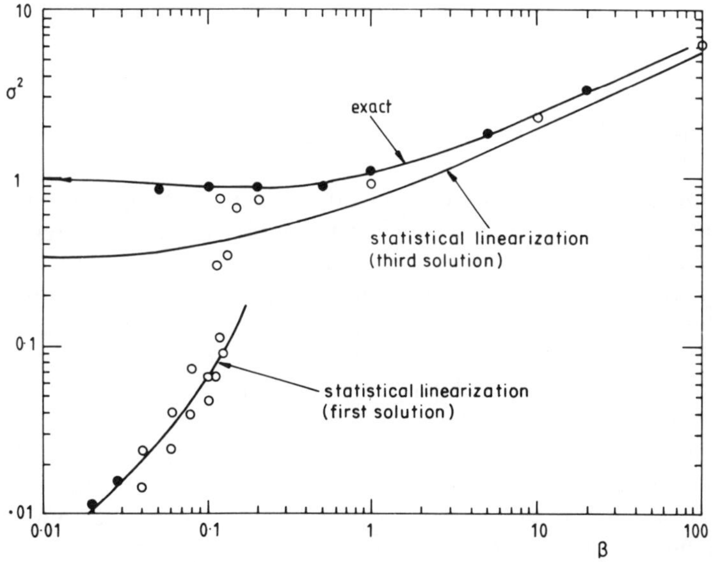

Figure 10.8
Variation of the mean square response with β, according to statistical linearization, for an oscillator with two static equilibrium positions. Comparison with simulation estimates (○ 1000 samples, ● 500 000 samples) and with the exact solution

Also shown in Figure 10.8 are the first and third solutions generated by the method of statistical linearization (see Figure 5.9b). The second solution is omitted since, as pointed out in Section 5.3.8, it is not physically realistic. Also shown in the figure is the exact solution for this problem, as computed from the FPK equation. It is noted that the asymptotic exact result

$$\sigma^2 \to 1 \quad \text{as } \beta \to 0 \tag{10.79}$$

can be deduced from simple physical reasoning, and does not require a consideration of the FPK equation. Thus, as $\beta \to 0$, the probability density function of the response will approach the form of two delta functions, situated at the positions of static equilibrium. Taking the scaling implicit in σ^2 into account, the asymptote of equation (10.79) is readily found.

It is seen from Figure 10.8 that at very low values of β both long duration (LD) and short duration (SD) simulation estimates are in agreement with the first solution. This corresponds to the situation where the probability of the response escaping from one of two potential wells is very small and the response is thus quasi-stationary, to a very good approximation. As β increases the LD simulation estimates show an early abrupt jump, whereafter they closely follow the exact solution, and are reasonably close to the third solution, obtained from statistical linearization. This jump, at $\beta \sim 0.04$, is due to the fact that, in the region of this level of excitation, the probability of the response staying within one well changes rapidly from a high value to a low value. Thus, beyond the critical value of β, the sample functions jump from one well to the other, with a reasonably high probability (see Figure 5.8c), within the duration of the sample function. Both the exact solution, and the third solution from statistical linearization, relate to this long-term behaviour, which is symmetric about a mean value of zero.

For the shorter duration sample functions the probability of not leaving one particular well stays high, over a wider range of β values. This trend is clearly shown by the results in Figure 10.8. Thus, the SD simulation estimates follow the first solution from statistical linearization up to a value of β of approximately 0.1, before there is a jump up towards the long-term solution. This critical value of β is, in fact, fairly close to that predicted by the statistical linearization method ($\beta = 1/6 \sim 0.167$). Again, at high values of β the SD estimates are in good agreement with the exact solution.

The lack of agreement between the exact solution and the third, long-term, statistical linearization at low values of β, is readily accounted for by the highly non-Gaussian nature of the response in this regime. As pointed out earlier, the density function for the response approaches the form of two symmetrically disposed delta functions, as $\beta \to \infty$. This represents an extreme deviation from the Gaussian form. As one would expect, the two solutions tend to converge as β becomes large, since the response becomes more Gaussian. For very large β values the cubic non-linear stiffness term dominates over the negative linear stiffness term. Therefore, the error of the statistical linearization result for σ approaches the value of 7.6% found earlier (see Table 10.1, where here $\nu = 3$).

It can be concluded from the results shown in Figure 10.8, that the theoretical long-term limit given by equation (10.79) is, in practice, unobtainable. No matter how long the sample functions are, which are used to generate estimates of σ^2, there will be a strong tendency to stay within one potential well, if β is low enough. Thus σ^2, as estimated from simulation results, will tend to follow the first statistical linearization solution, rather than the exact solution, for sufficiently low values of β, the range of β values for which the first solution is valid being dependent on the duration of the sample function used to estimate σ^2.

It is noted that simulation results for the mean of the response, as estimated from simulated sample functions, are also in close agreement with predictions from statistical linearization theory (see Figure 5.9a).

10.4.4 Oscillators with hysteresis

Earlier, in Chapter 8, some theoretical results were presented for the standard deviation of the response of a bilinear oscillator to white noise excitation, calculated according to the averaging method, due to Caughey, and according to the extended differential equation (EDE) method (see Figures 8.16 and 8.17). It was pointed out that the averaging method can be expected to be accurate only in circumstances where the response process is narrow-band in character.

In Figures 10.9a to 10.9d these, and other similar results, are compared with corresponding digital simulation estimates of the standared deviation of the response, σ. As in Figures 8.16 and 8.17, $\sigma/(S_0)^{1/2}$ is plotted against S_0, where S_0 is the constant spectral level of the excitation. The simulation results were obtained by numerically integrating the governing equations of motion for the third-order system, as given by equations (8.119) to (8.121). Each simulation estimate shown was derived from a single sample function of the response, consisting of 500 000 samples spaced at a time step, $\Delta\tau$, of 0.02.

It is observed that for $\alpha = 0.02$ and $\beta = 0$ (Figure 10.9a) the simulation estimates are in closer agreement with the EDE theory, than with the averaging theory, in the middle range of input strength, as measured by S_0 ($0.01 < S_0 < 0.5$). In this range the response is wide-band in character and one can not expect the averaging method to yield accurate results. The simulation estimates indicate, however, that whilst the EDE method gives a significant improvement in accuracy, there is still a substantial error, of approximately 40%.

As the input strength parameter, S_0, becomes large the simulation estimates show that the EDE theory becomes progressively less accurate. In contrast, the accuracy of the averaging method is seen to improve rapidly. This latter effect is simply due to the fact that the response becomes progressively more narrow-band as S_0 increases.

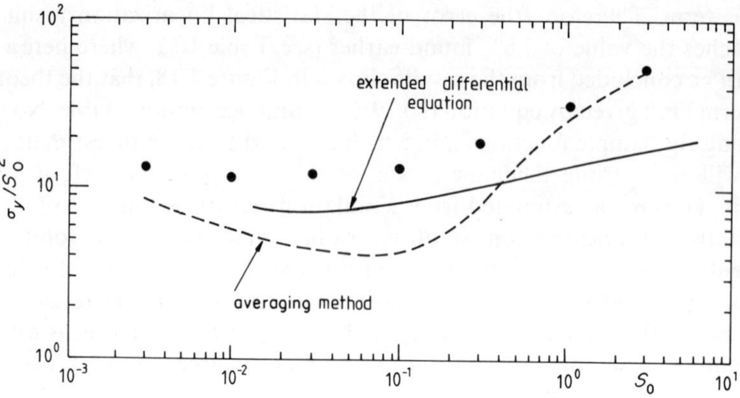

extended differential equation
averaging method
$\sigma_y/S_0^{1/2}$
10^2
10^1
10^0
10^{-3}
10^{-2}
10^{-1}
10^0
10^1
S_0

(a)

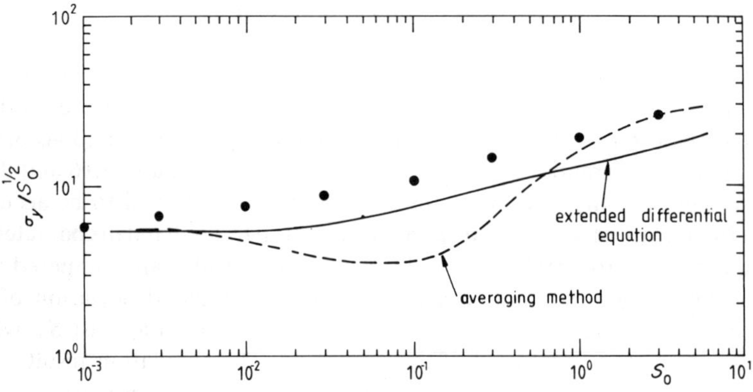

extended differential equation
averaging method
$\sigma_y/S_0^{1/2}$
10^2
10^1
10^0
10^{-3}
10^{-2}
10^{-1}
10^0
10^1
S_0

(b)

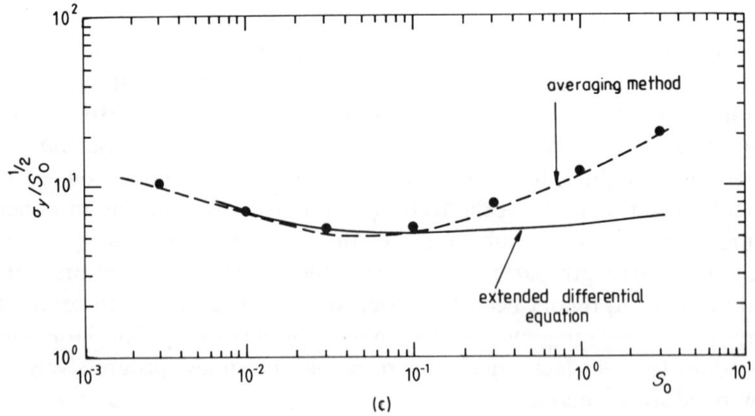

averaging method
extended differential equation
$\sigma_y/S_0^{1/2}$
10^2
10^1
10^0
10^{-3}
10^{-2}
10^{-1}
10^0
10^1
S_0

(c)

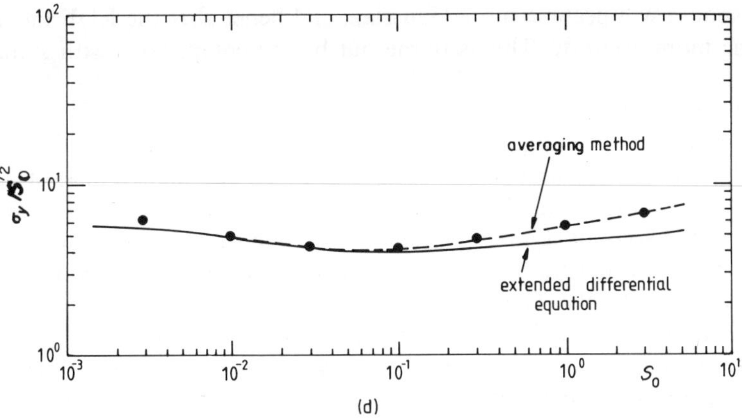

Figure 10.9
Variation of the standard deviation of the displacement response with input level for an oscillator with bilinear hysteresis. Comparison of results from two statistical linearization methods with simulation estimates (●). (a) $\alpha = 0.02$, $\beta = 0$. (b) $\alpha = 0.02$, $\beta = 0.1$. (c) $\alpha = 0.5$, $\beta = 0$. (d) $\alpha = 0.5$, $\beta = 0.1$

The response process also tends to become narrow-band at very low levels of excitation, as measured by S_0. Thus, not surprisingly, the theoretical curve derived from the averaging method is seen to become more accurate at very low values of S_0. There is also a tendency for the two theoretical estimates of σ to become closer, as S_0 is reduced to very low values. Numerical difficulties have, however, prevented EDE results being obtained at very low values of S_0 (see Chapter 8).

The rather poor accuracy of the EDE statistical linearization approximation, in this particular case, is attributable to the discontinuous nature of the non-linearity function, $G(\dot{y}, z)$ which appears in equation (8.121). As the analysis in Section 8.3.1 shows, the evaluation of the equivalent linear system parameters here, involves the expectation of derivatives of $G(\dot{y}, z)$, with respect to $\dot{y}$ and z. The expressions for these derivatives contain delta functions with the consequence that the values of equivalent linear system parameters (especially k (see equation (8.174)) are very sensitive to the shape of the assumed joint probability density function for $\dot{y}$ and z. Thus, pronounced deviations from the Gaussian form assumed in the EDE theory, which are certainly present in the case under discussion, will introduce significant errors in the evaluation of the parameters in the equivalent linear system, and hence in the approximate estimation for σ. As shown in Chapter 9, the statistical linearization result will lead to exact values of σ if the exact distribution of the response state variables is used. Thus errors in the estimation of σ, from the EDE theory can, in fact, only be attributable to the adoption of a Gaussian response distribution.

If viscous damping is introduced into the model then one can expect that the response will become more Gaussian, and hence that the EDE theory will become more accurate. This is borne out by the comparison with simulation estimates shown in Figure 10.9b, where $\alpha = 0.02$, as before, but now the viscous damping parameter β is 0.1. Again, the averaging method gives inaccurate results in the middle range of input strength level but gives results in good agreement with the simulation estimates, at both low values, and high values, of S_0.

The discontinuities in the bilinear hysteresis model become effectively less severe as the secondary slope parameter, α, is increased. Increasing α also has the effect of making the response more narrow-band, in the middle range of input strength levels. The comparison with simulation estimates shown in Figures 10.9c and 10.9d shows that the averaging method gives an accurate estimation of σ over the whole range of S_0, if $\alpha = 0.5$, for both $\beta = 0$ and $\beta = 0.1$. At low to intermediate levels of excitation the EDE results also agree with both simulation estimates, and with the result from the averaging method, and can therefore be judged to be accurate. However, once again the accuracy of the EDE theoretical result is seen to deteriorate at very high levels of excitation, particularly in the case where $\beta = 0$.

For smoother types of hysteresis it is reasonable to expect that the errors arising from the adoption of a Gaussian response distribution will be substantially smaller. This is borne out by comparisons between simulation estimates the EDE theoretical results for the case of curvilinear hysteresis,

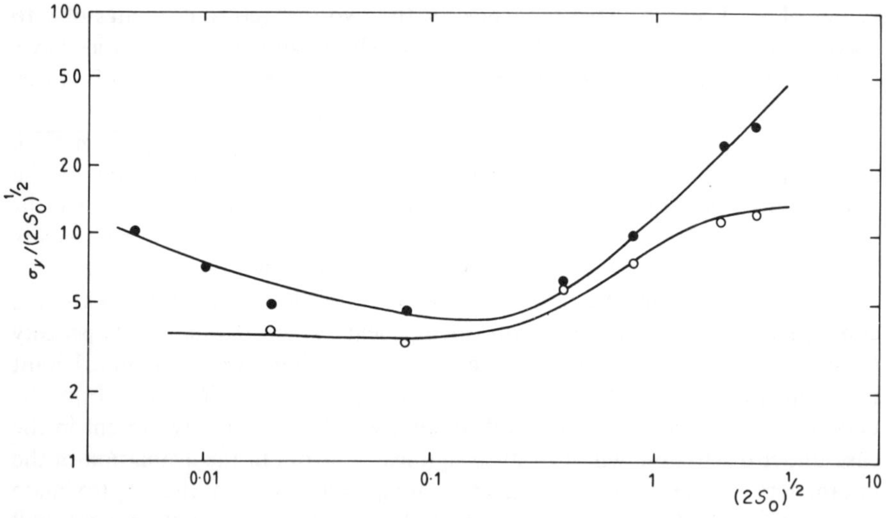

Figure 10.10
Variation of the standard deviation of the displacement response with input level for an oscillator with curvilinear hysteresis. Comparison of statistical linearization results (———) with simulation estimates (● $\beta = 0$, ○ $\beta = 0.1$) (reproduced from Wen 1980) by permission of the American Society of Mechanical Engineers).

where the appropriate $G(\dot{y}, z)$ function is given by equation (8.122). Figure 10.10 shows a typical such comparison between simulation estimates of σ_y and predictions from the EDE theory, over a range of S_0 values (see Wen (1980)). Here $\gamma = \nu = 0.5$, $A = 1$ and $n = 1$. It is seen that there is excellent agreement at two levels of viscous damping, corresponding to $\beta = 0$ and $\beta = 0.1$. It can be concluded that the EDE theory is much more accurate for this type of hysteresis, than in the bilinear case.

10.4.5 Multi-degree of freedom systems with hysteresis

The accuracy of the EDE statistical linearization method for analysing hysteretic systems has also been assessed by a number of authors in cases where there are several degrees of freedom. Again, the basis for this assessment is a comparison with simulation estimates.

Figure 10.11 shows a typical such comparison, for a system in the form of four degree of freedom shear beam structure incorporating hysteretic elements of the curvilinear type (see Baber and Wen (1981) for details). The methodology for analyzing this type of structure, by the EDE statistical linearization method has been described earlier, in Section 8.3.6. The four curves shown in Figure 10.11

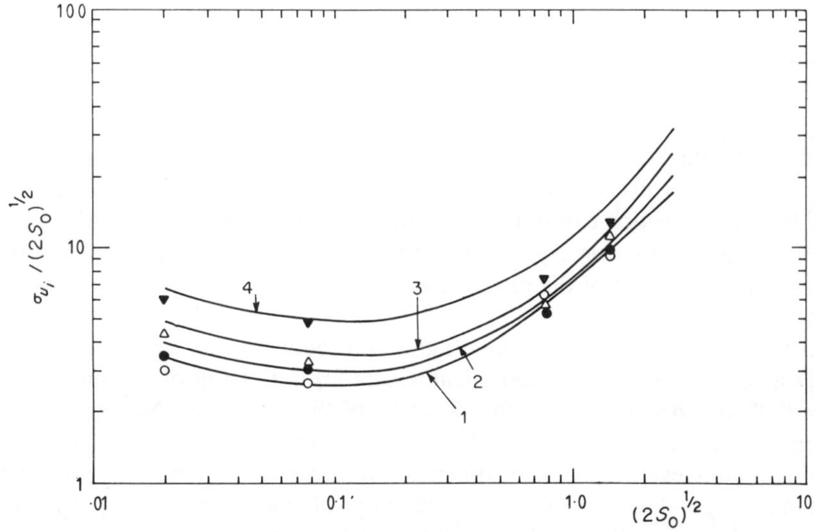

Figure 10.11
Variation of the standard deviation of the displacement response with input level for a four degree of freedom hysteretic system. Comparison of statistical linearization results with simulation estimates (○, 1; ●, 2; Δ, 3; ▼, 4) (reproduced from Baber and Wen (1981) by permission of the American Society of Civil Engineers).

(see also Figure 12 of Baber and Wen (1981)) show the variation of the standard deviation of the quantities u_i ($i = 1, 2, 3, 4$) with the constant spectral level of the white noise excitation, where u_i is the relative displacement between the ith and $(i+1)$th stories of the structure. The simulation estimates show excellent agreement with the theoretical curves, for each relative displacement, testifying to the accuracy of the theoretical method, when dealing with hysteresis of the curvilinear type.

10.4.6 Non-stationary response

In Chapter 7 it was shown that the statistical linearization method can be extended fairly readily to deal with non-stationary random excitation. The extension is particularly easy if the excitation can be modelled as stationary white noise, modulated by a deterministic function of time. Again, the accuracy of the results so obtained may be assessed through a comparison with digital simulation estimates. In this case the estimates must be derived by ensemble averaging over a set of generated sample functions, in the manner described earlier.

As an example of such a comparison, the problem defined by equations (7.84) and (7.85) will be examined again. Figure 10.12a and 10.12b show the estimated variation of the mean and standard deviation of the response process, $q(t)$, according to the statistical linearization theory, for the case where the modulation function, $\alpha(t)$, is of the form (Spanos, 1978, 1981b)

$$\alpha(t) = \exp(-0.025t) - \exp(-0.25t) \tag{10.80}$$

The results relate to low levels of viscous damping, as measured by the parameter $\beta = 2\zeta$, and to a non-linearity parameter ε equal to 2. Clearly this value of ε yields a system which can not be considered as weakly non-linear. Also shown in these figures are corresponding digital simulation estimates, derived from 300 sample functions of the response and drawn as continuous variations with time.

It is observed that there is good agreement between the simulated data and the statistical linearization theoretical results, for both levels of viscous damping ($\beta = 0.05$ and $\beta = 0.20$), over the duration of the excitation. Not only are the proper trends exhibited, such as the system rise time decreasing with increasing damping, but even the actual numerical values given by the two procedures are in quite good agreement. In fact, the range of relative error is approximately 0 to 10%. It is noted that the larger type of fluctuation in the simulation estimates of the mean response, compared with the fluctuation pertaining to the estimates of the response standard deviation is in line with equations (10.77) and (10.78). Moreover, the magnitude of the fluctuation is in agreement with the dispersion levels indicated by equations (10.77) and (10.78).

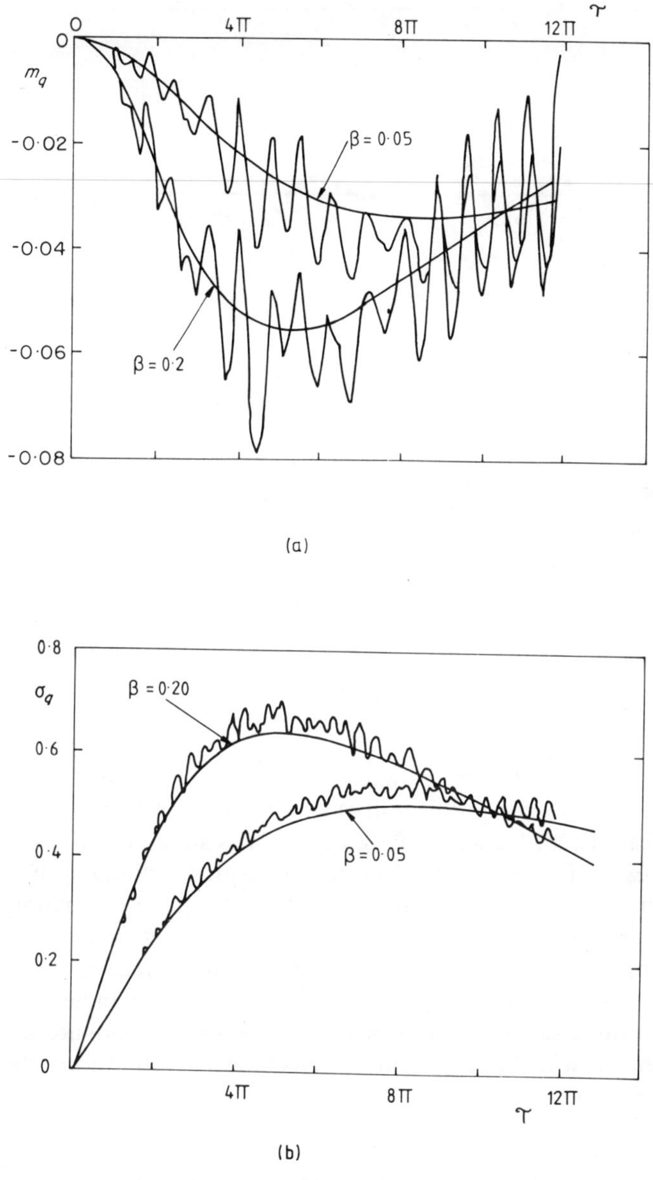

Figure 10.12

Variation of the standard deviation of the displacement response with time for an asymmetric Duffing oscillator with modulated white noise excitation. Comparison between statistical linearization results (———) and simulation estimates (~~~~~). $\beta = 0.05$ and 0.2. (a) Variation of the mean of the response with time. (b) Variation of the standard deviation of the response with time

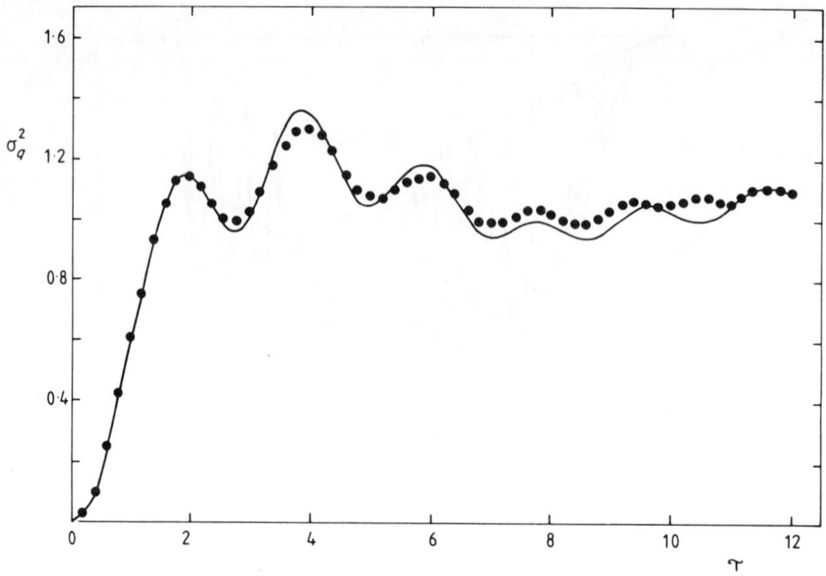

Figure 10.13
Variation of the mean square of the response with time for a Duffing oscillator with non-white excitation. $\lambda = 0.3$. Comparison of the statistical linearization approximation (———) with simulation estimates (●) (reproduced from Sakata and Kimura (1980) by permission from Academic Press)

As a final comparison between simulation estimates and statistical linearization results, the problem discussed in detail in Section 7.4.3 will once again be considered. Here a Duffing oscillator is subjected to non-stationary, non-white excitation, modulated by a deterministic function $a(t)$ of exponential form. Figure 10.13 shows a comparison of mean square simulation estimates with statistical linearization predictions, for the case of transient response to suddenly applied, stationary excitation. The parameters relevant to these results are the same as those for Figure 7.5 ($\lambda = 0.3$), and are given in Section 7.4.3. The simulation estimates, obtained by Sakata and Kimura (1980) are seen to be in close agreement with the approximate, theoretical result, during the transient phase of the response.

10.5 CONCLUDING REMARKS

The preceding analytical and Monte Carlo results and a variety of other studies (e.g. see Spanos, 1981) attest to reasonable reliability of the statistical linearization method in predicting the first and second statistical moments of the response of non-linear systems excited by stochastic process.

In fact, as was demonstrated by the example of a Duffing oscillator excited by white noise, there are cases for which the expression for the desired response statistics obtained by statistical linearization schemes can be proved to agree with the exact solution to the first order in terms of ε, the coefficient which is used to quantify the strength of the non-linearity. Therefore, the realiability of the method for systems with 'weak' non-linearities is established. However, this fact has been a source of difficulty to some users of the method in the sense that meaningful results are expected only for weakly non-linear dynamical systems. It is important to realize, however, that the statistical linearization method is not a perturbation technique, and it does not provide expressions for the response statistics which are necessarily in the form of polynomials in terms of the non-linearity parameter. In fact the method has been found surprisingly reliable even for the estimation of the response statistics of strongly non-linear structural systems (Spanos, 1981a). In this regard, values in the range 0 to 20% should be deemed as representative of the relative error of the derived approximate solutions, except when extreme forms of non-linearity are involved.

Generally, the extensively documented accuracy of the statistical linearization technique for estimating the statistical moments for system response, for a broad class of non-linear problems of engineering practice, should be deemed as a reasonable basis for attempting to use the method for a new problem of interest. Nevertheless, indiscriminate dependence on the results of the method is inadvisable. A more logical approach could involve the prior verification of a representative set of response statistics derived by the method of statistical linearization by comparisons with Monte Carlo simulations. Then, the method can be used repeatedly as the basis of critical technical estimations regarding a particular problem of interest.

From a practical viewpoint it is believed that the method of statistical linearization will continue to be widely used for treating old and new problems in a variety of technical fields. From a theoretical viewpoint it is anticipated that improved methods will be developed, which are more accurate in certain specific problems. However, such improvements must be reasonably versatile and tractable, if they are to be adopted in engineering practice. In this regard, the concept of statistical "quadratization", with possible extensions (Donley and Spanos, 1990), merits attention.

Appendix A

EVALUATION OF EXPECTATIONS

In applying the statistical linearization method it is often required to evaluate expectations of the form

$$E\{dg/dx\} = \int_{-\infty}^{\infty} (dg/dx) f(x)\, dx \tag{A1}$$

Table A1
x with zero mean

$g(x)$	$E\{dg/dx\}$	$E\{g(x)\}$
$\operatorname{sgn} x$	$\left(\frac{2}{\pi}\right)^{1/2}\sigma^{-1}$	0
$x\lvert x\rvert$	$\left(\frac{8}{\pi}\right)^{1/2}\sigma$	0
x^3	$3\sigma^2$	0
$x^{2n+1}\lvert x\rvert$	$\frac{(2\sigma^2)^{n+2}}{(2\pi)^{1/2}\sigma^3}\Gamma(n+2)$	0
x^{2n+1}	$\frac{(2\sigma^2)^{n+3/2}}{(2\pi)^{1/2}\sigma^3}\Gamma(n+\frac{3}{2})$	0
$(x+a)\lvert x+a\rvert$	$\left(\frac{8}{\pi}\right)^{1/2}\sigma\exp\left(\frac{-a^2}{2\sigma^2}\right)$	$(\sigma^2+a^2)\operatorname{erf}\left(\frac{a}{\sqrt{2}\sigma}\right)$
	$+2a\operatorname{erf}\left(\frac{a}{\sqrt{2}\sigma}\right)$	$+\left(\frac{2}{\pi}\right)^{1/2}a\sigma\exp\left(-\frac{a^2}{2\sigma^2}\right)$

$$\Gamma(v) = \int_0^{\infty} \exp(-x)x^{v-1}\, dx$$

$$\operatorname{erf}(x) = \frac{2}{\sqrt{\pi}}\int_0^x \exp(-u^2)\, du$$

Table A2
x with non-zero mean

$g(x)$	$E\{dg/dx\}$	$E\{g(x)\}$
$\operatorname{sgn}(x)$	$\left(\frac{2}{\pi}\right)^{1/2}\sigma^{-1}\exp\left(-\frac{\lambda^2}{2}\right)$	$1+\exp\left(\frac{\lambda}{\sqrt{2}}\right)$
x^2	$2\sigma\lambda$	$\sigma^2(1+\lambda^2)$
x^3	$3\sigma^2(1+\lambda^2)$	$\sigma^3(3\lambda+\lambda^3)$

$\lambda = \frac{m}{\sigma}$

and

$$E\{g(x)\} = \int_{-\infty}^{\infty} g(x)f(x)\,dx \tag{A2}$$

where x is a Gaussian random variable, with mean m and standard deviation σ. The symbol $f(x)$ is the probability density function of x. Table A1 gives some useful standard results, for various specific functions, $g(x)$, for the case where $m = 0$. Some limited results are given in Table A2 for the case where $m = 0$.

In evaluating these expectations the standard Gaussian form for $f(x)$ is taken (see equation (3.74)).

Appendix B

A USEFUL INTEGRAL FOR RANDOM VIBRATION ANALYSES

In calculating response statistics of oscillating systems to random excitations with rational power spectra it is necessary to calculate integrals of the form

$$I_m \equiv \int_{-\infty}^{\infty} \frac{\Xi_m(\omega)\,\mathrm{d}\omega}{\Lambda_m(-\mathrm{i}\omega)\Lambda_m(\mathrm{i}\omega)} \tag{B1}$$

where

$$\Xi_m(\omega) = \xi_{m-1}\omega^{2m-2} + \xi_{m-2}\omega^{2m-4} + \cdots + \xi_0 \tag{B2}$$

$$\Lambda_m(\mathrm{i}\omega) = \lambda_m(\mathrm{i}\omega)^m + \lambda_{m-1}(\mathrm{i}\omega)^{m-1} + \cdots + \lambda_0 \tag{B3}$$

and ξ_r, λ_r are arbitrary constants. This appendix provides a formula which can be used in determining I_m in a closed form. This formula is derived indirectly by considering certain properties of the spectrum of the stationary output of a linear time-invariant system to white noise input (Spanos, 1983). Useful generalizations for finite intervals of integration, and a broader class of integrands, are available (Spanos, 1987).

Equations for correlation and crosscorrelation functions

Consider the output of a linear system of order m to white noise input described by the equation

$$[\lambda_m D^m + \lambda_{m-1}D^{m-1} + \lambda_{m-2}D^{m-2} + \cdots + \lambda_0]x(t) = w(t) \tag{B4}$$

In equation (B4) D^r $(r = 0, \ldots, m)$ denotes the rth-order differential operator, λ_i, $(i = 1, 2, \ldots, m)$ are time invariant constants, and $w(t)$ represents a white noise process with correlation function

$$R_w(\tau) \equiv E\{w(t)w(t+\tau)\} = 2\pi\delta(\tau) \tag{B5}$$

where $\delta(\tau)$ is a two-sided Dirac delta function. It is assumed that the Routh–Hurwicz criterion is satisfied and the characteristic equation that corresponds to equation (B4) has roots with negative real parts. That is, the homogeneous part of equation (B4) is stable. Herein, it is assumed that $x(t)$ eventually becomes a stationary process with a correlation function $R_x(\tau)$ and a spectral density function $S_x(\omega)$. Using the results of Sections 4.3 and 4.4 it is easily shown that $S_x(\omega)$ is given by the equation

$$S_x(\omega) = \frac{1}{\Lambda_m(s)\Lambda_m(-s)} \qquad s = i\omega \tag{B6}$$

Furthermore, denote by $R_{xw}(\tau)$ the cross-correlation function of $x(t)$ and $w(t)$. Then, two ordinary differential equations governing the dependence of $R_x(\tau)$ and $R_{xw}(\tau)$ on the time lag variable τ can be derived. For this, the following formulas can be used

$$E\{x(t)D^r x(t-\tau)\} = (-1)^r D^r R_x(\tau) \tag{B7}$$

and

$$E\{w(t-\tau)D^r x(t)\} = D^r R_{xw}(\tau) \tag{B8}$$

These formulae can be readily proved by relying on the definition of the derivative of a real function (see Section 3.7). Replacing t by $t-\tau$ in equation (B4), then multiplying by $x(t)$ and ensemble averaging yields

$$[\hat{\lambda}_m D^m + \hat{\lambda}_{m-1} D^{m-1} + \cdots + \hat{\lambda}_0] R_x(\tau) = R_{xw}(\tau) \quad t \geqslant 0 \tag{B9}$$

where

$$\hat{\lambda}_r = (-1)^r \lambda_r \tag{B10}$$

Similarly, multiplying equation (B4) by $w(t-\tau)$ and ensemble averaging yields

$$(\lambda_m D^m + \lambda_{m-1} D^{m-1} + \cdots + \lambda_0) R_{xw}(\tau) = \pi \hat{\delta}(\tau) \quad \tau \geqslant 0 \tag{B11}$$

where $\hat{\delta}(\tau)$ is a one-sided delta function. Strictly speaking, equation (B11) is only a formal expression of the differential equation

$$(\lambda_m D^{m-1} + \lambda_{m-1} D^{m-2} + \cdots + \lambda_0 D^{-1}) R_{xw}(\tau) = \pi \quad \tau \geqslant 0 \tag{B12}$$

Equation (B12) shows that the derivatives $D^r R_{xw}(\tau); r = 1, \ldots, m-1$ are finite for $0 \leqslant \tau < \infty$. Thus, equation (B9) can be differentiated r times $(0 \leqslant r \leqslant m-1)$ to yield

$$(\hat{\lambda}_m D^{m+r} + \hat{\lambda}_{m-1} D^{m+r-1} + \cdots + \hat{\lambda}_0 D^r) R_x(\tau) = D^r R_{xw}(\tau) \quad \tau \geqslant 0 \quad 0 \leqslant r \leqslant m-1 \tag{B13}$$

The initial conditions at $\tau = 0$, for $R_{xw}(\tau)$, are now determined. Toward this end consider the impulse response $h(t)$ of the linear system described by equation (B4). Then, the stationary system output can be expressed as

$$x(t) = \int_{-\infty}^{\infty} h(t-u) w(u) \, du \tag{B14}$$

where u is a dummy variable. Multiplying equation (B14) by $w(t-\tau)$ and ensemble averaging yields

$$R_{xw}(\tau) = \pi h(\tau) \quad \tau \geqslant 0 \tag{B15}$$

Clearly, $R_{xw}(\tau)$ is not an even function of τ; in fact $R_{xw}(\tau)=0$ for $\tau<0$. Furthermore, equation (B15) can be generalized in the following form

$$D^r R_{xw}(\tau)=\pi D^r h(\tau) \quad r=0,\ldots,m-1 \tag{B16}$$

Equation (B16) can be used in conjunction with equation (B8) to determine the crosscorrelation between $w(t-\tau)$ and any of the derivatives of the stationary system output. At zero time lag, using the properties of $h(t)$, equation (B16) yields

$$D^r R_{xw}(0)=0 \quad r=0,\ldots,m-2 \tag{B17}$$

and

$$D^{m-1} R_{xw}(0)=\frac{\pi}{\lambda_m} \tag{B18}$$

Spectral moments of the system output

Equation (B13) in conjunction with equations (B17) and (B18) can be used to derive a formula regarding the spectral moments of the stationary output of the system. Define the spectral moment

$$M_{2r} \equiv \int_{-\infty}^{\infty} \omega^{2r} S_x(\omega)\,\mathrm{d}\omega \tag{B19}$$

where $S_x(\omega)$ is given by equation (B6). Note that $S_x(\omega)$ is related to $R_x(\tau)$ by the equation

$$R_x(\tau)=\int_{-\infty}^{\infty} \exp(-s\tau) S_x(\omega)\,\mathrm{d}\omega \tag{B20}$$

Thus, differentiating equation (B20) up to $2m-1$ times and taking into consideration equation (B19) yields

$$M_{2r-1}=0 \quad 0 \leqslant r \leqslant m \tag{B21}$$

and

$$M_{2r}=(-1)^r D^{2r} R_x(0) \quad 0 \leqslant r \leqslant m-1 \tag{B22}$$

Clearly, $M_r, r \geqslant 2m$, is unbounded. Substituting equations (B21) and (B22) into equation (B13) and setting $r=m-1, m-2,\ldots,0$ yields

$$\begin{aligned}
\lambda_{m-1}M_{2m-2}-\lambda_{m-3}M_{2m-4}+\lambda_{m-5}M_{2m-6}-\cdots &= \frac{\pi}{\lambda_m}\\
-\lambda_m M_{2m-2}+\lambda_{m-2}M_{2m-4}-\lambda_{m-4}M_{2m-6}+\cdots &= 0\\
0-\lambda_{m-1}M_{2m-4}+\lambda_{m-3}M_{2m-6}-\cdots &= 0\\
\cdot\ \cdot\ \cdot \qquad \cdots &= 0\\
\cdot\ \cdot\ \cdot \qquad -\lambda_2 M_2+\lambda_0 M_0 &= 0
\end{aligned} \tag{B23}$$

It is seen that the spectral moments M_{2r}; $r = 0, \ldots, m-1$ satisfy m linear algebraic equations. Thus, they can be determined by the classical Cramer's formula. For example the moment M_{2r} is determined by replacing the $(m-r)$th column of the coefficients determinant by the right-hand side of equation (B23). That is

$$M_{2r} = \frac{\begin{vmatrix} \lambda_{m-1} & -\lambda_{m-3} & \lambda_{m-5} & -\lambda_{m-7} & \cdots \pi/\lambda_m & \cdot & \cdot & \cdot \\ -\lambda_m & \lambda_{m-2} & -\lambda_{m-4} & \lambda_{m-6} & \cdots\ 0 & \cdot & \cdot & \cdot \\ 0 & -\lambda_{m-1} & \lambda_{m-3} & -\lambda_{m-5} & \cdots\ 0 & \cdot & \cdot & \cdot \\ 0 & \lambda_m & -\lambda_{m-2} & \lambda_{m-4} & \cdots\ 0 & \cdot & \cdot & \cdot \\ \cdot & \cdot & \cdot & \cdot & \cdots\ \cdot & \cdot & \cdot & \cdot \\ 0 & 0 & \cdot & \vdots & \cdots\ 0 & \cdot & -\lambda_2 & \lambda_0 \end{vmatrix}}{\begin{vmatrix} \lambda_{m-1} & -\lambda_{m-3} & \lambda_{m-5} & -\lambda_{m-7} & \cdots & \cdot & \cdot & \cdot \\ -\lambda_m & \lambda_{m-2} & -\lambda_{m-4} & \lambda_{m-6} & \cdots & \cdot & \cdot & \cdot \\ 0 & -\lambda_{m-1} & \lambda_{m-3} & -\lambda_{m-5} & \cdots & \cdot & \cdot & \cdot \\ 0 & \lambda_m & -\lambda_{m-2} & \lambda_{m-4} & \cdots & \cdot & \cdot & \cdot \\ \cdot & \cdot & \cdot & \cdot & \cdots & \cdot & \cdot & \cdot \\ 0 & 0 & \cdot & \cdot & \cdots & \cdot & -\lambda_2 & \lambda_2 \end{vmatrix}}$$

$$r = 0, \ldots, m-1. \qquad \text{(B24)}$$

Formula for I_m

Examining equations (B1) to (B3) and equation (B19) it is recognized that

$$I_m = \xi_{m-1} M_{2m-2} + \xi_{m-2} M_{2m-4} + \cdots + \xi_0 \qquad \text{(B25)}$$

Substituting into equation (B25) the values for M_{2r} which are given by equation (B24), and manipulating yields

$$I_m = \frac{\begin{vmatrix} \xi_{m-1} & \xi_{m-2} & & & & & & \xi_0 \\ -\lambda_m & \lambda_{m-2} & -\lambda_{m-4} & \lambda_{m-6} & \cdots 0 & \cdot & \cdot & \cdot \\ 0 & -\lambda_{m-1} & \lambda_{m-3} & -\lambda_{m-5} & \cdots 0 & \cdot & \cdot & \cdot \\ 0 & 0 & & & \cdots 0 & \cdot & \cdot & \cdot \\ 0 & 0 & \cdot & \cdot & \cdots & \cdot & -\lambda_2 & \lambda_2 \end{vmatrix}}{\begin{vmatrix} \lambda_{m-1} & -\lambda_{m-3} & \lambda_{m-5} & -\lambda_{m-7} & \cdot & \cdot & \cdot & \cdot \\ -\lambda_m & \lambda_{m-2} & -\lambda_{m-4} & \lambda_{m-6} & \cdot & \cdot & \cdot & \cdot \\ 0 & -\lambda_{m-1} & \lambda_{m-3} & -\lambda_{m-5} & \cdot & \cdot & \cdot & \cdot \\ 0 & \lambda_m & -\lambda_{m-2} & \lambda_{m-4} & \cdot & \cdot & \cdot & \cdot \\ \cdot & \cdot & \cdot & \cdot & \cdot & \cdot & \cdot & \cdot \\ 0 & 0 & \cdot & \cdot & \cdot & \cdot & -\lambda_2 & \lambda_0 \end{vmatrix}} \frac{\pi}{\lambda_m} \qquad \text{(B26)}$$

Equations (B24) and (B26) can simplify significantly certain calculations that are necessary in conducting random vibration analyses of linear systems. For example, consider the case of a SDOF linear oscillator with natural frequency ω_0 and ratio of critical damping ζ, which is exposed to white noise with a two-sided spectral density equal to unity. Clearly, in this case $m = 2, \lambda_2 = 1, \lambda_1 = 2\zeta\omega_0$, and $\lambda_0 = \omega_0^2$. Using these values, the stationary variances σ_x and σ_x of the oscillatory response can be conveniently determined by relying on equations (B24) or (B26). Specifically, it is found that

$$\sigma_{\dot{x}}^2 = M_2 = \frac{\pi}{2\zeta\omega_0} \tag{B27}$$

$$\sigma_x^2 = M_0 = \frac{M_2}{\omega_0^2} = \frac{\pi}{2\zeta\omega_0^3} \tag{B28}$$

It is noted that traditionally equations (B27) and (B28) are derived by using the theory of residues of complex functions. In fact, this theory can also be used to determine the general integral I_m (James *et al.*, 1965).

References

Abramovitch, M. and Stegun, I. A. (1972). *Handbook of Mathematical Functions*, Dover, New York.

Ahmadi, G. (1980). Mean Square Response of a Duffing Oscillator to Modulated White Noise Excitation by the Generalised Method of Equivalent Linearization, *J. Sound Vib.*, **71**, 9–15.

Ahmadi, G. (1981). Non-Stationary Random Vibration of a Nonlinear System with a Set-Up Spring, *Acustica*, **48**, 50–53.

Ahmadi, G. (1983). Stochastic Earthquake Response of a Structure on a Sliding Foundation, *Int. J. Engng Sci.*, **21**, 93–102.

Arnold, L. (1973). *Stochastic Differential Equations: Theory and Applications*, Wiley Interscience, New York.

Assaf, S. A. and Zirkle, L. D. (1976). Approximate Analysis of Non-Linear Stochastic Systems, *Int. J. Control*, **23**, 477–492.

Astrom, K. J. and Eykhoff, P. (1971). System Identification: A Survey, *Automatica*, **7**, 123–169.

Atalik, T. S. and Utku, S. (1976). Stochastic Linearization of Multi-Degree-of -Freedom Non-Linear Systems, *Earthquake Engng. Struct. Dyns.*, **4**, 411–420.

Atherton, D. P. (1975). *Nonlinear Control Engineering*, Van Nostrand, London.

Augusti, G., Baratta, A. and Casciati, F. (1983). *Probabilistic Methods in Structural Engineering*, Chapman and Hall, London.

Baber, T. T. (1984). Nonzero Mean Random Vibration of Hysteretic Systems, *J. Engng Mech., ASCE*, **110**, 1036–1049.

Baber, T. T. and Noori, M. N. (1984). Random Vibration of Degrading Structures with Pinching Hysteresis, *Proc. ASCE Engng Mech. Div. Speciality Conf. on Probabilistic Mechanics and Structural Reliability, Berkeley, California*, 147–150.

Baber, T. T. and Noori, M. N. (1985). Random Vibration of Degrading, Pinching Systems, *J. Engng Mech., ASCE*, **111**, 1010–1026.

Baber, T. T. and Wen, Y.-K. (1981). Random Vibration of Hysteretic, Degrading Systems, *J. Engng Mech. Div., ASCE*, **107**, 1069–1087.

Baker, W. E., Woolam, W. E. and Young, D. (1967). Air and Internal Damping of Thin Beams, *Int. J. Mech. Sci.*, **9**, 743–766.

Bartels, R. H. and Stewart, G. W. (1972). Solution of the Matrix Equation AX + XB = C, *Algorithm 432 in Communications of the ACM*, Vol. 15, No. 9.

Beaman, J. J. and Hedrick, J. K. (1981). Improved Statistical Linearization for Analysis and Control of Nonlinear Stochastic Systems: Part 1; An Extended Statistical Linearization Technique, *J. Dyn. Syst. Meas. Control., ASME*, **103**, 14–21.

Beavers, A. N. Jr. and Denman, E. D. (1975). A New Solution Method for the Liapunov Matrix Equation, *SIAM J. Appl. Maths.*, **29**, 416–421.

Belokobylskii, S. V. and Prokopov, V. K. (1982). Friction Induced Self Excited Vibrations of Drill Rigs with Exponential Drag Law, *Prikland Naya Mekhanika*, **8**, 98–101.

Bendat, J. S., Enochson, L. D., Klein, G. H. and Piersol, A. G. (1962). Advanced Concepts of Stochastic Processes and Statistics for Flight Vehicle Vibration Estimation and Measurement, *Flight Dyn. Lab., Aero. Syst. Dir. Air Force System Command, Wright-Patterson Air Force Base, Ohio, Tech. Rep. No. ASD-TR-62-1973.*

Bishop, R. E. D. and Price, W. G. (1986). A Note on Hysteretic Damping of Transient Motions. In I. Elishakoff and R. H. Lyon, (eds), *Random Vibration-Status and Recent Developments*, Elsevier, Amsterdam, 39–45.

Bishop, R. E. D., Gladwell, G. M. L. and Michaelson, S. (1965). *The Matrix Analysis of Vibration*, Cambridge University Press, Cambridge, UK.

Blevins, R. D. (1977). *Flow-Induced Vibration*, Van Nostrand Reinhold, New York.

Bogoliubov, N. and Mitropolsky, A. (1963). *Asymptotic Methods in the Theory of Nonlinear Oscillations* (2nd ed), Gordon and Breach, New York.

Bolotin, V. V. (1969). *Statistical Methods in Structural Mechanics*, Holden-Day, San Franscisco.

Bolotin, V. V. (1984). *Random Vibrations of Elastic Systems*, Martinus Nijhoff, The Hague.

Booton, R. C. (1954). Non-Linear Control Systems with Random Inputs, *IRE Trans. Circuit Theory*, **CT-1**, 9–18.

Booton, R. C., Mathews, M. V. and Seifert, W. W. (1953). Nonlinear Servomechanisms with Random Inputs, Dyn. Anal. Control Lab., Rep. No. 70., MIT, Cambridge, Mass., USA.

Bouc, R. (1967). Forced Vibration of Mechanical Systems with Hysteresis, *Abtract; Proc. 4th Int. Conf. Nonlinear Oscillations, Prague, Czechoslovakia.*

Bover, D. C. C. (1978). Moment Equation Methods for Non-Linear Stochastic Systems, *J. Math. Anal. Appl.*, **65**, 306–320.

Brook, A. K. (1986). The Role of Simulation in Determining the Roll Response of a Vessel in an Irregular Seaway, *Proc. Int. Conf. on the Safeship Project: Ship Stability and Safety*, Royal Inst. Naval Architects, London, UK.

Bruckner, A. and Lin, Y. K. (1987). Generalization of the Equivalent Linearization Method for Non-Linear Random Vibration Problems, *Int. J. Non-Linear Mech.*, **22**, 227–235.

Budgor, A. B. (1976). Studies in Nonlinear Stochastic Processes, I. Approximate Solutions of Nonlinear Stochastic Differential Equations by the Method of Statistical Linearization, *J. Statit. Phys.*, **15**, 355–374.

Cai, G. Q. and Lin, Y. K. (1988a). A New Approximate Solution Technique for Randomly Excited Non-Linear Oscillators, *Int. J. Non-Linear Mech.*, **23**, 409–420.

Cai, G. Q. and Lin, Y. K. (1988b). On Exact Stationary Solutions of Equivalent Non-Linear Stochastic Systems, *Int. J. Non-Linear Mech.*, **23**, 315–325.

Casciati, F. (1982). Probabilistic Analysis of Inelastic Structures, *Nucl. Engng. Des.*, **71**, 271–276.

Caughey, T. K. (1959). Response of a Nonlinear String to Random Loading, *J. Appl. Mech., ASME*, **26**, 341–344.

Caughey, T. K. (1960a). Random Excitation of a System with Bilinear Hysteresis, *J. Appl. Mech., ASME*, **27**, 649–652.

Caughey, T. K. (1960b). Random Excitation of a Loaded Nonlinear String, *J. Appl. Mech., ASME*, **27**, 575–578.

Caughey, T. K. (1963). Equivalent Linearization Techniques, *J. Acoust. Soc. Am.*, **35**, 1706–1711.

Caughey, T. K. (1964). On the Response of a Class of Nonlinear Oscillators to Stochastic Excitation, *Proc. Coll. Int. du Centre Nat. de la Rechercher Scient., No. 148, Marseille, France*, 393–402.

Caughey, T. K. (1971). Nonlinear Theory of Random Vibrations, *Adv. Appl. Mech.*, **11**, 209–253.

Caughey, T. K. (1986a). On the Response of Non-Linear Oscillators to Stochastic Excitation, *Prob. Engng Mech.*, **1**, 2–4.
Caughey, T. K. (1986b). Exact Solutions in the Theory on Non-Linear Random Vibration and Their Applications, *23rd Ann. Mtg. Soc. Engng Sci., State Univ. New York at Buffalo, USA.*
Caughey, T. K. and Ma, F. (1982). The Exact Steady State Solution of a Class of Non-Linear Stochastic Systems, *Int. J. Non-Linear Mech.*, **17**, 137–142.
Chang, T.-P., Mochio, T. and Samaras, E. (1986). Seismic Response Analysis of Nonlinear Structures, *Prob. Engng. Mech.*, **1**, 157–166.
Chen, S. S. (1977). Flow-Induced Vibrations of Circular Cylindrical Structures: Part I; Stationary Fluids and Parallel Flow, *Shock Vib. Digest*, **9**, 25–38.
Chu, C. (1985). Random Vibration of Nonlinear Building-Foundation Systems, PhD Thesis, Illinois, USA.
Chung, K. L. (1979). *Elementary Probability Theory with Stochastic Processes* (3rd ed), Springer-Verlag, New York.
Clarkson, B. L. and Mead, D. J. (1973). High Frequency Vibration of Aircraft Structures, *J. Sound Vib.*, **28**, 487–504.
Clough, R. W. and Penzien, J. (1975). *Dynamics of Structures*, McGraw-Hill, New York.
Constantinou, M. C. and Tadjbakhsh, I. G. (1984). Response of a Sliding Structure to Filtered Random Excitation, *J. Struct. Mech.*, **12**, 401–418.
Cooper, G. R. and McGillem, C. D. (1986). *Probabilistic Methods of Signal and System Analysis* (2nd ed), Holt, Rinehart and Winston, New York.
Cramer, H. (1946). *Mathematical Methods of Statistics*, Princeton University Press, Princeton, NJ.
Cramer, H. (1966). On the Intersections between the Trajectories of a Normal Stationary Stochastic Process and a High Level, *Arkiv Math.*, **6**, 337–349.
Crandall, S. H. (1963). Perturbation Techniques for Random Vibration of Nonlinear Systems, *J. Acoust. Soc. Am.*, **35**, 1700–1705.
Crandall, S. H. (1964). The Spectrum of Random Vibration of a Nonlinear Oscillator, *Proc. 11th Int. Congr. App. Mech., Munich, West Germany.*
Crandall, S. H. (1977a). Nonlinear Problems in Random Vibration, Int. Conf. Nonlinear Oscillations, *Abh. Akad. Wissen Sch., DDR*, **2**, 215–224.
Crandall S. H. (1977b). On Statistical Linearization for Non-Linear Oscillators, *Problems of the Asymptotic Theory of Non-Linear Oscillation*, Acad. Sci. Ukranian SSR, Naukova Duma, Kiev, USSR.
Crandall S. H. (1978). Heuristic and Equivalent Linearization Techniques for Random Vibration of Nonlinear Oscillators, 6th Int. Conf. Nonlinear Oscillations (ICNO), Prague, Czechoslovakia.
Crandall, S. H. (1980). Non-Gaussian Closure for Random Vibration of Non-Linear Oscillators, *Int. J. Non-Linear Mech.*, **15**, 303–313.
Crandall, S. H. (1985). Non-Gaussian Closure Techniques for Stationary Random Vibration, *Int. J. Non-Linear Mech.*, **20**, 1–8.
Crandall, S. H. and Lee, S. S. (1976). Biaxial Slip of a Mass on a Foundation Subject to Earthquake Motion, *Ing. Arch.*, **45**, 361–370.
Crandall, S. H. and Mark, W. D. (1963). *Random Vibration in Mechanical Systems*, Academic Press, New York.
Crandall, S. H. and Zhu, W. Q. (1983). Random Vibration: A Survey of Recent Developments, *J. Appl. Mech., ASME*, **50**, 953–962.
Crandall, S. H., Khabbaz, G. R. and Manning, J. E. (1964). Random Vibration of an Oscillator with Nonlinear Damping, *J. Acoust. Soc. Am.*, **36**, 1330–1334.
Crandall, S. H., Lee, S. S. and Williams, J. H. (1974). Accumulated Slip of a Friction-Controlled Mass Excited by Earthquake Motions, *J. Appl. Mech., ASME*, **41**, 1094–1098.

Dashevskii, M. L. (1967). Approximate Analysis of the Accuracy of Non-Stationary, Non-Linear Systems, Using the Method of Semi-Invariants, *Auto. Remote Control.*, **28**, 63–74.

Dashevskii, M. L. and Lipster, R. S. (1967). Application of Conditional Semi-Invariants in Problems of Non-Linear Filtering of Markov Processes, *Auto. Remote Control.*, **28**, 63–74.

Davenport, A. G. and Novak, M. (1976). Vibration of Structures Induced by Wind, In C. M. Harris and C. E. Crede, (eds) *Shock and Vibration Handbook*, McGraw-Hill, New York. Chap. 29.

Davies, H. G. and Nandlall, D. (1986). Phase Plane for Narrow Band Random Excitation of a Duffing Oscillator, *J. Sound Vib.*, **104**, 277–283.

Davis, H. F. (1963). *Fourier Series and Orthogonal Functions*, Allyn and Bacon, Boston, M.A.

Den Hartog, J. P. (1956). *Mechanical Vibrations* (4th ed), McGraw-Hill, New York.

Dimentberg, M. F. (1970). Oscillations of a System with a Nonlinear Cubic Characteristic under Narrow-Band Excitation, *Mekhanica Tverdogo Tela*, **6**, 150–154.

Dimentberg, M. F. (1988). *Statistical Dynamics of Nonlinear and Time-Varying Systems*, Research Studies Press, Taunton, UK.

Donley, M. G. and Spanos, P. D. (1990). *Introduction to Statistical Quadratization with Applications to Compliant Offshore Structures*, Springer-Verlag Lecture Notes in Engineering, New York.

Doob, J. L. (1953). *Stochastic Processes*, Wiley, New York.

Dunne, I. F. and Wright, J. H. (1985). Predicting the Frequency of Occurrence of Large Roll Angles in Irregular Seas, *Proceedings of the Royal Institution of Naval Architects*, **127**, 233–245.

Einstein, A. (1905). On the Movement of Small Particles Suspended in a Stationary Liquid Demanded by the Molecular-Kinetic Theory of Heat, *Ann. Phys.*, **17**, 549–569.

Elishakoff, I. (1983). *Probabilistic Methods in the Theory of Structures*, Wiley, New York.

Fang, T. and Wang, Z. N. (1986a). A Generalisation of Caughey's Normal Mode Approach to Nonlinear Random Vibration Problems, *AIAA J.*, **24**, 531–534.

Fang, T. and Wang, Z. N. (1986b). Complex Modal Analysis of Random Vibrations, *AIAA J.*, **24**, 342–343.

Faravelli, L., Casciati, F. and Singh, M. P. (1987). Stochastic Equivalent Linearization Algorithms, presented at the IUTAM Symp. on Nonlinear Stochastic Dynamic Systems, Innsbruck, Austria.

Feller, W. (1971). *An Introduction to Probability Theory and its Applications*, 2 Vols., (2nd ed), Wiley, New York.

Foster, E. T. (1968). Semilinear Random Vibrations in Discrete Systems, *J. Appl. Mech., ASME*, **35**, 560–564.

Foster, E. T. (1970). Model for Nonlinear Dynamics of Offshore Towers, *J. Engng Mech. Div., ASCE*, **96**, 41–67.

Froude, W. (1955). *The Papers of W. Froude*, Royal Institute Naval Architects, London.

Gawthrop, P. J. (1984). Parameter Estimation from Non-Contiguous Data, *Proc. Inst. Elect. Engng*, **131**, 261–266.

Gawthrop, P. J., Kountzeris, A. and Roberts, J. B. (1988). Parametric Identification of Nonlinear Ship Motion from Forced Roll Data, *J. Ship Research, SNAME*, **32**, 101–111.

Gelb, A. and Van Der Velde, W. E. (1968). *Multiple-Input Describing Functions and Nonlinear System Design*, McGraw-Hill, New York.

Gnedenko, B. V. (1962). *The Theory of Probability*, Chelsea, New York.

Goodman, T. R. and Sargent, T. P. (1961). Launching of Airborne Missiles Underwater;

Part XI: Effect of Nonlinear Submarine Roll Damping on Missile Response in Confused Seas, *Appl. Res. Assoc. Inc. Doc. No. ARA-964.*

Goodwin, G. C. and Payne, L. R. (1977). *Dynamic System Identification: Experiment Design and Data Analysis*, Academic Press, New York.

Goto, H. and Iemura, H. (1973). Linearization Techniques for Earthquake Response of Simple Hysteretic Structures, *Proc. Jap. Soc. Civ. Engng*, **212**, 109–119.

Grigoriu, M. and Allbe, B. (1986). Response of Offshore Structures to Waves, *J. Engng Mech.*, **112**, 729–744.

Grimmett, G. and Stirzaker, D. (1982). *Probability Theory and Random Processes*, Clarendon Press, Oxford, UK.

Gumestad, O. T. and Connor, J. J. (1983). Linearisation Methods and the Influence of Current on the Non-linear Hydrodynamic Drag, *Appl. Ocean Research*, **5**, 184–194.

Hagedorn, P. and Wallaschek, J. (1987). On Equivalent Harmonic and Stochastic Linearization for Nonlinear Shock-Absorbers. In F. Ziegler, G. I. Schuëller (eds) *IUTAM Symp. Nonlinear Stochastic Dynamic Systems*, Springer-Verlag, Berlin, Germany, 23–32.

Harrison, R. F. and Hammond, J. K. (1986). Approximate Time Domain Non-Stationary Analysis of Stochastically Excited Non-Linear Systems with Particular Reference to the Motion of Vehicles on Rough Ground, *J. Sound Vib.*, **105**, 361–371.

Hennig, K. and Roberts, J. B. (1986). Averaging Methods for Randomly Excited Non-Linear Oscillators. In I. Elishakoff and R. H. Lyon (eds) *Random Vibration-Status and Recent Developments*, Elsevier, Amsterdam, 143–161.

Ibrahim, R. A. (1985). *Parametric Random Vibration*, Research Studies Press, Taunton, UK.

Ibrahim, R. A. and Soundararajan, A. (1985). An Improved Approach for Random Parametric Response of Dynamic Systems with Non-Linear Inertia, *Int. J. Non-Linear Mech.*, **20**, 309–323.

Ibrahim, R. A., Soundararajan, A. and Heo, H. (1985). Stochastic Response of Nonlinear Dynamic Systems Based on Non-Gaussian Closure, *J. Appl. Mech., ASME*, **52**, 965–970.

Iwan, W. D. (1973). A Generalisation of the Concept of Equivalent Linearization, *Int. J. Non-Linear Mech.*, **8**, 279–287.

Iwan, W. D. (1974). Application of Nonlinear Analysis Techniques, *Appl. Mech. Earthquake Engng, Appl. Mech. Div., ASME, AMD*, **8**, 135–161.

Iwan, W. D. and Lutes, L. D. (1968). Response of the Bilinear Hysteretic System to Stationary Random Excitation, *J. Acoust. Soc. Am.*, **43**, 545–552.

Iwan, W. D. and Mason, A. B. Jr. (1980). Equivalent Linearization for Systems Subjected to Nonstationary Random Excitation, *Int. J. Non-Linear Mech.*, **15**, 71–82.

Iwan, W. D. and Patula, E. J. (1972). The Merit of Different Error Minimization Criteria in Approximate Analysis, *J. Appl. Mech., ASME*, **39**, 257–262.

Iwan, W. D. and Yang, I.-M. (1971). Statistical Linearization for Nonlinear Structures, *J. Engng Mech. Div., ASCE*, **97**, 1609–1623.

Iwan, W. D. and Yang, I.-M. (1972). Application of Statistical Linearization to Nonlinear Multidegree-of-Freedom Systems, *J. Appl. Mech., ASME*, **39**, 545–550.

Iyengar, R. N. (1975). Random Vibration of a Second Order Nonlinear Elastic System, *J. Sound Vib.*, **40**, 155–165.

Iyengar, R. N. (1988). Higher Order Linearization in Non-Linear Random Vibration, *Int. J. Non-Linear Mech.*, **23**, 385–391.

Jacobsen, L. S. (1930). Steady Forced Vibration Influenced by Damping, *J. Appl. Mech. ASME*, **52**, 169–181.

James, H. M., Nichols, N. B. and Phillips, R. S. (1965). *Theory of Servomechanisms*, Dover, New York.

Jazwinskii, A. H. (1970). *Stochastic Processes and Filtering Theory*, Academic Press, New York.

Kanai, K. (1961). An Empirical Formula for the Spectrum of Strong Earthquake Motions, *Bull. Earthquake Research Inst., Univ. of Tokyo, Japan*, **39**.

Kaplan, P. (1966). Lecture Notes on Nonlinear Theory of Ship Roll Motion in a Random Seaway, *Proc. 11th Towing Tank Conf., Tokyo, Japan*, 393–396.

Kazakov, I. E. (1954). Approximate Method for the Statistical Analysis of Nonlinear Systems, *Trudy VVIA*, No. 394.

Kazakov, I. E. (1955). Approximate Probability Analysis of Operational Precision of Essentially Nonlinear Feedback Control Systems, *Auto. Remote Control*, **17**, 423–450.

Kazakov, I. E. (1965a). Generalization of the Method of Statistical Linearization to Multidimensional Systems, *Auto. Remote Control*, **26**, 1201–1206.

Kazakov, I. E. (1965b). Statistical Analysis of Systems with Multi-Dimensional Nonlinearities, *Auto. Remote Control*, **26**, 458–464.

Khasminskii, R. Z. (1968). On the Averaging Principle for Stochastic Differential Itô Equations, *Kibernetika*, **4**, 260–279 (in Russian).

Kimura, K. and Sakata, M. (1981). Non-Stationary Responses of a Non-Symmetric, Non-Linear System Subjected to a Wide Class of Random Excitation, *J. Sound Vib.*, **76**, 261–272.

Kimura, K., Yagasaki, K. and Sakata, M. (1983). Non-Stationary Responses of a System with Bilinear Hysteresis Subjected to Non-White Random Excitation, *J. Sound Vib.*, **91**, 181–194.

Kirk, C. L. (1974). Application of the Fokker-Planck Equation to Random Vibration of Non-Linear Systems, *Cranfield Report Aero No. 20, Cranfield Institute of Technology, UK*.

Kobari, T., Minai, R. and Suzuki, Y. (1973). Statistical Linearization Techniques for Hysteretic Structures with Earthquake Excitations, *Bull. Disaster Prevention Inst., Kyoto University, Japan*, **23**, pts. 3–4, No. 215, 111–135.

Kolovskii, M. Z. (1966). An Accuracy Estimate of Solutions Obtained by the Statistical Linearization Method, *Auto. Remote Control*, **27**, 1692–1701.

Kolmogorov, A. N. (1956). *Foundations of the Theory of Probability*, Chelsea, New York.

Kozin, F. (1988). The Method of Statistical Linearization for Non-linear Stochastic Vibrations. In F. Ziegler, and G. I. Schuëller (eds) *IUTAM Symp. Nonlinear Stochastic Dynamic Systems*, Springer-Verlag, Berlin, 45–56.

Kramers, H. A. (1940). Brownian Motion in a Field of Force and the Diffusion Model of Chemical Reactions, *Physica*, **7**, 284–304.

Kree, P. and Soize, C. (1986). *Mechanics of Random Phenomena: Structural Dynamics*, Rydll.

Krylov, N. and Bogoliubov, N. (1937). Introduction a la Mechanique Nonlineaire: les Methodes Approaches et Asymptotices, *Ukr. Akad. Nauk. Inst. de la Mecanique, Chaire de Phys. Math. Ann.* (Translated by S. Lefshetz in *Ann. Math. Studies*, No. 11, Princeton, NJ, USA, 1947).

Kuznetsov, P. I., Stratonovitch, R. L. and Tikhonov, V. I. (1965). *Non-linearTransformations of Stochastic Processes*, Pergamon Press, Oxford, UK.

Langley, R. S. (1988a). Application of the Principle of Detailed Balance to the Random Vibration of Non-Linear Oscillators, *J. Sound Vib.*, **125**, 85–92.

Langley, R. S. (1988b). An Investigation of Multiple Solutions Yielded by the Equivalent Linearization Method, *J. Sound Vib.*, **127**, 271–282.

Lazan, B. J. (1968). *Damping of Materials and Members in Structural Mechanics*, Pergamon Press, Oxford, UK.

Leira, B. J. (1987). Multidimensional Stochastic Linearization of Drag Forces, *Appl. Ocean Research*, **9**, 150–162.

Lemaitre, J. (1971). Response of Nonlinear Systems to Random Loads—Bibliographic Analysis, *Office Nat. d'Etudes et de Recherches Aero., Note Technique*, No. 186 (in French).

Lennox, W. C. and Kuak, Y. C. (1976). Narrow Band Excitation of a Nonlinear Oscillator, *J. Appl. Mech., ASME*, **43**, 340–344.

Lin, Y. K. (1967). *Probabilistic Theory of Structural Dynamics*, McGraw-Hill, New York.

Lin, Y. K. and Cai, G. Q. (1988). Exact Stationary Response Solution for Second Order Nonlinear Systems Under Parametric and External White Noise Excitations: Part II, *J. Appl. Mech.*, **55**, 702–705.

Lin, Y. K. and Wu, W. F. (1984). Applications of Cumulant Closure to Random Vibration Problems, In T. C. Huang and P. D. Spanos (eds) *Proc. Symp. Random Vibrations, ASME Winter Ann. Mtg., New Orleans*, AMD, Vol. 65.

Lin, Y. K., Yong, Y., Cai, G. Q. and Bruckner, A. (1988). Exact and Approximate Solutions for the Response of Nonlinear Systems Under Parametric and External White Noise Excitations. In F. Ziegler and G. I. Schuëller, *IUTAM Symp. Nonlinear Stochastic, Dynamic Engineering Systems*, Springer-Verlag, Berlin, 323–333.

Loeve, M. (1977). *Probability Theory* (4th ed), Springer, Heidelberg.

Lutes, L. D. (1970a). Equivalent Linearization for Random Vibration, *J. Engng Mech. Div.*, **96**, 227–242.

Lutes, L. D. (1970b). Approximate Technique for Treating Random Vibration of Hysteretic Systems, *J. Acoust. Soc. Am.*, **48**, 299–306.

Lutes, L. D. and Takemiya, H., (1974), Random Vibration of a Yielding Oscillator, *J. Engng Mech. Div., ASCE*, **100**, 343–358.

Lyon, R. H., Heckl, M. and Hazelgrove, C. B. (1961). Response of a Hard-Spring Oscillator to Narrow-Band Excitation, *J. Acoust. Soc. Am.*, **33**, 1404–1411.

McCallion, H. (1973). *Vibration of Linear Mechanical Systems*, Longman, London.

Manning., J. E. (1975). Response Spectra for Nonlinear Oscillators, *J. Engng. Ind., ASME*, **97**, 1223–1226.

Maymon, G. (1984). Response of Geometrically Nonlinear Elastic Structures to Acoustic Excitation—An Engineering Orientated Computation Procedure, *Comps. Structs.*, **18**, 647–652.

Mei, C. and Paul. D. B. (1986). Nonlinear Multi-Mode Response of Clamped Rectangular Plates to Acoustic Loading, *AIAA J.*, **124**, 643–648.

Mei, C. and Wolfe, H. F. (1986). On Large Deflection Analysis in Acoustic Fatigue Design. In I. Elishakoff, and R. H. Lyon, *Random Vibration-Status and Recent Developments*, Elsevier, Amsterdam, 279–302.

Morison, J. R., O'Brien, M. P., Johnson, J. W. and Schast, S. A., (1950). The Force Exerted by Surface Waves on Piles, *Petroleum Trans., ASME*, **189**, 149–154.

Nakamizo, T. (1970). On the State Estimation of Non-linear Dynamic Systems, *Int. J. Control*, **11**, 683–695.

Newland, D. E. (1984). *An Introduction to Random Vibrations and Spectral Analysis* (2nd ed), Longman, London.

Nigam, N. C. (1983). *Introduction to Random Vibration*, MIT Press, Cambridge, MA.

Noguchi, T. (1985).The Response of a Building on Sliding Pads to Two Earthquake Models, *J. Sound Vib.*, **103**, 437–442.

Noori, M., Choi. J-D. and Davoodi, H. (1986). Zero and Nonzero Mean Random Vibration Analysis of a New General Hysteresis Model, *Prob. Engng. Mech.*, **1**, 192–201.

Orabi, I. I. and Ahmadi, G. (1987). A Functional Series Expansion Method for the Response Analysis of a Duffing Oscillator Subjected to White Noise Excitations, *Int. J. Non-linear Mech.*, **22**, 451–465.

Osinki, Z. (1971). Stochastic Processes in Nonlinear Vibrations, *Zagad. Drgan Niel.*, **12**, 101–111.
Pace, I. S. and Barnett, S. (1972). Comparison of Numerical Methods for Solving Liapunov Matrix Equations, *Int. J. Control*, **15**, 907–915.
Papoulis, A. (1984). *Probability, Random Variables and Stochastic Processes* (2nd ed), McGraw-Hill, New York.
Park, Y. J., Wen, Y-K. and Ang, A. H. S. (1986). Random Vibration under Bi-directional Ground Motions, *Earthquake Engng. Struct. Dyns.* **14**, 543–557.
Parzen, E. (1962). *Stochastic Processes*, Holden-Day, San Francisco.
Piszczek, K. and Niziol, J. (1986). *Random Vibration of Mechanical Systems*, PWN-Polish Scientific Pub.,Warsaw, Poland, and Ellis Harwood, Chichester, UK.
Pivovarov, I. and Vinogradov, O. G. (1987). One Application of Bouc's Model for Non-Linear Hysteresis, *J. Sound Vib.*, **118**, 209–216.
Press, H. and Houbolt, J. C. (1955). Some Applications of Generalized Harmonic Analysis to Gust Loads on Airplanes, *J. Aero. Sci.*, **22**, 17–26.
Priestley, M. B. (1981). *Spectral Analysis and Time Series*, 2 Vols, Academic Press, New York.
Rayleigh, J. W. S. (1877). *The Theory of Sound*, reprinted by Dover, New York 1945.
Reid, J. G. (1983). *Linear System Fundamentals: Continuous and Discrete, Classic and Modern*, McGraw-Hill, New York.
Rice, S. O. (1944). Mathematical Analysis of Random Noise In N. Wax (ed), *Selected Papers in Noise and Stochastic Processes*, Dover, New York, 1954, 133–294.
Richard, K. and Anand, G. V. (1983). Nonlinear Resonance in Strings under Narrow-Band Random Excitation—Part 1: Planar Response and Stability, *J. Sound. Vib.*, **86**, 85–98.
Roberts, J. B. (1966). On the Response of a Simple Oscillator to Random Impulses, *J. Sound Vib.*, **4**, 51–61.
Roberts, J. B. (1974). Probability of First Passage Failure for Stationary Random Vibration, *AIAA J.*, **12**, 1636–1643.
Roberts, J. B. (1976a). First Passage Time for the Envelope of a Randomly Excited Linear Oscillator, *J. Sound Vib.*, **46**, 1–14.
Roberts, J. B. (1976b). First Passage Probability for Non-Linear Oscillators, *J. Engng. Mech. Div., ASCE*, **102**, 851–866.
Roberts, J. B. (1977). Stationary Response of Oscillators with Non-Linear Damping to Random Excitation, *J. Sound. Vib.*, **50**, 145–156.
Roberts, J. B. (1978a). The Response of an Oscillator with Bilinear Hysteresis to Stationary Random Excitation, *J. Appl. Mech., ASME*, **45**, 923–928.
Roberts, J. B. (1978b). First-Passage Time for Oscillators with Nonlinear Damping, *J. Appl. Mech., ASME*, **45**, 175–180.
Roberts, J. B. (1978c). First Passage Time for Oscillators with Non-Linear Restoring Forces, *J. Sound Vib.*, **56**, 71–86.
Roberts, J. B. (1978d). The Energy Envelope of a Randomly Excited Non-Linear Oscillator, *J. Sound Vib.*, **60**, 177–185.
Roberts, J. B. (1981a). Response of Nonlinear Mechanical Systems to Random Excitation: Part I; Markov Methods, *Shock Vib. Digest*, **13**, 17–28.
Roberts, J. B. (1981b). Response of Nonlinear Mechanical Systems to Random Excitation: Part II; Equivalent Linearization and Other Methods, *Shock Vib. Digest*, **13**, 15–29.
Roberts, J. B. (1981c). Nonlinear Analysis of Slow Drift Oscillations of Moored Vessels in Random Seas, *J. Ship Research, SNAME*, **25**, 130–140.
Roberts, J. B. (1982). A Stochastic Theory for Non-Linear Ship Rolling in Irregular Seas, *J. Ship Research, SNAME*, **26**, 229–245.

Roberts, J. B. (1984). Techniques for Nonlinear Random Vibration Problems, *Shock Vib. Digest*, **16**, 3–14.
Roberts, J. B. (1985) Estimation of Non-Linear Ship Roll Damping from Free-Decay Data, *J. Ship Research, SNAME*, **29**, 127–138.
Roberts, J. B. (1986) First Passage Probabilities for Randomly Excited Systems: Diffusion Methods, *Prob. Engng. Mech.*, **1**, 66–81.
Roberts, J. B. and Dacunha, N. M. C. (1985). The Roll Motion of a Ship in Random Beam Waves: Comparison between Theory and Experiment, *J. Ship Research, SNAME*, **29**, 112–126.
Roberts, J. B. and Dunne, J. F. (1988). Nonlinear Random Vibration in Mechanical Systems, *Shock Vib. Digest*, **20**, 16–25.
Roberts, J. B. and Spanos, P. D. (1986). Stochastic Averaging: An Approximate Method for Solving Random Vibration Problems, *Int. J. Non-Linear Mech.*, **21**, 111–134.
Roberts, J. B. and Yousri, S. N. (1978). An Experimental Study of First-Passage Failure of a Randomly Excited Structure, *J. Appl. Mech., ASME*, **45**, 917–922.
Roberts, J. B., Ellis, J. and Hosseini-Sianaki, A. (1990). The Determination of Squeeze Film Dynamic Coefficients from Transient Two-dimensional Experimental Data (to be published in J. Tribology, ASME).
Robson, J. D. (1963). *An Introduction to Random Vibration*, Edinburgh University Press, Edinburgh.
Rothschild, D. and Jameson, A. (1968). Comparison of Numerical Methods for Solving Liapunov Matrix Equations, *Int. J. Control*, **11**, 181–198.
Roy, R. V. and Spanos, P. D. (1990). Wiener-Hermite Functional Representation of Non-linear Stochastic Systems (to be published in *J. Structural Safety*).
Rubinstein, R. Y. (1981). *Simulation and the Monte Carlo Method*, Wiley, New York.
Sakata, M. and Kimura, K. (1980). Calculation of the Nonstationary Mean Square Response of a Non-Linear System Subjected to Non-White Excitation, *J. Sound Vib.*, **73**, 333–344.
Sakata, M., Kimura, K. and Utsumi, M. (1984). Non-Stationary Response of Non-Linear Liquid Motion in a Cylindrical Tank Subjected to Random Base Excitation, *J. Sound Vib.*, **94**, 351–363.
Sawaragi, Y., Sugai, N. and Sunahara, Y. (1962). *Statistical Studies of Non-Linear Control Systems*, Nippon, Osaka, Japan.
Scheurkogel, A. and Elishakoff, I. (1988). An Exact Solution of the Fokker-Planck Equation for Nonlinear Random Vibration of a Two Degree of Freedom Systems. In F. Ziegler and G. I. Schuëller (eds), *IUTAM Symp. Nonlinear Stochastic Dynamic Engineering Systems*, Springer-Verlag, Berlin, 285–299.
Schiehlen, W. O. (1985). Random Vehicle Vibrations. In I. Elishakoff and R. H. Lyon (eds) *Random Vibration—Status and Recent Developments*, Elsevier, Amsterdam, 379–388.
Schmidt, A. X. and Marlies, C. A. (1948). *Principles of High Polymer Theory and Practice*, McGraw-Hill, New York. p. 573.
Shinozuka, M. (1972). Monte Carlo Solution of Structural Dynamics, *Comps. Structs.*, **2**, 855–874.
Sinitsyn, I. N. (1976). Methods of Statistical Linearization (Survey), *Auto. Remote Control*, **35**, 765–776.
Smith, H. W. (1966). *Approximate Analysis of Randomly Excited Nonlinear Controls*, MIT Press, Cambridge, MA.
Soni, S. R. and Surrendran, K. (1975). Transient Response of Nonlinear Systems to Stationary Random Excitation, *J. Appl. Mechs., ASME*, **42**, 891–893.

Soong, T. T. (1973). *Random Differential Equations in Science and Engineering*, Academic New York.

Spanos, P-T. D. (1976). Linearization Techniques for Non-linear Dynamical Systems, *Report EERL 76-04, Earthquake Engineering Research Laboratory, California Institute of Technology, Pasadena, CA, USA.*

Spanos, P-T. D. (1978). Stochastic Linearization Method for Dynamic Systems with Asymmetric Nonlinearities, *Report EMERL 1126, Engineering Mechanics Research Laboratory, University of Texas at Austin, USA.*

Spanos, P-T. D. (1980a). Formulation of Stochastic Linearization for Symmetric or Asymmetric MDOF Nonlinear Systems, *J. Appl. Mech., ASME*, **47**, 209–211.

Spanos, P-T. D. (1980b). Numerical Simulations of a Van der Pol Oscillator, *Comps. Maths. with Appls.*, **6**, 135–145.

Spanos, P-T. D. (1981a). Stochastic Linearization in Structural Dynamics, *Appl. Mech. Revs., ASME*, **34**, 1–8.

Spanos, P-T. D. (1981b). Monte Carlo Simulations of Responses of Non-Symmetric Dynamic Systems to Random Excitations, *Comps. Structs.*, **13**, 371–376.

Spanos, P-T. D. (1983). ARMA Algorithms for Ocean Wave Modeling, *J. Energy Resources Technology, ASME*, **105**, 300–309.

Spanos, P. D. (1986). Filter Approaches to Wave Kinematics Approximation, *Appl. Ocean Research*, **8**, 2–7.

Spanos, P. D. (1987). An Approach to Calculating Random Vibration Integrals, *J. Appl. Mech., ASME*, **54**, 409–413.

Spanos, P. D. and Agarwal, V. K. (1984). Response of a Simple Tension Leg Platform Model to Wave Forces Calculated at Displaced Position, *J. Energy Resources Techn., ASME*, **106**, 437–443.

Spanos, P-T. D. and Chen, T. W. (1981). Random Response to Flow-Induced Forces, *J. Engng. Mech. Div., ASCE*, **107**, 1173–1190.

Spanos, P-T. D. and Iwan, W. D. (1978). On the Existence and Uniqueness of Solutions Generated by Equivalent Linearization, *Int. J. Non-Linear Mech.*, **13**, 71–78.

Spanos, P. D. and Lutes, L. D. (1986). A Primer of Random Vibration Techniques in Structural Engineering, *Shock Vib. Digest*, **18**, 3–10.

Spanos, P. D. and Mignolet, M. D. (1989). ARMA Monte Carlo Simulation in Probabilistic Structural Analysis, *Shock Vib. Digest*, **21**, 3–14.

St. Denis, M. and Pierson, W. J. (1953). On the Motions of Ships in Confused Seas, *Trans. Soc. of Naval Architects and Marine Engineers (SNAME)*, **61**, 1–30.

Stoker, J. J. (1950). *Non-linear Vibrations*, Interscience, New York.

Stoker, J. J. (1968). *Non-linear Elasticity*, Nelson, New York.

Stratonovich, R. L. (1964). *Topics in the Theory of Random Noise*: 2 Vols, Gordon and Breach, New York.

Suzuki, Y. and Minai, R. (1987). Application of Stochastic Differential Equations to Seismic Reliability Analysis of Hysteretic Structures. In Y. K. Lin and R. Minai (eds) *Leçture Notes in Engng.*, No. 32, Springer, Berlin.

Tajimi, H. (1960). A Statistical Method of Determining the Maximum Response of a Building Structure During an Earthquake, *Proc. 2nd. World Conf. Earthquake Engng., Vol. II, Tokyo and Kyoto, Japan*, 781–798.

Takemiya, H. (1973). Equivalent Linearization for Randomly Excited Bilinear Oscillators *Proc. Jap. Soc. Civ. Engng.*, No. 219, 1–13.

Takemiya, H. and Lutes, L. D. (1977). Stationary Random Vibration of Hysteretic Structures, *J. Engng. Mech. Div., ASCE*, **103**, 673–688.

Takizawa, H. and Aoyama, H. (1976). Biaxial Effects in Modelling Earthquake Response of R/C Structures, *Earthquake Engng. Struct. Dyns.*, **4**, 523–552.

Thompson, J. M. T. and Hunt, G. W. (1973). *A General Theory of Elastic Stability*, Wiley-Interscience, London.

Timoshenko, S. P. and Woinowsky-Krieger, S. (1959). *Theory of Plates and Shells* (2nd ed), McGraw-Hill, New York.
To, C. W. S. (1984). The Response of Nonlinear Structures to Random Excitation, *Shock Vib. Digest*, **16**, 13–18.
To, C. W. S. (1987). Random Vibration of Nonlinear Systems, *Shock Vib. Digest.*, **19**, 3–9.
Vanmarke, E. H. (1976) Structural Response to Earthquake. In C. Lomnitz and E. Rosenblueth (eds) *Seismic Risk and Engineering Decisions*, Elsevier, Amsterdam, Chapter 8.
Vassilopoulos, C. (1971). Ship Rolling at Zero Speed in Random Beam Seas with Nonlinear Damping and Restoration, *J. Ship Research, SNAME*, **15**, 289–294.
Wedig, W. 1984. Critical Review of Methods in Stochastic Structural Mechanics, *Nucl. Engng. Des.*, **79**, 281–287.
Wen, Y-K. (1980). Equivalent Linearization for Hysteretic Systems under Random Excitation, *J. Appl. Mech. ASME*, **47**, 150–154.
Wen, Y-K. (1986) Stochastic Response and Damage Analysis of Inelastic Structures, *Prob. Engng Mech.* **1**, 49–57.
Wen, Y-K. (1989). Methods of Random Vibration for Inelastic Structures, *Appl. Mech. Revs., ASME*, **42**, 39–52.
Whittaker, E. T. (1937). *A Treatise on the Analytical Dynamics of Particles and Rigid Bodies* (4th ed), Cambridge University Press, Cambridge, UK.
Williams, J. H. Jr, (1973). Designing Earthquake Resistant Structures, *Technol. Rev.*, **76**, 37–43.
Wu, W. F. and Lin, Y. K. (1984). Cumulant-Neglect Closure for Non-Linear Oscillators under Random Parametric and External Excitations, *Int. J. Non-Linear Mech.*, **19**, 349–362.
Yaglom, A. (1962). *Stationary Random Functions*, Prentice-Hall, Englewood Cliffs, NJ.
Yang, C. Y. (1986). *Random Vibration of Structures*, Wiley, New York.
Yong, Y. and Lin, Y. K. (1987). Exact Stationary Response Solutions for Second Order Nonlinear Systems under Random Parametric and External White-Noise Excitations, *J. Appl. Mech., ASME*, **54**, 414–418.
Zhu, W-Q. (1983). Stochastic Averaging of the Energy Envelope of Nearly Liapunov Systems", in K. Hennig (ed), *IUTAM Symp. Random Vibrations and Reliability*, Frankfurt/Oder (GDR), Akademie-Verlag, Berlin, East Germany, 347–357.
Zhu, W-Q. (1989). Exact Solutions for Stationary Responses of Several Classes of Nonlinear Systems to Parametric and/or External White Noise Excitations, in: A Collection of Papers Presented at XVIIth International Conference on Theoretical and Applied Mechanics by Chinese Scholars, Beijing: Beijiang University Press, China.
Zhu, W-Q. and Lei, Y. (1989). A Stochastic Theory of Cumulative Fatigue Damage, *presented at the ICCOSAR meeting on Structural Reliability, San Francisco.*
Zhu, W-Q. and Yu, J-S. (1989). The Equivalent Nonlinear System Method, *J. Sound Vib.*, **129**, 385–395.

Author index

Subject index